ELECTRONICS
A General Introduction
for the Non-Specialist

ELECTRONICS

A General Introduction for the Non-Specialist

G. H. OLSEN

B.Sc., C.Eng., A.M.I.E.R.E., A. Inst. P.

*Senior Lecturer, Physics Dept., Rutherford
College of Technology*

Springer Science+Business Media, LLC

First published by
Butterworth & Co. (Publishers) Ltd.

ISBN 978-1-4899-6249-2   ISBN 978-1-4899-6535-6 (eBook)
DOI 10.1007/978-1-4899-6535-6

©
The Author
1968
Originally published by Plenum US in 1968.

Softcover reprint of the hardcover 1st edition 1968

Suggested U.D.C. number: 621·38

Library of Congress Catalog Card Number 68 – 25338

PREFACE

There can be few scientific workers nowadays who do not use electronic equipment in their work. For those who are not specialists in electronic engineering the field of electronics must indeed seem bewildering; every year sees the invention of new devices or the development of existing ones, and the pace at which electronics is expanding is so rapid that even electronic engineers feel that they are being overwhelmed. This book has therefore been written as a general introduction to the subject for those who find formal or examination texts to be unsuitable for their needs.

The author is indebted to several firms for their willingness to allow publication of data and circuits from their commercial literature: Mullard Ltd., Ferranti Ltd., Marconi Ltd., Texas Instruments Ltd., SGS-Fairchild Ltd., Philco International Ltd., Tektronix Ltd., Solartron Instruments Ltd., Standard Telephones and Cables Ltd., Hilger and Watts Ltd., Baldwin Instruments Ltd., Evans Electroselenium Ltd., and Mallory Batteries Ltd. Thanks are also due to the Editors of *Electronic Engineering*, *Wireless World*, *Instrument Practice* and Philips Technical Library as well as the Institution of Electronic and Radio Engineers for permission to publish circuit diagrams taken from various papers referred to in the text.

The author is grateful to the staff of Butterworth and Co. Ltd. for their help in the preparation of the manuscript; he also wishes to thank Miss Rosalind Lowe who typed the manuscript with meticulous care and who cheerfully suffered the many alterations in the handwritten work.

Finally the author must express his great appreciation to his wife, Mary. Her fortitude and self-denial as well as her assistance in reading and commenting upon the manuscript have not gone unnoticed. It is probably true to say that without her encouragement and support this book would never have been written.

CONTENTS

CONTENTS

Appendices

9

1

INTRODUCTION

The average reader usually avoids reading the preface of a textbook because such an introduction is often little more than a statement of the author's aims in writing the book, together with numerous acknowledgments to those who have helped him in his work. Understandably this can be dull for those who are keen to delve immediately into the subject matter. It is in the hope of arresting the attention of such readers that this short introductory chapter has been written. In it an attempt has been made to ensure that all readers know what the aims of the book are and what they can expect from a study of the text.

Paradoxically, let us begin by stating what this book is not intended to do. It is not an examination textbook that supplies the information and circuit analyses that have to be regurgitated when answering what are often stereotyped examination questions. Examinations, one supposes, are a necessary part of our education system and many excellent texts have been published that cater for the examination candidate. Unfortunately, such books do not meet the needs of many scientists who wish to gain only a working knowledge of the operation of some of the electronic equipment they use. It is not often that non-specialists in electronics are required to design electronic equipment from basic principles, but they may find it useful to be able to modify or extend the capabilities of their instruments, or be in a position to suggest such modifications or extensions to others. It is also useful to have sufficient knowledge to read specifications, to buy equipment wisely and to approach manufacturers and others knowledgeably. The author has enjoyed the privilege of teaching many such people who required a course that started right from the beginning assuming no knowledge whatever of the subject. Too many courses, these students felt, were aimed at the professional physicist and electronic or electrical engineer. These latter workers often require a detailed analytical approach and usually bring to the lectures a considerable background knowledge appropriate to electronics; many of the available textbooks, understandably, are written for such specialists. For the graduate chemist, biologist, medical doctor, mechanical engineer and mathematician, however, there has often been insufficient time in their studies to include work on electronics, and since electronic techniques are now widely used in almost every kind of measurement and control, many feel the inadequacy of their knowledge of electronics. This book has therefore been written for those who want an introductory account of the subject that is qualitative, informative, and is not overburdened with mathematics and circuit analyses. Although it is assumed that the reader has no previous knowledge of electronics, some acquaintance with certain aspects of physics and mathematics must be taken for granted. The reader is expected to have heard of Ohm's law, to be able to manipulate algebraic expressions, to perform very simple differentiations and to know what is meant by a simple integral. The mathematical content has, however, been restricted, but where necessary sufficient instruction is given. (The only exception to this is the section dealing with operational amplifiers.) Particular attention has been paid to the sections dealing with the j-operator and complex numbers in a.c. theory, both of which are extremely easy to understand.

In avoiding Laplace transforms and mathematical aspects of the quantum theory, wave mechanics and solid-state physics, it has been necessary to attempt qualitative descriptions of the processes involved. Such 'mechanistic' pictures as have been drawn rely heavily on analogies. Whilst such interpretations are not, one hopes, incorrect, the reader should be aware that they have limitations, and not press the analogies too far. Bearing this in mind, useful pictures of device behaviour can be constructed that will enable the reader to approach the use of such devices with confidence.

Of those who attend the author's introductory course on electronics, chemists have been in the majority and it is for this reason that the book is biased towards their needs. The subject of electronics is the same, however, for all students and so the major part of the book should prove satisfactory for workers in other fields. Undergraduate electrical engineers and physicists may also welcome a qualitative first approach.

A good deal of attention has been paid in the early chapters to the basic components, such as resistors, capacitors, inductors, valves and transistors. Readers are then introduced to combinations of these components that form the fundamental circuits from which most electronic equipment is built. An attempt has been made to keep a balance between thermionic valve circuits and those that use transistors and allied semiconductor devices. Although transistors are rapidly replacing thermionic valves in most applications, it would be wrong, in the author's view, to regard thermionic valves as of no account and to discount them as obsolete, especially in measurement applications. Several practical aspects of electronics are discussed and some useful circuits are given with details of the components used. Very often textbooks intended for examination work pay little attention to the practical details of the circuits they analyse.

Many electronic devices seem complex at first sight, but it should be realized that all electronic circuits, no matter how complex they may seem, can be regarded as the combination of a comparatively small number of basic units each performing a specific function. It follows that an understanding of the operation of the basic units leads to an understanding of the complete equipment. The basic units may be assembled in a variety of ways to form quite different composite equipment. For example, a cathode-ray oscilloscope incorporates amplifiers, an oscillator and two types of power supply, as well as the cathode-ray tube itself. A counting unit contains amplifiers, power supplies, perhaps an oscillator for calibration purposes as well as dekatron counting tubes. Some of the basic units are common to both items of equipment, although both perform quite different functions. The main differences to be found in electronic apparatus lie in the various transducers connected to the input (e.g. photocells, glass electrodes, strain gauges or temperature sensing elements) and in the indicator used at the output (e.g. a moving-coil meter, a cathode-ray tube or a pen recorder). This book, therefore, deals mainly with the basic units, although applications and complete equipment are discussed where appropriate.

Valves and transistors are classified as active devices because they modify the power supplied to them. The sources of the power in electronic circuits are batteries or more conveniently the electric generators that feed power into the public electricity supply. The ultimate source of power is the sun or radioactive materials; nowadays, it is possible to utilize power from the sun by converting the radiant energy directly to electrical energy with the aid of solar cells. Power can be modified to produce electrical oscillations or to amplify small signals. Although valves and transistors are the agents by which such modifications are made, they do not actually supply power to a circuit. In spite of this, we often find it convenient to regard such devices as doing so, and bear in mind that they are energized from a power supply. For this reason, these components are called active devices. Associated with valves and transistors are those components that consume power or otherwise control the flow of energy in some way. These components (resistors, capacitors and inductors) are regarded as passive elements.

There is no doubt that practical periods in the laboratory constitute one of the attractive features of the courses given by the author. Many people like to get down to building apparatus for themselves. The author has no way of knowing what facilities the reader has for practical work, therefore no details of experiments are given. No amount of reading alone, however, will make a person competent to deal with electronic equipment; practical experience is the only solution. The reader should therefore take every opportunity to do practical work. If necessary he should buy a kit and build a transistor radio or make an audio amplifier for record reproduction purposes. The end products will have more than merely educational value. On no account must the reader become discouraged if the apparatus he builds does not work the first time it is switched on, or when the source of trouble with faulty apparatus cannot be located swiftly. (The author still has difficulties in this respect.) To develop practical skill, experience and perseverance are essential. The final chapter deals briefly with some practical aspects of fault finding and the building of apparatus.

2

PASSIVE COMPONENTS—RESISTORS, CAPACITORS AND INDUCTORS

Electric current may be regarded as a flow of electrons. The flow may be along a wire, a carbon rod, through a gas, a vacuum, a transistor or a thermionic valve. Electronics is the subject which deals with the way in which this flow can be controlled and modified so as to produce useful results. Each electron carries a small negative charge ($1{\cdot}602{\times}10^{-19}$ coulombs) and when many millions of them flow along a wire a charge q will pass a particular point during a specified interval of time. If a small charge, dq, passes a point in a small interval of time, dt, then dq/dt is the rate at which charge is passing. This rate is called electric current, i.e.

$$i = \frac{dq}{dt}$$

When q is in coulombs and t is in seconds, i is in amperes. One ampere is thus equivalent to the passage of one coulomb per second past a given point.

Materials vary enormously in their ability to allow the passage of current when an electric pressure, i.e. voltage, is applied. Silver and copper for example present very little opposition to the flow of electrons and are, therefore, termed conductors. Short, thick wires made from these materials will pass currents of several amperes when the voltage between the ends of the wire is only a fraction of a volt. Mica, quartz, polythene and porcelain, however, pass practically no current when voltages are applied, unless such voltage s are above a breakdown value, in which case the material disintegrates. Such materials are known as insulators. Intermediate between conductors and insulators is the class of materials known as semiconductors. Many elements and compounds are semiconductors, germanium and silicon being perhaps the best known since transistors are made from these elements.

When a material is made into a rod or wire to form an electronic component, it is generally true to say that the current passing through the component depends upon the applied voltage and the temperature. If, for a given temperature, the current is directly proportional to the applied voltage the component is said to obey Ohm's law. Such components are called resistors (strictly, linear resistors) because a graph of current against voltage is a straight line. A component that does not meet this requirement is termed a non-linear

15

resistor. We shall use the term 'resistor' to mean 'linear resistor' and stress non-linearity in other cases by using the term 'non-linear resistor'.

Frequent reference to graphs will be made when studying electronic components since pictorial representations of data make easier the task of understanding the behaviour of resistors, valves, transistors, photocells, etc. The graph of current through a device against applied voltage is often the most useful one to consider; such graphs are called characteristic curves.

RESISTORS

The characteristic curve for a resistor is particularly simple and is shown in *Figure 2.1*. It is obvious that a linear relationship exists. From the graph we may deduce that voltage, V, is proportional to current, I, therefore, V/I is constant. The constant ratio is defined as the resistance, R, and is measured

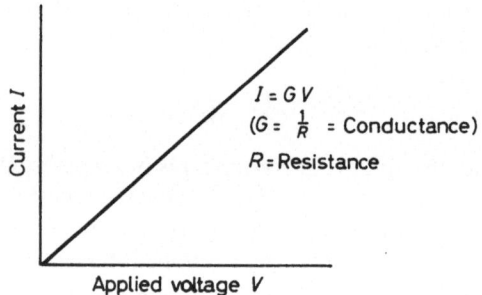

Figure 2.1. Graph of current against voltage for a linear resistor

in ohms when V is in volts and I in amperes. $V/I = R$ is the mathematical embodiment of Ohm's law. Equally we may say that $I/V = G$ where G is a constant known as the conductance. $G = 1/R$ so the units of G are reciprocal ohms or mhos (ohm backwards). Alternatively, G may be expressed in amperes per volt. *Figure 2.1* shows current being represented by the ordinate (*y-axis*) since it is usual to regard current as being the dependent variable, i.e. its value depends upon the voltage (which is represented by the abscissa or *x*-axis). This arrangement is conventional when considering valves and transistors.

Resistors are used in electronic circuits to provide specific paths for electric currents and to serve as circuit elements that limit the current to some desired value. They provide a means of producing voltages as, for example, in a voltage amplifier. Here variations of valve or transistor currents are made to produce varying voltages across a resistor placed in series with the amplifying device. Resistors are also used to build various networks and filters and to act as voltage dividers. The many uses of resistors will become increasingly apparent as the reader progresses in his study of electronics.

Fixed Resistors

Figure 2.2 shows the circuit symbols for fixed value resistors, i.e. resistors whose value is not mechanically adjustable. The importance of the non-reactive type of fixed resistor will be made clear later in the text.

General symbol   Non-reactive type

Figure 2.2. Circuit symbols for fixed resistors

General-purpose resistors are usually made of a carbon composition. They are cheap to buy and serve most purposes, but they are not so stable as other types to be described. A drift in the resistance of the component during its life may be anything up to 10 or 15 per cent of the initial resistance. General-purpose resistors are used in non-critical positions in the circuit where the power dissipation does not exceed 2 W. Two forms are manufactured, the insulated type and non-insulated type. In the former type *(Figure 2.3)* carbon

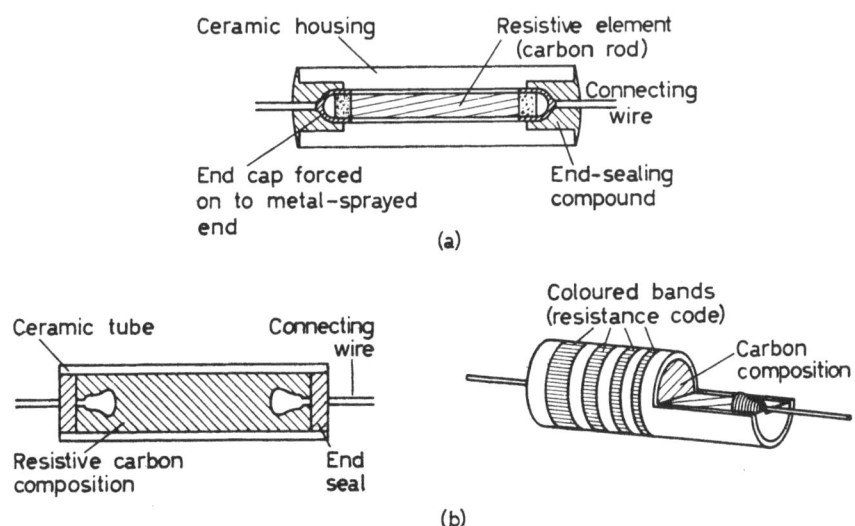

Figure 2.3. Two forms of insulated carbon composition resistors

black, a resin binder and a filler are mixed in the required proportions to produce a resistance of the desired value in the finished component. The mixture is then injected under pressure into a ceramic tube and cured in an oven. Subsequently the resistors are automatically sorted into batches before marking with paint rings according to the standard colour code. Each batch contains only those resistors whose resistance is within a few per cent of some specified nominal value. The advantage of the insulated type is that short-circuits are prevented when many components must occupy a small volume.

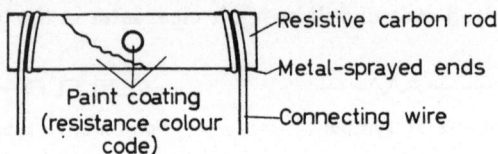

Figure 2.4. An uninsulated carbon rod resistor

The open or uninsulated type *(Figure 2.4)* permits better heat dissipation so that for the same wattage rating they are physically smaller than the corresponding insulated type. More detailed information on resistors and other components may be found in G.W.A. Dummer's book *Modern Electronic Components*[1].

Resistance values for carbon composition resistors are indicated by means of a standard colour code. Each colour represents a digit, as follows:

Black	0	Blue	6
Brown	1	Violet	7
Red	2	Grey	8
Orange	3	White	9
Yellow	4	Gold	$\pm5\%$ tolerance
Green	5	Silver	$\pm10\%$ tolerance

The band system is used with insulated resistors. Coloured bands are painted on the ceramic tube towards one end *(Figure 2.5a)*. The first digit of the resistance value is indicated by the colour of the band nearest the end; the next digit is indicated by the colour of the next band; the colour of the third band indicates the power of the decimal multiplier, i.e. the number of noughts. For example orange represents 10^3 i.e. three noughts. If the third ring is gold the multiplier is 10^{-1}; if it is silver the multiplier is 10^{-2}. As an example if the bands are orange, orange and gold the resistance is 3·3 ohms. Sometimes the first digit is represented by a band somewhat wider than the others. If the fourth band is absent the actual resistance value is within $\pm20\%$ of the nominal value. If the fourth band is silver or gold the tolerance is $\pm10\%$ or

18

±5% respectively. For uninsulated resistors the 'body-tip-spot' system is used. The body colour represents the first digit, the tip colour the second digit and the spot colour the number of noughts *(Figure 2.5b)*.

The so-called 'preferred value' system for resistance values needs explaining because of the seemingly odd numbers used. Before the 1939–45 war the main standard values were 10, 25 and 50 with their multiples of ten. The

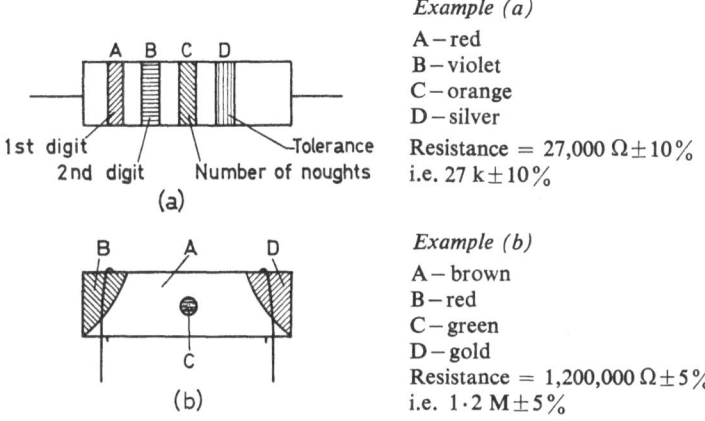

Example (a)
A – red
B – violet
C – orange
D – silver
Resistance = 27,000 Ω ± 10%
i.e. 27 k ± 10%

Example (b)
A – brown
B – red
C – green
D – gold
Resistance = 1,200,000 Ω ± 5%
i.e. 1·2 M ± 5%

Figure 2.5. The colour coding system for carbon composition resistors

manufacture of resistors to cover the intervening values resulted in the production of many more resistance values than was necessary. A further disadvantage arose in the possibility of finding resistors with given nominal markings having resistances greater than some components bearing a higher nominal marking. The overlapping is illustrated by *Figure 2.6a* and shows a state of affairs which can not be tolerated during war-time. The most efficient system is based on a logarithmic scale. For a given tolerance spread (20% in *Figure 2.6b*) the minimum overlapping occurs when the nominal values are 10, 15, 22, 33, 47, 68, etc. A complete table is given in the appendices.

In specifying the values of resistance up to 999 ohms the number of ohms is used (e.g. 470 Ω); for resistance values between 1,000 and 999,999 the number of thousands followed by k is used (e.g. 680 kΩ, 1·2 kΩ); from one megohm upwards the number of megohms is used followed by M (e.g. 4·7 MΩ). In electronic texts the symbol Ω for ohms is frequently dropped in connection with resistances, and will be here also.

Where the disadvantages of instability, noise and comparatively low-wattage dissipation are troublesome, types of resistors other than the carbon composition variety must be used.

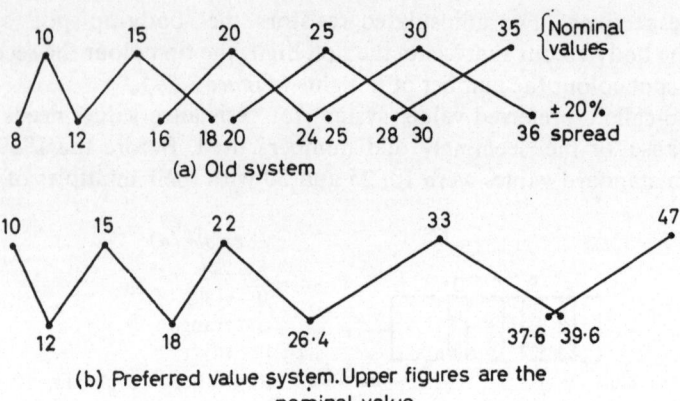

Figure 2.6. (a) shows the degree of overlapping with the old system whilst (b) shows the tolerance spreads using the preferred value system. The tolerance in each case is ±20%

Stability is concerned with the change in resistance under working conditions or whilst being stored on the shelf. If high stability (which is not the same as high accuracy) is required the cracked-carbon or metal-oxide type of resistor may be used.

Cracked-carbon resistors are made by coating ceramic rods with conducting carbon. This is achieved by passing the rods through heaters to bring their temperature to a sufficiently high value for pyrolytic action to take place. A hydrocarbon, usually methane, previously purified by removing water vapour, oxygen and carbon monoxide, is then passed into a reaction tube. Decomposition, or cracking, then occurs at the surface of the ceramic rod and carbon is deposited. For the best electrical properties the carbon coating should be thick. High resistances are obtained by spiralling the coating, such spiralling being carried out by machines which automatically produce the desired resistance value. End-caps are then force-fitted to the rod and the carbon film protected from damage and moisture by the application of numerous coatings of suitable varnishes and paints. The nominal value of high-stability resistors is often painted on to the resistance body in figures and words, although some manufacturers use the colour code system. An additional band of salmon pink is then used to indicate that the resistor is a high-stability type. Various wattage dissipations are available up to a usual figure of 2 W. The tolerances are usually ±5% or ±1%.

Metal-oxide film resistors are now displacing cracked-carbon high-stability types because of their increased stability, greater reliability and smaller size for equal power rating. Since they have been available in quantity, the cost compares favourably with that of other forms of resistor. The resistor is formed by spraying a solution of stannic chloride and antimony trichloride

20

onto a glass or porcelain rod. On increasing the temperature to red heat, hydrolysis occurs and a vitreous layer of oxide is formed. By varying the composition of the solution and the thickness of the oxide layer, various resistance values and wattage dissipations are possible. Spiralling as for the cracked-carbon types is also used to achieve the required resistance.

Metal film and metallic fibre resistors are manufactured for their special properties. Details of these types can be found in the reference already given[1].

Where very high values of resistance are required, say up to 10^{12} Ω, special precautions must be taken. Such resistors may be required in the radio-chemistry laboratory for use in connection with ionization chambers and vibrating-reed, quartz-fibre or valve electrometers; such instruments are employed in the detection and measurement of currents as small as 10^{-15} A. The resistors used are usually of the carbon composition film type. The ceramic rod on which the film is deposited is hermetically sealed within a glass tube. Since the glass is coated with silicone lacquer to minimize the effects of moisture, care must be taken not to handle the body of the glass envelope.

Before the choice of a resistor for a particular application can be made the user needs to know more than merely the ohmic value of the resistance. The wattage rating, i.e. power-handling capacity, must be known. This is obtained from a knowledge of the maximum current that the resistor will be required to pass, I_{max}, or alternatively the maximum voltage, V_{max}, likely to be applied. The power rating of the resistor must then exceed $I_{max}^2 R$ or V_{max}^2/R so as to allow a margin of safety. Knowledge of the stability, accuracy or manufacturing tolerance, maximum operating temperature and maximum operating voltage may also be required. In general, the voltage across any insulated 1 W carbon composition or cracked-carbon resistor should not exceed 500 V. Some extension is permissible if the resistor is an uninsulated type, but it is usually safer to adhere to the lower limit. For $\frac{1}{2}$W resistors the figure should not exceed 300 V. Where the voltage across two points to be connected by a resistor exceeds 500 V, two or more resistors of suitable value must be connected in series. For wirewound resistors the permissible maximum voltage may be about 2,500 V, but it would be wise when the voltage is high to consider the manufacturer's data.

Where higher wattage dissipations are encountered, say from 3 W upwards to 100 W or more, wirewound resistors are generally used. These resistors are wound with resistance wire on ceramic formers and the whole assembly is protected. The protection may take several forms; cement, lacquered or vitreous enamelled coatings are the usual choice. If the wattage dissipation is really high it may be necessary to adopt an open-wound construction and protect the wire with a metallic grid or housing. Since this type of resistor is used for purposes not requiring high precision, 10% and 20% tolerances are usual.

Where the highest precision is required together with good long-term stability, precision types of wirewound resistors are used. Typical applications are in the field of instrumentation and in the construction of standard resistance boxes. Very close tolerances are possible, a figure of $\pm 0.1\%$ being common. Wire (usually Nichrome or Manganin) is carefully selected and wound onto bobbins or sectionalized spools. After artificial ageing and stabilizing the resistors are vacuum impregnated with varnish or some other suitable sealing material. Methods of winding vary, but the aim is to reduce inductance and capacitance associated with the resistor to a minimum.

Variable Resistors

For many purposes it is necessary to alter at will the resistance value of a resistor in a given circuit location, or to vary the position of the tapping along a potential divider (e.g. in a volume control where it is necessary to vary the amplitude of an input signal to an audio amplifier). Where continuous (rather than discrete) variations of resistance are required the ordinary type of 'volume control' is used. This type of variable resistor is often called a potentiometer although the use of this term is incorrect since measurement of potential is rarely involved. A better term to describe this component is 'potential divider'. The circuit symbol for a potential divider is shown in *Figure 2.7*. Electrical connections are made to the two ends of the resistive track element and brought out to solder tags. A mechanically adjustable wiper arm makes contact with the resistive element. The connection to the wiper arm is brought out to a third tag. By moving the position of the arm it is possible to select a suitable fraction of the total resistance. For potential division the total voltage is applied across the resistive element. A variable fraction of this voltage is then available between the wiper arm contact and either of the remaining tags.

For locations where negligible or very low values of current are involved the resistance element is of the carbon-track variety. There are several meth-

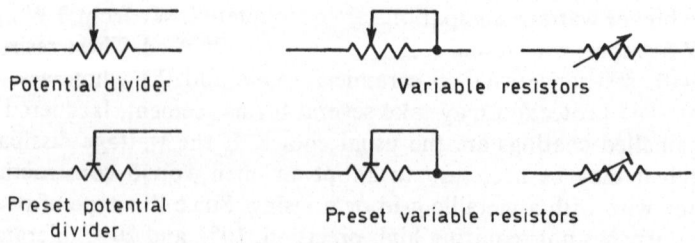

Potential divider Variable resistors

Preset potential divider Preset variable resistors

Figure 2.7. Circuit symbols for variable resistances

22

ods of producing the track, a common one being to spray the carbon suspension onto a plastic strip and cure it at high temperatures. The tracks are then stamped out and formed into an arc subtending an angle of something over 300°. Alternatively, the track may be formed on a circular disc. This method is employed when carbon composition moulded tracks are involved. The power dissipation with these types is not often greater than 1 W. For greater power dissipation wirewound elements are used. In one common type a rectangular strip of insulating material is wound with resistance wire and the former is then bent into a circular shape. A contact arm makes electrical connection with the wire at the edge of the strip. For high wattage dissipations open types of resistor elements must be used.

Various so-called 'laws' of these devices are available. A linear potential divider is one in which the carbon track or wire is uniform throughout the element. Equal increments of angular rotation bring equal increments of resistance into the circuit. Many other 'laws' are possible depending upon the application for the component. In volume controls it is desirable that equal increments of angular rotation of the shaft bring about equal changes in subjective loudness. This is achieved by having a logarithmic law, i.e. the resistance along the track varies in a logarithmic manner. In carbon tracks this is produced by selective spraying of the solution through suitable masks. In scientific apparatus the log. law produces a roughly linear scale in decibels (a term explained on page 216). Other laws such as semi-log, linear-tapered, inverse log, inverse semi-log, etc., are available for special applications.

It is possible to have two components ganged together so that a single spindle operates both wiper arms simultaneously.

There are occasions when intermediate resistor ratios need to be accurately known and reproducible. It is then advisable to have discrete variations of resistance. A suitable potential divider can be made by connecting several resistors in series and bringing the intermediate junctions to a suitable switch. A potential divider of this kind is one form of attenuator. An attenuator in electronics is a device for reducing the amplitude of a voltage or current. Most resistive attenuators used for potential division are more complicated than the one described above because problems of matching the load to the generator are involved. Those readers who are interested in designing attenuators are referred to the book *Telecommunications* by W. Frazer (Macdonald, 1957). Attenuators need not always be constructed with resistors; capacitors or inductors may also be used where appropriate.

Series and parallel connections

Resistors are said to be in series when they are connected together to form a chain. If the ends of the chain are connected to a source of electric power, such as a battery, then the same current passes through each resistor, i.e. the magnitude of the current is the same at every point in the circuit. Across any given resistor in the chain a voltage must exist. It is the voltage (or potential difference) between the ends of the resistor that causes the current to flow through it. The magnitude of the voltage, V, is given by Ohm's law, viz. $V = IR$, I being the current and R the resistance. If around a series circuit we

$R = R_1 + R_2 + R_3 = 1\,k + 5\,k + 4\,k = 10\,k$

$I = \dfrac{V}{R} = \dfrac{20}{10 \times 10^3} = 0.002 \text{ A} = 2 \text{ mA}$

(1 milliamp (mA) = 0.001 A = 10^{-3}A)

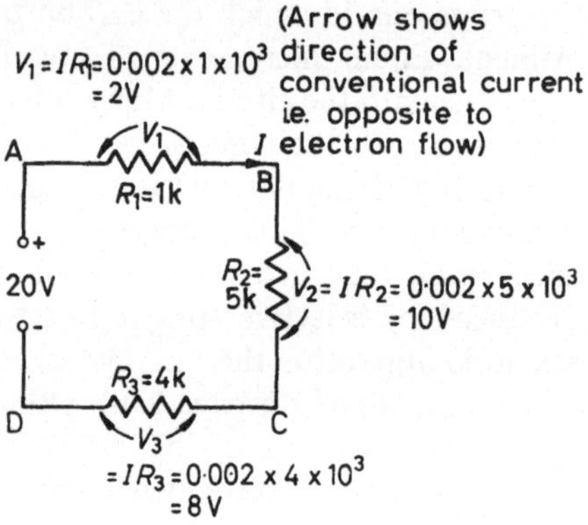

(Arrow shows direction of conventional current ie. opposite to electron flow)

$V_1 = IR_1 = 0.002 \times 1 \times 10^3 = 2\text{V}$

$R_1 = 1\,k$

20V

$R_2 = 5\,k$

$V_2 = IR_2 = 0.002 \times 5 \times 10^3 = 10\text{V}$

$R_3 = 4\,k$

$V_3 = IR_3 = 0.002 \times 4 \times 10^3 = 8\text{V}$

Figure 2.8. Voltage and current calculations for a simple series circuit

add together the voltages across each resistor the sum of the voltages must be equal to the applied voltage. In *Figure 2.8*

$$V = V_1 + V_2 + V_3$$
$$= IR_1 + IR_2 + IR_3$$
$$= I(R_1 + R_2 + R_3) = IR$$

The voltage source is, therefore, presented with a total resistance R given by

$$R = R_1 + R_2 + R_3$$

It is important for the beginner to understand what is meant by the terms 'voltage', 'a rise or fall of voltage in a circuit' and 'voltage drop'. The voltage (or potential difference) between two points in a circuit is a measure of the amount of work required to move unit positive charge from one point to the other at a higher potential. Let us appeal to an analogy in an attempt to clarify the position. If we were on a hillside a certain amount of work would

24

be necessary to raise a load of material from one point to another higher up the hill. The amount of work would depend upon the weight of the load, mg (where m is the mass and g the acceleration due to gravity), and the difference in height between the two points. If we take unit weight, e.g. let $mg = 1$ pound weight, then the work done in ft. lb. in raising the load would be numerically equal to the vertical difference in height when heights are measured in feet. The work done in raising bodies to various heights may be recovered by allowing the bodies to fall back to their original positions. The work done may be evident as heat, or, alternatively, the kinetic energy of motion may be made to turn a turbine and produce electricity or do other useful work. Bodies at various heights on the hillside all have, therefore, different potential energies, and such energies may be calculated in terms of their relative heights. In measuring height it is convenient to have some zero reference level. Conventionally, this is taken as mean-sea-level. Associated with every point on the hill is thus a potential relative to our mean-sea-level. Similarly in the electric circuit of *Figure 2.8* we have a battery which is a source of electromotive force (e.m.f.). This force can make charges move in a circuit. The battery thus establishes a potential gradient in the circuit. If the point D is chosen as our reference zero of potential then the points C, B and A are at progressively higher potentials. Current will, therefore, flow around the circuit as long as the battery can supply enough energy to sustain a voltage gradient. In resistors, this energy is evident as heat, but if the resistors were replaced by electric motors, the energy would be experienced as mechanical movement. There is thus no need for the beginner to wonder why current is not 'used up' as coal is on a fire, for example. The current is no more used up than is the water which, when falling down a gradient, is made to do work by turning a turbine.

The difference in potential between two points is measured practically by a voltmeter. The voltmeter is connected between the two relevant points and a small (usually negligible) current passes through the meter and does work in turning a pointer across a scale. (Details of how voltmeters operate are given later in the book.)

Work done, W, equals potential difference, V, times the charge, Q.

$$\text{i.e.} \quad W = VQ$$

$$\therefore \frac{dW}{dt} = V\frac{dQ}{dt} = VI$$

that is to say the rate of doing work, i.e. the power, P, in watts is equal to the product of the voltage and the current. Using Ohm's law we have

$$P = VI = I^2R = V^2/R$$

Since the current is the same through all the resistors, it must be limited to the least maximum value of current that can safely be passed through any

25

single resistor. Hence in *Figure 2.9* if I_A, I_B and I_C are the maximum permissible currents for R_A, R_B and R_C the value of the applied voltage V is limited to that which causes the least of these currents to flow.

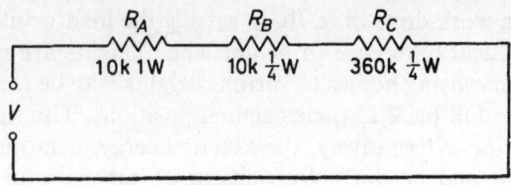

Figure 2.9

For resistor R_A

$$I_A^2 R_A = 1$$

$$\therefore I_A = \sqrt{\frac{1}{10^4}} = 0\cdot01 \text{ A} = 10 \text{ mA}$$

For resistor R_B

$$I_B^2 R_B = \frac{1}{4}$$

$$\therefore I_B = \sqrt{\frac{1}{4\times10^4}} = 5 \text{ mA}$$

For resistor R_C

$$I_C = \sqrt{\frac{1}{4\times36\times10^4}} = 0\cdot83 \text{ mA}$$

Hence in the chain the current must not exceed 0·83 mA.

Resistors are said to be connected in parallel when they are connected as shown in *Figure 2.10*. Since the same voltage, V, exists across each resistor then the current drawn by R_1, R_2 and R_3 is V/R_1, V/R_2 and V/R_3, respec-

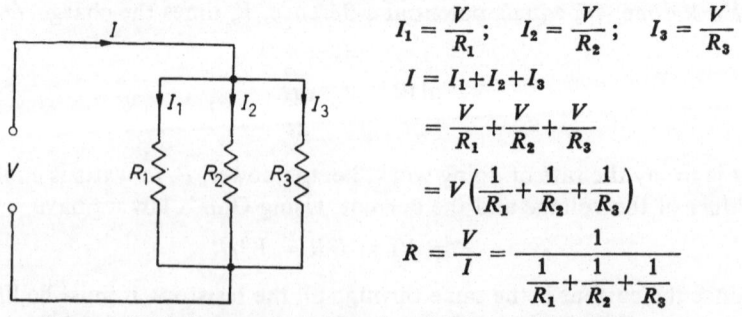

$$I_1 = \frac{V}{R_1}; \quad I_2 = \frac{V}{R_2}; \quad I_3 = \frac{V}{R_3}$$

$$I = I_1 + I_2 + I_3$$

$$= \frac{V}{R_1} + \frac{V}{R_2} + \frac{V}{R_3}$$

$$= V\left(\frac{1}{R_1} + \frac{1}{R_2} + \frac{1}{R_3}\right)$$

$$R = \frac{V}{I} = \frac{1}{\frac{1}{R_1} + \frac{1}{R_2} + \frac{1}{R_3}}$$

Figure 2.10. Voltages and currents in a simple parallel circuit

tively. The supply current I therefore divides into I_1, I_2 and I_3.

$$I = I_1 + I_2 + I_3$$

$$= \frac{V}{R_1} + \frac{V}{R_2} + \frac{V}{R_3} = V\left(\frac{1}{R_1} + \frac{1}{R_2} + \frac{1}{R_3}\right)$$

In a parallel circuit, as in a series circuit, the effective resistance, R, is equal to the applied voltage divided by the total current.

$$\therefore \frac{V}{I} = R = \frac{1}{\dfrac{1}{R_1} + \dfrac{1}{R_2} + \dfrac{1}{R_3}}$$

The effective resistance is always less than the lowest resistance in the circuit.

Reduction of actual circuits to an equivalent circuit is a useful procedure as we shall see later when we deal with transistors and valves. It is left to the reader to confirm the steps illustrated by *Figure 2.11*.

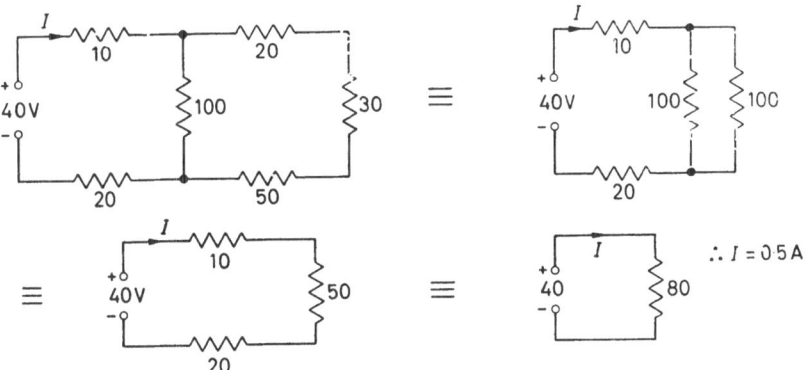

Figure 2.11. Reduction of a network to a simple equivalent circuit

The Wheatstone bridge network is one of the most important measuring networks we have, especially when it is extended to measuring techniques in the alternating current field. If, in *Figure 2.12* we apply a voltage between A and C then a special case arises if the resistors bear such a relation to each other that the fall in potential from A to the point D is the same as from A to B. Under these conditions D and B are at the same potential and hence a detector between D and B draws no current. The resistances are then related by the equation

$$\frac{R_1}{R_2} = \frac{R_3}{R_4}$$

27

If any three are known, the value of the fourth may be calculated. R_3 and R_4 are usually standard ratio arms and R_2 a standard variable resistor. An unknown resistor is then placed in the R_1 position. The bridge is said to be brought into balance when adjustment of R_1, R_3 and R_4 reduces the current in the detector arm to zero.

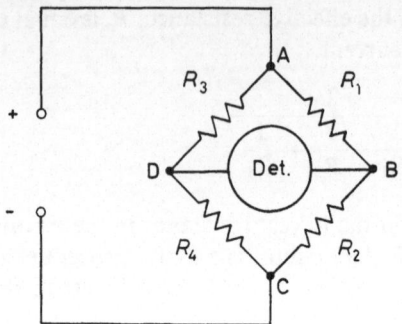

At balance the potentials at D and B are equal. The detector (Det.) shows a null reading. Then

$$\frac{R_1}{R_2} = \frac{R_3}{R_4}$$

Figure 2.12. The basic Wheatstone bridge

Non-linear Resistors

These resistors fall into two classes—the temperature-sensitive type and the voltage-sensitive type. The temperature-sensitive types are often known as thermistors[4]. They consist of the sintered oxides of manganese and nickel with small amounts of copper, cobalt or iron added to vary the properties. The physical shape is usually a bead or rod. The circuit symbol for the device is shown in *Figure 2.13*. The resistance of such materials is given by

$$R = a \exp (b/T) \tag{2.1}$$

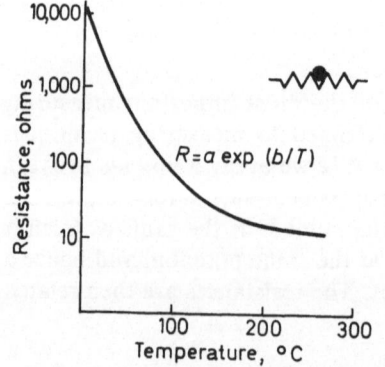

 Circuit symbol. Black dot shows negative temperature coefficient.

$$\text{Temperature coefficient} = \frac{\text{Rate of change of resistance with temp.}}{\text{Resistance}}$$

$$\frac{dR/dT}{R} = \frac{1}{R} a \exp (b/T) \left(-\frac{b}{T^2}\right) = -\frac{b}{T^2}$$

Figure 2.13. The resistance/temperature characteristic for a typical thermistor

where a and b are constants depending upon the composition and physical size, and T is the temperature in °K.

The commonest application is for surge suppression, especially in television sets. These sets are usually built without mains transformers and the heaters of the valves are connected in series across the mains. On first switching on, the filaments are cold and thus their resistances are low. A large current surge would therefore result. However, by including a thermistor in the chain, this surge is prevented because thermistors have a large resistance when cold and a small resistance when hot (which is opposite to the effect in valve heaters). A typical graph of resistance against temperature is shown in *Figure 2.13*. *Figure 2.14* shows the plot of voltage against current. The passage of current

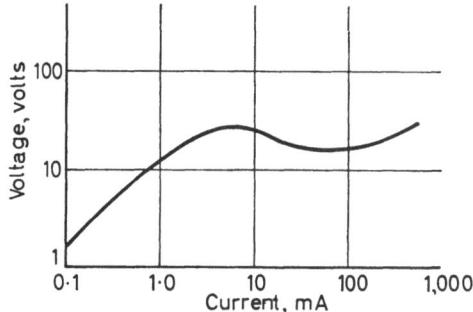

Figure 2.14. The voltage/current characteristic for a typical thermistor

through the thermistor gives rise to a heating effect. Thermal equilibrium is reached when the heat lost to the surroundings is equal to the heat generated by the current. Each point on the graph represents the current for a given voltage after thermal equilibrium has been achieved. For low values of current the heating effect is very small. For a constant ambient temperature, therefore, the thermistor behaves as a normal linear resistor and has a straight-line voltage/current characteristic. At a certain point (when the current is 1 mA in the example shown) the heating effect is sufficient to cause a fall in resistance, and there is a consequent departure of the graph from linearity. For a certain current value (about 5 mA for our thermistor) the voltage reaches a maximum. Further increases in current heat the thermistor sufficiently to cause such a drop in resistance that the voltage across the thermistor falls. Above about 100 mA in our example the resistance has fallen to its minimum value. Any further increases in current are then accompanied by increases in voltage across the device.

Care must be taken to ensure that in any circuit the voltage across a thermistor is not held at a value above that of the maximum shown in the graph.

29

(It will be necessary to consult the appropriate curve for any given thermistor.) If we persist in holding voltage above the critical value the current will rise to a value which will be excessive for the device.

Because the temperature coefficient of thermistors is large at moderate temperature, an obvious application of the device in the chemistry laboratory, or works, is in temperature measurement and control[5]. A simple arrangement of thermistor and meter can be used as a thermometer, but for best results a bridge circuit is used with a thermistor forming one arm.

Thermistors have such widespread applications as thermometers that it will pay the reader to consult the Standard Telephones and Cables Ltd. Application Report on *The Design of Thermistor Bridge Networks for Direct Reading*

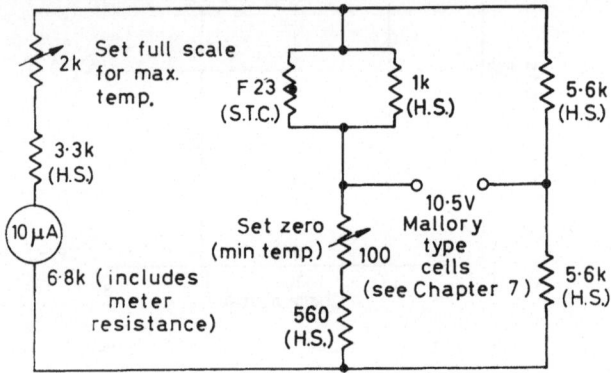

Figure 2.15. A circuit suitable for use as a clinical thermometer to read temperatures over the range 35 to 45°C. The F23 thermistor of Standard Telephones and Cables Ltd. is a small bead type in the end of a suitable glass holder. Fixed resistors should be high stability (H.S.) types

Thermometers. This report deals with the circuit design of bridges where the unbalance is exhibited on a meter or pen recorder as an indication of the temperature which is being measured. *Figure 2.15* gives one such bridge network. The function of the 1 k resistor in parallel with the type F23 thermistor is to linearize the graph of current against temperature. For the circuit given there is good linearity over the temperature scale 35 to 45°C. The corresponding meter currents are 0 to 10 μA. A disadvantage of these direct reading bridges is their dependence on supply voltages, which must be very stable. *Figure 2.16* gives other examples of thermistor bridges useful over the range 0 to 250°C.

The measurement of differences in temperature under varying ambient conditions requires two thermistors connected in a bridge circuit. The change of resistance with temperature of a thermistor, dR/dT, depends upon the magnitude of T, i.e. on the ambient temperature; also there is a wide variation of the

30

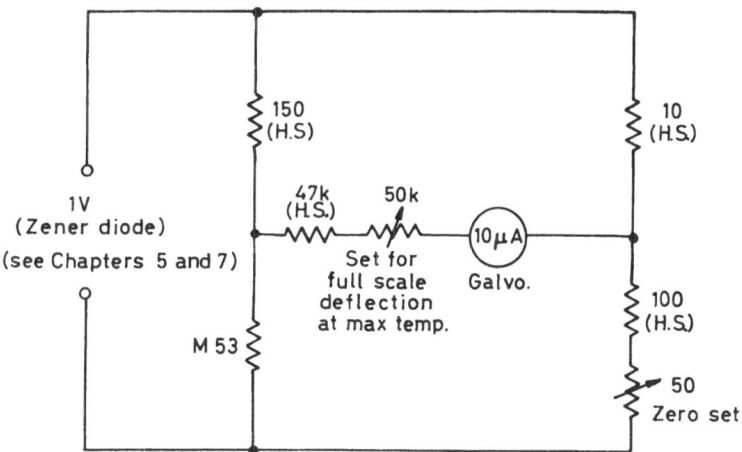

Figure 2.16a. A circuit suitable for measuring temperatures over the range 0–250°C. The M53 is a small bead set upon a disc about 5 mm in diameter. This type is particularly suited to measuring surface temperatures. A Pye Scalamp is a suitable galvanometer

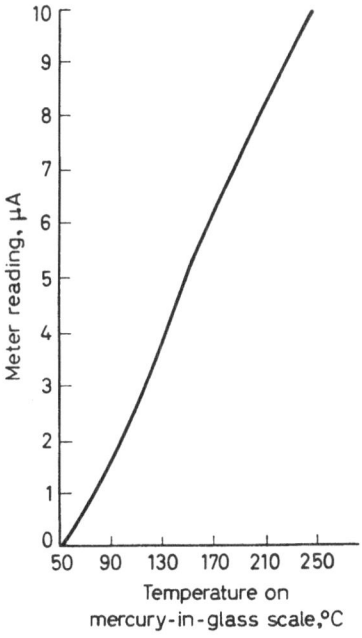

Figure 2.16b. Calibration curve for the circuit of *Figure 2.16a*

31

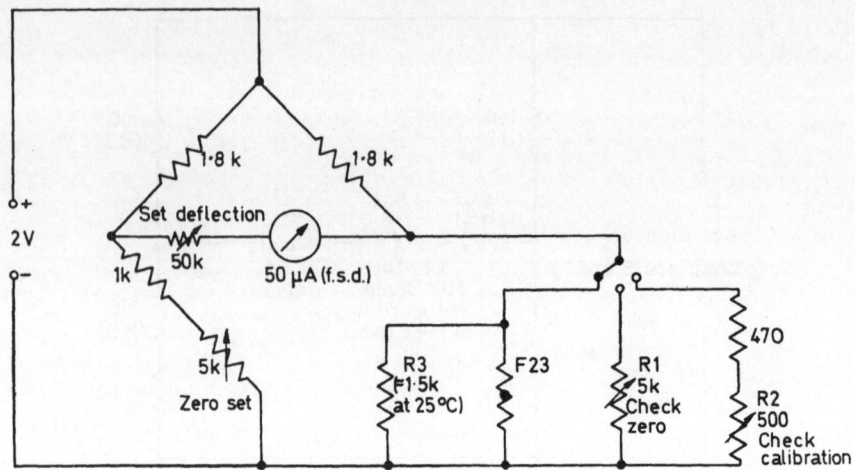

Figure 2.16c. Circuit for measuring temperature in the range 0–50°C (after Atkins and Settering[5]; see also Scarr[4]). R1 and R2 eliminate the need for a stabilized supply. R1 is preset to a value equal to the parallel combination resistance of R3 and the resistance of the thermistor at the desired starting temperature. R3 is equal to the resistance of the thermistor at that temperature which is midway between the maximum and minimum values (i.e. 25°C when the desired range is 50°C). R2 is preset so that its value, in series with the 470 Ω resistor, gives a total resistance equal to the parallel combination of R3 and F23 at some desired calibration temperature (e.g. 50°C corresponding to full-scale deflection)

constants a and b in equation 2.1 even among thermistors of the same type. However, the costly selection of matched components for the bridge can be avoided by using the method described by Nordon and Bainbridge[3]. *Figure 2.17* shows the conventional bridge circuit. The linearizing resistors r_1 and r_2 are usually equal to the resistances of the respective thermistors at the middle of the temperature range. (The analysis of this is given in the S.T.C. booklet mentioned above.) The errors associated with the non-linearity have been further reduced by Nordon and Bainbridge by making the linearizing resistance, r, equal to the expression

$$r = R_T \cdot \frac{(b-2T)}{(b+2T)} \tag{2.2}$$

where T the absolute ambient temperature, b the exponent in equation 2.1, and R_T the resistance of the thermistor at the middle of the temperature range. The error in linearization for a range of 50 degrees (Celsius) is about 2 per cent for one sample when $r = R_T$, but only 0·5 per cent when equation 2.2 is used.

The resistances of R_3 and R_4 should be large compared with the parallel arrangements in the remaining arms. This then keeps the current flowing through the thermistors almost constant when the thermistor resistance varies. A good practical arrangement is to have R_3 and R_4 each about 100 times the resistances in the other arms. The variable resistor brings the bridge into balance and thus sets the zero; the adjustment is necessary to allow for variations between thermistor samples. Once the bridge is balanced with the thermistors each at the same temperature, it will remain so at other ambient temperatures provided the temperatures of both thermistors remain equal.

$$R_T = a \exp (b/T)$$

$$r_1 = R_{T1} \frac{(b-2T)}{(b+2T)}$$

where R_{T1} is the resistance of the thermistor Th1 at a temperature of $T°K$.

$$r_2 = R_{T2} \frac{(b'-2T)}{(b'+2T)}$$

where b' is the constant in the exponent for thermistor Th 2

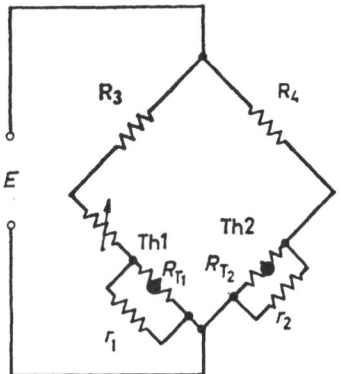

Figure 2.17. The use of unmatched thermistors for the measurement of temperature difference under varying ambient conditions (Nordon and Bainbridge[3])

In vacuum systems the thermal conductivity of the gas is a function of the gas pressure. By ensuring a constant wattage dissipation in the thermistor the temperature of the thermistor, and thus its resistance, is a measure of the gas pressure. Using a bridge circuit a range of 10^{-5} mm to 10 mm of mercury can be measured (Dummer[1], p. 93).

Thermistors with positive temperature coefficients are available. In a circuit diagram they are represented by the symbol shown in *Figure 2.18.* Their main use is to protect current-carrying circuits from damage due to excessive rises in temperature. For example an electric motor may be subjected to dangerous rises in temperature in the event of sustained overloading, a locking of the rotor arm or blocked ventilation ducts. Such a temperature rise may be prevented by including in the power lead a thermistor with a positive temperature coefficient. At normal temperatures the thermistor's resistance is low and presents little interference with the current flow. Above a certain temperature, known as the Curie temperature, the resistance rises sharply, thus preventing excessive current flow.

Positive temperature coefficient (p.t.c.) resistors are made from barium titanate. Below the Curie temperature the resistance is almost constant; but once the Curie temperature has been exceeded there is a rapid rise of resistance. *Figure 2.18* shows the curves for two typical examples. The Curie, or switch, temperature is determined by the manufacturing techniques used. A range of switching temperatures is available. For example Mullard market p.t.c. resistors in the form of discs fitted with connecting leads and protected by an insulating lacquer. The switching temperatures for these types are 35, 50, 80 and 110°C.

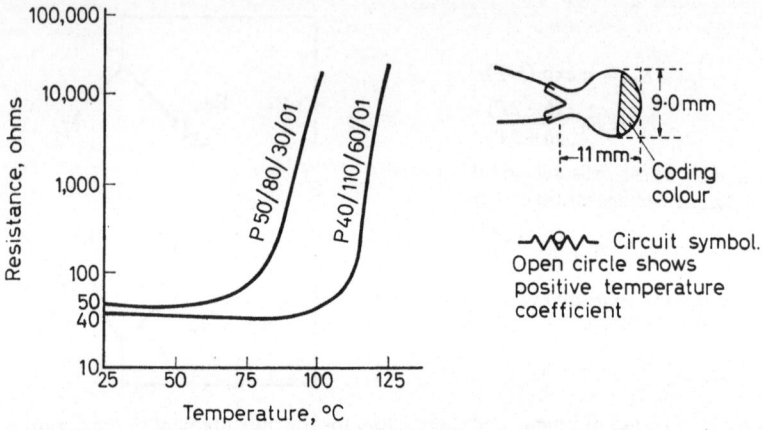

Figure 2.18. Resistance/temperature curves for two Mullard positive temperature coefficient thermistors

The voltage-sensitive type of non-linear resistor is a ceramic-like resistance material with the unusual property of being able to change its resistance in accordance with the applied voltage. Metrosil is the trade name used by A.E.I. Ltd. for this type of resistor. It is formed by dry-pressing silicon carbide with a ceramic binder and firing at about 1,200°C. The usual shape takes the form of discs or rods. The relationship between the applied voltage and current can be stated

$$I = KV^\alpha$$

where K is a constant equal to the current in amps when $V = 1$ volt. Its value depends upon the physical dimensions and composition of the material and also on the manufacturing process. α is a constant which is independent of the physical size of the resistor; its value depends only on the composition and manufacturing process. Common values of α lie between 2 and 6. The value of

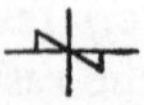

Figure 2.19. Circuit symbol for a voltage-sensitive resistor

34

K is usually between 3×10^{-2} and 10^{-5}. The circuit symbol for this type of resistor is shown in *Figure 2.19*. The component is used as a surge limiter, voltage stabilizer, and also for producing special waveforms and for changing the operating performance of a circuit.

CAPACITORS

Capacitors are components that have the ability to store electric charge. A capacitor consists of two conductors in close proximity separated by an insulator called the dielectric. The circuit symbol for a fixed value capacitor is shown in *Figure 2.20*.

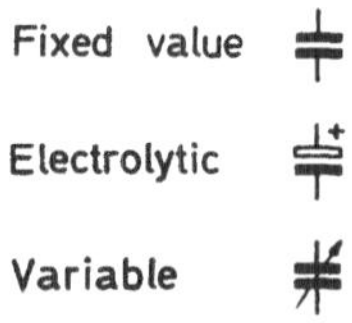

Figure 2.20. Circuit symbols for capacitors

Fixed value

Electrolytic

Variable

If a potential difference, V, is established across the dielectric by connecting the conductors to the terminals of a battery or some other generator of steady voltage, a certain charge, q, is stored. Doubling the voltage increases the charge by a factor of 2; and, in fact, the ratio of charge to potential difference is constant, i.e.

$$\frac{q}{V} = C \quad \text{(a constant)}$$

The constant C is known as the capacitance. When q is in coulombs and V in volts, C is in farads. The farad, F, is a very large unit of capacitance for electronic purposes so the microfarad, μF, which is one millionth of a farad, is used. The picofarad (pF) is one millionth of a microfarad so $1 \text{ pF} = 10^{-6}\mu\text{F} = 10^{-12} \text{ F}$.

Since

$$\frac{q}{V} = C \quad \text{i.e.} \quad q = CV$$

$$\frac{dq}{dt} = C\frac{dV}{dt}$$

We have already seen that dq/dt (a rate of change of charge with time) is the current flowing, therefore

$$i = C\frac{dV}{dt} \tag{2.3}$$

The simple Ohm's law relationship for resistors cannot therefore be applied to capacitors since the current is proportional not to the applied voltage, but to the rate of change of the voltage. An alternative way of expressing equation 2.3 is

$$V = \frac{1}{C} \int i \, \mathrm{d}t$$

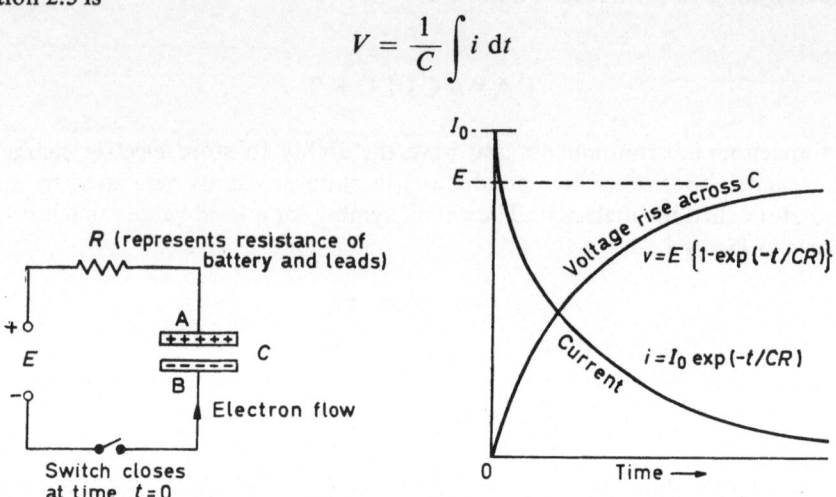

Figure 2.21. The charging of a capacitor from a source of steady voltage. Note that the current is a maximum when the voltage across C is zero, i.e. when C is uncharged. The current falls to zero when the voltage rises to a maximum. v and i are the instantaneous values of voltage and current respectively. I_0 is the current flowing immediately the switch is closed

The application of a steady voltage to the plates of a capacitor may be understood by referring to *Figure 2.21.* Let the switch be closed at time $t = 0$. At this instant there is no charge on either plate. Electrons from the upper plate, A, are attracted to the positive side of the battery and flow into that terminal. Since the current in a series circuit is the same everywhere, the number of electrons flowing out of the upper plate equals the number of electrons repelled by the negative terminal on to the lower plate, B. The upper plate, therefore, becomes positively charged because of a deficiency of electrons, and the lower plate becomes negatively charged. The electrons from the upper plate come from the atoms of the material from which the plate is made. (In a conductor at room temperature, there are many millions of valence electrons that are detached from their parent atoms and are in free random motion unless acted upon by an electric or magnetic field.) After a short time, depending upon the size of the capacitor and the resistance of the battery and lead wires, the voltage between the capacitor plates rises, for all practical purposes, to that of the battery. The rise is an exponential one, as shown in the formula given in *Figure 2.21.*

36

If now the switch is opened the capacitor remains charged because there is no way for the electrons on plate B to reach plate A. The capacitor thus stores the energy received from the battery and holds it until needed. (We are, of course, assuming that the dielectric is a perfect insulator. In practice the dielectric is not perfect and some leakage of charge from B to A occurs.) If a conductor is placed across the plates of a charged capacitor, electrons will flow

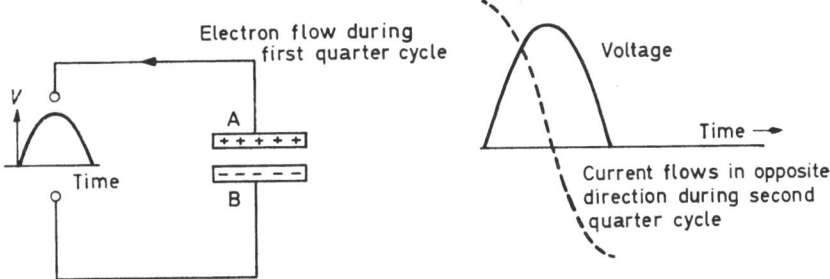

Figure 2.22a. Application of a sinusoidal voltage to a capacitor. After the regular ebbing and flowing has been established (i.e. the so-called steady-state condition) the voltage and current waveforms are as shown on the right. When the voltage is zero the current is a maximum. At maximum voltage the current is zero and the capacitor is fully charged. As the applied voltage falls, the voltage on the capacitor causes current to be returned to the source

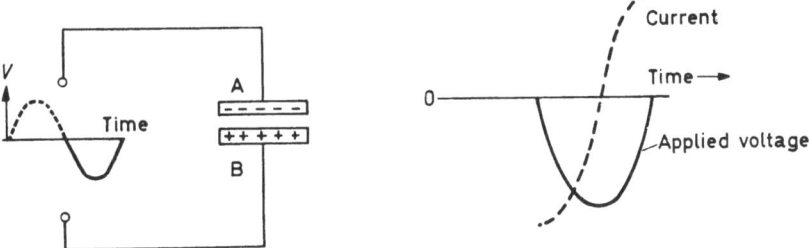

Figure 2.22b. Current and voltage waveforms during the next half-cycle of applied voltage

from B to A, thus returning the capacitor to its original uncharged state. The energy will be dissipated in the form of heat in the conductor since the latter has resistance.

In considering the charging action, it will be realized that the voltage built up across the capacitor plates is a counter voltage which opposes the battery voltage. This counter voltage is at a maximum when the capacitor is fully charged and the current has dropped to zero.

If the battery is replaced by some source of alternating voltage then the supply voltage will be constantly changing. If the change is a sinusoidal one, as it is on the alternating current mains supply, then the capacitor will be constantly charging and discharging as is shown in *Figure 2.22*. There will be a constant ebbing and flowing of electrons in the wires from the supply up to the capacitor and thus an alternating current will be flowing. In the d.c. case, once the initial charging is over no further current flows. The capacitor thus acts as an open circuit in that it offers infinite resistance to the steady voltage supply. However, in the alternating current (a.c.) case we obtain an alternating current as long as the supply is connected. No electrons, of course, actually pass through the dielectric of the capacitor; the latter is charging to one polarity, discharging and then charging to the reverse polarity, but so far as the rest of the circuit is concerned an alternating current flows just as it would if the capacitor were replaced by a resistor. We shall develop this further in the next chapter when considering some aspects of alternating current theory. For the moment, the reader should note that the maximum voltage across the capacitor occurs at a time when the current falls to zero. The maximum current flows when the voltage between the plates is zero, i.e. when the capacitor is uncharged. We see, therefore, that the current into the capacitor and the voltage across the plates do not achieve their maximum values at the same time as they would do if the capacitor were replaced by a resistor. The current and voltage are thus out of step in a time sense; they are said to be out of phase.

Three factors affect the capacitance of a capacitor; they are the area of the plates, the distance between the plates, and the dielectric material. The greater the area of the plates, the greater will be the capacitance. For a given dielectric and plate area the closer the plates are to each other, the greater will be the capacitance. The type of material used between the plates also affects the capacitance. Associated with this insulating material is a constant known as the dielectric constant. For air (strictly, vacuum) this constant is 1 whilst for mica and paraffin-waxed paper it is 6 and 2·2 respectively. The greater the dielectric constant the greater the capacitance. The choice of dielectric is not made by considering the dielectric constant alone. The dielectric strength (i.e. the voltage per given thickness before electrical breakdown occurs) also must be considered. Economic and other factors are also involved; mica for example is an excellent dielectric, but the cost of making a capacitor of several hundred microfarads would be prohibitive. No dielectric is perfect and some leakage is inevitable. For mica this leakage is very small, but in electrolytic capacitors leakage currents may be as high as several milliamps. In some circuit positions, e.g. in power supplies, leakage is not important, but in other locations leakage may be of paramount importance.

The physical appearance and properties of a capacitor vary a good deal depending upon the nature of the conductors and the dielectric material. Values of capacitance from 1 pF to several thousands of microfarads are readily available. One classification of capacitors depends upon the dielectric used. The dielectric materials in common use are wax-impregnated paper, mica, high-permittivity ceramics (permittivity is another name for dielectric constant), electrolytic oxide films and plastics such as polystyrene, polythene and polytetrafluoroethylene.

If several capacitors with differing dielectrics were given equal charges, the latter would leak away at different rates and hence the potential difference between the plates would drop at different rates. For polystyrene dielectrics the leakage time is several days, for impregnated paper several hours, and for ordinary plain-foil electrolytics only seconds need elapse. The ultimate dielectric strength is measured in terms of the voltage needed to puncture or rupture a dielectric of given thickness. Thickness and the nature of the dielectric therefore determine the maximum voltage that can be applied. When selecting a capacitor for a particular purpose therefore a knowledge of the capacitance is not sufficient; the maximum voltage likely to exist between the plates, and hence the working voltage of the capacitor, must also be specified.

General purpose capacitors use wax- or oil-impregnated paper as the dielectric. Two long rectangular aluminium foils separated by a slightly larger strip of the impregnated paper are rolled up like a Swiss Roll. They may then be inserted into an insulated cylinder and sealed at the ends. Connecting wires to each foil are brought out separately from each end; alternatively, the foil rolls may be encapsulated in some plastic insulating material. This type of capacitor is relatively cheap and has a reasonable capacity-to-volume ratio. The range of capacitance available is from about 100 pF to 1 or 2 μF. Working voltages vary, but values up to 500 V are common. The value of the capacitance is clearly stated on the body of the component together with the tolerance (usually ±20%). At one end there may be an ink or paint band passing right round the protecting cylinder. The wire from that end is connected to the outer foil, and in the circuit should be connected to 'earthy' points, e.g. the h.t. negative or to chassis or to the point with a potential nearest that of the chassis. Irrespective of the d.c. working voltage the a.c. voltage should never exceed about 300 V r.m.s. at 50 Hz. (Hz is now the accepted way of expressing a frequency and is equivalent to c/s, i.e. cycles per second. Hz is an abbreviation of Hertz.) If alternating voltages in excess of this are used, Dummer[1] has pointed out that burning of the paper may result due to repeated ionization of air bubbles trapped during the manufacturing process. The construction of the capacitor as rolls of foil makes it inevitable that some inductance is associated with the capacitance. This inductance can be troublesome at high frequencies.

Capacitors using mica dielectrics have a capacitance range from a few pF up to about 20 nF (1 nanofarad, nF = 1,000 pF = 10^{-9} F). They have a low power factor (which means that the losses are low), a high operating voltage and excellent long-term stability. Precision mica capacitors are used as sub-standards and can be adjusted to have values within 0·01% of the nominal value. For ordinary purposes silver films are formed by reduction of the oxide on to both sides of the selected mica plate.

High values of capacitance are obtained by stacking several metal foils interleaved with mica dielectric. Tolerances of 5%, 2% and 1% are commonly available. The mica capacitors are protected in one of two ways. They may

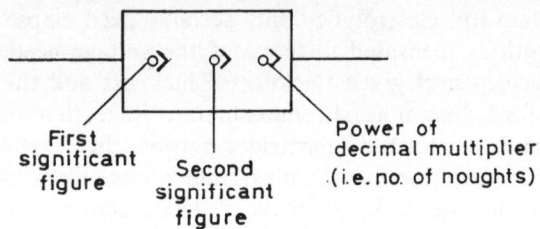

Figure 2.23. Colour coding for mica capacitors using the three-dot system. The colours represent the same numbers as in the resistance colour code. Tolerance is ±20% and the voltage rating is 500 V

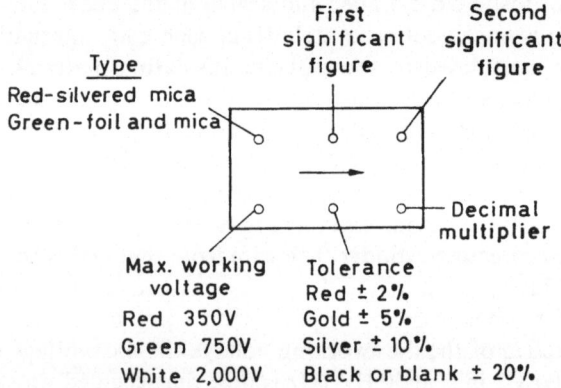

Figure 2.24. British six-dot system for coding mica capacitors. The upper row is read from left to right and the lower row from right to left.

be sandwiched between insulating boards and waxed to keep out the moisture, the capacitance value and tolerance being marked in ink or paint on the board in figures or words; alternatively, the capacitor may be housed in a moulding of insulating material. The values in picofarads and tolerances are indicated by colour codes. *Figures 2.23* and *2.24* show two frequently used arrangements of coloured codes.

40

Ceramic dielectric capacitors are made with three different values of permittivity. The low-permittivity types (often referred to as low-k types since k is a symbol often used to represent permittivity) are made from steatite and have an excellent high frequency performance. The medium-permittivity types have a negative temperature coefficient and are therefore used as temperature compensating capacitors in tuned circuits. The high-k types with dielectric

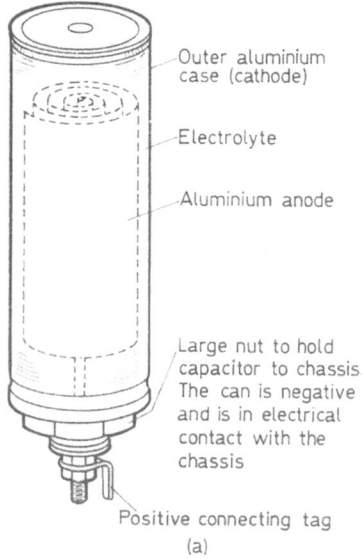

Outer aluminium case (cathode)

Electrolyte

Aluminium anode

Large nut to hold capacitor to chassis. The can is negative and is in electrical contact with the chassis

Positive connecting tag

(a)

Figure 2.25a. The construction of a 'wet type' of electrolytic capacitor

constants up to 1,200 are used where space is at a premium. The high-k value permits physically small capacitors to have a comparatively large capacitance. Unfortunately, this latter type is very temperature-sensitive.

The capacitance values of the ceramic dielectrics range from 1 pF to about 10 nF, such values being indicated by a 'band + 4 dot' system, a five-dot system or marking in figures. There is unfortunately no standard system used throughout the Commonwealth, America and European countries, and some confusion arises in reading the markings on many of these capacitors. In general, the first and second dots represent the significant figures, the third dot the decimal multiplier, and the fourth dot the tolerance. The band if present indicates the temperature coefficient. The best way out of the difficulty is to buy capacitors from a known source using a known code.

Electrolytic capacitors are used when large values of capacitance are required with small physical size. Typical applications are in power supply smoothing filters, by-pass capacitors in bias circuits and in circuits incorporating transistors. In these locations the high leakages encountered are of little consequence. In the low-voltage range, capacitances up to thousands of microfarads are available. For the higher voltages in h.t. supplies (say up to 600 V) capacitance values of 8, 16, 32 and 50 μF are commonly used. Tolerances are fairly wide being usually —20% to +50%.

The ordinary plain foil types have a thin rectangular sheet of aluminium wound into a spiral *(Figure 2.25)* so as to be conveniently housed in a cylinder of similar metal. The spiral sheet is immersed in an electrolyte based on

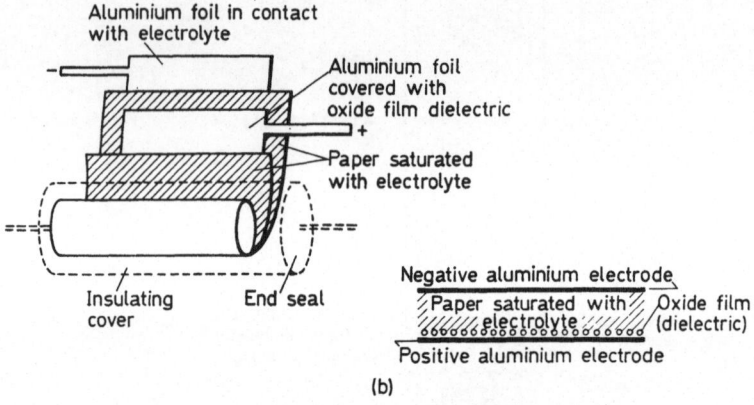

Figure 2.25b. The 'dry type' of electrolytic capacitor in which the electrolyte is in the form of a paste of glycol and ammonium tetraborate

ammonium tetraborate. The electrolyte is in contact with an outer housing cylinder of aluminium. An extremely thin layer of oxide is deposited on the aluminium spiral by anodic oxidation, the process being known as 'forming'. The large capacitance is a result of the oxide layer being so thin ($\sim 10^{-5}$ cm), and the effective area being much increased on etching. One plate of the capacitor is the aluminium spiral, and the other conductor is the electrolyte itself. The dielectric is the oxide layer, which is electrically very strong. If, however, breakdown does occur by the application of not too great an overload then on removing the voltage source the oxide layer will re-form. This self-healing process is a further advantage of electrolytic capacitors. Care must be taken to ensure that these capacitors are connected correctly in the circuit since, unlike previous types, they are polarized. The manufacturer clearly marks which connection must go to the positive potential. The circuit symbol for this type of capacitor is shown in *Figure 2.20* where the unfilled bar represents the positive end. The capacitance value is marked on the body in figures together with the maximum working voltage conditions. Needless to say, alternating voltages must not be applied across this type of capacitor. Variations superimposed on a steady d.c. level are allowed, but the maximum alternating current through the capacitor must not exceed the manufacturer's specified limit.

Tantalum-foil electrolytics are now well developed for use in transistorized equipment. They have the advantage of small physical size, long shelf-life (in contrast to other types of electrolytics which need re-forming if stored for long periods), and great reliability. The range available is mainly for low-voltage operation.

Variable capacitors take two forms. The first is probably best known to readers as the tuning agency in radio receivers. This type consists of a movable set of specially shaped plates or vanes that interleave with a set of fixed plates when the spindle is rotated. The dielectric most frequently used is air although thin sheets of flexible insulant are sometimes found in older capacitors and where space is at a premium. When used in radio sets, oscillators and filters, two or more capacitors are often ganged so that rotation of only a single shaft is required to alter capacitance values in several circuits simultaneously. The values lie usually in the range 50 pF to 500 pF.

Trimmers, or pre-set capacitors, are used when the capacitance needs changing only very infrequently. These may be miniature versions of the larger variable capacitors; alternatively, they may be of the compression type where, by turning a screw, the metal foils and dielectric are compressed to a greater or lesser degree, thereby changing the distance between the metal foils and thus altering the capacitance. A common range of value for a trimmer is 3–30 pF but maximum values of 50 pF to 750 pF are readily available.

Precision variable capacitors are in principle the same as the air-dielectric tuning capacitor, but the workmanship of standards made by firms like Sullivan Ltd. and Muirhead Ltd. is superb. The engineering of the gears and bearings involved is of a very high order and naturally the cost of such capacitors is very high. Quartz pillars are used to insulate the end-plates carrying the fixed plates of machined aluminium. Duralumin shafts carry the movable vanes and are mounted between hardened steel centres or jewelled bearings.

Capacitors in parallel

When capacitors are connected in parallel, the same voltage exists across each capacitor. Each capacitor, therefore, acquires a charge proportional to its capacitance. The total charge is the sum of all the charges. The supply, therefore, is presented with a total capacitance, C, equal to the sum of the individual capacitances C_1, C_2, etc.,

$$\text{i.e.} \quad C = C_1 + C_2 + C_3 \dots$$

Connecting capacitors in series results in a reduction of the total capacitance to a value less than that of the smallest capacitor used. The formula for the total capacitance may easily be shown to be given by

$$C = \frac{1}{\dfrac{1}{C_1} + \dfrac{1}{C_2} + \dfrac{1}{C_3}} \cdots$$

INDUCTORS

Inductance is that property of an electrical circuit which tends to prevent any change of current in that circuit. Devices having the primary function of introducing inductance into a circuit are called inductors. All inductors consist of coils of insulated wire wound on to a suitable bobbin or former. Although the core of the coil may be air, it is more usual to concentrate the magnetic flux in the core by using suitable ferromagnetic substances; magnetically soft iron and ferrites are used as described below. Since inductors oppose changes of current, one of their functions in electronic circuits is to provide a large opposition to the flow of alternating current whilst simultaneously presenting very little opposition to the flow of steady currents. A common example of this is discussed in Chapter 7 dealing with power supplies; in the process of converting alternating current from the public electricity supply to direct current for use in electronic equipment, it is found that a large alternating ripple is superimposed on the desired steady current. By using an inductor, it is possible to filter out the alternating ripple. A low impedance path is provided for the unwanted alternating component by means of a capacitor; the inductor is used to 'choke off' the alternating component. When used in this way an inductor is often referred to as a choke. Chokes, or inductors, for use in power supplies must present a large opposition to the flow of alternating currents whose frequency is low. The frequency of the supply is only 50 Hz (60 Hz in the U.S.A.) and we shall see in Chapter 7 that the ripple frequency is usually twice the supply frequency, namely, 100 Hz. To present a sufficiently large impedance at such low frequencies, the inductance must be large. (The unit of inductance, defined below, is the henry; inductors for use in power supplies must have inductances of between 3 and 25 H.) In order to achieve such a high inductance in coils of reasonable physical size, the coil is wound around a core of magnetically soft iron. The inductance of any coil can be shown to be proportional to the square of the number of turns, the cross-sectional area of the coil and the permeability of the core. (Permeability is a measure of the ease with which a material can be magnetized.) The inductance of a given coil in air can, therefore, be increased

44

hy a factor of 1,000 if an iron core is used, since the permeability of iron is about 1,000 times that of air. To minimize the losses due to eddy currents in the core, the latter consists of thin sheets of iron, called laminations; these laminations are coated with a thin layer of insulating material which prevents eddy current of any significant magnitude being established. *Figure 2.26* shows the typical shape of the laminations used in low-frequency inductors. Such shapes aid the manufacture of inductors by automatic means. The coil itself is wound on a suitable former with aid of a machine; the core is then assembled by inserting the laminations in an interleaved fashion into the coil. The latter is thus surrounded by an iron path for the magnetic flux.

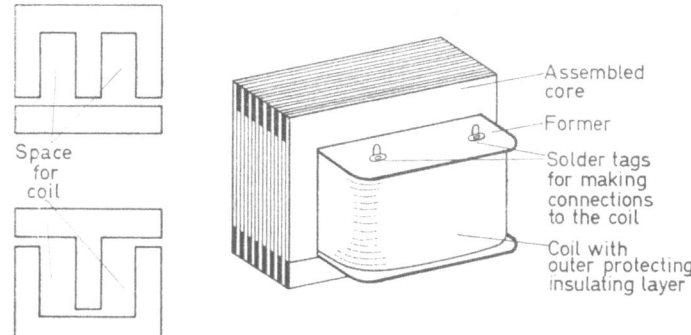

Figure 2.26. Construction of an inductor for power frequencies. The shape of the laminations is shown on the left. Either combination may be used

Such flux is confined almost entirely to the iron core and very little flux leakage exists. Since practically all the magnetic flux threads the coil, the inductor has its maximum inductance. Further details are given in Chapter **7** in which different types of low-frequency chokes are discussed.

To understand the way in which inductors present an opposition to alternating currents, but not to direct current, we must revise our ideas on electromagnetic induction. It is not possible here to go into any great detail; standard books on electricity and magnetism are the works to consult.

An electric current flowing through a conductor has a magnetic field associated with it. It is usual to visualize the magnetic field by inventing magnetic flux lines, i.e. the paths that would be taken by a fictitious isolated north pole. Such flux lines are shown in *Figure 2.27*. Whilst steady currents produce magnetic fields, Faraday showed that the converse was not true; to induce a current in a wire, it is necessary for the magnetic field to vary. Consider now the coil represented in *Figure 2.27*. If no current exists in the coil, and by some means a magnetic field is made to grow in the coil (e.g. by plunging a

45

bar magnet into the coil), an e.m.f. will be induced. The magnitude of the e.m.f. is given by

$$E = -\frac{dN}{dt}$$

where N is the magnetic flux and E is the e.m.f., i.e. the e.m.f. is equal to the rate of change of the magnetic flux. The minus sign arises because the direction of the induced e.m.f. is such as to oppose the change producing it (Lenz's law). This means that if an external circuit exists, currents will be produced in the coil which themselves give rise to a magnetic field; this latter field is

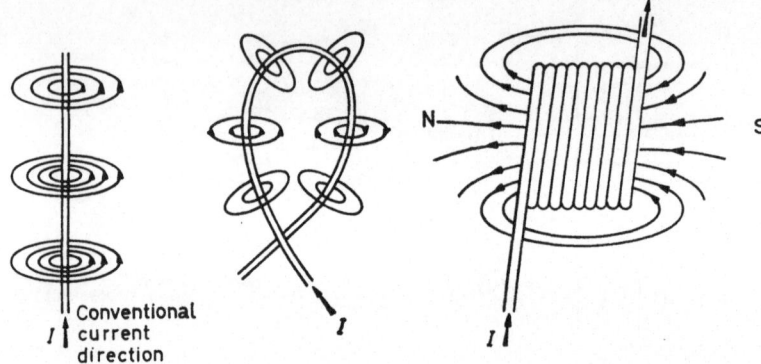

Conventional
I current
direction

I

I

Figure 2.27. Lines of force representing magnetic fields around current-carrying wires in which the current is steady. Steady magnetic fields associated with conductors do not however induce currents in the wires. To induce a current the field must be varying

opposite in sense to the magnetic field inducing the current. Instead of plunging a magnet into the coil let us apply a voltage to the ends of the coil. If the voltage is steady the only opposition to current is the resistance from which the coil is made. If the voltage is varying, however, there will be a continual change in the current which in turn produces a changing magnetic field. As the magnetic field grows, there will be increasing magnetic flux threading the coil. The rate of increase depends upon the frequency of the applied alternating voltage. Such a changing magnetic field induces an e.m.f. in the coil that opposes the applied e.m.f.; there is thus some opposition to the establishment of the magnetic field. When the applied e.m.f. is diminishing the magnetic field collapses. This collapsing field attempts to keep the current at its maximum value. Whether the applied e.m.f. is rising or falling therefore the inductor opposes changes of current in the circuit. For a given inductor the opposition increases as the frequency of the applied e.m.f. increases. In a sense, the inductor may be said to have 'electrical inertia'.

The back e.m.f. induced in the coil is proportional to the rate of change of flux which is itself proportional to the rate of change of current, therefore

$$e \propto -\frac{di}{dt}$$

where e is the back e.m.f. and i the instantaneous value of the current. Hence

$$e = -L\frac{di}{dt}$$

where L is the constant of proportionality known as the self-inductance of the coil. When i is in amperes, t in seconds and e is in volts, L is in henries (H). Thus if the current is changing in a coil of inductance one henry at the rate of one ampere per second then the back e.m.f. produced is one volt. Like the capacitor, no simple relationship exists between the applied voltage and the current; the implications of this are discussed in Chapter 3.

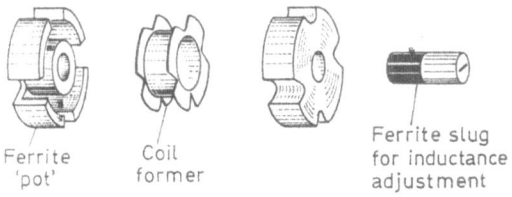

Ferrite 'pot'   Coil former   Ferrite slug for inductance adjustment

Figure 2.28. Essential parts of a 'pot-core' inductor

In Chapter 3 we shall discuss the combination of capacitors and inductors to form tuned or resonant circuits. Such tuned circuits enable a signal of one frequency to be isolated from other signals of different frequencies. Radio and television sets use many such circuits. Used in this way inductors must be designed for operation at frequencies much higher than 50 Hz. The losses associated with iron cores are too great to make operation at radio frequencies efficient; the cores of high frequency inductors are therefore made from ferrites, or iron-dust held in a suitable binding material. Ferrites are chemical compounds of the form MFe_2O_4 where M is a divalent metal commonly Mg, Mn, Zn or Ni. This non-metallic material combines reasonable permeability (from several hundreds up to about 1,200) with high resistivity. Eddy currents at high frequencies are therefore largely avoided.

Inductors for filter and other purposes are now based on what are called pot-cores (see *Figure 2.28*). Here the ferrites are cast as cylinders closed at one end. The coil of wire, wound on a suitable former, is placed within two of these cylinders so as to be completely shielded. The magnetic circuit is completed by a central cylindrical core. For a given inductor some variation of inductance is possible by adjusting the position of the central core or by varying the pressure at which the two ferrite cylinders are held together.

Mutual Inductance

If two coils are placed close to each other so that a varying magnetic field in one coil induces an e.m.f. in the second coil, the two coils are said to be inductively coupled. The changing magnetic flux due to the current in the first or primary circuit must interlink the secondary circuit in order to induce an e.m.f. in the secondary coil. The phenomenon is called mutual induction. The mutual inductance, M, between the two coils is measured in henries and depends upon (*a*) the number of turns in the primary coil, (*b*) the number of turns in the secondary coil, (*c*) the relative position of the coils, and (*d*) the permeability of the medium between the coils.

$$M = \frac{\text{Induced voltage } (e) \text{ in the secondary coil}}{\text{Rate of change of current in the primary}}$$

If the secondary coil forms part of a complete secondary circuit then the induced e.m.f. (*e*) will cause a current to flow in that circuit. This secondary current produces a flux that opposes the flux produced by the primary current, *i*.

$$e = -M\frac{\mathrm{d}i}{\mathrm{d}t}$$

where *i* is the primary current. The minus sign signifies that the induced voltage produces a secondary current, and hence flux, that opposes the primary flux.

When inductors are connected in series the total inductance is calculated in the same manner as that used for resistors in series. Since each inductor contributes to the opposition to change of current the total inductance, L, is equal to the sum of the individual inductors.

$$L = L_1 + L_2 + L_3 \ldots$$

If, however, mutual inductance exists between the individual coils then the total inductance will depend upon the relative connections of the coils. *Figure 2.29* shows the two possible connections for a pair of coils. If the magnetic fields are mutually assisting then the total inductance is given by $L = L_1 + L_2 + 2M$; whereas if the fields are opposing $L = L_1 + L_2 - 2M$.

Inductors in parallel can be represented by a total inductance given by

$$L = \cfrac{1}{\cfrac{1}{L_1} + \cfrac{1}{L_2} + \cfrac{1}{L_3}} \quad \cdots$$

assuming no magnetic coupling between the coils.

When two coils are electrically isolated, but connected by means of magnetic flux, they form a transformer. The principles of a mains transformer are discussed in Chapter 7. Radio frequency transformers use similar principles,

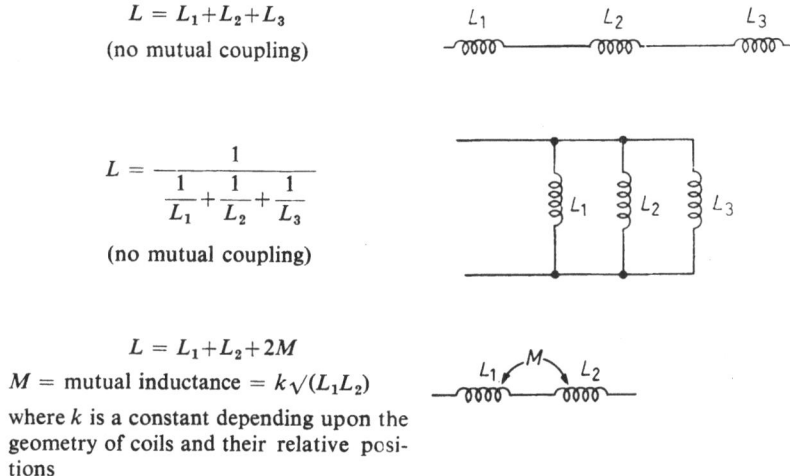

$L = L_1 + L_2 + L_3$

(no mutual coupling)

$$L = \cfrac{1}{\cfrac{1}{L_1} + \cfrac{1}{L_2} + \cfrac{1}{L_3}}$$

(no mutual coupling)

$L = L_1 + L_2 + 2M$

$M = $ mutual inductance $= k\sqrt{(L_1 L_2)}$

where k is a constant depending upon the geometry of coils and their relative positions

$L = L_1 + L_2 - 2M$

Figure 2.29. Formulae associated with combinations of coils

but are, of course, much smaller and use ferrite or iron dust cores; additionally the two coils are often associated with capacitors so as to form resonant circuits. The efficient transfer of power from one coil to another then takes place mainly at one frequency.

MINIATURIZATION

There has been a great deal of discussion and work done in recent years on the subject of microminiaturization and solid-state circuits. It is certain that in many applications the more familiar passive and active elements will, in their present form, be superseded. Using evaporation techniques, resistors, capacitors and inductors can be combined into single small circuit blocks. By using the solid-state circuit the passive elements can readily be integrated with transistors. A paper by Roberts and Campbell[6] gives some of the modern forms of R–C networks and shows by way of example a complete phase-shift oscillator in modern solid block form.

Extreme miniaturization is achieved by using epitaxial techniques in which a thin wafer of p-type or n-type silicon a few thousandths of an inch thick forms the base material. Layers are then grown on the surface in such a way as to preserve the single crystal lattice structure. By using photoengraving and etching processes semiconductor resistors and active devices (transistors and diodes) can be formed. (Chapter 5 on semiconductor devices explains the semiconductor terms used and the way in which transistors are fabricated.) The final result is that an entire electronic circuit can be formed on a small silicon chip; for example, manufacturers today have no difficulty in making an amplifier consisting of 16 to 20 transistors and the associated resistors on a silicon chip no larger than 75 thousandths of an inch square. Whole transmitters consisting of the necessary amplifiers, oscillators and mixers can easily be accommodated within a small metal can of the type now being used to house a single conventional transistor.

When this form of equipment is widely adopted servicing will be a thing of the past. On the one hand the units promise to be economic, very reliable, and to have a greatly extended useful life; on the other hand, if, for some reason, failure of a particular component occurs, the block or unit must be discarded since no repair is possible. It must be emphasized that these units are not 'just around the corner' so far as the bulk of general laboratory equipment is concerned. They are being developed primarily for satellite and space research. Whilst they will no doubt influence the form of future electronic equipment, it will be many years before we see the last of thermionic valve and transistor apparatus in its present form. Computers and digital control apparatus will be the first equipment to use monolithic, integrated circuits in large numbers. There are, in fact, already available some computing systems that incorporate these miniature circuits.

An alternative miniaturization construction is based upon thin-film circuits. The final arrangement is larger than the monolithic circuit, but production techniques are somewhat easier. For small runs the thin-film circuits arc cheaper to produce. The process involves the vacuum deposition of resistive material (e.g. nichrome) on insulating substrates. Masking, photoengraving and etching processes produce the final resistor. Small capacitors are formed from aluminium films separated by a silicon dioxide dielectric. The transistors are subsequently added to complete the circuit.

The whole subject of microminiaturization is a fascinating prospect for the future. Those interested in reading more about the subject should consult the papers listed in the references[6-9].

REFERENCES

1. Dummer, G. W. A. *Modern Electronic Components*. Published by Pitman & Sons Ltd. (second edition, 1966).
2. *The Design of Thermistor Bridge Networks for Direct Reading Thermometers.* Application Report of Standard Telephones & Cables Ltd. MK/167.
3. Nordon, P. and Bainbridge, N. W. 'The use of unmatched thermistors for the measurement of temperature difference under varying ambient conditions'. *J. Sci. Instrum.* 1962, **39**, 399.
4. Scarr, R. W. A. 'Thermistors, their theory, manufacture and application'. *Proc. Instn Elect. Engrs*, 1960, **107**, Part B. No. 35. See also 'Thermistor applications' by the same author in *Direct Current*, Dec. 1960.
5. Atkins, P. A., Settering, R. A. 'Thermistors—a survey of their applications in temperature measurement and control'. *Instrum. Pract.* 1959 (Oct.), 1042.
6. Roberts, D. H. and Campbell, D. S. 'Techniques of microminiaturization'. *J. Brit. Inst. Radio Engrs*, 1961, **22**, No. 4, p. 281.
7. Hittinger, W. C. and Sparks, M. 'Microelectronics'. *Sci. Amer.* 1965 (Nov.), 57.
8. Dummer, G. W. A. and Granville, J. W. *Miniature and Microminiature Electronics*. Pitman. 1962.
9. *Thin Film Circuits for Microminiaturization.* Mullard Publication, July 1963.

3

THE RESPONSE OF CIRCUITS
CONTAINING PASSIVE COMPONENTS

WAVEFORMS

An examination of a piece of electronic equipment under working conditions will reveal many different voltages and currents at various points throughout the circuit. The useful signal information is carried by voltages and currents that vary, so we frequently consider the graphs of the magnitudes of these electrical quantities as functions of time. The shape of a given graph is known as the waveform of the voltage or current being studied. In many cases the waveforms are periodic (i.e. they repeat themselves exactly in equal successive intervals of time) so the resulting graphs have a regular pattern. Experience in interpreting the waveforms in an electronic circuit often produces valuable information about the behaviour of the electronic apparatus or associated equipment. *Figures 3.1* and *3.2* show the essential features of several common waveforms and introduce some of the terms used in connection with them.

One of the important uses of electronic equipment in the laboratory is the processing of information. During his experiments the scientific worker must consider such quantities as temperature, pH, conductivity, colour, pressure, wavelength, etc. If these quantities are to be measured, recorded, or processed for a computer or control system, it is necessary to find suitable transducers which produce electrical voltages that represent quantitatively the quantities to be studied. Such transducers may be thermocouples for temperature measurements, glass electrodes for pH measurements, or photocells for wavelength or colour studies. The output from such transducers is rarely large enough to operate some recording or indicating device directly. It is the role of electronic equipment to amplify or modify the transducer outputs so as to produce signals of suitable magnitude or power.

All electronic apparatus, no matter how complicated it may seem when considered as a whole, may be regarded as the combinations of various fundamental circuit arrangements. Once we have studied these fundamental circuits, and realized that they are common to many seemingly totally dissimilar items of equipment, then we shall be in a position to understand the functioning of such diverse instruments as a pH meter, a cathode-ray oscilloscope,

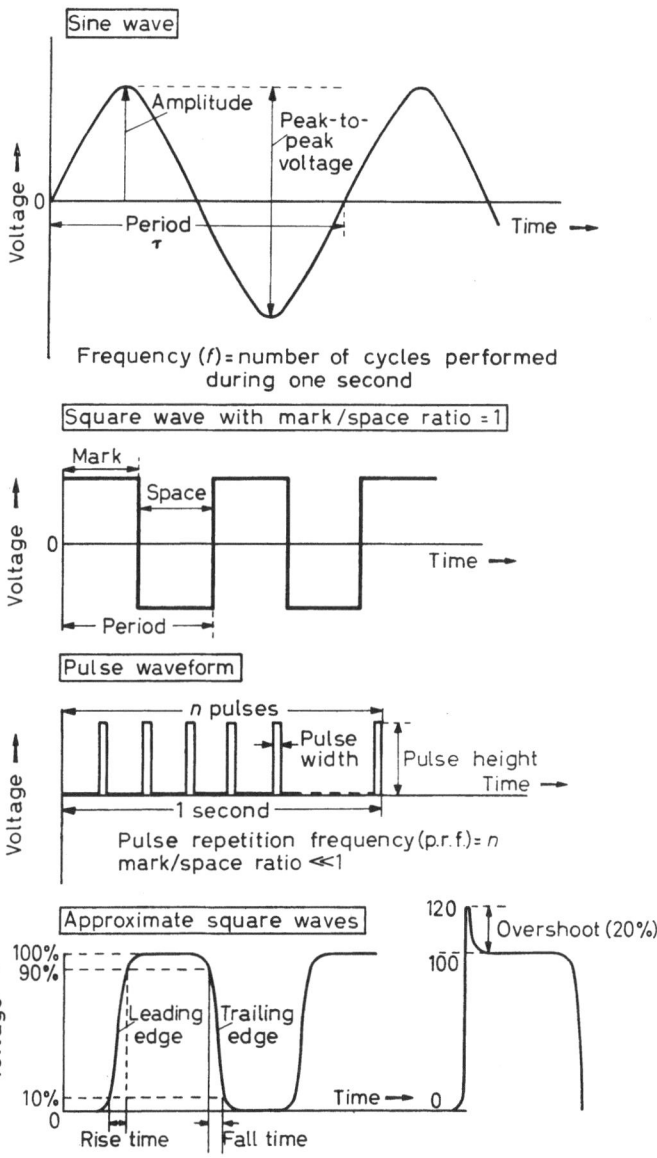

Figure 3.1. Some common waveforms with associated terms

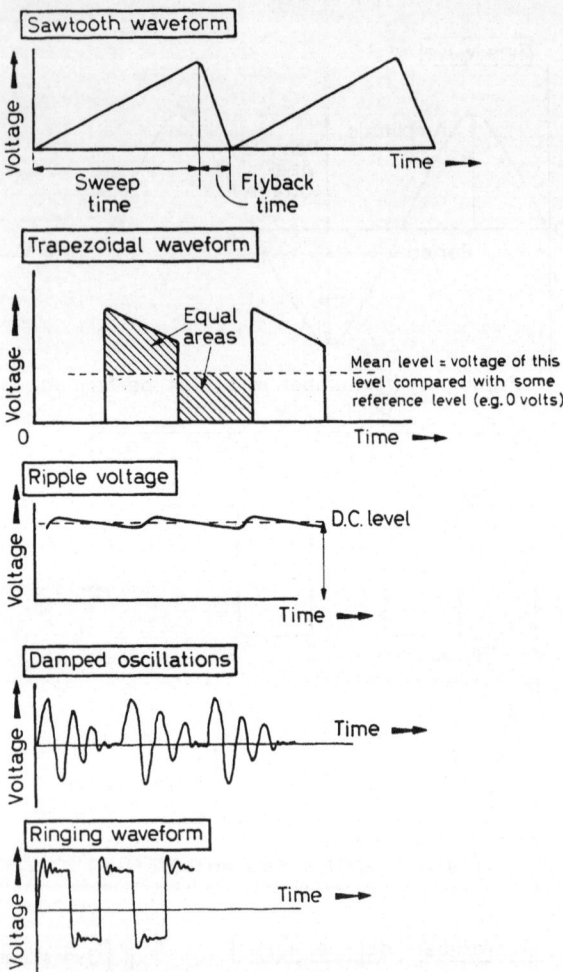

Figure 3.2. Further examples of waveforms

a valve-voltmeter, a temperature controller, etc. The electronic engineer must take care to design his equipment so as to reduce to a minimum any distortion of waveform in amplifying devices, or to produce a desired result with accuracy (e.g. an output voltage which is the integral of an input voltage or, say the sum of several input voltages). The equipment he designs is assembled from passive components—the resistors, inductors, capacitors already discussed—combined with the amplifying or active components such as thermi-

54

onic valves and transistors. This chapter is devoted to the discussion of the effects various standard arrangements of resistance, capacitance and inductance have on signal waveforms.

The Capacitor-Resistor (CR) Circuit

Let us consider first the circuit of *Figure 3.3*. If at time $t = 0$ the switch is moved to A and allowed to stay there, the accompanying graphs show the resultant voltages across C and R at any subsequent time. The sum of the capacitor voltage, V_C, and the voltage across the resistor, V_R, must equal the supply voltage, V. After a time equal to about six time constants the charging current will be practically zero; as a result $V_R = 0$, and the whole of the supply voltage appears across the capacitor. On moving the switch to B, the capacitor will discharge through the resistor. V_C therefore falls to zero. Since now $V_C + V_R = 0$, the voltage across the resistor will be $-V$ initially and subsequently will approach zero as the capacitor becomes discharged. The position

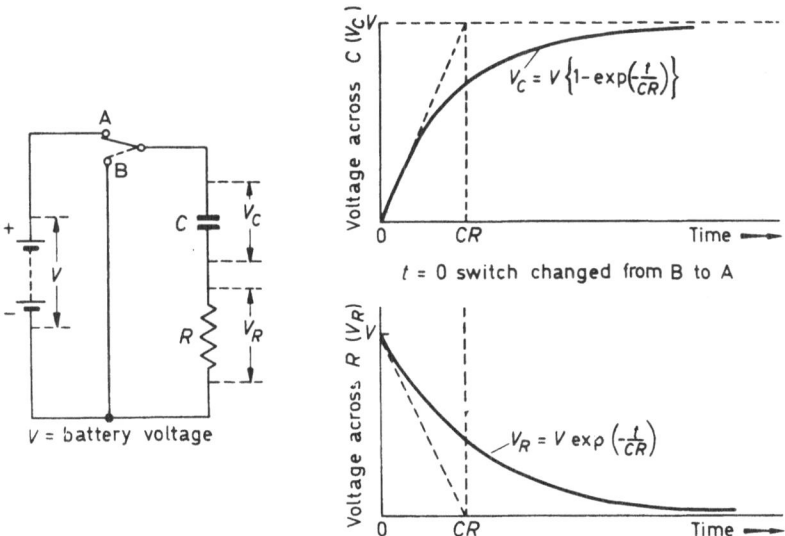

Figure 3.3. Voltage waveforms associated with C and R during charging. $V_C + V_R = V$. $CR =$ time constant, i. e. the time it would take for the voltage across the resistor to become zero if the original rate of charging the capacitor had been maintained. When $t = CR$, $V_R = V/e$ where e is the exponent of the natural logarithms equal to 2·718. An alternative definition for the time constant is therefore the time it takes for the voltage across R to fall from V to about 37% of V. In the same time the charging of the capacitor will have reached a stage where the voltage across the capacitor has become $\{1 - (1/e)\}\ V$, i.e. about 63% of V

is shown in *Figure 3.4*. If now the switch is operated continuously, staying as long in position A as in position B, the resulting waveform presented to the CR circuit will be rectangular with a mark-space ratio of 1. The waveforms across *C* and *R* will now depend on the period of the square wave and the values of *C* and *R*.

Figure 3.5 shows the resulting waveforms when a square wave with a period of about *CR* is applied to a capacitor-resistor circuit.

A simple differentiating circuit is formed when the time constant *CR* is much smaller than the period of the applied waveform. It is shown in Appen-

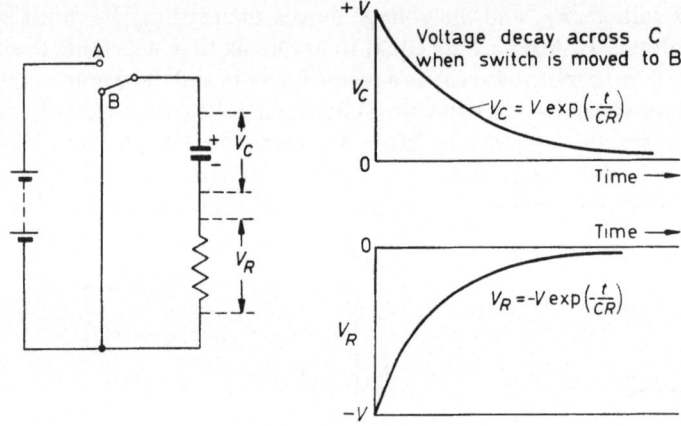

Figure 3.4. Voltage waveforms associated with *C* and *R* when the switch is moved to B. *C* then discharges through *R*. $V_C + V_R = 0$

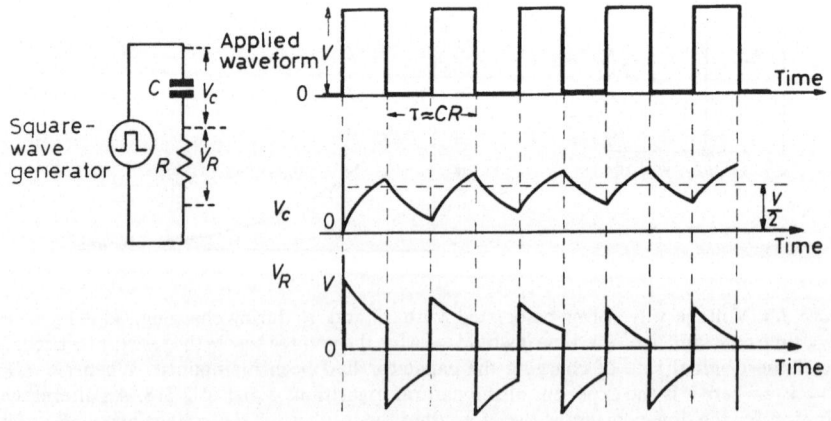

Figure 3.5. Waveforms resulting from the application of an input square wave with a period of about *CR*

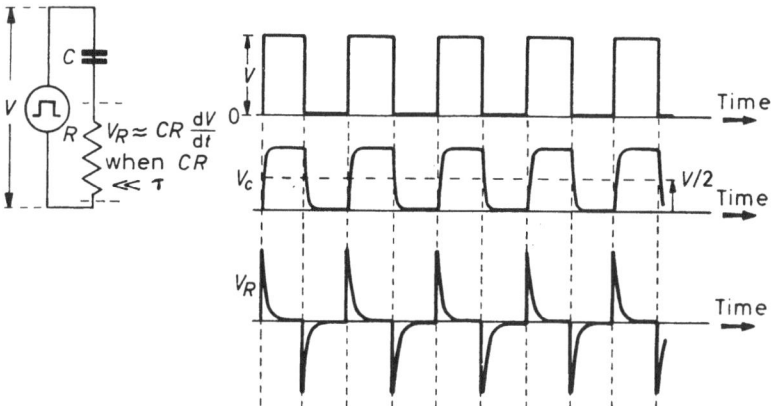

Figure 3.6. A simple differentiating circuit is formed when small values of *C* and *R* are used, i.e. *CR* ≪ the period of the applied waveform

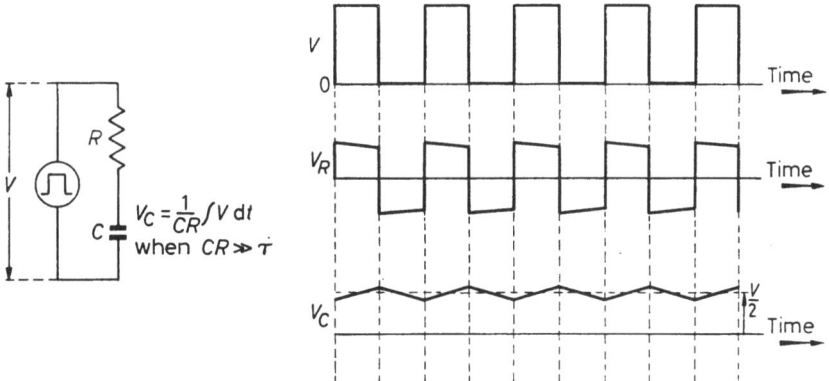

Figure 3.7. The simple integrating circuit. The output voltage across the capacitor is approximately proportional to the time integral of the input voltage when *CR* ≫ the period of the applied waveform

dix 1 that the voltage across the resistor is a close approximation to the differential coefficient of the input voltage with respect to time when *CR* is small. *Figure 3.6* shows the waveforms involved.

A simple integrating circuit is formed when the time constant *CR* is much longer than the period of the applied waveform. In this case, as Appendix 1 shows, the voltage across the capacitor is approximately the time integral of the applied voltage, *V*, provided *CR* is at least five or six times longer than the period of the applied voltage (see *Figure 3.7*).

The coupling circuit of *Figure 3.8* is extensively used to connect one stage of an amplifier to the next. The usefulness of this arrangement will become increasingly apparent as the reader progresses in his studies of electronics. The object of this circuit is to transmit only the signal waveform from one part of the circuit to the next, blocking out any steady component that the

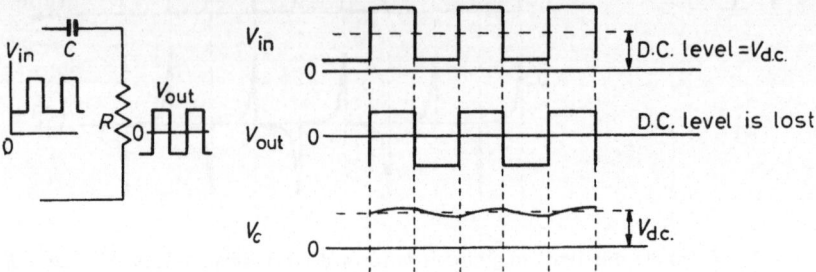

Figure 3.8. The coupling circuit. Here C and R are made large so that CR is a good deal greater than the period of the lowest frequency waveform it is desired to handle

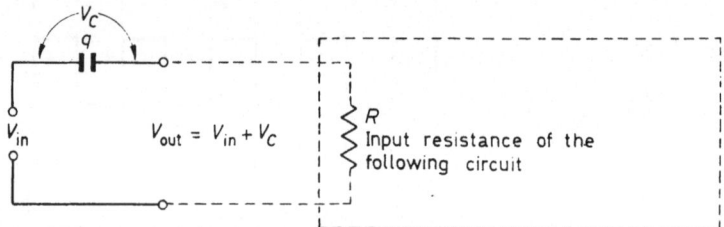

Figure 3.9. By keeping the charge on the capacitor, q, constant V_C is constant. V_{out} will vary only if V_{in} varies. If there is no leakage in C then the variations are transmitted without change in magnitude or waveform. Only the d.c. level is lost. In practice, the following circuit inevitably allows some change in the charge, q, but this change is usually kept to a minimum by using a large value of C and having a large value of input resistance in the following circuit

input signal contains. Provided the time constant is large, say greater than five or six times the period of the input signal, the voltage across the resistor has a waveform almost identical with that of the varying part of the input signal. It will be remembered from Chapter 2 that the voltage across a given capacitor depends upon charge stored. If that charge does not alter, the voltage across the capacitor will not change. We can arrange that very little charging or discharging of the capacitor takes place by making R large. For

58

example in a coupling circuit to be used with signals whose frequency is as low as 50 Hz, the value of R may well be 470 k with a capacitance value of 0·25 μF. The resulting time constant, CR, is $0·25 \times 10^{-6} \times 470 \times 10^3 \approx 0·12$ seconds, which is long compared with the period of a 50 Hz signal, i.e. 0·02 seconds. *Figure 3.9* further illustrates the point about the transmission of a variable voltage by a capacitor. It must be emphasized that the function of the capacitor in a coupling circuit is to block out the d.c. level and to transmit only variations of the input signal. This it does by *not* charging or discharging. In so far as charging and discharging occurs the capacitor is failing to transmit the variations properly. We shall see in later chapters that some charging and . discharging is inevitable, but we must arrange for this to be a minimum.

The Inductance-Resistance (LR) Circuit

Inductance in a circuit resists changes in current in that circuit, consequently if a series LR circuit is supplied with square waves, pulses, or other waveforms having fast rise times, the waveforms across the inductor or resistor will differ from that of the signal waveform. *Figure 3.10* shows the waveforms produced with an applied square wave.

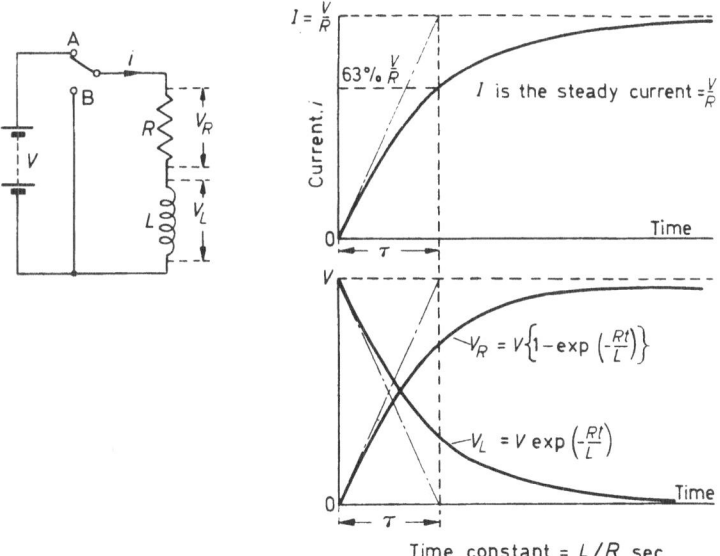

Figure 3.10a. The establishment of a steady current (V/R) is shown in the upper graph. Once a steady current is flowing there will be a steady magnetic field associated with the inductor. The lower graph shows the corresponding voltages across the two components

59

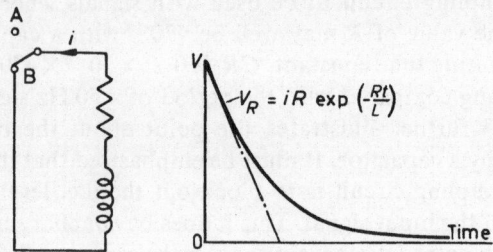

Figure 3.10b. Once the current in the circuit of *Figure 3.10a* is steady the voltage across the inductor is zero. The inductor stores energy ($\frac{1}{2}LI^2$) as a magnetic field. If the switch is now moved to B the field collapses, and the voltage across R decays exponentially

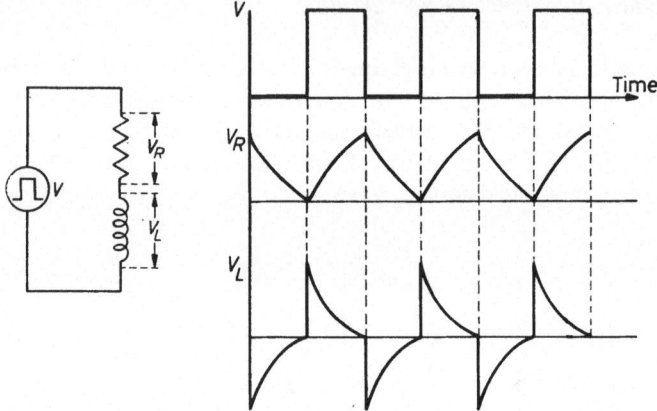

Figure 3.10c. Waveforms obtained after steady-state conditions have been achieved. The time constant is less than the period of the applied square wave

Fourier (1768–1830), a French mathematician, was one of the first to realize that all periodic waveforms could be synthesized by the combination of sine waves of the appropriate frequency, amplitude and phase. Conversely any periodic waveform may be analysed into its sine wave components as shown in *Figure 3.11*. We see therefore that voltages and currents that vary sinusoidally are of fundamental importance in electronics. Any analysis that is satisfactory for sine waves is usually satisfactory for other periodic waveforms. If, for example, an amplifier does not perform well when square waves are applied to the input terminals, it will be found that the performance of the amplifier is inadequate in some respects. This is not surprising when it is realized that a true square wave contains the fundamental and all the odd harmonics up to infinite frequency. It is, therefore, more than academic

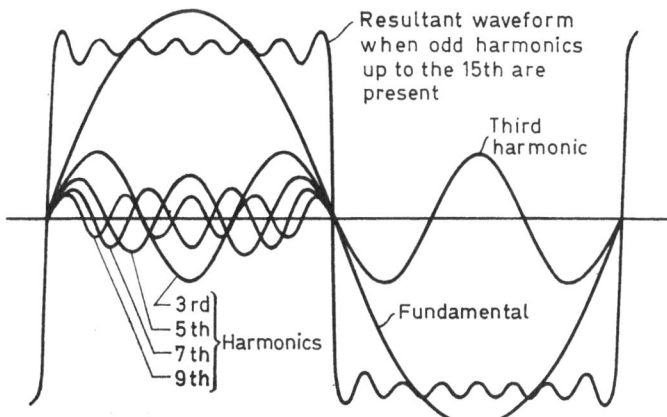

Figure 3.11. A square wave consists of the fundamental plus all the odd harmonics out to an infinite frequency. The sum of the odd harmonics up to 15 is shown together with the appropriate odd harmonics up to the 9th

interest that urges us to study the response of LCR networks to various applied sinusoidal voltages. The interpretation of the waveforms produced by an amplifier to which a square-wave signal is applied is dealt with in Chapter 11. The reader is referred to *Figure 11.13* and the related text.

Sine Waves

The generation of sine waves in a laboratory is usually carried out by a piece of electronic equipment called an oscillator. Such oscillators produce at their output terminals voltages that vary sinusoidally with time. Various forms of electronic oscillators are discussed in Chapter 11.

The alternating current supply mains uses a sinusoidal waveform, since this waveform is easy to generate and avoids design difficulties in the distribution equipment used for high powers. A simple sine-wave generator is shown schematically in *Figure 3.12*. The voltage induced in the rectangular coil is zero when the wires along the length of the coil, AB and CD, are travelling parallel to the direction of the lines of force. When AB and CD are moving at right angles to the direction of the lines of force, there is a maximum rate of cutting of these lines, and the voltage induced in the coil is a maximum. As the coil continues to rotate, the induced voltage falls to zero, and thereafter will increase in magnitude, but be of opposite polarity. Provided the coil is rotated at a constant angular velocity, ω, the voltage between the slip-rings varies sinusoidally with time.

61

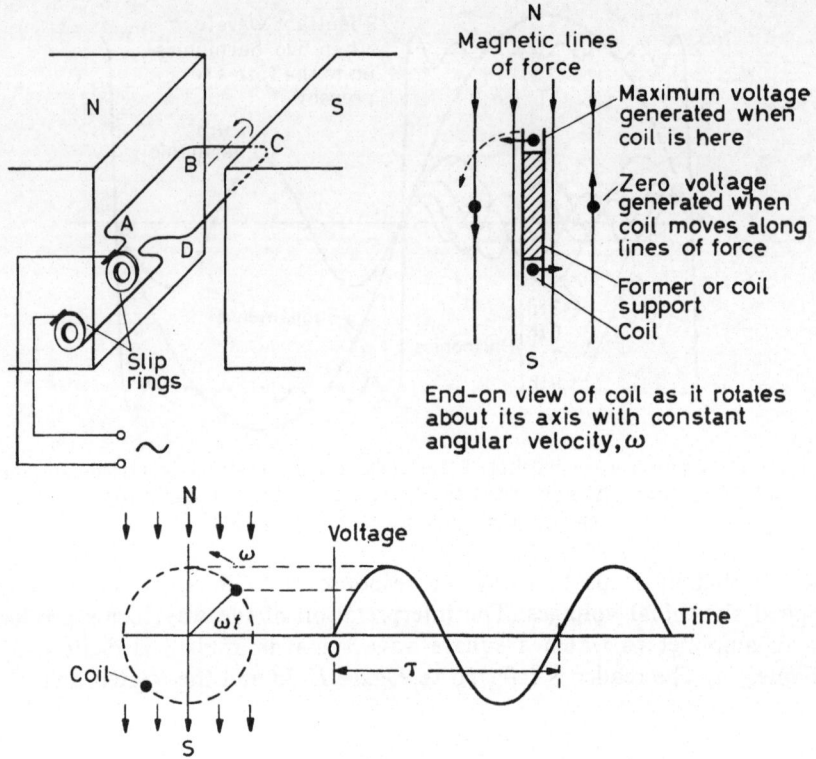

Figure 3.12. The generation of sine waves in a simple dynamo
1 revolution = rotation of 2π radians $\therefore \omega\tau = 2\pi$

$$\tau = \frac{1}{f} = \frac{2\pi}{\omega} \qquad \therefore \quad \omega = 2\pi f$$

One revolution, which is equivalent to a rotation of 2π radians (360°), produces one complete cycle of the waveform. The time taken to complete one cycle is known as the period or periodic time, τ; if the angular velocity of the coil is ω then $\tau = 2\pi/\omega$. If the coil makes f revolutions per second then the sine wave is said to have a frequency of f cycles per second. One cycle is therefore completed in $1/f$ seconds. Thus

$$\tau = \frac{1}{f} = \frac{2\pi}{\omega} \quad \text{and} \quad \omega = 2\pi f$$

Figure 3.13 illustrates the various terms associated with sine waves. The most important value associated with sinusoidal voltages and currents is the root-mean-square (r.m.s.) value. The majority of instruments used for measuring alternating voltages and currents are calibrated in r.m.s. values. The

62

need to use the r.m.s. value arises when we consider powers associated with this waveform. Suppose, for example, that we wish to know the power being supplied to an electric fire or some other heating equipment, or to an electric motor, then, in the d.c. case, the power, P, is given by multiplying the current, I, by the supply voltage, V, i.e. $P = VI$. With alternating current however, the voltage and current are continually varying, so we must find some figure that effectively represents the sinusoidal voltage for power calculations. Since

Mean value of half sine wave $= \dfrac{\displaystyle\int_0^{\pi} V_{max} \sin \omega t \, \mathrm{d}(\omega t)}{\pi}$

$= \dfrac{2V_{max}}{\pi} = 0{\cdot}637 \ V_{max}$

R.M.S. value $= \sqrt{\dfrac{\displaystyle\int_0^{2\pi} V_{max}^2 \sin^2 \omega t \, \mathrm{d}(\omega t)}{2\pi}}$

$= \dfrac{V_{max}}{\sqrt{2}} = 0{\cdot}707 \ V_{max}$

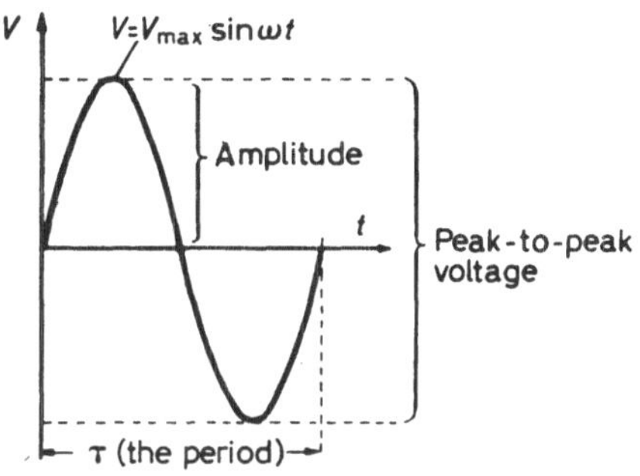

Figure 3.13. Terms associated with sine waves

power is proportional to the square of the voltage or current ($P = I^2R = V^2/R$) then we find the mean of the square of the alternating voltage (or current) and take the square root; hence we have the r.m.s. value. Integral calculus shows the r.m.s. value of a sinusoidal voltage (or current) to be 0·707 times the peak value.

The mean or average value of a sine wave is $2/\pi$ times the peak value (see *Figure 3.13*). This quantity is not so frequently used as the r.m.s. value; however, we shall need to know the definition of a mean value for later discussions on moving-coil meters and for the chapter on rectifiers and power supplies.

Sine Wave Response in LCR Circuits

The study of the application of sinusoidal signals to circuit arrangements of resistance, capacitance and inductance is of great importance, especially when it is remembered that all periodic waveforms may be resolved into a related set of sine waves.

When an alternating voltage is applied to the ends of a resistor, the results are easy to understand. We may draw directly on our experience with direct current and infer that a sinusoidal applied voltage produces a sinusoidal current through the resistor. Furthermore, at the instant when the voltage is a maximum, the current is also a maximum; when the voltage is zero, the current is zero. The current and voltage are said to be in phase (see *Figure 3.14a*). By dividing the r.m.s. value of the voltage by the r.m.s. value of the current we calculate the resistance, *R*. The r.m.s. values are chosen because these are the ones indicated by measuring instruments.

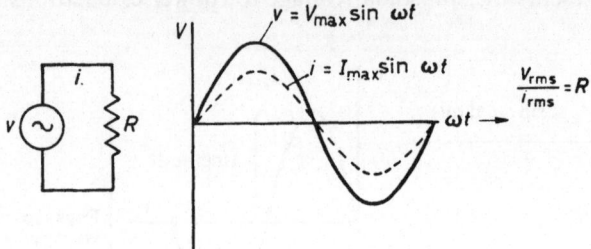

(*a*) Current and voltage in phase

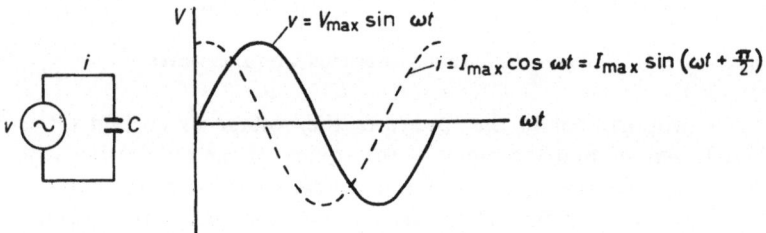

(*b*) Current leads voltage by 90° i.e. current reaches its maximum value a quarter of a period before the voltage reaches its maximum

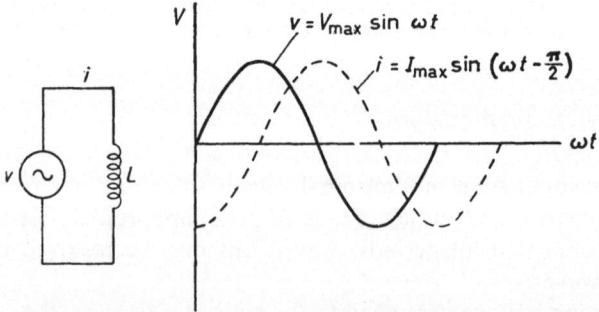

(*c*) Current lags voltage by 90° (alternatively voltage leads current by 90°)

Figure 3.14. The phase relationships between voltage and current for the basic passive components

The application of a sinusoidal voltage to a capacitor results in a rather more complicated behaviour than that experienced with a resistor. It will be recalled that during the charging and discharging of a capacitor maximum current flows when the charge on the capacitor, and hence the voltage across it, is zero. The maximum voltage exists at the time that the capacitor is storing its maximum charge. At this latter time however there is no movement of electrons and thus the current in the capacitor leads is zero. The facts may be summarized mathematically, as follows:

$$i = C \frac{dv}{dt}$$

where i is the instantaneous current value, C the capacitance and v the instantaneous value of the applied voltage. If a sinusoidal voltage is applied then $v = V_{max} \sin \omega t$ and

$$i = C \frac{d(V_{max} \sin \omega t)}{dt}$$

$$= \omega C V_{max} \cos \omega t \tag{3.2}$$

Now $\cos \omega t = \sin (\omega t + \pi/2)$ so the waveform of the current is sinusoidal. However, there is a phase displacement of $\pi/2$ or 90°, which tells us that the maximum current flows at the time the applied voltage is zero. *Figure 3.14b* shows the relevant waveforms. The current reaches its maximum at some time t (say) and a quarter of a period later the voltage reaches its maximum. The current and the voltage are out of phase; and we say that when a sinusoidal voltage is applied to a capacitor the current leads the voltage by 90°. The time t was deliberately selected as being some time when the voltage and current had achieved their regular waveforms, i.e. after the so-called 'steady-state' conditions had been established. Initially, of course, if the voltage is applied at time $t = 0$ the current and voltage waveforms do not follow the patterns already discussed, but go through a transient state before the steady-state condition is reached. These initial transients are usually of very short duration. It is not, therefore, worthwhile deriving the mathematical expressions for the transient waveforms here.

Equation 3.1 can be rearranged to give

$$\frac{V_{max} \cos \omega t \quad \text{(voltage)}}{i \quad \text{(current)}} = \frac{1}{\omega C} = \frac{1}{2\pi f C} = X_C$$

The term $1/(2\pi f C)$ is called the capacitive reactance of the capacitor and is often given the symbol X_C. The reactance of a capacitor is analogous to the resistance of a resistor since it is a measure of the opposition to the flow of alternating current. This opposition to flow depends upon the capacitance (the larger the capacitance the less is the opposition to the flow of current) but, unlike the resistor, the opposition also depends upon the frequency of the applied voltage. The greater the frequency the less is the opposition to the current. Conversely when the frequency is zero, i.e. when a steady voltage is applied to the capacitor, X_C is infinitely large and no current passes. This accords with our previous knowledge that direct current is prevented from flowing through a capacitor by the dielectric. When f is in cycles per second and C is in farads X_C is measured in ohms; for example, if an r.m.s. voltage of 10 volts at a frequency of 500 Hz were applied to a capacitance of 20 µF then the r.m.s. current that would flow in the leads to the capacitor would be V/X_C, i.e.

$$10/[1/(2\pi\ 500 \times 20 \times 10^{-6})] = 2\pi \times 10^{-1} = 0\cdot628 \text{ amps (r.m.s.).}$$

The application of a sinusoidal voltage to an inductor results in a steady-state current that lags the voltage by 90° (see *Figure 3.14c*). If the current through the inductor is given by $i = I_{max} \sin \omega t$ then

$$v = L \frac{di}{dt} = L \frac{d(I_{max} \sin \omega t)}{dt} = \omega L I_{max} \cos \omega t = \omega L I_{max} \sin \left(\omega t + \frac{\pi}{2}\right)$$

Since ωL is the reactance and $I_{max} \sin (\omega t + \pi/2)$ is the current term then Voltage = Reactance × Current. This statement represents the alternating current version of Ohm's law.

The opposition to the flow of alternating current is given in this case by X_L where X_L is the inductive reactance and is equal to $2\pi f L$.

When resistors, capacitors and inductors are connected together to form various series and parallel circuit arrangements, it is still possible to represent the alternating waves graphically by plotting their instantaneous values against time. Such graphs give us the mutual relationships between the various currents and voltages in the circuit, but they are tedious to draw. A much better method of representing the alternating quantities and their mutual relationships is to use what is called a 'phasor diagram'. Many books use the term 'vector diagram', but as the expressions 'vector' and 'vector analysis' have definite and well-defined meanings in physics and mathematics, it is better to avoid these terms. It is quite wrong, in the author's opinion, to represent quantities such as voltage and current by vectors. We may use phasors, which have the same mathematical properties as vectors, and thus avoid the physical anomaly of representing scalar quantities such as voltage and current by vectors.

In a phasor diagram, voltages and currents are represented by lines (called phasors), the lengths of which are proportional to the maximum values of the waves involved. The angles between the lines represent the phase angles between the waveforms. The phasors are, by definition, supposed to rotate about a fixed point, in an anticlockwise direction, at a constant angular velocity, ω. *Figure 3.15* should make the position clearer. Here we have a voltage and a current represented by the lines OA and OB, respectively. The length of OA represents the amplitude, or maximum value, of the voltage whilst OB represents the amplitude of the current. The angle AOB is the fixed phase angle between the voltage and the current and is given the symbol ϕ. As both phasors rotate about O at a constant angular velocity, ω, the projections of OA and OB on YY' represent the instantaneous values of the voltage and current at any time. If we plot the various projections at subsequent times we regain the original graphs which are to be superseded by the phasor diagram. It is customary to adopt the trigonometrical convention and take our reference direction along the x-axis from O.

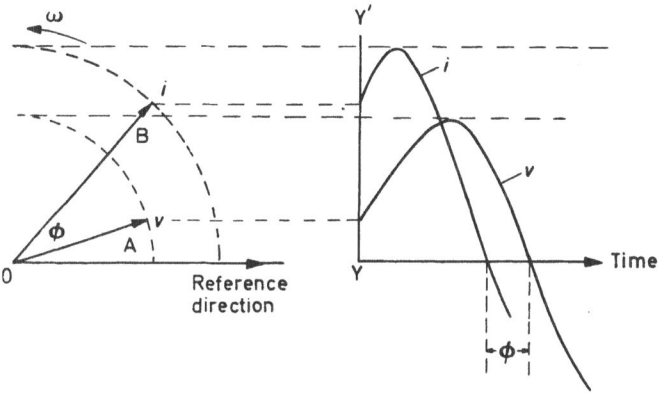

Figure 3.15. The representation of voltage and current graphs by means of phasors. ϕ is the angle representing the time or phase displacement between the two waveforms

In order to be able to construct and understand phasor diagrams there are three basic configurations to remember. These are given in *Figure 3.16*. In the case of the resistor, the angle between the phasors representing v and i is zero, showing that the current and the voltage are in phase. The diagram associated with the capacitor shows the current leading the voltage by 90°; for the inductor the diagram shows the current lagging the voltage by 90°. It is now much easier to represent the relationships between the phases and amplitudes of the various currents and voltages in a circuit containing combinations of resistance, capacitance and inductance.

In circuits that have the components in series, the current is the same at all points throughout the circuit at any given time. It is convenient, therefore, to take the current phasor as a reference. For example, in *Figure 3.17* we have a sinusoidal voltage applied to a circuit consisting of a capacitor in series with a resistor. The voltage across the resistor, v_R, will be in phase with the current, i, so a phasor representing v_R will point in the same direction as the current phasor. The phasor representing v_C is drawn at right angles to the reference direction to show a lag of 90°. If we measure v_C and v_R with meters and add the two readings, the sum will not be the same as the supply voltage v.

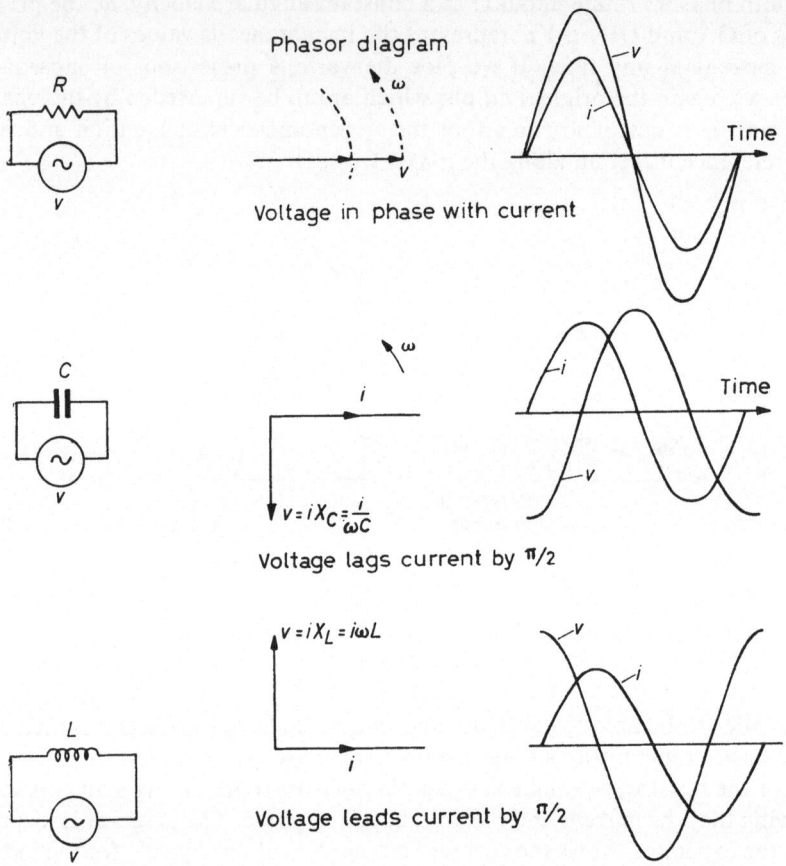

Figure 3.16. Phasor diagrams for the three basic passive components

This is because the meters are reading r.m.s. values and are taking no account of phase. We may add two voltages or currents correctly by using the phasor diagram. Since phasors obey the same mathematical laws as vectors, we add the phasors by completing the parallelogram (rectangle in this case) and measuring the appropriate diagonal. Thus $v = v_C + v_R$, but the addition is not an arithmetic one since it must obey the phasor addition rule. If we consider

68

magnitudes alone then

$$v = \sqrt{(v_R^2 + v_C^2)} \qquad \text{(Pythagoras' theorem)}$$

We may also extract phase information from the diagram noting that the supply current, i, leads the supply voltage by an angle

$$\phi = \cos^{-1} \frac{v_R}{v} = \cos^{-1} \frac{v_R}{\sqrt{(v_R^2 + v_C^2)}}$$

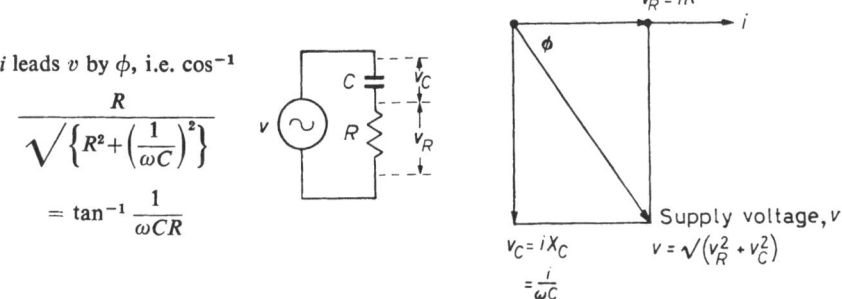

i leads v by ϕ, i.e. $\cos^{-1}$

$$\frac{R}{\sqrt{\left\{R^2 + \left(\dfrac{1}{\omega C}\right)^2\right\}}}$$

$$= \tan^{-1} \frac{1}{\omega C R}$$

Figure 3.17. Phase relations in a simple series RC circuit

The total opposition to the flow of alternating current by a circuit containing a reactive component (C and/or L) as well as resistance is known as the impedance of the circuit. Impedance is measured in ohms and is usually given the symbol Z.

$$Z = \frac{v}{i}$$

and from *Figure 3.17*

$$Z = \frac{\sqrt{(v_R^2 + v_C^2)}}{i} = \frac{\sqrt{\left\{i^2 R^2 + \left(\dfrac{i}{\omega C}\right)^2\right\}}}{i}$$

$$= \sqrt{\left\{R^2 + \left(\dfrac{1}{\omega C}\right)^2\right\}}$$

If the reader constructs a phasor diagram for a series combination of inductance and resistance he will find that $Z = \sqrt{\{R^2 + (\omega L)^2\}}$ and $\phi = \tan^{-1} (\omega L / R)$.

For parallel circuits the voltage is the same across each component, so in this case the reference direction is that of the phasor representing the supply voltage. In *Figure 3.18* we have a resistor in parallel with an inductor. The current through the resistor i_R is in phase with the supply voltage whilst i_L, the current through the inductor, lags the supply voltage by 90°. (The arrows show the direction of the current at some convenient instant. They are very frequently used in the circuit diagrams associated with phasor diagrams, even though we realize that the quantities involved are alternating ones. Their use arises in circuit analyses in which certain parallels are drawn between the

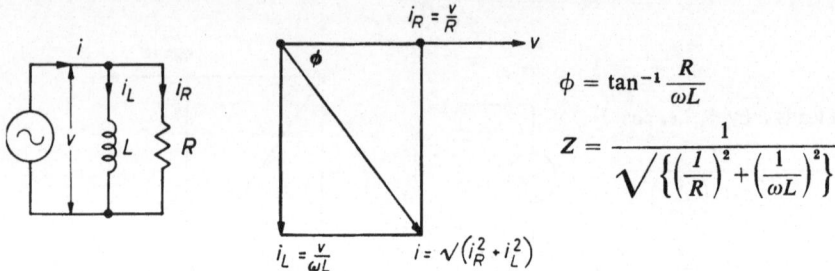

Figure 3.18. Phase relations in a simple parallel LR circuit

a.c. and d.c. case. Ohm's law and Kirchhoff's laws, for example, are valid in circuits carrying alternating currents.) The sum of i_L and i_R is the supply current, i, and here again we must observe the phasor addition rule. The circuit consisting of a resistor in parallel with an inductor has an impedance given by $Z = 1 \Big/ \sqrt{\left\{\left(\dfrac{1}{R}\right)^2 + \left(\dfrac{1}{\omega L}\right)^2\right\}}$.

Resonant circuits are those in which both inductance and capacitance are present. A series resonant circuit is one in which the capacitor, inductor and resistor are in series. In practice, the resistance is that of the wire used in winding the inductance and an actual resistor is usually absent. Nevertheless, a practical inductor in series with a capacitor is equivalent to a series arrangement of resistor, inductor and capacitor. During operation the voltage across the inductor leads the current through the circuit and has a magnitude of iX_L. The voltage across the capacitor, iX_C, lags the current by 90° and is, therefore, 180° out of phase with v_L. There is some frequency at which $X_L = X_C$. If the supply current is delivered at this frequency, then the voltage across the capacitor is equal in magnitude, but of opposite polarity, to the voltage across the inductor. The two voltages therefore cancel, and the supply current is limited only by the resistor. When the resistance is very small the current may rise to values considerably greater than those present at any other frequency. The phenomenon is known as resonance, and the special frequency involved is called the resonant frequency. At this frequency, f_r, the supply

70

current and the supply voltage are in phase. At frequencies removed from resonance the current will lead or lag the supply voltage depending upon whether the capacitive or inductive reactance is the dominant one. In general the impedance of a series circuit is given by $Z = \sqrt{\left\{R^2 + \left(\omega L - \dfrac{1}{\omega C}\right)^2\right\}}$

$$V = \sqrt{\{V_R^2 + (V_L - V_R)^2\}}$$

$$Z = \sqrt{\left\{R^2 + \left(\omega L - \dfrac{1}{\omega C}\right)^2\right\}}$$

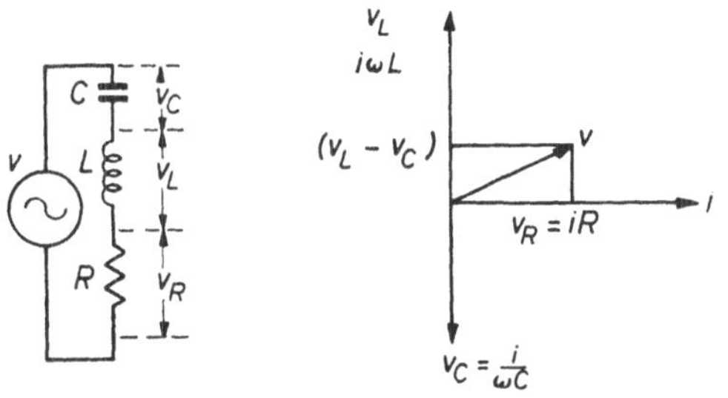

Figure 3.19a. Series resonant circuit

(Figure 3.19a). This is a minimum when $\omega_r L = \dfrac{1}{\omega_r C}$, where ω_r is the resonant angular frequency.

$$\omega_r = 2\pi f_r = \frac{1}{\sqrt{LC}}$$

$$\therefore f_r = \frac{1}{2\pi\sqrt{LC}}$$

In practical resonant circuits the resistance is usually low compared with the reactance of the coil since such resistance is almost wholly that of the wire from which the coil is wound. X_L is therefore greater than R and the voltage developed across the coil is greater than the supply voltage. This arises because at resonance the impedance of the circuit is simply R and so the current is given by $i = v/R$. The voltage across the inductor is iX_L, therefore $v_L = = vX_L/R$. The factor X_L/R, i.e. $\omega L/R$, is known as the magnification factor, and is given the symbol Q because this factor is a measure of the quality of the coil. In an ideal inductor, the resistance is zero and hence Q is infinitely large. In practical coils Q factors of 200–300 are common although many inductors, notably smoothing chokes, have a much lower figure. The high voltage across L is, of course, offset by the equally high voltage across C. The voltage across C is iX_C (which is equal to iX_L at resonance since $X_C = X_L$) and is of opposite polarity to v_L i.e. v_C is 180° out of phase with v_L. The magnified voltage across the inductor (and capacitor) is a very useful property of the series resonant circuit. Using this circuit, it is possible to select a particu-

71

lar frequency from a complex signal consisting of variations of voltage at many different frequencies. A particular example arises in radio where it is necessary to select only one frequency from many that are present in the aerial circuit *(Figure 3.19c)*. The voltage developed across the capacitor at the reso-

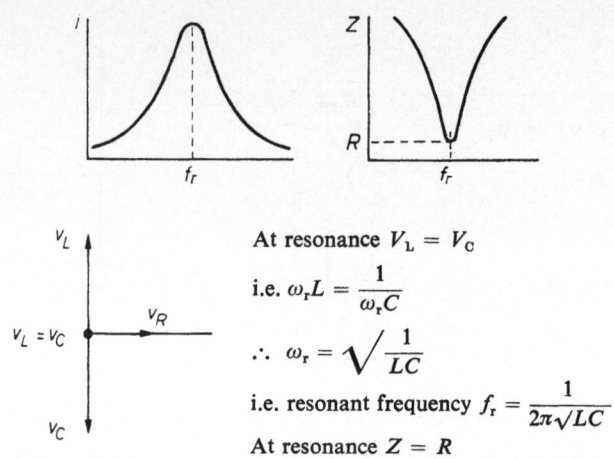

At resonance $V_L = V_C$

i.e. $\omega_r L = \dfrac{1}{\omega_r C}$

$\therefore \ \omega_r = \sqrt{\dfrac{1}{LC}}$

i.e. resonant frequency $f_r = \dfrac{1}{2\pi\sqrt{LC}}$

At resonance $Z = R$

Figure 3.19b. Conditions at resonance. Impedance is low at resonance

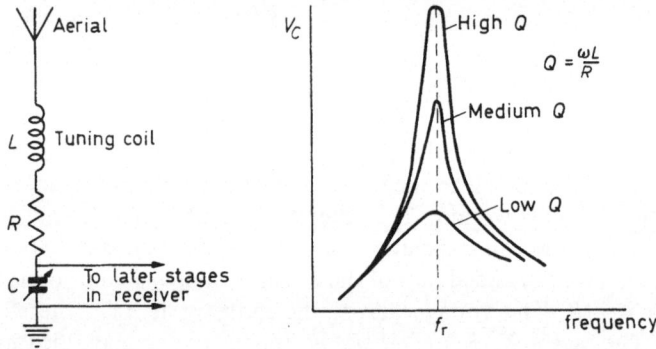

$Q = \dfrac{\omega L}{R}$

Figure 3.19c. The selection of a particular frequency using a series resonant circuit

nant frequency is many times greater than that developed at any other frequency. A plot of v_C against frequency gives us a graph known as a resonance curve. In *Figure 3.19c* the effect of different Q values on the sharpness of the peak is illustrated. When the resistance is low, the circuit is highly selective, whilst, on the other hand, a high resistance brings about poor selectivity because of the low Q-factor. Series resonant circuits should, therefore, always be fed from a low impedance generator.

When the generator impedance is high, then the selection of a particular frequency is best carried out by a parallel resonant circuit. The latter circuit has a high impedance at resonance. An additional advantage is that both the capacitor and inductor can each have one of their terminals at a common or earth potential. The parallel resonant circuit is shown in *Figure 3.20* along with the corresponding phasor diagram. Here it is convenient to take as our reference the phasor representing the current through the LR branch of the circuit. At resonance the supply voltage and supply current are in phase and the impedance of the circuit is purely resistive.

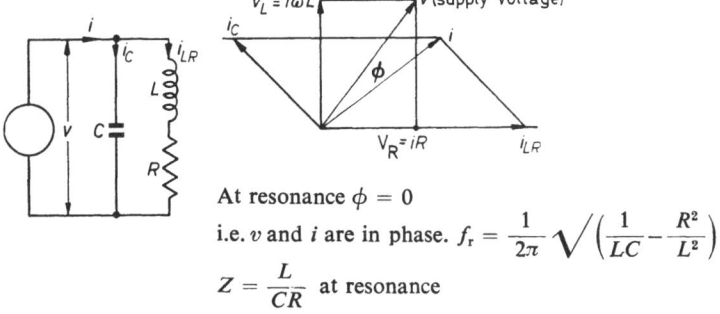

At resonance $\phi = 0$

i.e. v and i are in phase. $f_r = \dfrac{1}{2\pi}\sqrt{\left(\dfrac{1}{LC} - \dfrac{R^2}{L^2}\right)}$

$Z = \dfrac{L}{CR}$ at resonance

Figure 3.20. Phasor diagram for a parallel resonant circuit. Impedance is very high at resonance

It is shown later in the chapter that the resonant frequency of this circuit is given by

$$f_r = \frac{1}{2\pi}\sqrt{\left(\frac{1}{LC} - \frac{R^2}{L^2}\right)}$$

In order to resonate, the value of R must not exceed $2\sqrt{(L/C)}$. In practice a high-Q coil is used and so R is small in any case. The term R^2/L^2 is, therefore, much smaller than $1/(LC)$ and may usually be neglected in the expression for the resonant frequency. The resonant frequency for the parallel circuit is thus practically the same as that for the series circuit when the Q-factor is high.

At resonance the impedance of the parallel circuit is given by $Z = L/CR$. For a theoretical coil with no resistance, the impedance is infinitely large. As before the selectivity is reduced for finite values of R.

The j-notation and Circuit Analysis

For more complicated circuits than the parallel resonant circuit discussed above the construction of phasor diagrams becomes difficult or impossible. A much more elegant way of analysing a.c. circuits is to replace the phasor diagram by an algebraic technique. The possibility of doing this was first shown by Steinmetz (1893) who recognized the similarity between phasor diagrams and the 'Argand diagrams' used by mathematicians to represent diagrammatically what are called complex numbers. There is nothing difficult about the concept of a complex number; it is called complex because it is a number that requires two parts, unlike a simple arithmetic number such as 3 or 12 or a simple algebraic number like a, b, x, etc. Complex numbers arose in the solution to quadratic equations. It will be recalled that a quadratic equation of the form $ax^2 + bx + c = 0$ has solutions given by

$$x = \frac{-b \pm \sqrt{(b^2 - 4ac)}}{2a} = \frac{-b}{2a} \pm \frac{\sqrt{(b^2 - 4ac)}}{2a}$$

If $4ac$ is greater than b^2 then x is not a simple real number, but is made up of two parts, the so-called real part, $-b/2a$, and a second part called the imaginary part, $\pm \sqrt{\{(b^2 - 4ac)/2a\}}$. This second part is called imaginary because it is impossible to conceive of a real number which when squared will give a negative quantity, i.e. the square root of $b^2 - 4ac$ is not real when $4ac$ is greater than b^2. To overcome this mathematicians invented the number $i = \sqrt{-1}$. Since i can so easily become confused in electronics with the symbol for current, the letter j is used to denote $\sqrt{-1}$. Being uneasy about the new concept, Argand attempted successfully to represent the number in diagram form. The idea of representing a number by a diagram is not new and readers will be familiar with *Figure 3.21* in which real positive numbers are represented by

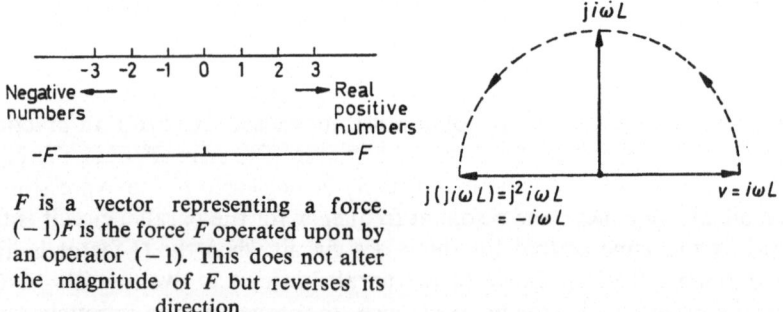

F is a vector representing a force. $(-1)F$ is the force F operated upon by an operator (-1). This does not alter the magnitude of F but reverses its direction

Figure 3.21. Illustrations of the effect of two operators viz. (-1) and j

distances along a line from the origin to the right. By operating on all of these numbers using the operator -1 we obtain the negative numbers to the left. Because we are so familiar with negative numbers, it is sometimes forgotten that -1 is an important mathematical invention not known to the ancients. The operator -1 is particularly important when dealing with vectors. It is a

74

matter of common experience that if a force F is acting upon a body, another force equal in magnitude, but opposite in direction will nullify the first force, i.e. the resultant of the two forces is zero. Representing this diagrammatically we choose a line of suitable length to represent the magnitude of the force and the direction of the line shows the direction of the force. Another line of the same length, but pointing in the opposite direction, represents $-F$. Since the vectorial sum of these two forces is zero, $\bar{F}+(-\bar{F}) = 0$ i.e. $\bar{F}-\bar{F} = 0$. The fact that common experience makes it impossible to conceive of a negative force (or velocity, or any other vectorial quantity) does not detract from the usefulness of defining $-F$ to mean a force of the same magnitude, but opposite in direction to F. Evidently then operating on a vector with the operator -1 has the effect of turning the vector through $180°$ without changing its magnitude. We can, therefore, manipulate vectors algebraically instead of diagrammatically by using the triangulation theorems and the operator -1.

Suppose now we define an operator j such that jF means a vector of the magnitude of F but rotated anticlockwise through $90°$. Operating on jF by j again will give us $j(jF)$ i.e. j^2F. This vector will now be pointing in the opposite direction to F, therefore $j^2F \equiv -F$. In other words, $j^2 \equiv -1$ and so $j = \sqrt{-1}$. Readers should avoid trying to conceive of $\sqrt{-1}$ in any way other than that of a vector operator rotating a vector through $90°$.

Since phasors obey the same mathematical laws as vectors, it is useful to apply the same operators to them. For example, if we have a phasor representing a voltage say $i\omega L$ then $ji\omega L$ represents a voltage of the same magnitude, but advanced through $90°$ *(Figure 3.21)*. Consider now the phasor diagram of *Figure 3.22*. The voltage v_R is equal to iR, and the phasor representing it is in

$$v = v_R + jv_L$$
$$= iR + ij\omega L$$
$$= i(R + j\omega L)$$
$$= iZ \text{ where } Z = R + j\omega L$$

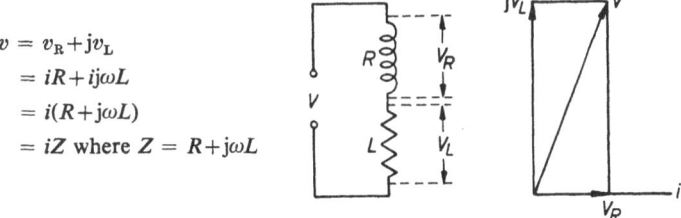

Figure 3.22. The replacement of a phasor diagram by complex numbers

the reference direction. The voltage v_L is equal to $i\omega L$, but to show that it is advanced by $90°$ we have $v_L = ji\omega L$. It is customary to keep the j and ω together so we would write $v_L = ij\omega L$. The supply voltage v is the phasor sum of v_R and v_L, i.e. $v = v_R + v_L = iR + ij\omega L = i(R + j\omega L)$. Dividing by i we have $v/i = R + j\omega L$. This is the impedance Z. The expression for Z is, therefore, a complex one made up of two parts, viz., a real part R and an imaginary part $j\omega L$. Such terms, although frequently used in textbooks on electronics, are not, in the author's opinion, satisfactory, since some confusion can arise in the mind of a beginner when confronted with the statement that the reactive component is imaginary. It is better, therefore, to call a complex quantity, or number, a general number, and to regard such a number as having an ordinary component and a quadrature component. The terms 'ordinary' and 'quadrature' replace 'real' and 'imaginary'. The quadrature component is so called because of the 90° shift from the reference direction.

The magnitude of a general number is found by applying Pythagoras' theorem. Thus, if Z is expressed as the general number $R+j\omega L$ then the magnitude, i.e. modulus, of Z is written with two lines bracketing the symbol, and is given by $|Z| = \sqrt{\{R^2+(\omega L)^2\}}$. When the relationship between the general number and corresponding phasor diagram is appreciated, information about phase angles is obtained in addition to a knowledge of the magnitudes of the quantities involved. For example, if the sum of two phasor voltages is given by $v = iR+ji\omega L$ then it follows immediately that $|v| = i\sqrt{\{R^2+(\omega L)^2\}}$ and that v leads i by an angle $\phi = \cos^{-1}R/\sqrt{\{R^2+(\omega L)^2\}}$.

On being first introduced to ordinary arithmetic numbers at school we are taught to add, subtract, multiply and divide them. Making progress into algebraic numbers the same manipulations must be mastered and new rules learnt. Understandably therefore before we can use general numbers successfully, we must learn to manipulate them correctly. All the ordinary rules of algebra apply except that when we come across the expression j^2 we replace it by -1. The procedures are best illustrated by examples. If we have two general numbers $(a+jb)$ and $(x+jy)$ then the rules are, as follows:

Addition: $(a+jb)+(x+jy) = (a+x)+j(b+y) = A+jB$

The ordinary components and quadrature components are merely added grouping them so that $a+x$ gives an ordinary number A and $jb+jy$ gives the quadrature component jB.

Subtraction: This is achieved by replacing the plus sign with the appropriate minus sign thus $(a+jb)-(x+jy) = (a-x)+j(b-y)$. The number $a+jb$ has a component a in the reference or 'easterly' direction and a quadrature component b in the 'north' direction since $+jb$ is by definition an operation on the phasor b rotating it anticlockwise through 90°. If we operate on jb with j we obtain $j^2b = -b$, and operating once more with j we obtain $-jb$. This is equivalent to rotating b in the opposite sense, i.e. in the clockwise direction, by 90°. The phasor $-jb$ lies, therefore, in the 'south' direction. The general number $a-jb$ thus lies in the fourth quadrant. A further example involving minus signs is

$$(a+jb)-(x-jy) = (a-x)+j(b+y)$$

Multiplication

$$(a+jb)(x+jy) = ax+jbx+jay+j^2by$$
$$= (ax-by)+j(bx+ay)$$
$$= C+jD \text{ (say)}$$

Division

The first process in dividing $a+jb$ by $x+jy$ appears to be difficult because the initial step in the division is to rationalize the denominator, i.e. to eliminate the implied square root sign in the denominator. The square root becomes obvious when the modulus of the denominator is considered. If $N = x+jy$ then $|N| = \sqrt{(x^2+y^2)}$. Square roots are awkward to handle especially when they are in the denominator of a fraction. The rationalization is achieved by multiplying both the numerator and the denominator by the conjugate of the denominator. (The conjugate of a general number is another general number with the same ordinary and quadrature components, but with the sign changed in front of the j term; the conjugate of $R+jX$, for example, is $R-jX$.) Multiplying both the numerator and the denominator of a fraction by the same factor leaves the factor unchanged. The complete division process is as follows:

$$\frac{a+jb}{x+jy} = \frac{(a+jb)}{(x+jy)} \times \frac{(x-jy)}{(x-jy)}$$
$$= \frac{(ax+by)+j(bx-ay)}{x^2-j^2y}$$
$$= \frac{(ax+by)+j(bx-ay)}{(x^2+y^2)}$$

Evidently by multiplying the denominator by its conjugate, the difference of two squares is obtained which on simplification yields an ordinary number for the denominator. The division is completed by obtaining the answer as a general number

$$\frac{a+jb}{x+jy} = \frac{ax+by}{x^2+y^2} + j\frac{(bx-ay)}{x^2+y^2}$$
$$= A+jB$$

where $A = (ax+by)/(x^2+y^2)$ and $B = (bx-ay)/(x^2+y^2)$

Some elementary applications of the j-notation—One of the great advantages in using the j-notation lies in the ease with which expressions for the impedances of the various branches of a network can be formulated. There is no need any longer to think consciously of phase since the j symbol automatically takes this into account for us. Three expressions must be remembered:

(a) Resistive terms have no j associated with them since the currents and voltages across a resistive circuit are in phase.

(b) For inductive terms the reactance is $j\omega L$. When a voltage v is involved $v = ij\omega L$; automatically we have the voltage expression showing a 90° phase lead of v on i.

(c) For capacitive terms the reactance is $-j/\omega C$ so that $v = \dfrac{-ij}{\omega C}$, showing that the voltage lags the current by 90°. This reactive term is not generally adopted, however, since it is of great benefit to keep the j and ω terms together. However

$$-\frac{j}{\omega C} = \frac{-j}{\omega C} \cdot \frac{j}{j} = \frac{1}{j\omega C}$$

The reader should, therefore, remember the expression for capacitive reactance as $1/j\omega C$.

When using these expressions, it must be realized that they are valid only when analysing circuits in which the variations of voltage and current are sinusoidal, and the so-called 'steady-state' conditions have been reached. Conclusions may also be drawn about periodic waves; but, for analyses involving transients (sharp or quick changes of voltage or current) and transient response, it is necessary to employ other techniques, e.g. Laplace transforms. These, however, are outside the scope of this book. Interested readers may care to consult the book by Day[1].

Figure 3.23 shows the two simple resonant circuits already discussed. The impedance Z of the series arrangements is $Z = R + j\omega L + \dfrac{1}{j\omega C} = R + j\left(\omega L - \dfrac{1}{\omega C}\right)$. We could obtain $|Z|$ by the rules already given, which yields the result $\sqrt{\left\{R^2 + \left(\omega L - \dfrac{1}{\omega C}\right)^2\right\}}$. However, by using the j-notation another very important advantage arises. At resonance the voltage and current are in phase. There can, therefore, be no j-term. If we examine any resonant circuit and obtain an expression for the impedance then, by equating the j-term to zero, we obtain the condition for resonance. For example, in the series resonant circuit, $Z = R + j\omega L + 1/(j\omega C)$. At resonance the j term is zero, therefore, $j\{\omega L - 1/(\omega C)\} = 0$ and hence the resonant frequency is $1/2\pi\sqrt{LC}$.

The power of the j-notation is in greater evidence when the parallel circuit of *Figure 3.23b* is considered. When components are in parallel it is better to deal with admittances rather than impedances, since the former are merely added. (An admittance, Y, is the reciprocal of an impedance, Z, just as conductance, G, is the reciprocal of resistance, R.) Without drawing any phasor

$$Z = R + j\omega L + \frac{1}{j\omega C}$$

$$= R + j\left(\omega L - \frac{1}{\omega C}\right) \text{ and}$$

$$|Z| = \sqrt{\left\{R^2 + \left(\omega L - \frac{1}{\omega C}\right)^2\right\}}$$

(a)

At resonance j term $= 0$ $\quad \therefore \quad \omega L = \frac{1}{\omega C}$

i.e. $f_r = \frac{1}{2\pi\sqrt{LC}}$

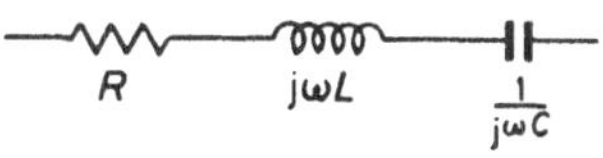

$$Y_1 = \frac{1}{Z_1} = j\omega C$$

$$Y_2 = \frac{1}{Z_2} = \frac{1}{R + j\omega L}$$

$$Y = Y_1 + Y_2 = j\omega C + \frac{1}{R + j\omega L}$$

$$= j\omega C + \frac{R - j\omega L}{R^2 + (\omega L)^2}$$

$$= \frac{R}{R^2 + (\omega L)^2} + j\frac{\omega C R^2 + \omega^3 L^2 C - \omega L}{R^2 + (\omega L)^2}$$

at resonance j term $= 0$

$\therefore \quad \omega C R^2 + \omega^3 L^2 C - \omega L = 0$

$$\therefore \quad f_r = \frac{1}{2\pi}\sqrt{\left(\frac{1}{LC} - \frac{R^2}{L^2}\right)}$$

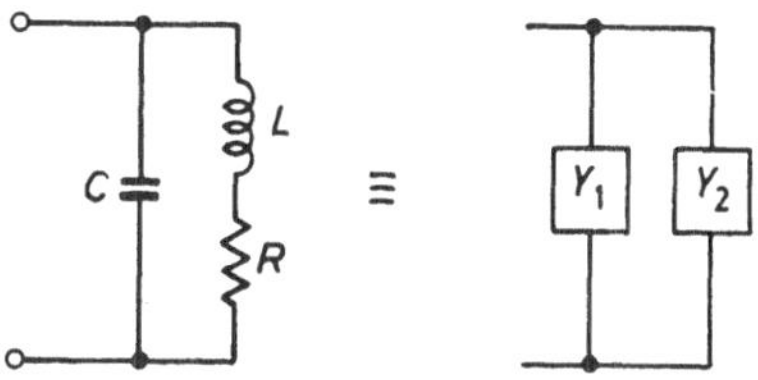

(b)

Figure 3.23. The j notation as applied to series and parallel resonant circuits

diagrams the symbolic (i.e. algebraic) analysis yields quite simply the resonant frequency formula $f_r = \frac{1}{2\pi}\sqrt{\left(\frac{1}{LC} - \frac{R^2}{L^2}\right)}$. The resonant frequency for an inductance and capacitance in parallel is thus slightly different from that when the components are in series. The difference is very small however for high-Q coils.

REFERENCES

1. Day, W. D. *Introduction to Laplace Transforms for Radio and Electronic Engineers.* Iliffe, 1960.

4

ACTIVE DEVICES 1 —
THERMIONIC VALVES

The signals produced by thermocouples, pH electrodes, photocells, piezoelectric crystals, etc., are rarely powerful enough to operate meters and recording devices directly. Such signals, therefore, need to be amplified before they are fed into the indicating device. Amplification, which is the most common of all electronic operations, cannot be performed by using networks of resistors, capacitors and inductors. It is necessary to energize such networks by using some active device. The two major groups of active devices are (a) thermionic valves and (b) semiconductor devices. This chapter deals with thermionic valves.

The earliest type of electronic amplifying device depended upon the controlled flow of electrons through a region which had been pumped as free as possible from air and other gases. The system of enclosing the necessary electrodes in a glass envelope or tube, and evacuating the tube as highly as possible, is still today the standard way of manufacturing thermionic valves.

THE DIODE

Although not an amplifying device, the thermionic diode provides a convenient starting point because all thermionic valves are diodes into which other electrodes have been introduced for specific purposes. We shall discuss in this and later chapters the role of the diode in power supplies and in the performance of electronic functions such as demodulation or detection, waveshaping and voltage clamping.

The electrons in thermionic devices are produced by a process known as thermionic emission. When a wire such as tungsten is heated to incandescence, sufficient energy is available to allow the electrons to leave the wire and form a cloud around it. This cloud of electrons is known as the space charge. When the wire is coated with a mixture of barium and strontium oxides, the cloud is formed at a much lower temperature. This is because the work function of the surface has been reduced by the presence of the oxides. The work function is a measure of the work needed to 'evaporate' the electrons from the surface. Equilibrium is established when the rate of 'evaporation' is equal to

80

the rate at which electrons return to the wire; the equilibrium is, therefore, a dynamic one. When the wire is held in an evacuated tube, the cloud of electrons extends for a considerable distance from the wire. By placing a second electrode within the tube, it is possible to draw electrons from the space charge to this electrode. This is achieved by making the second electrode positive with respect to the wire. The flow of electrons from the emitting surface, known as the cathode, to the collecting electrode, known as the anode (or plate in American terminology) constitutes an electric current through the device. The latter is known as a diode since two electrodes are involved. The electrons can be made to flow in one direction only (hence the term valve). On making the anode negative with respect to the cathode, the electric field between the electrodes is such as to prevent the flow of electrons to the anode. *Figure 4.1* shows the construction of the diode together with its characteris-

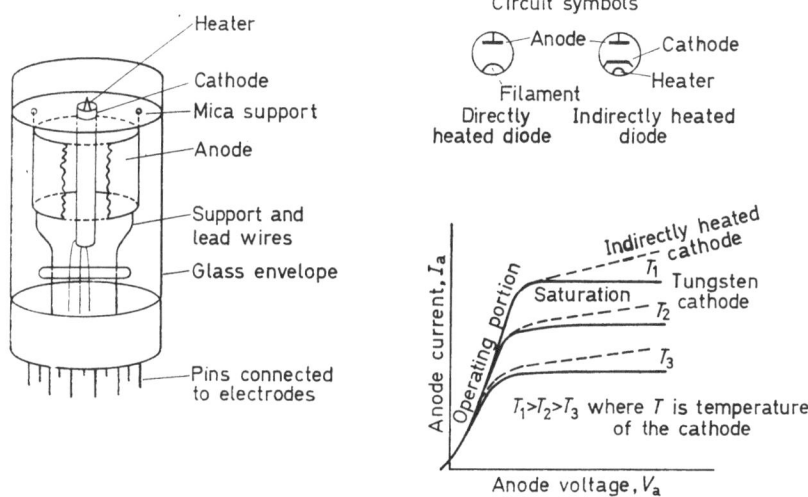

Figure 4.1. Details of a small thermionic diode

tic curves. These curves or graphs are a useful way of representing the behaviour of the device. The anode current is represented by I_a and the voltage between anode and cathode by V_a. The small finite value of current with zero and small ($<$ about 400 mV) negative values of V_a is of no importance in practice; many graphs are published showing the curve passing through the origin, although this is not strictly correct. The small current at zero V_a is explained by the finite energy of the electrons on being emitted from the cathode.

The cathodes for small-power diodes (passing current up to about 300 mA) are nickel cylinders coated with barium and strontium oxides. The necessary temperature is obtained by having a heating element enclosed within the tube, but insulated electrically from the cathode by a refractory material such as alumina. The heating element can then be supplied with alternating current from a secondary winding on a mains transformer. It is necessary to avoid high voltages between the heater and the cathode otherwise breakdown will occur. The maximum heater/cathode voltage is specified by the manufacturer and is commonly about 50 volts. When large powers are involved (as for example in transmitters) the anode voltage may be in excess of 1 kV. At this voltage the ionic bombardment of the cathode by positive ions would damage an indirectly heated cathode in a short time. Such ions arise from the residual gases within the tube. It is therefore necessary to employ a wire or filament type of cathode made from tungsten or thoriated tungsten. Such a wire can withstand ionic bombardment, but unfortunately the efficiency of electron emission is low.

Diode Applications

The most important application of thermionic diodes is in power supplies where it is necessary to convert alternating voltages from the mains to steady voltages for feeding electronic circuits. As this is the subject of a separate chapter however, no discussion will be undertaken here except to explain the elementary principles of rectification and demodulation.

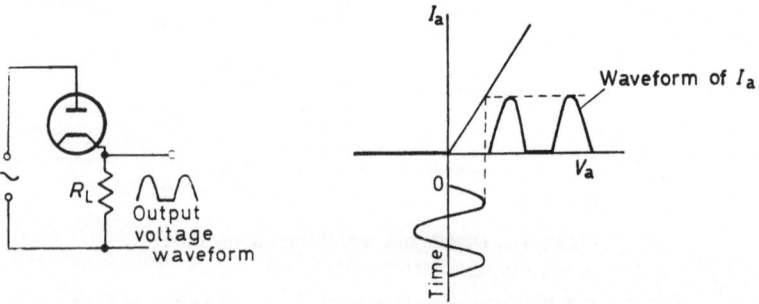

Figure 4.2. Basic principles of rectification

Figure 4.2 shows a thermionic diode and a series resistor load R_L, the combination being supplied with an alternating sinusoidal voltage. This voltage may be obtained from an oscillator or the secondary winding of a transformer. When the applied voltage is such that the anode is positive with respect to the cathode, current flows in the circuit. If we neglect the resistance of the diode (which is non-linear, and therefore varies with V_a) then the

82

current waveform is sinusoidal during the first half-cycle. When the polarity of the voltage is reversed, the anode becomes negative with respect to the cathode and no conduction takes place. *Figure 4.2* shows the position in graphical form. The voltage across the resistor, R_L, is proportional to the current, and is therefore a series of half-sine-waves. The alternating voltage has thus been converted to a series of unidirectional sinusoidal pulses, a process known as rectification.

Apart from rectifying the mains to supply power at a steady voltage, the rectifying properties of a diode are most frequently used in the detection or demodulation of an amplitude-modulated wave. It frequently happens that a signal of low frequency cannot be handled efficiently. For example, in radio communication the audio signals associated with speech and music could not be efficiently radiated as electromagnetic waves at their fundamental frequencies, since aerials many hundreds of miles long would be needed. A high-frequency carrier wave is generated, and this can be radiated by aerials of realistic size. The signal information is then impressed upon the carrier wave by varying or modulating the amplitude of the carrier wave in sympathy with the signal. After efficient radiation and pick-up by the receiving aerial the signal information is recovered from the carrier wave by a process known as demodulation or detection. A pair of earphones or audio amplifier and loudspeaker system could not respond to the frequency of the carrier wave, so this wave is first rectified (i.e. detected) and the audio information recovered by means of a capacitor-resistor circuit as shown in *Figure 4.3*.

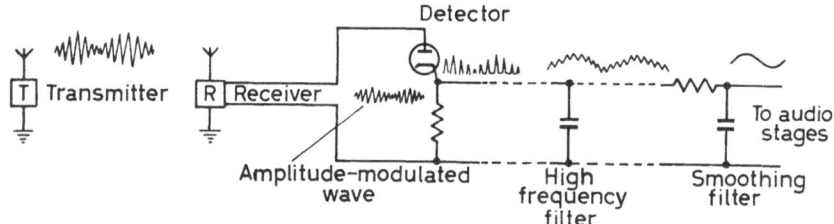

Figure 4.3. Basic demodulation of an amplitude-modulated wave in telecommunications

Exactly the same principle may be used in a laboratory for studying slowly varying quantities. The measurement of a slowly varying temperature by a thermocouple is an example. The small thermocouple output will need to be amplified. If the variations are very slow (less than 1 cycle per second) then an expensive d.c. amplifier will be required to amplify the signal directly. However, if only the variations of temperature are important, a much cheaper a.c. amplifier may be used yielding accurate results. The signal is made to modulate the amplitude of the voltage from an oscillator generating voltages at a

6

frequency of say 400 Hz. This signal is easily amplified by an a.c. amplifier. The output of the amplifier can then be fed into a detector and an amplified version of the original thermocouple voltage variations obtained. We need not at the moment consider the circuitry of the arrangement shown in block diagram form in *Figure 4.4*. The individual units making up the complete apparatus will be discussed at the appropriate stage.

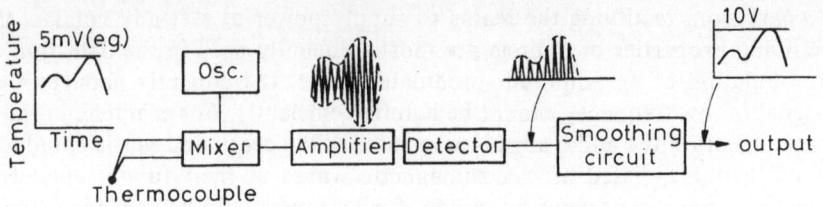

Figure 4.4. Method of amplifying a slowly varying voltage

The Diode as a Clamp and Limiter

In electronic circuits the need may arise for the voltage at some point in the circuit to be restricted within specified limits. For example, the output of a servo amplifier driving a motor may be satisfactory during normal operation, but sometimes overloads within the amplifier cause the output to be excessively high, which may result in damage to the motor. By limiting the output, such damage can be avoided. Damage to valves or transistors may occur when the input potentials to such devices are outside the normal working limits. In such circumstances it is highly desirable to incorporate a protecting circuit in the form of a clamping or limiting diode.

To understand the limiting action, it must be remembered that when the anode of a diode is negative with respect to the cathode no conduction takes place and the diode then acts like an open circuit and can be ignored. If however conditions arise that make the anode positive with respect to the cathode, conduction occurs and, in a thermionic diode, the resistance of the device is only a few hundred ohms. During conduction therefore the potential of the anode is practically the same as that of the cathode. (For a semiconductor diode, the resistance is much lower and the clamping is even better than that using a thermionic diode.) Consider now the circuits of *Figure 4.5*. In *Figure 4.5a* as soon as the input goes positive, the diode conducts and acts like a near short circuit to the output terminals. The output voltage is therefore nearly zero. During the next half cycle when the input is such as to make the anode negative with respect to the cathode, no conduction occurs; the diode is open circuit and the output waveform is identical to the input waveform if no loading is applied to the output terminals. In *Figure 4.5b* the insertion of a low resistance voltage source biases the diode. The input voltage must now exceed $+5$ V before conduction occurs. Thereafter the output voltage is held at $+5$ V since the voltage drop across the diode is negligible. As soon as the input voltage falls below $+5$ V conduction ceases and the input and output waveforms are identical. In *Figure 4.5c* two diodes are used to limit the output voltage

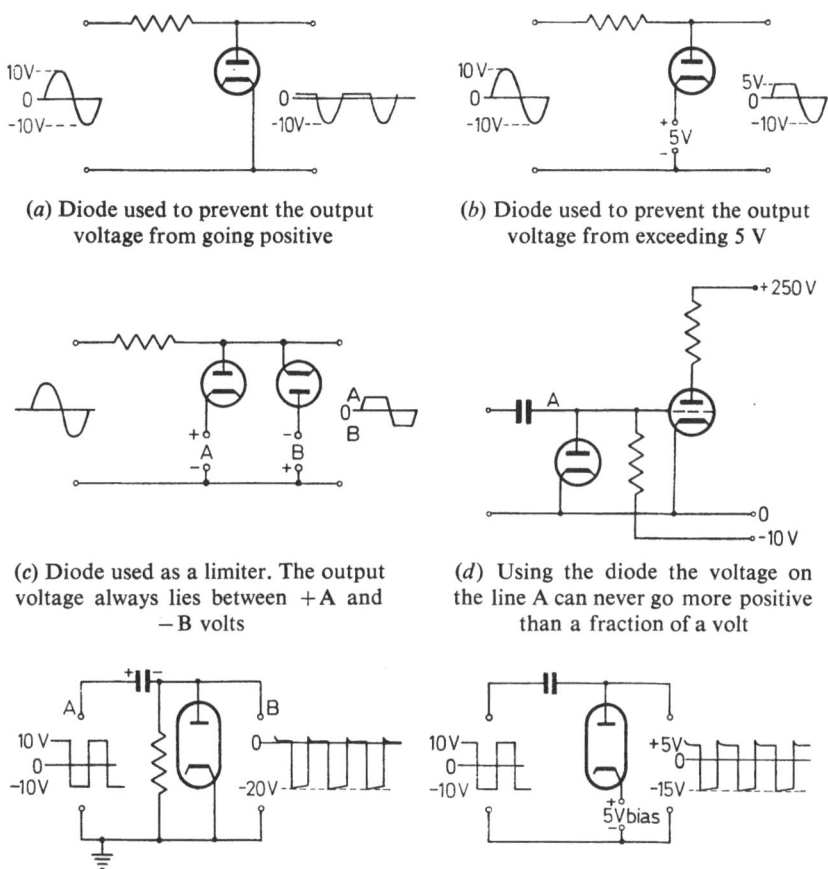

(a) Diode used to prevent the output voltage from going positive

(b) Diode used to prevent the output voltage from exceeding 5 V

(c) Diode used as a limiter. The output voltage always lies between $+A$ and $-B$ volts

(d) Using the diode the voltage on the line A can never go more positive than a fraction of a volt

(e) The diode clamp without and with bias. In the left-hand diagram, when A goes positive the diode conducts and the capacitor charges to practically 10 V with the polarity shown. The output terminal B is at zero potential at this time. When A goes negative the diode becomes non-conducting and the output voltage at B is then the sum of the supply voltage and the capacitor voltage. These circuits can be referred to as d.c. restoration circuits. Any steady component of a signal is lost in the CR coupling network. The diode is used to restore this steady component to the waveform

Figure 4.5. The use of the diode as a voltage limiter, and as a clamp or restoration circuit

within certain specified limits determined by the biasing voltages. *Figure 4.5d* shows the application of a diode clamping circuit to the input of a triode valve. We shall see later in the chapter that the grid of the triode must never be positive with respect to the cathode otherwise an excessive current would flow. The diode prevents this, the operation of the circuit being similar to *Figure 4.5a*. *Figure 4.5e* shows how to use a diode as a clamp to hold a waveform at some desired voltage level.

THE TRIODE

Although the diode is useful for the purposes described in the previous section, such a device is unable to perform the function of amplification. If we place a simple series circuit consisting of a diode and resistor arcross the terminals of a signal source, such as an oscillator, then it can readily be seen that neither the voltage across the diode nor that across the resistor can be greater than the supply voltage. In order to be able to amplify a signal the diode must be modified by interposing a wire grid between the cathode and the anode. The resulting three-electrode device is known as a triode.

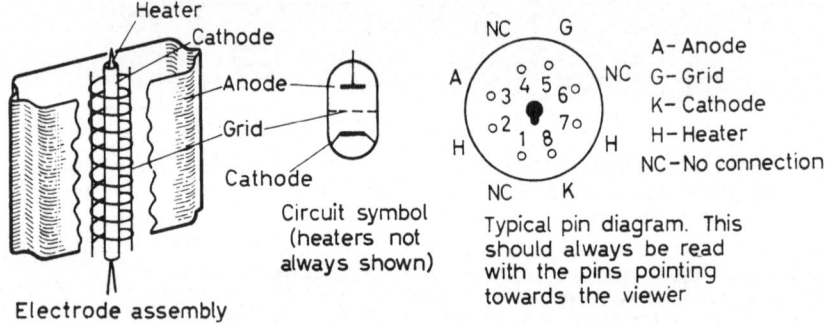

Figure 4.6. The main constructional features of a thermionic triode. For any particular valve the manufacturer's data sheet should be consulted for the pin connections and type of base used, e.g. IO (international octal), B7G, B9A, etc.

The main constructional features are shown in *Figure 4.6*. The electrons emitted from the cathode pass through the grid and are collected by the anode. The flow is controlled by the combined effect of the electrostatic fields due to the grid and the anode. Since the distance between the grid and cathode is much smaller than that between the anode and cathode, small variations of grid voltage have the same effect on the anode current as much larger variations of anode voltage. It is therefore possible to apply a small signal voltage between the grid and cathode and obtain an amplified version of the signal

between anode and cathode. Before discussing the amplifying action, however, we must first consider the fundamental properties of the triode. These
properties may be understood by considering the families of characteristic
curves associated with the device.

Since the anode current, I_a, is a function of (i.e. depends upon) two independent variables, namely the anode voltage, V_a, and the grid voltage, V_g, it
is not possible to summarize the behaviour of the triode by a set of curves on
a single pair of axes. Two sets of curves must be drawn, one showing the variation of anode current with grid voltage for fixed values of anode voltage (i.e.
the I_a/V_g characteristics) and another showing the variation of anode current
with anode voltage for fixed grid voltages (i.e. the I_a/V_a characteristics). These
sets of curves are known respectively as the static mutual characteristics and
the static anode characteristics. They are static characteristics because the
graphs are obtained under steady conditions, no signal variations being present. From these curves the behaviour of the triode can be predicted and
amplifiers can be designed to meet given specifications.

Wherever possible we associate with electronic components a certain quantity or parameter that gives us a quantitative estimation of the component's
performance. In the case of a resistor, for example, the graph of current
against applied voltage—the I/V curve—is a straight line that leads to the definition of resistance, R. The symbols C and L are used when estimating the
performance of capacitors and inductors. With a triode a single parameter
would be insufficient to predict the performance of the valve in a given circuit.
It is necessary therefore to define the minimum number of parameters that
adequately describe the behaviour of the device. In the case of a triode there
are three important parameters associated with the characteristic curves. They
are defined as follows:

(a) *Amplification factor*, μ. This parameter enables the voltage gain of a
given circuit to be predicted. The actual voltage gain is never, in practice,
equal to the amplification factor for reasons we shall see when dealing with a
simple amplifier circuit.

$$\mu = \frac{\text{Small change in anode voltage } (\Delta V_a)}{\text{Small change in grid voltage necessary to bring the anode current back to its original value } (\Delta V_g)}$$

$$= -\left(\frac{\Delta V_a}{\Delta V_g}\right)_{I_a = \text{constant}} = -\frac{\partial V_a}{\partial V_g}$$

Referring to *Figure 4.7a*, if the anode current is increased from B to C by increasing the anode voltage from V_{a1}, to V_{a2}, then in order to bring the current
back to its original value the grid voltage will have to be decreased, going from
its value at C (and B) to the more negative value at A. μ therefore tells us how

87

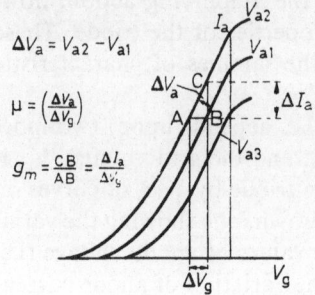

$$\Delta V_a = V_{a2} - V_{a1}$$

$$\mu = \left(\frac{\Delta V_a}{\Delta V_g}\right)$$

$$g_m = \frac{CB}{AB} = \frac{\Delta I_a}{\Delta V_g}$$

(a) Static mutual characteristics

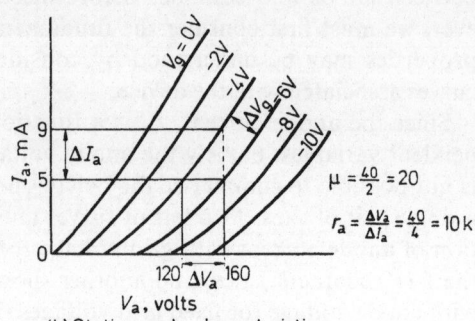

$$\mu = \frac{40}{2} = 20$$

$$r_a = \frac{\Delta V_a}{\Delta I_a} = \frac{40}{4} = 10\,k$$

V_a, volts

(b) Static anode characteristics

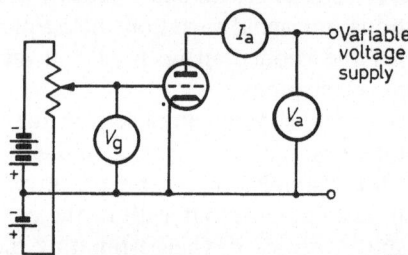

(c) Circuit for obtaining the static characteristics

Figure 4.7. Characteristics and parameters for a typical triode

much greater the grid voltage is in its effect on the anode current than the anode voltage. It should be noted that rises in anode voltage are offset by falls of grid voltage. The ratio $\Delta V_a/\Delta V_g$ is therefore negative. The amplification factor is however a positive quantity and so the definition of μ must include a minus sign.

(b) The a.c. or anode slope resistance, r_a. This parameter is a measure of the internal resistance of the triode to small variations of anode current and may therefore be regarded as the dynamic resistance. This is not, of course, the absolute resistance of the valve obtained by dividing the anode voltage by the corresponding anode current. We are concerned only with the amplification of signals and hence in variations of current and voltage. This is why the incremental or a.c. resistance is taken. Its value is the effective one in amplifier calculations.

$$r_a = \frac{\text{Small change in anode voltage}}{\text{Small change in anode current}} \quad \text{for a constant } V_g$$

$$= \left(\frac{\Delta V_a}{\Delta I_a}\right)_{V_g = \text{constant}} = \frac{\partial V_a}{\partial I_a}$$

88

When V_a is in volts and I_a in milliamps, r_a is in kilohms.

(c) *Mutual conductance*, g_m. This is the third parameter and is defined by

$$g_m = \frac{\text{Small change in anode current}}{\text{Small change in grid voltage}} \quad \text{for a constant } V_a$$

$$= \left(\frac{\Delta I_a}{\Delta V_g}\right)_{V_a = \text{constant}} = \frac{\partial I_a}{\partial V_g}$$

It is usual to have I_a in milliamps and so the units of g_m are most frequently given in milliamps per volt. In America the term transconductance is used instead of mutual conductance the units being millimhos (mho is a reciprocal ohm unit, therefore millimhos are the same as milliamps per volt).

From the graphs it will be seen that the values of μ, r_a and g_m depend upon the positions at which the parameters are taken. This is why it is wrong to refer to the parameters as valve constants. Over a given operating region the parameters are reasonably constant, but one must always take into account a change in value when working away from this region. Nevertheless a parameter is a useful measure or gauge of performance under specified conditions. Manufacturers always state the conditions (I_a, V_a, etc.) under which their published parameters were determined.

The three parameters are not independent. It can be shown that $\mu = r_a g_m$ and so if any two are known the third may be calculated.

The Triode as an Amplifier

The application of a varying signal voltage to the grid of a triode results in variations of anode current. To obtain voltage amplification it is necessary to pass the anode current through a resistor so that variations of current through the resistor give rise to variations of voltage across it. The resistor placed in the anode lead for this purpose is known as a load resistor. The ratio between the alternating voltage across the resistor to the alternating voltage applied between grid and cathode is known as the stage gain, A. Quantitatively, the stage gain is given by

$$A = \frac{\mu R_L}{r_a + R_L}$$

where R_L is the load resistance.

It can be seen that the stage gain is never as large as the amplification factor. The explanation lies in the fact that some of the voltage necessary for driving the alternating component of anode current through the load resistor is lost in the internal a.c. impedance, r_a, of the device.

89

It is often a matter of difficulty for the beginner to sort out the different volt-age and current components and symbols. Whilst for correct operation of a thermionic valve it is necessary to have steady h.t. voltages, grid bias voltages and steady currents, it is really only the variations of anode current, grid volt-age and anode voltage that are of final interest. It is therefore convenient to devise a circuit that is equivalent, so far as variations are concerned, to the complete circuit. Such an equivalent circuit is shown in *Figure 4.8*, and re-

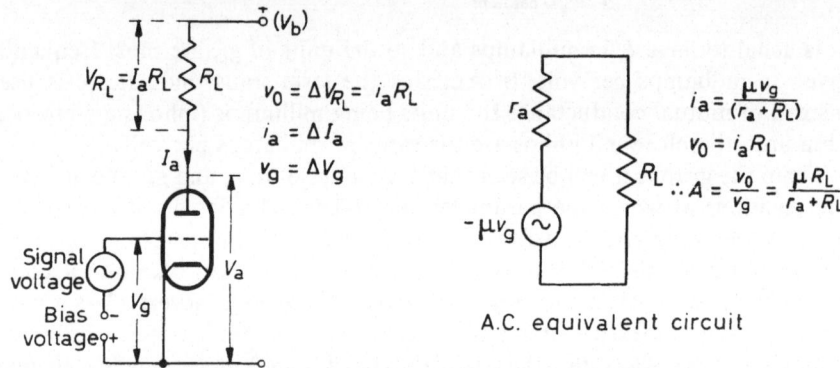

Figure 4.8. Equivalent circuit for a triode amplifier

presents a triode amplifier. The equivalent circuit is derived from mathemati-cal analyses that are given in most textbooks. By convention the alternating quantities are represented by small letters of the alphabet, whereas the total or absolute quantities are represented by capital letters; thus I_a symbolizes the actual anode current and i_a represents only the variation or alternating component of this current. By adopting an equivalent circuit many mathema-tical analyses can proceed without reference to the steady operating condi-tions. Convenient though an equivalent circuit is in the derivation of formulae for stage gain, input and output impedances, frequency of oscillation, etc., the designing of electronic apparatus must take the steady operating condi-tions into consideration.

Before discussing the specific details of the design of a simple amplifier the term 'bias voltage' must be explained. In *Figure 4.9* the dotted lines show the static mutual characteristics. When the grid is sufficiently negative with re-spect to the cathode, the valve is cut off. In the absence of a load resistor and with an h.t. voltage, V_b, of 250 V, increases in grid voltage from the cut-off value give rise to increases of anode current; such increases in anode current are predicted by the dotted curve $V_a = 250$ V. When however a load resistor is placed in the anode line, increases in anode current bring about decreases in the anode voltage; this is because of the voltage dropped across the load resistor. The dynamic characteristic cannot therefore be coincident with the dotted line $V_a = 250$ V, but must drop progressively to positions representing lower anode voltages. Plotting the new curve point-for-point yields a dynamic characteristic. In *Figure 4.9* the solid line is an example of a dynamic characteristic. The position of the dynamic characteristic depends upon the supply voltage and the load resistance. For a given anode voltage and load resistor the dynamic characteristic shows the value of the anode current for any given grid voltage.

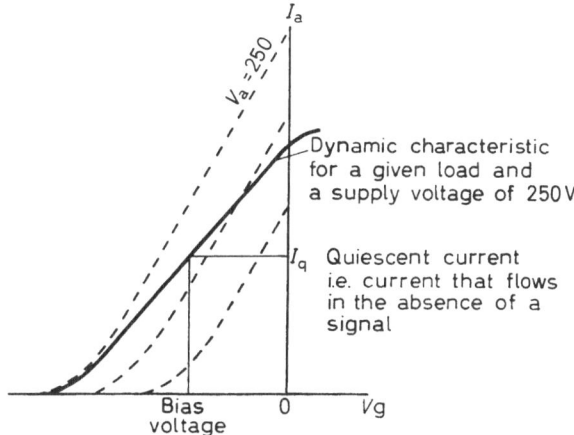

Figure 4.9. The dynamic characteristic

The dynamic characteristic is not straight throughout its length, but curves in the region of the cut-off voltage and also in the region where the grid voltage is positive. If faithful amplification of the signals is to be obtained, operation must be confined to the straight portion of the characteristic. This is achieved by ensuring that the signal voltage applied to the grid is superimposed upon a steady negative voltage. The value of this steady voltage is determined by observing that value of grid voltage that corresponds to the centre of the straight portion of the dynamic characteristic. This steady voltage is known as the grid bias voltage. At one time the bias voltage was supplied by a grid bias battery, but nowadays it is customary to use automatic grid bias or to obtain a fixed bias voltage from the power supply.

There is an additional point to consider when distortion-free amplification is required. The amplitude of the signal must not be too large otherwise, even with the correct value of grid-bias voltage, operation will extend into the curved regions of the characteristic.

Automatic biasing is achieved by using the circuit of *Figure 4.10*. The necessary bias voltage is produced by placing a resistor in the cathode line. When the quiescent current I_q (which is the current obtained in the absence of a signal) passes through the bias resistor in the cathode line there is necessarily a voltage developed across this resistor. The magnitude of the voltage is determined by the product of the quiescent current and the bias resistance. The method of estimating the values of I_q and the bias resistance are described below. A consideration of *Figure 4.10* shows that the end of the resistor con-

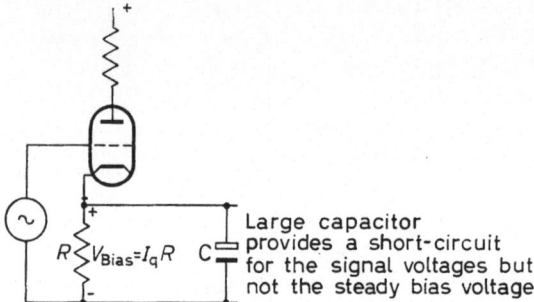

Large capacitor provides a short-circuit for the signal voltages but not the steady bias voltage

Figure 4.10. Circuit for automatic grid bias. The large capacitor C stabilizes the bias voltage when signals are being applied

nected to the cathode becomes positive with respect to the chassis or earth line. This must be so since the electrons (which are negatively charged) move along the resistor in the direction of the cathode, and subsequently travel to the anode and the h.t. positive terminal of the power supply. With no input signal, i.e. with the input terminals of the amplifier effectively shorted, the potential of the grid must be that of the chassis or earth line, and hence must be negative with respect to the cathode. If the shorting link is removed and a signal applied, the alternating signal voltage must be superimposed on the bias voltage. Unfortunately the presence of the bias resistor prevents the full signal voltage from being applied between the grid and cathode. As a result there is a fall in the gain of the amplifier, which is not always desirable. This point is further discussed in Chapter 9.

To avoid the fall in gain, a short-circuit path across the bias resistor must be provided for the signal voltage, but not for the bias voltage. A capacitor of large capacitance provides such a path since its impedance at zero frequency is infinitely large, whereas at nearly all practical signal frequencies the impedance is very low.

Amplifier Design

It would be impossible for manufacturers to publish the dynamic characteristics for all possible combinations of supply voltage and load resistance to
be used with any given valve. Since the dynamic characteristics are not available, and it would be tedious to plot them out, amplifier designs are made on
the published anode characteristics.

$$V_a + I_a R_L = V_b$$

$$\therefore \quad I_a = -\frac{1}{R_L}V_a + \frac{V_b}{R_L}$$

This represents a straight line with

a slope $-\dfrac{1}{R_L}$ and intercept on I_a

axis of V_b/R_L. The bias voltage is

ignored because it is so small

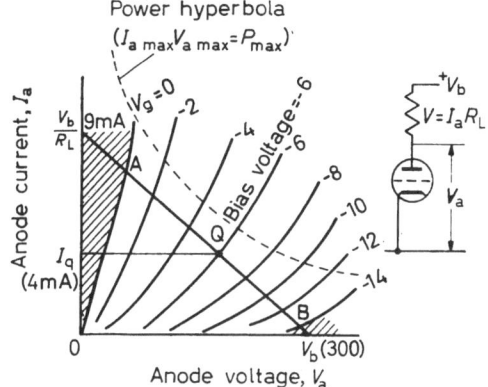

Figure 4.11. Construction of the load line

The first step in the design is to construct a load line on the characteristics.
From *Figure 4.11* we note that $V_b = V_a + I_a R_L$ and so $I_a = -(1/R_L)V_a +$
V_b/R_L. This is the equation of a straight line. When $I_a = 0$, $V_a = V_b$, that
is to say, with zero anode current, there is no voltage drop across the load
resistor and the anode voltage equals the supply voltage. This is one point on
the line. When $V_a = 0$, $I_a = V_b/R_L$, which gives the second point. A straight
line, known as the load line, is drawn between the points. The slope of the
line is evidently $-1/R_L$.

Figure 4.12 shows how the load line can be derived from the dynamic characteristic. Normally when designing an amplifier the dynamic characteristic
need not be considered, but it may be convenient for readers to compare the
two graphs and thus understand how a load line drawn on a set of static anode
characteristics can be used to design and predict the behaviour of a single-
stage amplifier. The hyperbola drawn is obtained from the manufacturer, who
specifies the maximum power, P_{max}, that may safely be dissipated at the anode.
This power is the product of $I_{a, max}$ and $V_{a, max}$. Plotting $I_{a, max}$ against $V_{a, max}$
will give a hyperbola. In a properly designed amplifier the load line must lie
below the power hyperbola. For reasons already given, operation at positive
grid voltage should not be permitted. The hatched area *(Figure 4.11)* between

the I_a ordinate and the $V_g = 0$ characteristic is therefore forbidden. For very low anode currents, it will be noticed that the characteristics are not evenly spaced, but become quite crowded at values of grid voltage near the cut-off voltage. It is usual to avoid operation where the crowding of the characteristics is rather severe. This is equivalent to avoiding operation on the curved part of the dynamic characteristic. This leaves a certain length of load line,

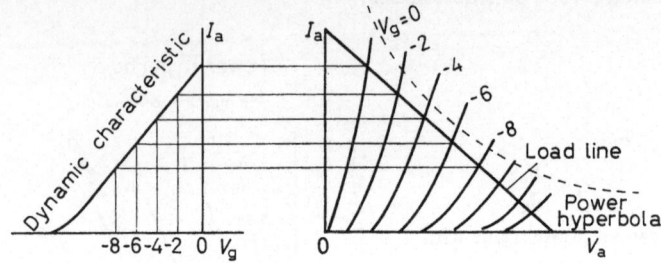

Figure 4.12. The derivation of the load line from the dynamic characteristic

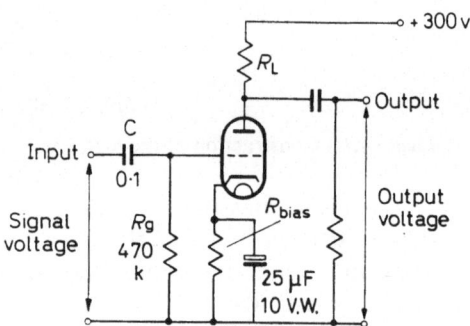

Figure 4.13. A complete single-stage amplifier. For calculation of component values see text. C prevents any upset in bias conditions by blocking any steady voltage from the signal source. R_g prevents the grid from becoming inoperative by allowing a leakage path for the discharge of C

AB, along which operation is permitted. The operating point Q is chosen to be at the centre of AB. Having determined the position of Q, the required bias voltage and the quiescent current can be estimated.

Let us suppose that we are required to design a single-stage amplifier *(Figure 4.13)* with a gain of 16 to operate over most of the audio band (say signals with frequencies from 40 Hz to 20 kHz). The supply (h.t.) voltage, V_b, is 300 V. The design proceeds as follows:

(1) Select a valve with a suitable amplification factor (μ). The gain must be 16, therefore from the gain formula, we realize that μ must be greater than 16. Let us choose a triode with μ of 20 and $r_a = 8$ k.

94

(2) Calculate the value of R_L from

$$A = \frac{\mu R_L}{r_a + R_L} \qquad \text{i.e.} \qquad 16 = \frac{20R_L}{8\,k + R_L}$$

$$\therefore \quad 4R_L = 8 \times 16\,k$$

$$\therefore \quad R_L = 32\,k$$

Choosing the nearest preferred value $R_L = 33\,k$.

(3) Construct the load line on the static anode characteristics to pass through the points ($I_a = 0$, $V_a = 300\,V (= V_b)$)) and ($I_a = V_b/R_L = 300/33\,k = 9\,mA$, $V_a = 0$.)

(4) Choose the operating point Q and read off the quiescent current (4 mA) and the bias voltage ($-6\,V$) (see *Figure 4.11*).

(5) Calculate the bias resistance from $V_{bias} = I_q \times R_{bias}$. In our case $R_{bias} = 6/(4 \times 10^{-3}) = 1\cdot5\,k$.

(6) We note that the quiescent current is 4 mA and, therefore, the dissipation in the anode resistor is $(4 \times 10^{-3})^2 \times 33 \times 10^3 = 0\cdot528$ watts. To leave an adequate safely margin the anode load should be a 1 W resistor.

(7) The time constant of $C_1 R_g$ must be long enough so that a reasonable response to low frequencies is obtained. $0\cdot1\ \mu F$ and 470 k are adequate values for audio-frequency work. The bypass capacitor across the bias resistor should be 25 to 50 μF. The working voltage need only be about the bias voltage. A 10 V working would be adequate, giving us a margin of safety.

MULTIELECTRODE VALVES

When a triode is being used as an amplifier of low-frequency signals, the effect of the capacitances that exist in the valve between the various electrodes may be neglected. The anode and grid, being conductors in fairly close proximity separated by a dielectric, form a small capacitor, and when attempting amplification at high frequencies this capacitance is troublesome. More will be said about this when we discuss amplifiers in Chapter 9, but for the present it must be pointed out that voltages may be fed back to the grid from the anode circuit. These voltages lead to instability and oscillation in the amplifier, thus rendering it useless for amplification purposes.

The first attempt to overcome the problem involves interposing a second grid between the anode and the control or signal grid. This second grid serves as an electrostatic screen and reduces the anode-to-grid capacitance from about say 5 pF to 0·001 pF. The resulting valve is known as a screened grid valve. In order to maintain the flow of electrons through the valve the screen grid must be kept at a high potential relative to the cathode. If the anode voltage were 250 V, for example, the voltage on the screen may be about 200 V. Unfortunately, when the electrons reach the anode, they have considerable speed and hence kinetic energy. Such energy is sufficient to dislodge electrons from the anode material. The electrons so emitted from the anode are known as secondary electrons, the phenomenon being called secondary emission. It is the secondary emission that accounts for the kink in the characteristic shown in *Figure 4.14*. For low anode voltages, the electrons are collected by

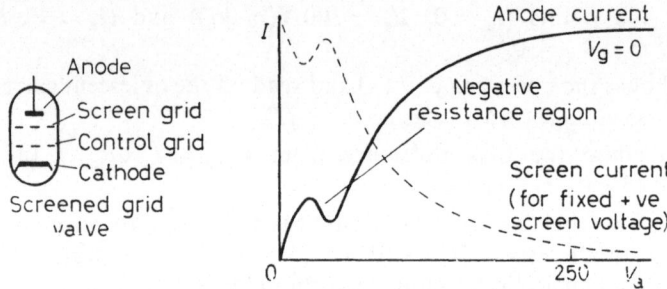

Figure 4.14. Characteristic curve for a screened grid valve

the screen grid. The anode current then falls. For higher anode voltages, the secondary electrons cannot reach the screen grid and hence there is the usual rise of anode current with anode voltage. The kink however, is very inconvenient, since operation on the portion that has a negative slope means in effect that the valve is acting as a negative resistance. Because the valve was invented for use at high frequencies (radio frequencies), the loads are not resistors, but tuned circuits consisting of inductors and capacitors. Operation on the negative resistance portion of the characteristic may neutralize the positive resistance of the tuned circuit, thus reducing the effective losses to zero. The amplifier will then burst into oscillation and will no longer perform as an amplifier.

Secondary emission can be prevented by slowing the electrons on their journey from the screen grid to the anode. This is achieved by introducing a third grid into the valve and placing it between the screen grid and the anode. The potential of this third grid is low; in the majority of cases the grid is connected directly to the cathode. This third grid is known as the suppressor grid since it suppresses the secondary emission. The five-electrode valve that results is known as a pentode. Stage gains with this type of valve are usually many times higher than those possible with triode valves. *Figure 4.15* shows the circuit symbol for a pentode together with a typical set of anode characteristics. It will be noted that the characteristics are almost horizontal beyond the knee. This means that the anode current is almost independent of the anode voltage, a fact that we shall use in future discussions on the cathode-ray oscilloscope and constant current generators. Flat curves of this type are characteristic of high impedance devices, i.e. constant-current generators. The equi-

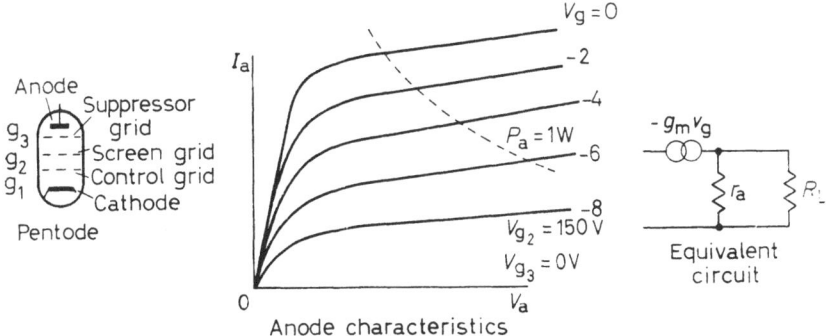

Figure 4.15. Characteristics of a pentode valve. Published curves are for some specified value of voltage for g_2. The suppressor voltage is usually zero, i.e. g_3 is connected to the cathode. Figures given are only examples

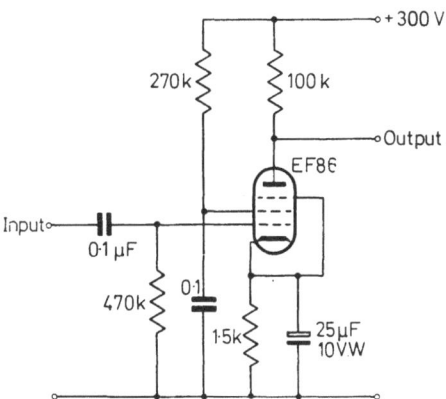

Figure 4.16. Circuit of a single-stage pentode amplifier with a voltage gain of about 100. Note the decoupling capacitor between the screen grid and the h.t. −line. This is to prevent the voltage on the screen grid from varying. Fixed screen grid voltages give greater gain

valent circuit of a triode is not, therefore, convenient to use. Instead the equivalent circuit shown in *Figure 4.15* is adopted. The actual circuit of a single-stage audio amplifier is shown in *Figure 4.16*. The EF86 is a special low-noise pentode in which particular attention has been paid to the construction of the heater and rigidity of the electrodes. The output from the valve has a very low hum content—i.e. the output is almost free from 50 Hz signals induced by the a.c. feeding the heater. Variations of anode current with mechanical vibration (known as microphony) are also minimized in this type of pentode.

When very large output powers are required, as opposed to voltage swings, the multielectrode structure of the pentode poses difficulties associated with heat dissipation. The kinkless or beam tetrode largely overcomes the difficulties *(Figure 4.17)*. In this type of valve the suppressor grid is dispensed with,

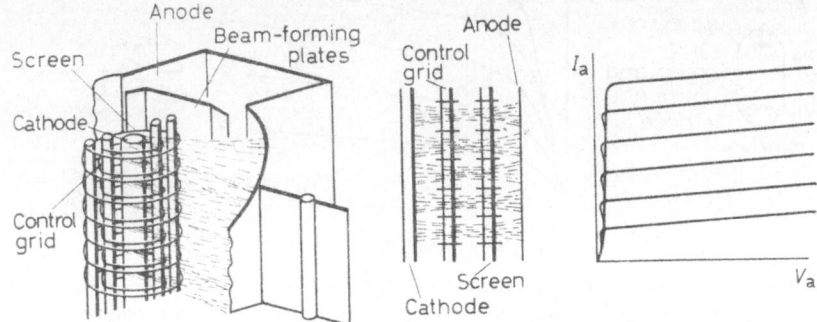

Figure 4.17. Details of the beam tetrode. The screen and control grids are aligned so as to concentrate the beams. The density of electrons at the anode is so great that secondary emission is effectively suppressed

and secondary emission is suppressed by concentrating the electrons into beams. The concentration is achieved by aligning the control-grid wires with the screen-grid wires, by the use of beam-forming plates, and by constructing specially shaped anodes. The knee of the characteristic is much closer to the ordinate than that of a pentode, therefore more of the load line is available for operation. The greater variations of anode voltage and current that are permitted lead to an increase of efficiency of this type of valve over that obtainable with triodes and pentodes.

In an attempt to reduce the physical size of apparatus, manufacturers often enclose more than one valve assembly within a single glass envelope. The electronic functions are not altered in any way. The circuit symbols used for common multi-valve arrangements are shown in *Figure 4.18*.

It is not possible to mention every coding system used by valve manufacturers. One system however, devised by Mullard Ltd., is being used by several manufacturers. In this code the letters have the following significance:

(a) First letter C—200 mA heater
               D—0·5 V to 1·5 V heater
               E—6·3 V heater
               P—300 mA heater
               U—100 mA heater

(b) The second and subsequent letters

A—single diode

B—double diode

C—triode

D—output triode

E—tetrode

F—voltage amplifying pentode

K—heptode or octode

L—output pentode or tetrode

Y—half-wave rectifier

Z—full-wave rectifier

(c) The first figure represents the type of base, the most important being 3—Octal base, 8—B9A (noval) base and 9—B7G base. The second figure indicates the order of development.

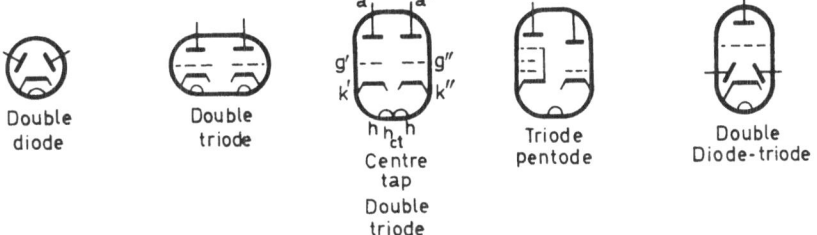

Double diode   Double triode   Centre tap Double triode   Triode pentode   Double Diode-triode

Figure 4.18. Common examples of circuit symbols used to represent multi-valve arrangements within a single envelope

Thus an ECC81 is a double triode with a 6·3 V heater using a B9A base. A PCF80 is a triode and pentode within the same glass envelope. A B9A base is used and the heater current is 300 mA.

In the United States two systems of labelling valves are used. The older system uses numbers alone, but this has been superseded by designations that involve a number followed usually by two letters and then a further number, e.g. 6SL7. The first number indicates the approximate heater or filament voltage. The figure 1 indicates a voltage of 1·4 V, 6 means 6·3 V, 12 means 12·6 V and so on. The letters indicate the type of valve. The letter S stands for single-ended, as opposed to a double-ended valve in which some of the leads are brought out of one end and the remaining leads brought out of the the other. The top of a double-ended valve usually has only a single lead to a grid or an anode. In double-ended valves the S is omitted, in which case a single letter alone may be used, e.g. 6A8. The last figure refers to the number of useful elements brought out to an external connection. The four-element designation may be followed by a hyphen and further letters, e.g. 6SJ7–GT. The suffix after the hyphen denotes the type of construction used, G representing a large glass envelope and GT a small straight-sided glass tube. The letter M indicates that the outside of the bulb has been metallized. This metallized coating acts as an electrostatic screen. Connection to the coating is made via one of the base pins.

Valves used for H.M. Forces may have CV numbers. Tables are readily available to enable a CV number to be identified with a manufacturer's coding.

5

ACTIVE DEVICES 2–
SEMICONDUCTORS

In 1948 the American physicists, J. Bardeen and W. H. Brattain announced the invention of the transistor[1], a new type of amplifying device made from semiconducting crystals. Very few at that time could have foreseen the revolutionary developments that were to follow, developments so important and far-reaching as to change the whole outlook of the science and technology of electronics. The physical principles involved in transistor action[2] had been worked out in conjunction with their colleague, W. Shockley; and in recognition of their work the three physicists were awarded jointly the 1956 Nobel Prize for Physics.

From the early days of the Second World War until 1950 a very great increase was made in our knowledge of the properties of semiconductors. Semiconductors had, of course, been known for a long time; Faraday in 1833 performed experiments with galena and carborundum; the rectifying properties of certain solids were discovered some two years later. A silicon rectifying diode was known as early as 1906 with the invention of a detector by Pickard, and most readers will be aware of the 'cat's whisker' detectors of early radio sets that made use of the rectifying properties of a metal-to-semiconducting crystal contact. The crystal in this case was usually impure galena, and many a tricky domestic situation arose in the selection and maintaining of the best location of the 'cat's whisker' on the crystal. The invention and development of thermionic diodes and triodes overshadowed the crystal diode, rendering it temporarily obsolete, but prior to and during the last war a great deal of work was done on germanium diodes for telecommunication purposes. It now seems likely that transistors and allied semiconductor devices will replace thermionic tubes in almost every type of electronic equipment.

The term 'transistor' arises from a combination of the italicized portions of the words *trans*former and re*sistor*, since the device is made from resistor material and transformer action is involved in the operation of a transistor. The first transistors were of the point contact variety, but two or three years later the junction transistor made its appearance. In the early days the point-contact transistor performed better at higher frequencies than the junction types; but improved manufacturing techniques have enabled the junction transistor to establish its superiority in this respect. The vulnerability of point contact transistors to mechanical shock has precluded the use of this device for all practical purposes; the junction transistor is now so firmly established that we will consider this type alone.

Transistors have several advantages over their thermionic counterparts. They are small and light, being usually less than one gram in weight and less than 1 cm³ in volume. Most of the bulk is made up of the container and silicone grease or other filler. The actual transistor itself is very small; in micromodule integrated circuits several hundreds of transistors can be fabricated on a single silicon chip about 0·1 inch square, although the single transistor in normal use is somewhat larger than the micromodule type. Transistors can withstand mechanical vibrations and shocks that would ruin a thermionic valve. This is because a transistor is a single crystal device and therefore has no delicate electrode structure to damage. In addition there is no vacuum to preserve within a vulnerable glass envelope. Heaters are not required in transistors and thus there is a substantial saving of power; a small amplifying thermionic valve usually requires about 2 watts of heater power. Since no heaters are required, transistors operate immediately the supply is connected, no warming up period being necessary. The supply voltages required are low and rarely exceed 40 V. Voltages from 6 V to 24 V are common. As far as can be judged, the life of a transistor that is properly installed and operated within its ratings should be indefinitely long. It is therefore an ideal device for use in remote locations such as space satellites and submarine telecommunication cables that need booster or repeater amplifiers along the line.

Problems associated with limited power output, noise, frequency response and operating temperatures are all yielding to intensive research. Nowadays there are very few applications in which a thermionic device is clearly superior to a transistor.

CONDUCTORS, INSULATORS AND SEMICONDUCTORS

The devices discussed in this chapter depend for their action on the controlled flow of electric charges through a suitably prepared crystal. To understand the rectifying and amplifying properties of crystal devices a certain amount of preliminary solid-state physics needs to be known.

Our present concept of the nature of matter is that all materials are made of atoms. These atoms are basically electrical and have a small, massive, positive nucleus (consisting of protons and neutrons) which is surrounded by a cloud of electrons. In a neutral atom the number of electrons equals the number of protons. The charge on an electron is equal to that on the proton, but of opposite sign, the electron being negative and the proton positive. After the discovery of the electron by J. J. Thomson in 1897, later attempts to elucidate the nature of matter resulted in the concept of a structure in which the electrons occupied certain shells about the central nucleus. The theories depended heavily on the work of spectroscopists; Bohr was the first to invoke the quantum theory and show that the electrons were not free to occupy shells at any arbitrary distance from the nucleus. Most readers will be familiar with his theory of the hydrogen atom in which the single electron could occupy only certain discrete levels. Intermediate energy levels are forbidden to the electron. Energy can be acquired by the electron only in discrete quanta. The acquisition of energy results in a transition to another shell, and the atom is then in an unstable state. When the electron drops to its original level, energy is given out in the form of electromagnetic radiation of frequency v where $E = hv$, E being the energy involved, and h being Planck's constant. The transitions from the various permitted levels result in radiations of a set of specific frequencies, and these are manifest in the characteristic line spectra for the element.

If sufficient energy is available the electron is able to break away from the parent nucleus and the atom is then said to be ionized. Two charges are then available, the negative one on the electron and a positive one of the same magnitude, but opposite sign, on the ion which remains. The electrons are comparatively mobile because of their small mass. The mobility of the positive ions is much lower since the mass is almost that of the neutral atom. The amount of energy needed to ionize an atom is measured in electron-volts. Voltage is a measure of the work needed to move unit charge against the electric field between two specified points. When 1 joule of work is expended in moving 1 coulomb, the voltage between the points is 1 volt. When the charge is that on an electron, the work involved is an electron-volt. The ionization energy is therefore measured in electron-volts, but since the only charge involved is an electron, it is customary to refer to ionization potential. In the case of hydrogen this potential is 13·6 V and hence it requires 13·6 electron-volts to ionize a hydrogen atom when the electron is originally in the ground state, i.e. lowest energy level. The various levels associated with the atom may be represented by a 'thermometer' type diagram *(Figure 5.1)* in which the horizontal width has no significance.

When the atom is more complex than hydrogen the spectrum is more complicated, as might be expected. Provided the atoms of the element are widely spaced (as in the gaseous state for example) line spectra are obtained. As the atoms become more densely packed interaction between the outer electrons of the atoms occurs. Instead of obtaining a single energy line, two atoms will produce two lines and n atoms will produce n lines. The single line, therefore, degenerates into a band, and a band spectrum is produced. The form of the 'thermometer' energy level diagram will depend upon the inter-atomic distance of the atoms after they have come together into a crystal

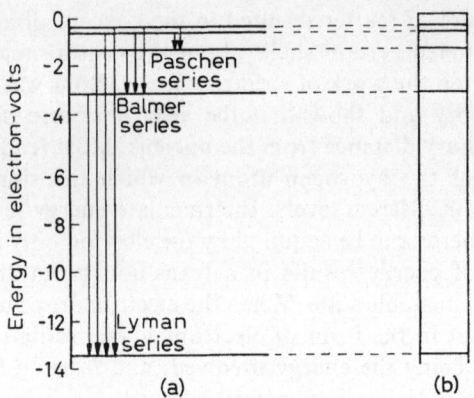

Figure 5.1. Method of representing discrete energy levels for hydrogen. (*a*) Conventional diagram. (*b*) 'Thermometer' type diagram

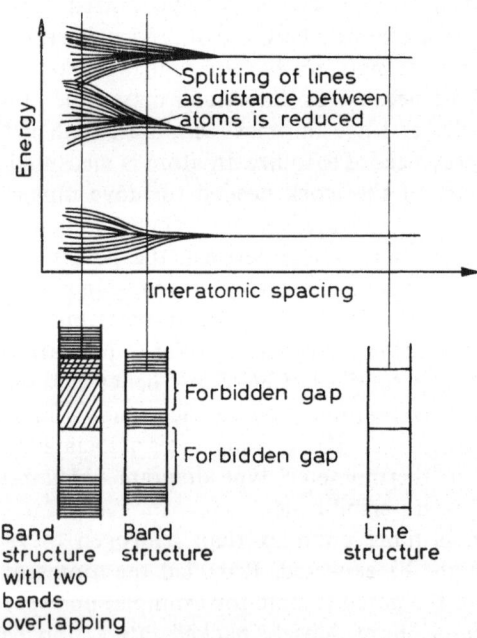

Figure 5.2. Energy level diagrams for various interatomic distances

lattice. *Figure 5.2* shows two possibilities. So far as we are concerned, we may ignore all the energy levels except those associated with the outermost or valence electrons. In the crystals with which we are involved the bonding is

104

covalent and results from a sharing of the valence electrons. Readers will be well aware that the chemical properties are determined by the valence electrons, all other inner electrons playing no part in chemical unions. It is the transition between two particular bands (namely, the valence band and the conduction band) that will be our main concern. Quantum theory predicts that a forbidden region of energies may exist between the two bands. No valence electron can occupy an energy level in this forbidden region or gap. The allowed band of energies is strictly of fine line structure, but the lines are too close to be resolved. Each line represents the energy that can be occupied by an electron. Since in a crystal there are many millions of valence electrons, there will be many millions of lines in the band. Diagrammatically, it is possible to show only a few.

Some materials like quartz and mica will not conduct electricity, whereas copper and silver conduct current very easily. The energy-level diagram represents one way of interpreting these facts. For every material there is a characteristic valence band structure and a given forbidden gap. Conduction in a material can occur only if electrons are available as charge carriers. In so far as they are locked in a valence bond or other trap, they are not available for conduction. To become available they must break away from the bond by the acquisition of sufficient energy from thermal or other sources (e.g. light or x-rays). To remain as free charge carriers, the acquisition of energy must take them from their original level to an higher unoccupied energy level. The Pauli exclusion principle dictates that if the upper level is already filled or falls into the forbidden gap the electron cannot remain a free charge carrier. The picture we have of an insulator therefore is one in which all the possible levels in the valence band are occupied by electrons. In addition the forbidden gap is large so that the probability of electrons acquiring sufficient energy from thermal sources, or from an electric field, to move up into the conduction band is extremely low. No electrons can therefore be freed for conduction. A conductor on the other hand may be represented in more than one way depending upon the element. The diagram may show a forbidden gap that is large, but only half of the valence band occupied. Very little energy is necessary to move the electron up the valence band. Such energy is always available at normal temperatures. Many charge carriers are present and the material is a good conductor. Alternatively the valence band may be filled, but if the forbidden gap is very small, or absent because of overlapping between the filled valence band and the empty conduction band above it, then again very little energy is required to move a valence electron to a higher energy level. Copper is an atom where the outer electron is extremely easy to detach from the parent atom because of the elongated elliptical orbit of the valence electron. This is interpreted as an overlapping of the valence and conduction bands. In addition the valence band is only half-filled.

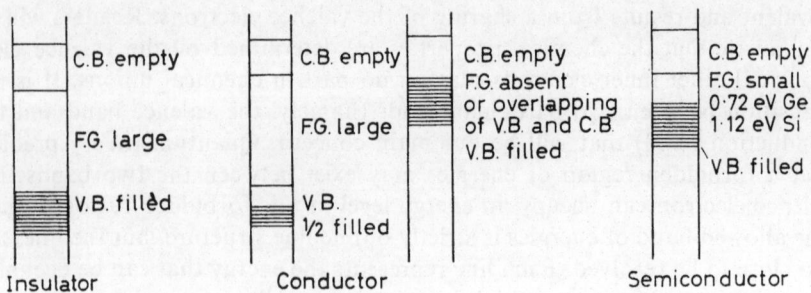

Figure 5.3. Energy-level diagrams for an insulator, conductor and semiconductor (C.B., conduction band; V.B., valence band; F.G., forbidden gap)

Between the two extremes, there is a range of elements and compounds where the valence band is filled and the forbidden gap is not too large, hence the material is neither an insulator or a conductor; it is a semiconductor. The two best known elements are germanium and silicon because transistors and allied devices are made from them. The forbidden gap for germanium is 0·72 eV, whilst that for silicon is 1·12 eV. Diamond, which is a reasonably good insulator, has a forbidden gap of 5·3 eV, whilst the gaps for mica and quartz are much larger. The line of demarcation between the classes is not well defined. The resistivities of quartz and marble are respectively about 10^{18} and 10^{10} ohm cm. Both are classified as insulators, although marble is one hundred million times a better conductor than quartz. Silver and mercury have resistivities of the order of 10^{-6} and 10^{-4} ohm cm. Semiconductors are in the range of about 1 ohm cm to 10^5 ohm cm. The figures for intrinsic

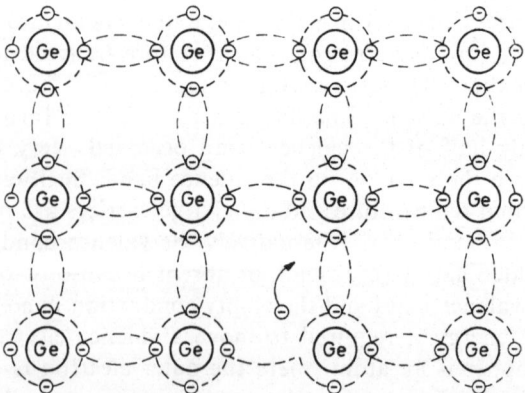

Figure 5.4. Diagrammatic representation of intrinsic germanium showing bonding of the valence electrons and the formation of an electron-hole pair. The charge carriers produced, namely holes and electrons, are responsible for intrinsic conduction

106

(i.e. pure) germanium and silicon are 47 ohm cm and $2 \cdot 3 \times 10^5$ ohm cm respectively at 300°K.

Germanium and silicon are both tetravalent elements. When techniques are used to produce transistor material, extremely pure single crystals are made in which the atoms take up a regular tetrahedron pattern (i.e. the diamond structure). In our diagrams it will be convenient to represent the crystal as a flat array as in *Figure 5.4*, bearing in mind the true structure in its three-dimensional lattice.

For intrinsic germanium, where the impurity content is less than 1 part in 10^8, some conduction can occur at room temperatures because the forbidden gap is not large and sufficient energy is available from thermal sources. There is always a possibility at any point in the crystal that a covalent bond will be ruptured. The electron is released into the crystal. The absence of an electron is in effect a positive charge since the germanium atom has lost a negative electron. The absence of an electron is called a 'hole'. It is convenient to regard a hole as a separate entity having a positive charge equal in magnitude to the charge on the electron. An analogous concept arises when we refer to the absence of water in a tube as a bubble. As the bubble floats up the tube we say the bubble has moved, although we realize that a more complicated movement of water molecules is involved. In the same way, when an electron drops into a hole (and the valence bond is re-established) the region from which the electron came must be positively changed. It is convenient to regard this as the movement of a hole. The rupture of a bond results therefore in the formation of an electron-hole pair.

The conduction that occurs in pure crystals is called intrinsic conduction, and, in so far as it occurs, it is a nuisance. If intrinsic conduction were absent semiconductor devices would be superior to those now available. The reason for this will be made clear later in the chapter. At room temperatures germanium contains approximately 2×10^{13} conduction electrons per cm³. At elevated temperatures the number is considerably higher. Silicon, with a larger energy gap than germanium, contains only about 10^{10} electrons per cm³ at room temperature. Silicon devices are therefore superior to germanium ones when operation at room and elevated temperatures is considered. Germanium devices are satisfactory up to about 70 or 80°C whereas silicon devices will work happily at 150°C.

Impurity Conduction

One way of increasing the conductivity of a crystal is to add a controlled amount of certain impurities. The conduction that occurs then is called impurity conduction and is of paramount importance in the operation of semiconductor devices.

When small amounts (1 part in 10^7 approximately) of a pentavalent impurity such as arsenic or phosphorus are added during crystal formation, the impurity atoms lock into the crystal lattice since they are not greatly different in size from a germanium atom, and the crystal is not unduly distorted. Four of the impurity valence electrons form valence bonds with adjacent germanium atoms, but the fifth is very easily detached at normal room temperatures, and becomes free within the crystal. The pentavalent impurity donates an

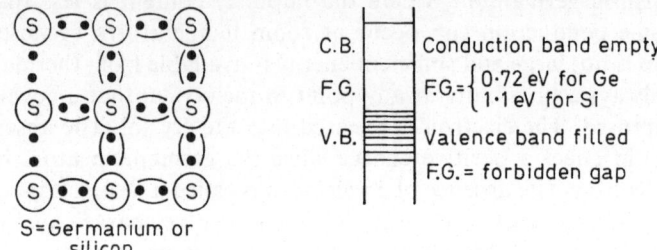

Figure 5.5a. Intrinsic semiconductor with the corresponding energy-level diagram

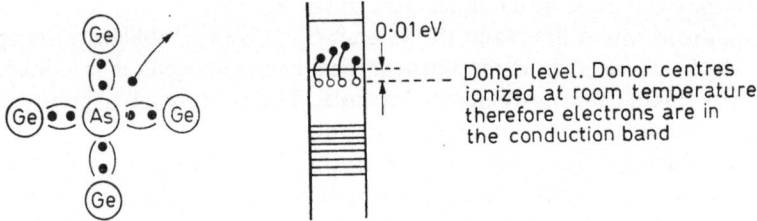

Figure 5.5b. Representation of *n*-type germanium. Conduction is mainly by electrons since they are in the majority. Some holes are formed upon the breakdown of valence bonds but they are in the minority

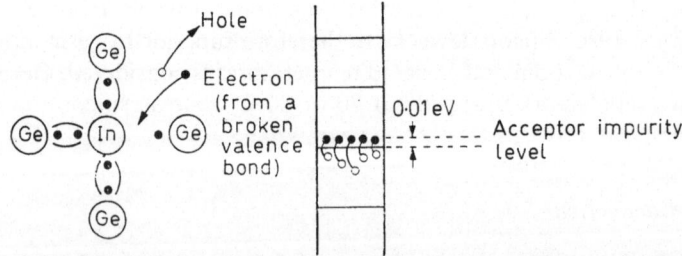

Figure 5.5c. Representation of *p*-type germanium. At room temperatures sufficient stray electrons from electron-hole pair formation are available to fill acceptor centres. Holes (or vacancies) are therefore created in the valence band. Conduction is mainly by holes since these are in the majority. The acceptor centre is thus negatively charged

108

electron to the crystal and is therefore called a donor impurity. The material is then known as n-type germanium *(Figure 5.5b)*. The crystal as a whole is of course still electrically neutral, but the number of available charge carriers has been considerably increased. Since only about 0·01 eV is required at room temperatures to release the fifth electron into the conduction band, all the donor centres are ionized.

Instead of adding a pentavalent impurity to an otherwise pure crystal a trivalent impurity such as indium, boron or aluminium may be included. The three valence electrons enter into covalent bonding, but a region of stress exists where the fourth bond ought to be, and any stray electron in the vicinity can be trapped by the impurity centre which then becomes negatively charged. The region from which the electron came is positively charged. In effect the trivalent impurity has injected a hole into the crystal. Such an impurity is called an acceptor impurity since it accepts electrons into the impurity centre. Germanium or silicon doped in this way has many holes as charge carriers and is therefore termed a p-type semiconductor *(Figure 5.5c)*.

In an n-type semiconductor electrons are in the majority and are called majority carriers. Some holes exist owing to the formation of electron-hole pairs at room temperature; these are called minority carriers. Alternatively in a p-type semiconductor, holes are the majority carriers and electrons the minority carriers. The number of charge carriers available determines the resistivity. Most of the germanium and silicon used in devices has a resistivity of 0·1 to 10 ohm cm after the impurities have been added.

THE pn JUNCTION RECTIFIER

If, during manufacture, a crystal is doped (i.e. impurities added to modify the characteristics) so as to be p-type at one end and n-type at the other, a very important practical semiconductor device is obtained. It is the concern of the technologist to ensure that the change from p-type to n-type material occurs abruptly at a plane or other well-defined boundary called a pn junction. After formation the electrons diffuse from the n-type material and fill some of the acceptor centres. The result is that a layer of negative charge lies along the junction on the p-side. A corresponding diffusion of holes to the n-side creates a layer of positive charge along the junction on the n-side. The final result *(Figure 5.6)* is the creation of a potential gradient across the junction. The electric field in the junction region is between fixed charged centres, viz. the acceptors, which are negatively charged, being on the p-side, whilst the fixed donor centres, positive because of the loss of an electron, are on the n-side. The potential gradient increases with diffusion until eventually it is great enough to prevent any further migration of electrons from the n-side and holes from the *p*-side. A dynamic equilibrium is then established. The potential gradient is often represented, as in *Figure 5.6*, by a small internal fictitious battery with the polarity as shown.

The application of a potential difference across the whole crystal can be considered by referring to *Figure 5.7*. If the positive pole of the external

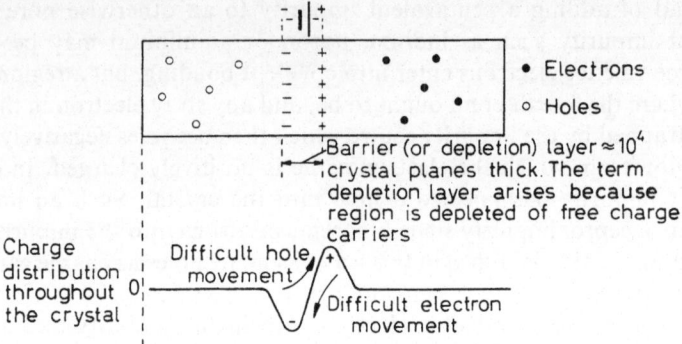

Figure 5.6. The *pn* junction. Upon formation the electrons diffuse from the donor impurities leaving the centres positively charged. The electrons are captured by the acceptor centres. Two layers of charge are formed along the junction boundary. Diffusion ceases when the barrier potential (represented by a fictitious internal battery) is high enough to prevent holes diffusing from the *p*-side, and electrons from the *n*-side

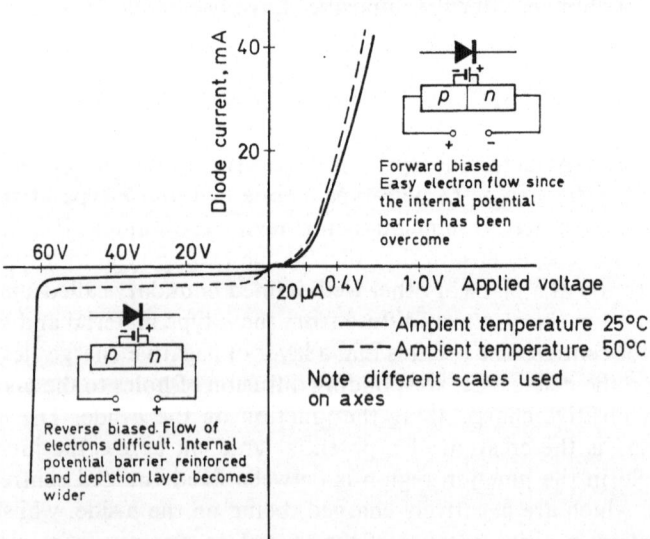

Figure 5.7. The behaviour and characteristics of a *pn* junction when subject to external applied voltages

110

e.m.f. source is connected to the p-side, and the negative pole is connected to the n-side, the electrons and holes will be drawn away from the junction and the potential barrier that exists there will be lowered. When the external voltage exceeds a few hundred millivolts, the internal 'battery' is totally overcome. Electrons and holes then flow freely across the junction, and the whole crystal is then a good conductor of electricity. When however the polarity of the external voltage source is reversed, so that the negative terminal is connected to the p-type region and the positive terminal is connected to the n-type region, then the potential barrier within the crystal is increased. Electrons are repelled towards the junction on the p-side and holes are repelled towards the junction on the n-side. A comparatively thick barrier layer is

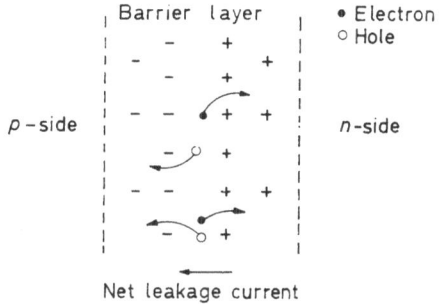

Figure 5.8. At room temperatures electron-hole pairs are formed in the depletion layer constituting the junction. The holes are swept to the p-side and the electrons to the n-side. The net leakage current is from the n-region to the p-region

formed in which there are very few free charge carriers. Such a layer, known also as a depletion layer, is a very good insulator, and thus very little current flows through the crystal. The *pn* junction device is therefore a rectifier, conducting electricity well in one direction, but not in the other.

Across the depletion layer there is a strong electric field. If a covalent bond in the junction region is ruptured by the acquisition of energy by the valence electron, then the hole formed will be swept to the p-side and the electron to the n-side. This constitutes a leakage current *(Figure 5.8)*. We see now why the intrinsic conduction should be low. Ideally the current should be zero when diode is reverse biased. This is why silicon diodes are preferred to germanium ones. At elevated temperatures the leakage current of a silicon device is many times smaller than that of a corresponding germanium diode. For example, at 60°C and with an inverse voltage of 50 V (i.e. a voltage applied so that the diode is in the non-conducting or reverse-biased state) a typical small germanium diode may have a leakage current of 20 μA.

111

The leakage current for a corresponding silicon device under the same conditions may be only about 0·05 μA.

If the reverse voltage is increased sufficiently, the few free charge carriers in the depletion layer gain sufficient energy to break down many covalent bonds. The reverse current then increases enormously. More will be said about this when we discuss zener diodes. The breakdown of the junction is not a permanent damaging of the device. When the reverse voltage is reduced the barrier layer returns to its normal state, provided the maximum reverse current has not been exceeded. For rectification purposes a diode is chosen whose reverse breakdown voltage exceeds any peak inverse voltage likely to be encountered.

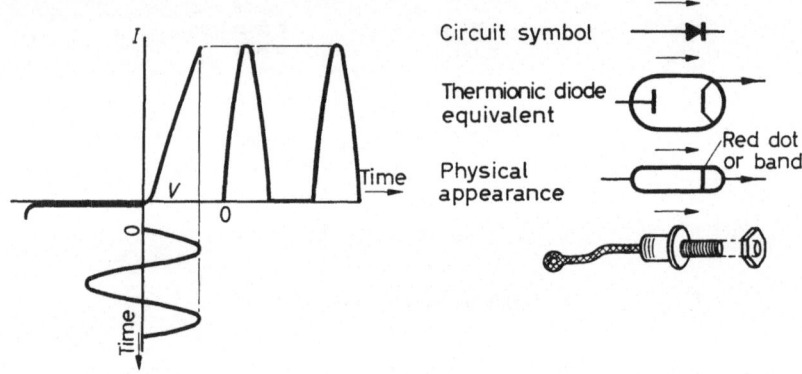

Figure 5.9. The diode used as a rectifier. The arrow in each case shows the direction of conventional current when the conduction is easy

Semiconductor diodes are now replacing thermionic valves for all rectification purposes *(Figure 5.9)*. We have previously seen how thermionic diodes can be used as rectifiers, demodulators, clamps and voltage limiters. Exactly the same circuits may be used when the thermionic diode is replaced by a semiconductor diode. In replacing a thermionic diode in any circuit the practical points to consider are (a) the maximum forward current to be drawn, and (b) the peak inverse voltage likely to be encountered. Semiconductor diodes have better forward characteristics than their thermionic equivalent. The forward resistance is low and diode losses during conduction are consequently reduced. The efficiency is further increased in that no heater power is required.

The subject of power rectification in power supplies for electronic equipment is discussed in Chapter 7 where details of circuits that use crystal diodes for power rectification are given.

112

ZENER DIODES

When a *pn* junction diode is biased in the reverse direction, the majority carriers (holes in the *p*-side and electrons in the *n*-side) move away from the junction. The barrier or depletion layer becomes thicker, and the absence of charge carriers is manifest as a very high resistance region across which current transfer becomes very difficult. The presence of minority carriers, and the electron-hole pair formation in the barrier layer, account for the flow of a very small leakage current. This leakage (i.e. reverse) current remains very small for all reverse voltages up to a certain value. Once this value has been exceeded there is a sudden and substantial rise in the reverse current. The voltage at which the rise in current occurs is called the breakdown voltage, although the breakdown is non-destructive provided the current is insufficient to cause the power dissipation to exceed the maximum allowed for the device. Under this condition, the breakdown is a reversible process. Over the operating range of reverse current, it is found that the voltage across the diode remains nearly constant. For this reason the most common application of zener diodes is in voltage regulation and for use as voltage reference elements (see Chapter 7). The range of applications is, however, very large, and these diodes are being used as overvoltage protection devices, clippers and limiters, square-wave generators and in coupling and biasing circuits.

The term zener diode arises because it was thought at one time that the mechanism responsible for the behaviour of the diode during breakdown was that proposed by Dr. Carl Zener in 1934[8]. He attempted to account for the breakdown in solid dielectrics. It is now believed for voltages above about 5 V the breakdown is an avalanche effect corresponding to the Townsend discharge effect observed during the breakdown in gaseous dielectrics (Chapter 7). The true Zener effect accounts for the breakdown below about 5 V.

Although the avalanche effect is now accepted, the term 'zener diode' is retained to cover both types. Other names in use are avalanche, reference, and stabilizing diodes.

The characteristics of diodes specifically manufactured for zener or avalanche operation are shown in *Figure 5.10*. For true zener diodes it is arranged that the depletion layer is extremely narrow. This is achieved by having very high impurity concentrations on both sides of a *pn* junction with abrupt boundaries. Since the bulk of the crystal has a comparatively low resistivity, nearly all the applied voltage exists across the narrow depletion layer. The electric field within the depletion layer is very high. If, for example, the applied voltage is 2 V and the depletion layer is 2×10^{-6} cm thick we have a field strength of $2/(2 \times 10^{-6})$ volts per cm, i.e. 1,000 kilovolts per cm. In this very large field electrons are excited from the valence band into the conduction band. The excitation is a quantum mechanical process in which electron-hole pairs are formed. The source of the required energy is the electric field. The number of covalent bonds ruptured is very large and rises rapidly with small increases of voltage. The resultant current increases exponentially with voltage. The depletion layer in a true zener diode is so narrow that the electrons pass through without ever striking an atom of the crystal lattice. It will be noticed that the 'knee' of the characteristic is more rounded than that for avalanche diodes.

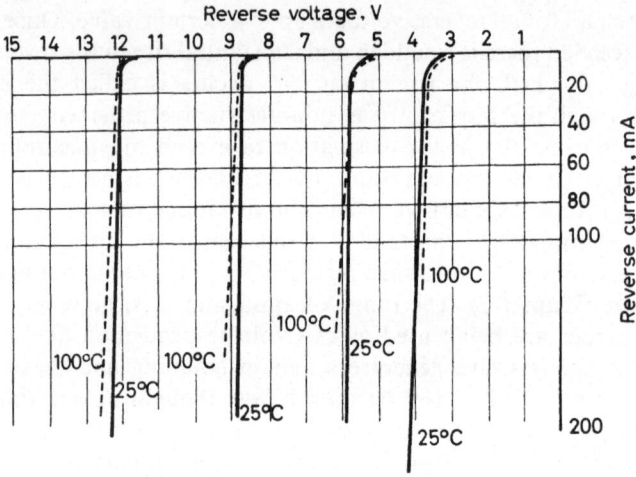

Figure 5.10. Typical characteristics of a set of silicon zener diodes

In the avalanche type of diode the depletion layer is somewhat thicker, being of the order of 10^{-4} to 10^{-5} cm. The thicker depletion layer is produced by using higher resistivity material than that used in a true zener diode. This is achieved by introducing a lower concentration of impurities into the intrinsic material. By controlling the resistivity of the material during manufacture, any desired value of breakdown voltage can be achieved. The mechanism of the breakdown is different from the zener effect. In the depletion layer of a *pn* junction thermal energy is responsible for the formation of electron-hole pairs. The leakage current is due to the movement of electrons accelerated in the field across the barrier layer. For avalanche diodes that are to be used in the breakdown condition, the depletion layer is deliberately made narrower than that in normal *pn* junctions. As the field across the depletion layer is increased it eventually reaches a critical value. The applied voltage to produce this critical value may be anywhere in the range 8 V to one or two

114

hundreds of volts for silicon devices. Once the critical or breakdown voltage has been reached, sufficient energy is gained by the thermally released electrons to enable them to rupture covalent bonds on collision with a lattice atom. The released electrons then go on to release further electrons, and so on in a chain or avalanche reaction. The depletion layer must be thick enough to allow this mechanism to proceed and this is the reason for increasing the layer thickness beyond the value found in true zener diodes.

For silicon diodes that are manufactured to have a breakdown in the range 5 V to 8 V, the current rise is due to a combination of the two mechanisms.

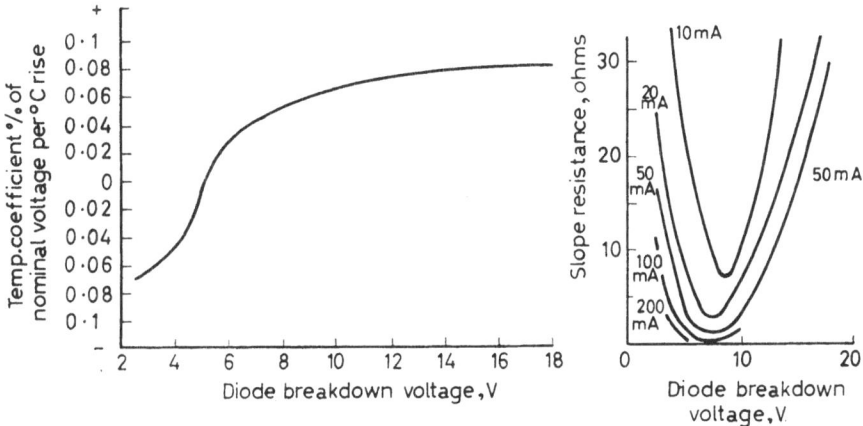

Figure 5.11. Temperature coefficient of voltage and the variation of slope resistance with reverse current for the Z2A series of silicon zener reference diodes (Standard Telephones and Cables)

One difference between the two mechanisms is that the temperature coefficient of a zener diode is negative, whereas that of an avalanche diode is positive. *Figure 5.11* shows how the temperature coefficient varies with diode breakdown voltage. As a consequence of this it is possible to combine a true zener diode with an avalanche diode so that the combination has a specified breakdown voltage and zero temperature coefficient.

Zener diodes are available at present in $\pm 5\%$, $\pm 10\%$ and $\pm 20\%$ tolerances. The same preferred numbers for the nominal breakdown voltages are used as those already in use for resistors, e.g. 4·7 V, 15 V, 22 V etc.. Since such a wide range of breakdown voltages are available, zener diodes are preferred to gas-discharge tubes for purposes of voltage stabilization. There is no discontinuity in the transition from the 'on' to 'off' state and no special arrangements are required for 'firing'. The devices are of course smaller and more robust than gas-discharge tubes and have an almost limitless life.

Apart from the working voltage and temperature coefficient the potential user should consider the slope resistance. The a.c. slope resistance is related to the working voltage and the reverse current. It is, in fact, the slope of the reverse voltage/reverse current characteristic at constant junction tempera-

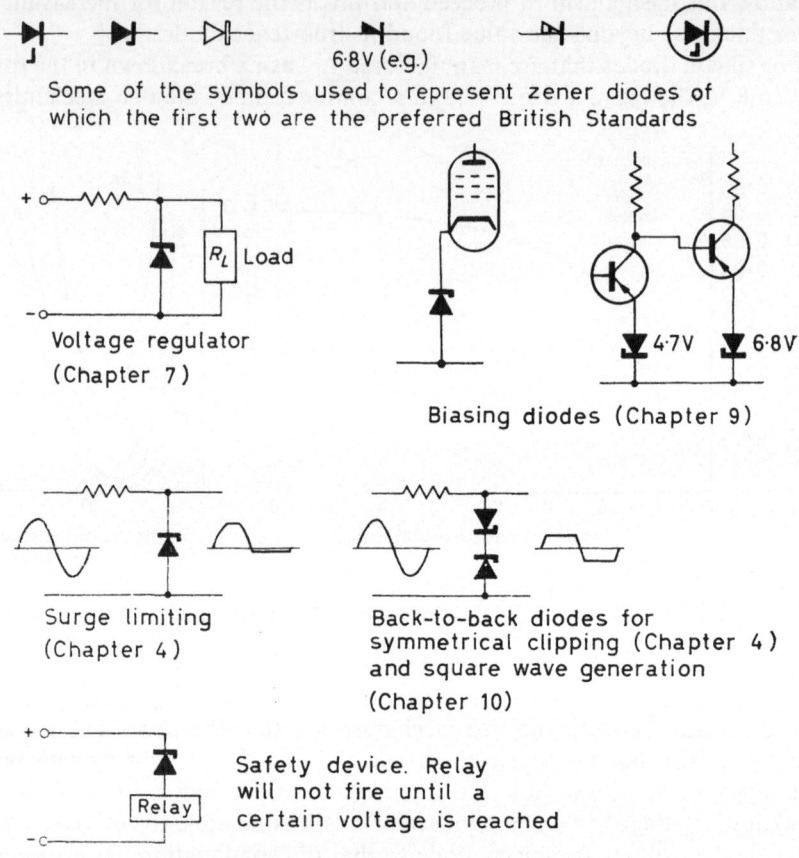

Some of the symbols used to represent zener diodes of which the first two are the preferred British Standards

Voltage regulator (Chapter 7)

Biasing diodes (Chapter 9)

Surge limiting (Chapter 4)

Back-to-back diodes for symmetrical clipping (Chapter 4) and square wave generation (Chapter 10)

Safety device. Relay will not fire until a certain voltage is reached

Figure 5.12. Some of the applications of zener diodes

ture. When the characteristic is plotted slowly enough for the junction temperature to become steady between each measurement the slope is the d.c. slope resistance. For zero temperature coefficient conditions the a.c. and d.c. slopes are identical. The lower the slope resistance the more constant is the operating voltage with changes in current. Graphs or sets of typical values are available from manufacturers for the diodes they manufacture. *Figure 5.11* shows the temperature coefficient curve and slope resistance curves for the Z2A range of silicon zener reference diodes of Standard Telephone and Cables Ltd.

Common applications for zener diodes are shown in circuit form in *Figure 5.12*. The diagrams indicate the chapters in which the relevant applications are discussed further.

TUNNEL DIODES

In 1958 Esaki described a new type of diode[4] with unusual electrical characteristics. The tunnel, or Esaki, diode is a *pn* junction in which both the *p* and *n* regions are heavily doped. Donor and acceptor concentrations of the order of 10^{19} impurity atoms per cubic centimetre are used. At these high concentrations the depletion layer is extremely thin, being only about 100 Å across.

For reverse voltages the device acts like a zener diode with zero breakdown voltage. Since the charge carrier density on both sides of the junction is so high, and the depletion layer is so thin, small forward voltages (up to 200 mV) cause the electrons to 'tunnel' through the potential barrier instead of going over it. This quantum mechanical tunnelling is a rather complex concept, requiring for its understanding a considerable background knowledge of solid-state physics. As this book is limited to introductory qualitative descriptions we shall content ourselves here with a brief pictorial account of the mechanism.

The characteristics and behaviour of any semiconductor device depend upon the concentration of free charge carriers. Such carriers are produced by thermal excitation of electrons from the valence bonds and from donor and acceptor impurity centres. The number of free electrons in the conduction band depends upon the number of free energy levels in the conduction band, the total number of electrons available and the probability of an electron acquiring sufficient energy, at a particular temperature, to be raised into the conduction band. By making some basic assumptions Fermi (and Dirac independently) calculated how the energy, on average, would be distributed among the free electrons in a crystal. Their statistics led to the expression

$$p(E) = \frac{1}{1 + \exp{(E - E_f)}\, kT}$$

$p(E)$ is the probability that an electron will be occupying any particular energy level E at an absolute temperature of T; k is Boltzmann's constant, equal to $8 \cdot 6 \times 10^{-5}$ eV per deg K; E_f is the Fermi level. From the expression, when $E = E_f$, $p(E) = 1/(1+1)$ i.e. 1/2. The Fermi level is thus the boundary between energy states which are more than half-filled and those which are more than half empty. The Fermi level is, of course, a mathematical concept

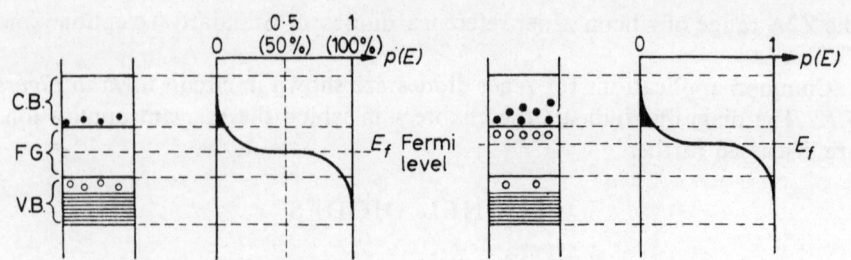

(a) Plot of Fermi probability function for intrinsic germanium (or silicon) at a temperature above 0°K (e.g. room temperature)

(b) Plot of $p(E)$ for n-type material. Note the raising of the Fermi level. Temperature is the same as for (a)

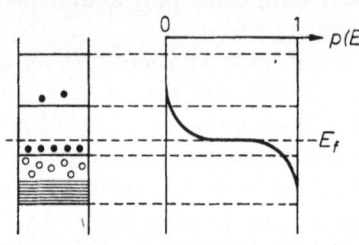

(c) Plot of $p(E)$ for p-type material. E_t is lower than half-way between the valence and conduction bands

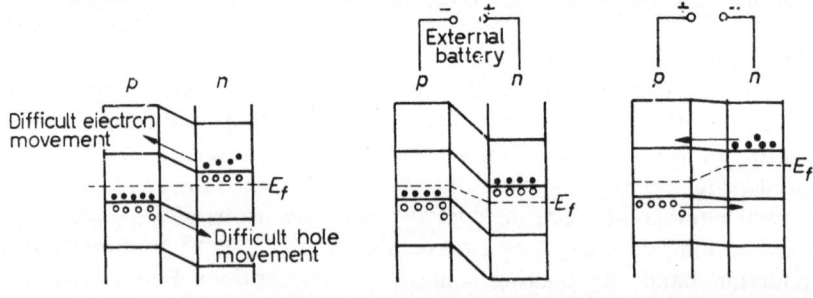

pn junction, with no applied bias, showing formation of a potential hill

pn junction with reverse bias. Movement of charges even more difficult

pn junction with forward bias. Movement of charges is easy

(d) Behaviour of normal pn junction under various bias conditions

Figure 5.13. Energy-level diagram interpretation of pn junction behaviour

and must not be confused with the energy band structure of allowed and forbidden bands in any particular crystal arrangement.

Figure 5.13a shows the energy level diagram for intrinsic germanium, and alongside is a plot of the probability function $p(E)$. For temperatures above

0°K there is some probability that a valence electron will be excited to the lower levels of the conduction band. At high levels within the conduction band the probability of an electron occupying such levels is zero. In the valence band some of the upper levels will be vacant because of a loss of electrons to the conduction band. The probability of these levels being filled is not, therefore, quite 1. The Fermi level is situated in the middle of the forbidden gap. For n-type germanium *(Figure 5.13b)* the number of electrons in the conduction band is increased over that for the intrinsic semiconductor. The probability of these levels in the conduction band being occupied is increased. The curve $p(E)$ therefore moves up and the Fermi level rises. Provided the temperature remains the same as in the previous intrinsic case, it will be seen that the probability of the upper levels of the conduction band being occupied has increased. The presence of donor impurities evidently decreases the number of holes present in the material; in other words, the formation of electron-hole pairs in the intrinsic material has been suppressed. *Figure 5.13c* shows the lowering of the Fermi level in p-type germanium.

When p- and n-type material are together in a single crystal, a *pn* junction is formed. In the absence of applied voltages, the Fermi level must be the same on both sides of the junction and so a shifting of levels as in *Figure 5.13d* takes place. This diagram is the energy level interpretation of the formation of a potential hill. When the p-side is connected to the negative terminal of a battery and the n-side to the positive terminal the energy of the electrons in the p-side is raised whilst that of the electrons in the n-side is lowered. The potential hill therefore increases and it is increasingly difficult for current to traverse the junction as the external reverse voltage is raised. If the connection to the battery is reversed however, the energy level movements are reversed and there is a lowering of the potential hill. Electrons can now easily pass from the n-side to the p-side.

In tunnel diodes the concentrations of impurities is so high that the Fermi level is depressed into the valence band for p-type material and raised into the conduction band for n-type material. The semiconductors are said to be degenerate since so many majority carriers are present that the upper energy levels at the top of the valence band in p-type material are vacant, and the lower levels in the conduction band of n-type material are filled. Even without any external applied voltage there is an overlapping of the bands as in *Figure 5.14a*. When a reverse bias voltage is applied there is the zener breakdown already described, and a current flows. The depletion layer is so thin that this effect is evident at extremely low reverse voltages. When a forward voltage of up to about 100 mV is applied the position is as shown in *Figure 5.14b*.

The filled levels in the conduction band on the n-side become level with the empty levels of the valence band on the p-side. The electrons on the n-side instead of moving up the large potential gradient to the conduction band of the p-side, tunnel through the barrier to occupy the upper levels of the p-side valence band. In other words at high impurity concentrations, and for small applied voltages, the junction acts as an ohmic one and current can pass readily in both directions. As the forward bias voltage is increased the position is that shown by *Figure 5.14c*. There is no longer coincidence of the

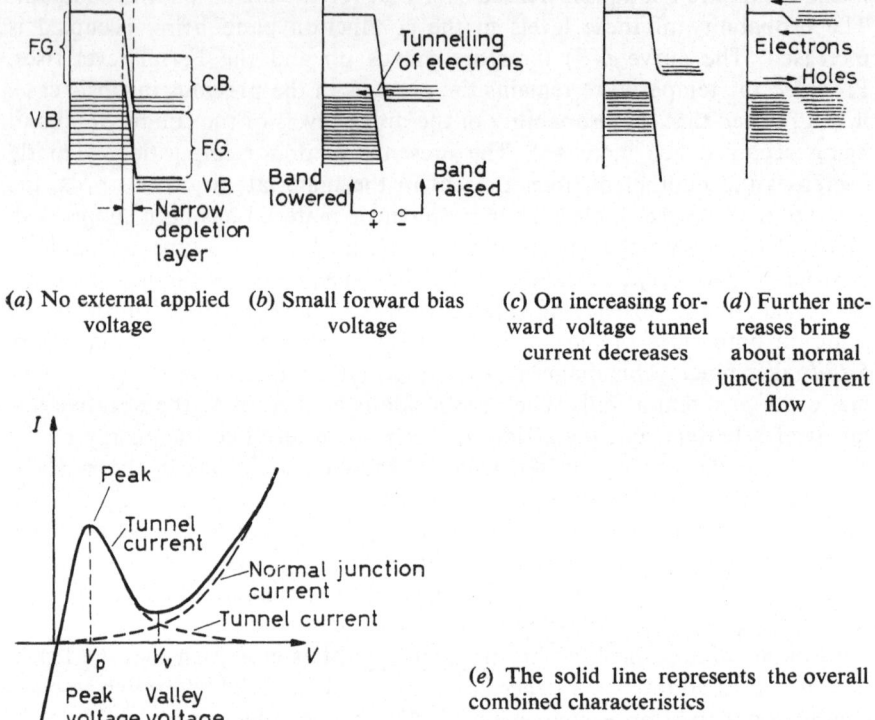

(a) No external applied voltage   (b) Small forward bias voltage   (c) On increasing forward voltage tunnel current decreases   (d) Further increases bring about normal junction current flow

(e) The solid line represents the overall combined characteristics

Figure 5.14. Pictorial representation of tunnel diode behaviour

relevant levels, and the tunnel current falls. Further increases of forward voltage bring into play the normal *pn* junction behaviour and the normal junction current is evident. The combined effect is shown in *Figure 5.14e*.

The significant part of the characteristic is that portion with a negative slope. Over this range increases of voltage are accompanied by decreases of current. In this region therefore the device exhibits a negative resistance. We shall see in Chapter 10 how this characteristic can be exploited when using a tunnel diode as an oscillator.

120

The other important application of a tunnel diode is its use as a switch. *Figure 5.15* shows the operation of the switching action. A static load line is chosen to cut the characteristic close to the peak and the valley as shown. The switching action is between the two stable operating points A and B. Suppose a start is made at position B. The application of a negative-going pulse at the input will cause a fall in the voltage across the diode. This is accompanied by a rise in current and a shift to the operating point A. The diode is then in the 'off' condition. A positive-going pulse will cause a rise in the voltage across the diode and the diode will revert to its original condition. The diode is then said to be 'on'.

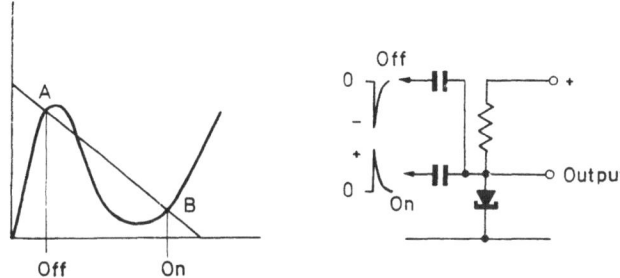

Figure 5.15. Tunnel diode used as a switch

Solid-state switches of this type may be operated at the rate of 1,000 million times per second. They are therefore ideal for incorporation into digital computers, and much research is being undertaken in this connection. So far, factors which need not concern us here have prevented large-scale construction of tunnel diode computers. We shall see later in the book how these diodes can be used as switching elements in logic circuits. Logic circuits are used in automatic control of factory and chemical processes, in data processing, and in counting.

THE JUNCTION TRANSISTOR

The *pn* junction, like its thermionic counterpart, can perform the function of rectification, but is unable to amplify a signal. By setting up a system of two closely-spaced *pn* junctions within a single crystal of germanium or silicon a transistor is formed. Such a device, as we shall see, can be made to act as an amplifier when incorporated in a suitable circuit. The transistor may be constructed in several different ways, but as an introduction to the device let us first consider the alloyed or fused-junction transistor, which is usually based on germanium. After reducing the germanium dioxide to the

121

metal in an atmosphere of hydrogen, the impure polycrystalline bar is subject to a process of zone refining. Molten zones, produced by high-frequency heating coils, are made to pass along the bar. In its passage along the bar the zone sweeps out the impurities. It is a triumph of modern technology that the impurities that remain are as low as 1 part in 10^{10}. The bar is then melted

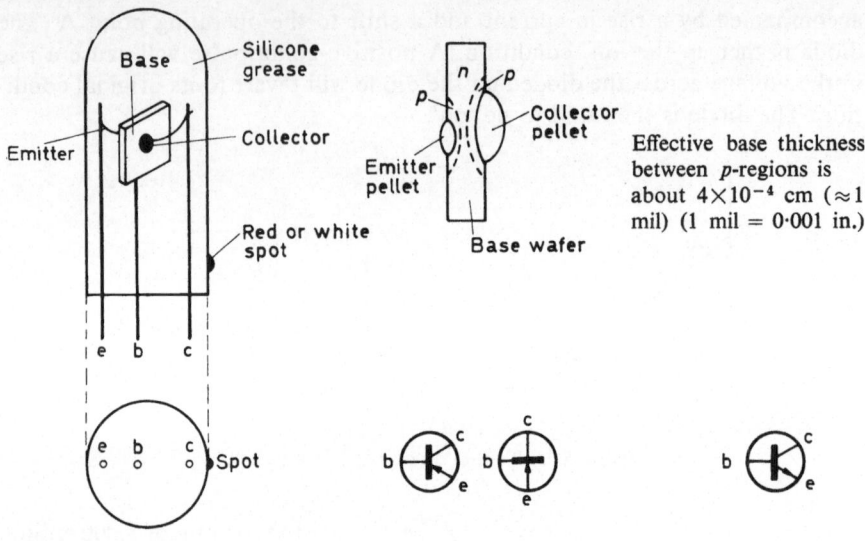

Looking at base with leads pointing to viewer. The base lead is nearer to the emitter lead than to the collector lead

Two recognized circuit symbols for a *pnp* transistor. The arrow points to the base. The left-hand symbol is recommended by British Standards Institution

Circuit symbol for an *npn* transistor. The arrow points away from the base

Figure 5.16. The main constructional features of an alloyed junction transistor

in a suitable crucible and a controlled amount of impurity is added. For purpose of illustration we shall assume that *n*-type germanium results. A single crystal seed is then made to touch the melt and is then slowly withdrawn. The result is a single large crystal of *n*-type germanium. This crystal is subsequently sawn into wafers approximately one tenth of an inch square. After etching and washing two small indium pellets are alloyed to the *n*-type germanium, one on either side. During subsequent heat treatment the indium diffuses into the germanium converting it to *p*-type. The diffusion is stopped just before the *p*-zones meet. An extremely thin section of *n*-type germanium (about 4×10^{-4} cm thick) is thus sandwiched between two layers of *p*-type germanium. The result is a *pnp* junction transistor. After washing and drying, ohmic contacts are made to the *n*-type wafer, known as the base, and also to the two indium pellets, the larger of which is called the collector, and the smaller, the emitter *(Figure 5.16)*. The whole assembly is then encapsulated within a metal or glass container. In the case of the glass container a coat of black lacquer is applied to the outside walls to prevent light from reaching

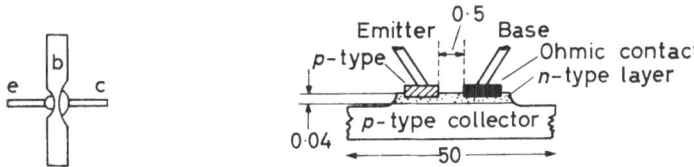

(*a*) Microalloy type. Effective base thickness ≈ 0.1 mil

(*b*) Mesa transistor. *n*-type layer produced by diffusion or epitaxial growth. Dimensions in mils. 'Table' is 6×8 mils. The emitter and base are parallel strips about 1 mil wide and 7 mils long

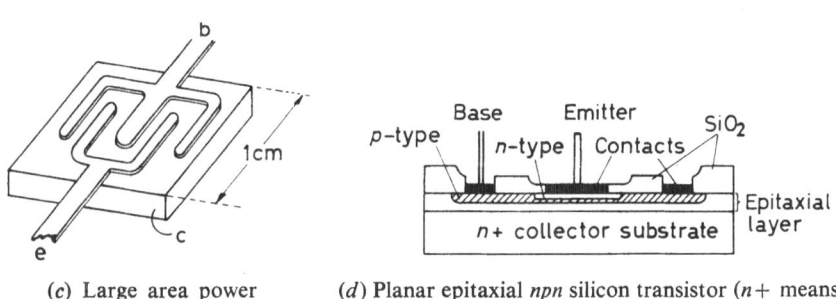

(*c*) Large area power transistor

(*d*) Planar epitaxial *npn* silicon transistor (*n*+ means heavily doped *n*-type material)

Figure 5.17. Diagrams of alternative transistor constructions

the transistor. If light were to reach the base region the transistor action would be modified in an undesirable way. Light, as well as heat, can supply energy for the rupture of covalent bonds. The resulting carriers are unwanted, especially if their numbers are modulated by varying light intensities from, say, a fluorescent lamp.

By starting with *p*-type germanium, and alloying in donor impurities, an *npn* junction transistor is formed. The difference is explained later in the chapter.

Attempts to improve the frequency response and power handling capabilities have led to other forms of construction. The frequency response is limited by the time it takes for the charge carriers to cross the base region. The technique of making alloyed transistors does not allow bases to be made

123

much thinner than about 0·3 mil (1 mil = 0·001 in). To fabricate thinner bases a different form of construction leads to the surface barrier transistor. Tiny jets of etching liquid are played on the each side of the base wafer. By allowing an infra-red beam to pass first through one jet, then the thin base region and finally on to a detector, it is possible to control the final base thickness to close limits. The emitter and collector junctions are made by a microalloying process yielding a microalloy transistor (MAT) *(Figure 5.17a)*. Alternatively, diffusion techniques may be used whereby an impurity in gaseous form is allowed to diffuse into a prepared substrate.

When larger amounts of power have to be handled than is possible with the types previously described, a mesa construction may be employed. (*Mesa* is a Spanish word meaning 'table-topped hill' and is of Latin origin). The collector is laid down on a thick metal base-plate which serves as a connection as well as heat-dissipating agent. If the collector region is *p*-type, a layer of *n*-type material is produced upon it as a very thin layer by a gaseous diffusion process. An ohmic contact (i.e. a non-rectifying contact) is eventually made to this *n*-layer which acts as the base. An emitter *p*-region is then formed close to the base contact *(Figure 5.17b)* and a high-frequency *pnp* transistor results that can also dissipate more power than the alloyed type of transistor. For large power outputs larger emitter and base areas are used *(Figure 5.17c)*.

Epitaxial techniques are now being used to produce transistors (*epi* = = upon; *taxis* = arrangement; from the Greek). When atoms from a fluid phase condense on a crystalline substrate, the orientation of the final crystal is controlled by the substrate, and there is a continuation of the crystal lattice structure. The deposited layer is then said to be epitaxial. The layers can be given desired properties by introducing suitable doping agents into the vapour phase. Relatively thin active volumes are involved and abrupt or graded junctions may be produced at will. The process leads to the most versatile and reliable transistors presently available. The diffusion and photo-lithographic processes involved are most amenable to quanity production techniques. Silicon planar epitaxial transistors, as well as being reliable and versatile, are therefore being produced at competitive prices. *Figure 5.17d* shows diagrammatically a typical planar epitaxial transistor. A low-resistivity single crystal substrate acts as the collector region. Upon this substrate is grown an epitaxial layer of high resistivity. Losses in the bulk of the collector crystal are low owing to the low resistance, but the base-to-collector junction, being formed in the high resistivity material, can withstand reverse voltages as high as 50 volts or more (companed with 10 or 15 V for the forms of construction previously discussed). An efficient collector region in a transistor that will operate at high voltages is therefore obtained. A layer of silicon dioxide, with suitable holes photoengraved in it, covers the surface of the epitaxial layer. Impurities are then allowed to diffuse downwards and underneath the insulating SiO_2 layer to form the base and emitter regions. Suitable ohmic contacts are then made to the regions. The electrical performance of transistors made in this way is excellent. One of the great advantages of using the epitaxial techniques is that resistors, capacitors and transistors can all be fabricated on a single silicon chip. This results in what is called a monolithic integrated circuit. Physically small amplifiers, oscillators, transmitters and computer circuits are now possible and available.

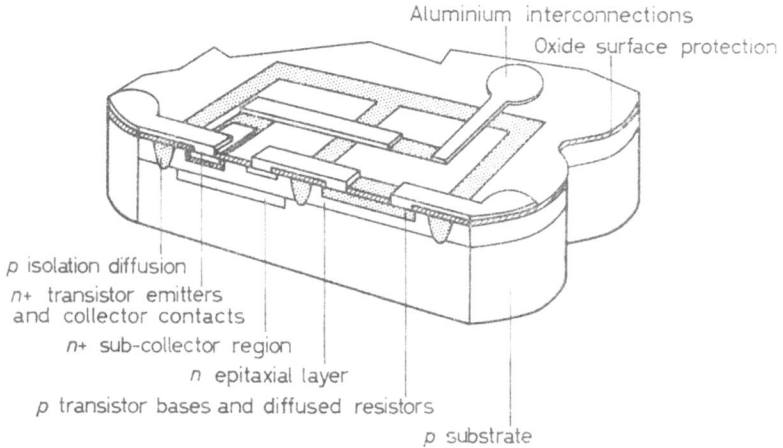

Figure 5.18. Sectional view of a Fairchild planar-epitaxial integrated circuit (reproduced by courtesy of SGS-Fairchild Ltd.)

Figure 5.18 shows a sectional view of a Fairchild planar-epitaxial integrated circuit. In a typical Fairchild example we may have a small silicon chip some 70 or 100 thousandths of an inch square upon which an entire d.c. amplifier consisting of 9 transistors and associated resistors is fabricated. Such a microminiaturized module is easily housed in a small can similar to that used to contain a single conventional transistor.

Transistor Operation

Since a transistor consists of two *pn* junctions within a single crystal, transistor action can be explained with reference to *Figure 5.19*. For diagrammatic purposes the base region is shown fairly thick, but in fact the *pn* junctions are very closely spaced and the active portion of the base is very thin. In the absence of any external applied voltages the collector and emitter

125

depletion layers are about the same thickness, the widths depending upon the relative doping of the collector, emitter and base regions. During normal transistor operation the emitter-base junction is forward biased so that current flows easily in the input or signal circuit. The bias voltage is about 200 mV for germanium transistors and about 400 mV for silicon devices. *Figure 5.19* shows the bias voltage being obtained from a battery. In practice, as we shall see in Chapter 9 on amplifiers, this bias is obtained in other ways. The collector-base junction is reverse-biased by the main supply voltage. Often, as in transistor radios for example, this voltage comes from a battery.

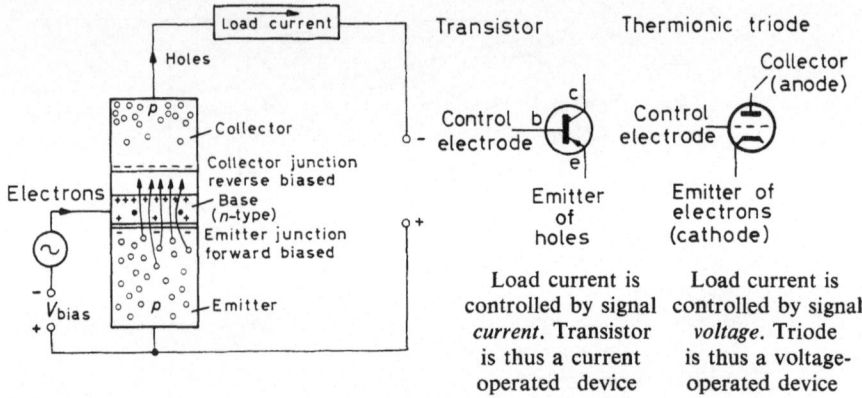

Figure 5.19. Diagrammatic representation of the amplifying action of a transistor

Voltages of 4·5 V, 6 V and 9 V are common. The collector junction is therefore heavily reversed-biased and the depletion layer there is quite thick. The voltage gradient in the depletion layer is high because nearly all of the battery voltage appears across it. There is little loss of voltage in the bulk of the collector portion of the crystal because most of this bulk is made from low resistivity material. In epitaxial transistors this is achieved by using heavily doped material; in alloyed types the collector pellet is considerably larger than the emitter pellet. The voltage drop across the emitter junction is low and can be ignored.

The injection of an electron into the base region by a signal source will now be considered. Once in the base, the electron attracts a hole from the emitter region. Combination of the hole and electron is not likely to occur however. The base region is lightly doped compared with the emitter region and so the lifetime of the hole in the base region is quite long. In addition the base is extremely thin so the hole, instead of combining with the signal electron or with an electron of the *n*-type base material, diffuses into the collector-base junction. The hole then comes under the influence of the

126

strong field there and is swept into the collector and hence into the load circuit. In a good transistor many holes pass into the collector region before eventually the signal electron is eliminated by combination with a hole. A small signal current can thus give rise to a large load current, and so current amplification has taken place. In practical transistors for every electron injected into the base 50 to 250 holes may be influenced to flow into the collector region. The current gain or amplification is therefore 50 to 250. This current gain is given the symbol α', β or h_{fe}. The significance of the symbols is explained in the chapter on amplifiers. (The figures given cover most of the common types of transistors, but transistors are available with a β outside this range.)

An *npn* transistor behaves in a similar fashion except that holes are injected into the base and electrons flow from the emitter into the collector. To maintain the correct bias conditions the polarity of the external batteries must be reversed.

When using a transistor in the way described above it will be noticed that the emitter lead is common to both the signal circuit and the load circuit. Operation is then said to be in the common-emitter mode. In the U.S.A. this is referred to as the grounded-emitter mode because the emitter is connected to the grounded (i.e. earthed or chassis) line. This is the most frequently used voltage amplifier arrangement because, compared with the common-base mode described below, little current is drawn from the signal source. An alternative way of viewing this is to say that the input resistance of the common-emitter amplifier is higher than that of the common-base mode.

There is a range of input base currents over which the load current is directly proportional to the base current. This is the operating range. Since the load current is controlled by a signal current the transistor is regarded as a current-operated device. It will be remembered that for a thermionic triode the load current is directly proportional to the grid voltage within the proper operating range. The thermionic triode is therefore a voltage-operated device. The collector current in a transistor does, of course, depend upon the base-emitter voltage, but since the relationship is very non-linear, it is not profitable to consider this aspect of transistor behaviour.

Figure 5.20 shows the three basic transistor arrangements. The common-emitter mode is the most commonly used arrangement for voltage amplification because, as has already been pointed out, very little current is required from the signal source.

The common-base mode of operation is also capable of voltage amplification is spite of the fact that in this mode the collector current is less than the emitter current by an amount equal to the base current. The change in collector current, i_c, divided by the change in the emitter (i.e. signal) current, i_e, is known as the current amplification and is given the symbol α. For this mode the current gain is therefore less than unity. Voltage amplification is achieved by the use of high values of load resistor. The transistor is able to maintain the current through the load because the device is a good constant current generator, that is to say it can deliver a current that is substantially independent of the load resistor provided variations of this resistor fall

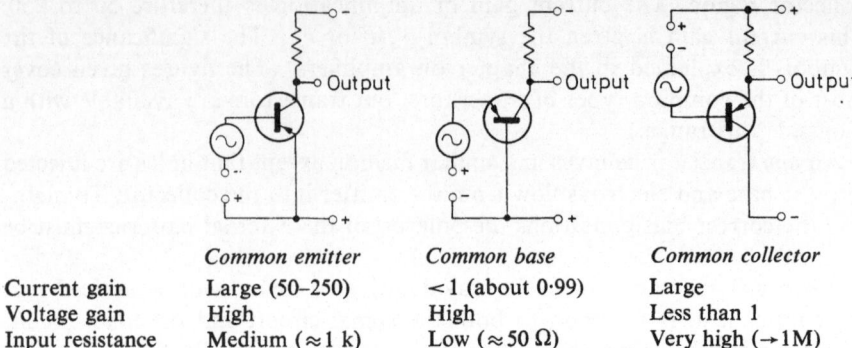

	Common emitter	Common base	Common collector
Current gain	Large (50–250)	<1 (about 0·99)	Large
Voltage gain	High	High	Less than 1
Input resistance	Medium ($\approx$1 k)	Low ($\approx$ 50 Ω)	Very high ($\to$1M)

Figure 5.20. The three basic amplifier arrangements together with some of their properties. The figures are given only as a guide to the magnitudes involved

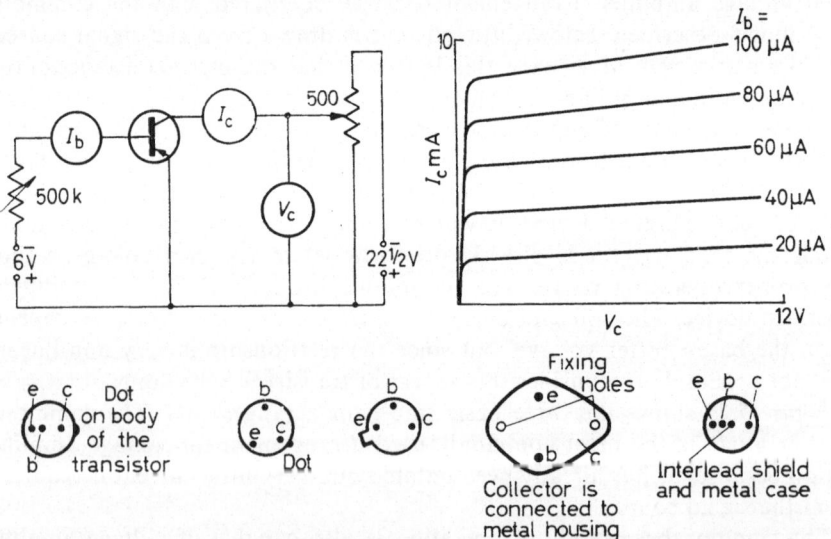

Figure 5.21. Characteristics of a transistor in the common-emitter mode. Figures given are typical of a small low-power transistor. Modifications to the circuit will be required for other types (e.g. high-voltage or power transistors). Some connection diagrams are given for transistors in common use. In all cases the wires are pointing from the transistor to the viewer

128

within reasonable limits A typical set of characteristic curves for the common-emitter mode are shown in *Figure 5.21*. The characteristics for the common-base mode are similar in shape, but I_c is plotted against V_c for several emitter currents instead of base currents. It will be realized that for a given emitter current the collector current is essentially independent of the collector voltage over the working range. This is typical of a device with a high internal resistance (e.g. the pentode). In the case of the transistor the high resistance is associated with the wide depletion layer at the base-collector interface. With such a high internal resistance variations of external load resistance do not affect the collector current very much. The output voltage available from this amplifier is the product of the load resistor, R_L, and i_c. The input voltage is the product of R_i and i_e where R_i is the input resistance to the amplifier. The voltage amplification is therefore given by $A = i_c R_L / i_e R_i$. Although the collector current, i_c, is less than the emitter current, i_e, the two are approximately equal. By making R_L much greater than R_i considerable voltage amplification can be achieved. R_i in this mode is typically about 50 ohms plus the internal impedance of the signal source; R_L is several kilohms.

The circuit was preferred to the common-emitter arrangement because of its better frequency response and freedom from thermal runaway (a term explained in Chapter 9). However the development of silicon transistors that can operate well at high frequencies, and which do not exhibit the phenomenon of thermal runaway in correctly designed circuits, have made the common-base amplifier less popular. The low input impedance of the common-base arrangement is a serious disadvantage that contributes to the unpopularity of this circuit.

The common-collector circuit has a voltage gain of less than unity and so is useless as a voltage amplifier. However this circuit has very important impedance matching properties that are to be discussed later in Chapter 9.

The method of obtaining transistor characteristics is shown in *Figure 5.21*. Only the common-emitter characteristics are shown since it will be obvious enough how those for the common-base arrangement are obtained.

THE UNIJUNCTION TRANSISTOR

This type of transistor consists of an n-type silicon bar with ohmic contacts (called base one, b_1, and base two, b_2) at opposite ends. A single rectifying contact is made between the two base contacts usually closer to b_2 than b_1. This rectifying contact is called the emitter. During normal operation b_2 is held positive with respect to b_1 via a suitable load resistor *(Figure 5.22)*. Provided the emitter junction is reversed biased, there is no emitter current and the n-type silicon bar acts as a simple potential divider, the voltage at the emitter contact being a fraction, β, of the voltage from b_2 to b_1. In order to maintain this condition the emitter junction must remain reverse biased; this is ensured by keeping the emitter voltage, V_e, less than βV, where V is the voltage from b_2 to b_1. When V_e is allowed to exceed βV the junction becomes forward biased and current flows in the emitter-to-b_1 region. The emitter injects holes into this region, and as they move down the bar there is an increase in the number of electrons in the lower part of the bar. The result is

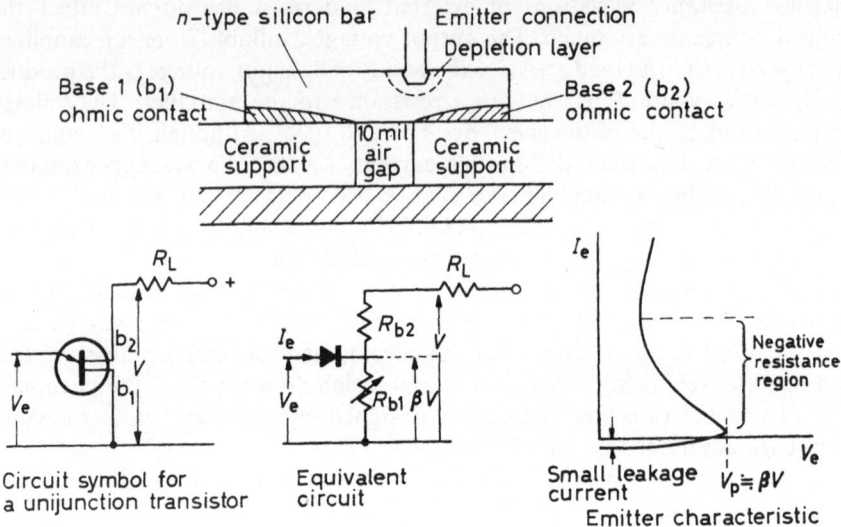

Figure 5.22. The unijunction transistor of the General Electric Co.

a substantial decrease in the bar resistance from the emitter to b_1. This decrease in emitter-to-b_1 resistance brings about a fall of emitter voltage. A condition arises whereby increases of emitter current are accompanied by decreases of emitter voltage, resulting in a negative resistance characteristic.

Unijunction transistors are low in cost and are characterized by a stable triggering voltage which is a fixed fraction of the voltage between b_1 and b_2. They have a negative resistance characteristic which is uniform among units of the same type and which is stable with temperature and throughout the unit's life. Apart from its use in oscillators and time delay circuits, the major use of a unijunction is as a pulse generator to fire silicon-controlled-rectifiers. As this latter application is the subject of Chapter 8 no more need be said here.

130

SILICON TEMPERATURE SENSORS

During investigations into silicon strain gauges it has been discovered that by a proper selection and cutting of a suitably prepared silicon crystal, a high-sensitivity solid-state temperature sensor could be produced. The resistance of such a sensor was found to vary in a linear way with temperature. The device is stable and has an extremely rapid response to changes of temperature *(Figure 5.23)*.

The silicon temperature sensor is a single crystal of almost pure silicon. Unlike the thermistor, with its compounded oxide structure, the silicon unit can be manufactured with a high degree of reproducibility. The almost linear temperature/resistance graph is a distinct advantage over the logarithmic curve of the thermistor.

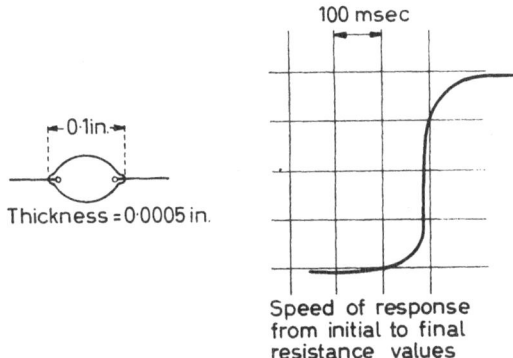

Figure 5.23. Dimensions of a silicon temperature sensor together with the graph showing the speed of response

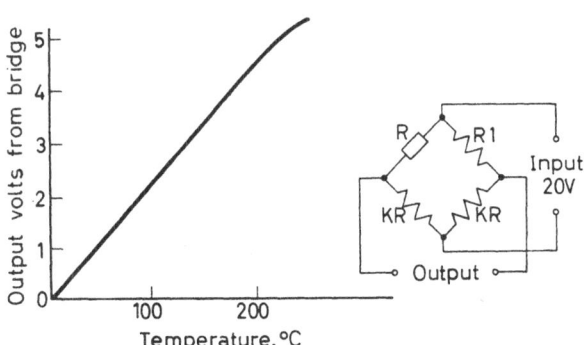

Figure 5.24. Calibration curve for the silicon sensor when bridge connected. R1 balances for zero output at the lower temperature (e.g. 0°C). For linearity $K = 5$

The sensors are available with an operating temperature range of about −170 to +300°C. They are normally calibrated over the range −10 to +170°C with an accuracy better than ±0·5°C. The theoretical response time is about five times faster than platinum for similar geometries. The relative resistance change over the range −10 to +250°C is from 40 to 50 per cent of the maximum resistance. The high sensitivity means that the output from the bridge circuit of *Figure 5.24* is high enough to drive milliammeters, measuring and control devices directly. An intermediate amplifier is often unnecessary. These sensors are available from Scientific Furnishings Ltd. of Poynton, Cheshire.

FIELD-EFFECT TRANSISTORS (F.E.T.s)

Although it has been possible for more than thirty years to make field-effect devices in the laboratory, it is only recently that the significant advances in semiconductor technology have made possible the manufacture of reliable units with useful characteristics. Basically there are three types of field-effect transistor, namely, the reverse-biased *pn* junction type, the insulated gate device based on a single crystal, and the insulated-gate version that uses a polycrystalline layer of semiconducting material such as cadmium sulphide. The descriptive account that follows relies heavily on one of the author's articles in *Wireless World*[3], and thanks are due to the Editor of that journal for kindly giving permission to reproduce some of the work here.

The Reverse-biased Diode Field-Effect-Transistor

This type of device was first proposed by Shockley, who called it a unipolar field-effect transistor because only one type of charge carrier is used to carry the current. This is different from a conventional bipolar transistor in which both majority and minority charge carriers are involved.

Figure 5.25 shows schematically the construction of such a device. A bar of *n*-type material has *p*-type impurities introduced into opposite sides. These *p*-type regions form the control electrode known as the gate. Between the gate electrodes there exists a channel of conducting material extending to ohmic contacts at the ends. One end is called the source and the other end the drain. Majority carriers (electrons in this case) may then flow along the channel from source to drain between the gate electrodes.

For very small source-to-drain voltages (i.e. voltages much less than the gate-to-source voltage) the device acts as an ohmic resistor whose resistance depends upon the effective width of the channel between the gate electrodes.

132

The gate-channel junctions are operated as reverse-biased *pn* junctions. As the reverse voltage on the gate is increased, the depletion layer extends into the body thus reducing the effective channel width and hence the conductance. For somewhat larger drain-to-source voltages the profiles of the depletion layers are not parallel to the centre line through the device, but converge in the region towards the drain as shown in *Figure 5.25*. Near to the drain the reverse bias is larger because of the voltage gradient along the channel. Provided the convergence is not excessive, however, the resistance of the channel, and hence the drain current, is modulated by varying the gate voltage. For drain-to-source voltages of the order of the gate voltage and above, the

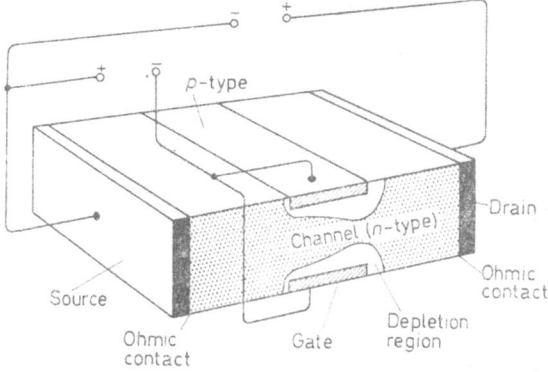

Figure 5.25. The Shockley-type field-effect device

convergence is such that the channel approaches what is termed the 'pinch-off' condition. It might be expected that the channel conductance falls to zero as the depletion layers from each side meet, thus cutting off the drain current. This does not happen, however, owing to the screening effect of the charge carriers at the centre of the channel. We find in fact that once this condition has been reached the drain current is almost independent of the drain voltage. The channel thus loses its ohmic properties and a saturation effect is observed. For any given gate voltage therefore the drain current rises linearly as the drain voltage is increased from zero to a small value. Thereafter further increases in drain voltage produce little further increase in the drain current, and an approximate saturation condition obtains. For a more negative value of gate voltage a similar mechanism operates, but the saturation current is lower because the whole effective channel is narrower. *Figure 5.26* shows a typical set of characteristic curves for the device.

Although improvements in transistor technology have modified the physical arrangement used in the Shockley transistor, the principle of operation re-

133

mains unaltered. The early Shockley types could not be made in commercial numbers with the techniques available in the mid-1950s. Now that the industry has mastered masking, diffusion and epitaxial techniques for silicon devices, it is possible to manufacture f.e.t.s with a reasonable degree of reproducibility. *Figure 5.26* shows the physical form of a modern f.e.t. taken from a Ferranti report[4]. The drain characteristics for one of their commercially available devices are also given.

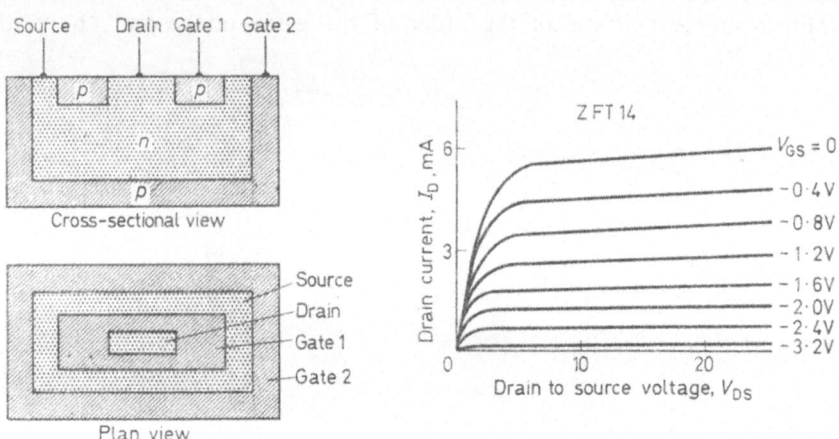

Figure 5.26. A modern form of construction due to Ferranti Ltd. For many applications gate 1 and gate 2 are connected together in the external circuit

Fundamentally, the devices are cheap to make; with the fast-increasing use of large numbers of f.e.t.s it is expected that the cost will compare favourably with that of conventional transistors.

The outstanding advantage of this, and other types of field-effect transistor is the high input impedance. The reverse-biased f.e.t. has an input resistance of about 10^{10} ohms whilst the insulated gate types now under development have input resistances approaching 10^{15} ohms with input capacitances of under 5 pF. The conventional transistor has an input resistance of the order of only thousands of ohms. Even with special circuitry this input resistance cannot be raised much above a megohm or two. This means that it is possible to combine the desirable features of the thermionic valve with those of a conventional transistor in a device that could be a superior substitute for both in many circuit applications. Some of these circuits appear later in the book.

134

The Metal-oxide-semiconductor Transistor (M.O.S.T.)

An attempt to increase further the input resistance of field-effect devices has resulted in a return to an early idea due to O. Heil, whereby the gate electrode is electrically insulated from the conducting channel. The construction and mode of operation is therefore significantly different from the Shockley reverse-biased diode type. Although only in the development stage

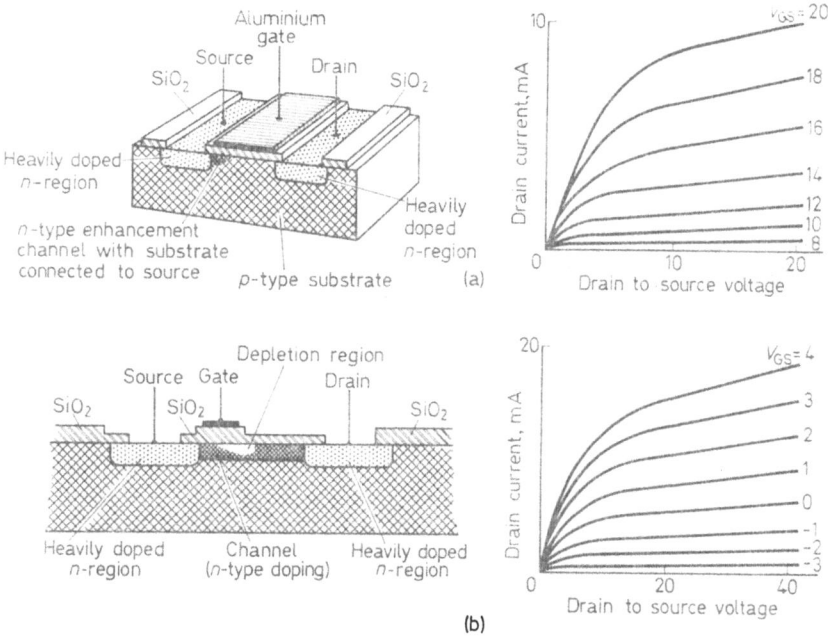

Figure 5.27. The constructions of Hofstein and Heiman showing (*a*) an enhancement type unit and (*b*) a depletion type device

at the moment, it seems that this latest device may give a much improved performance over the diode type.

There are several ways in which the insulated gate type of transistor may be constructed. Let us consider first a prototype model of the kind made by Hofstein and Heiman[5]. Once this type has been understood, modifications can be easily appreciated. The main constructional features are shown in *Figure 5.27a*. A *p*-type silicon body is used as a substrate upon which are diffused two heavily doped *n*-regions in closely spaced parallel strips along the body. A layer of silicon dioxide some 1,000 Å thick is then thermally

135

grown or evaporated on to the surface using a mask to leave the n-type regions uncovered. On the surface of the silicon dioxide insulating layer, and between the n-type regions, an aluminium layer is deposited, which acts as the gate electrode. This method of insulating the gate is preferred to the use of a discrete wafer of insulating material partly because of the thinness that can be achieved, and also because a thermally grown silicon dioxide layer passivates the silicon surface (i.e. it reduces very considerably the density of surface traps). Ohmic contacts are made to the n-regions (one of which acts as the sink and the other the source) and also to the gate. By making the gate positive with respect to the source, a positive bias exists between the gate and the p-type body in the region of the course. Positive charge carriers are repelled into the body and negative charge carriers are attracted to the surface. At the body-silicon dioxide interface there is thus induced an n-type layer of mobile charge carriers. This layer connects the drain and source resistively; it is often referred to as an inversion layer because, on increasing the gate voltage from zero, the channel, originally p-type, becomes intrinsic and then finally an n-type layer is formed. Further increases in gate voltage increase the number of electrons in the channel thus reducing the resistance between the source and drain. If a voltage is applied between the source and drain, a drain current, I_D, will flow. The magnitude of the drain current can be varied by applying varying voltages to the gate. Although the gate potential is positive relative to the source potential, no current is taken by the gate, since the silicon dioxide acts as an excellent dielectric.

Input resistances of the order of 10^{12} ohms have been achieved in available British units; whilst the Americans are claiming up to 10^{15} ohms for some of their transistors. There is thus available a solid-state device that is a close equivalent to a triode insofar as it is a voltage-operated device with a very large input impedance.

Insulated gate field-effect devices may be operated in one of two ways, namely, the enhancement mode or in the depletion mode, depending upon the form of construction used. In the enhancement mode there is an n-type channel between heavily doped n-type regions with the gate extending across the entire channel as in *Figure 5.27a*. The gate is forward-biased enhancing the number of electrons in the channel and reducing the source-to-drain resistance. At zero gate voltage the number of charge carriers in the channel is very low and so the drain current is effectively zero. One of the disadvantages of the enhancement type unit is the large capacitance associated with the gate electrode. To overcome this an offset gate that does not cover the whole of the channel is used. Normally this would produce a very high resistance in the channel region not influenced by the gate. However, by suitable doping, a channel may be produced that has appreciable conductivity at zero gate voltage. Such a transistor is a depletion type and has the

136

drain, source and channel regions all of the same conductive-type material although the drain and source regions are still heavily doped. The gate voltage must then be driven to some negative value before the drain current is zero. *Figure 5.27b* shows the cross-section of this type of unit together with typical characteristics. The pinch-off voltage, V_p, for a given transistor may be positive, zero or negative depending upon the construction. In practice, it is difficult to determine just when the drain current is zero so V_p is defined as that voltage that reduces the drain current to some specified low value (say 10 to 20 μA).

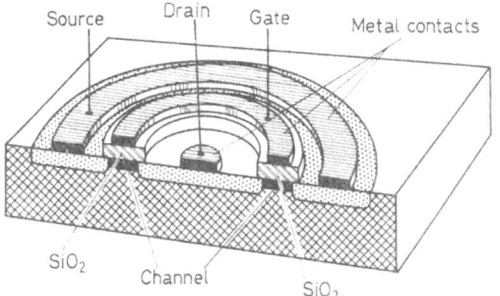

Figure 5.28. Alternative geometry for an insulated-gate f.e.t. The heavily doped regions are shown dotted

Figure 5.28 shows an alternative geometry. Some manufacturers (e.g. Ferranti and Mullard) make a fourth connection to the substrate creating a four-terminal device. Many workers are now exploiting f.e.t.s, and since several applications have been published, they will not be repeated here (see for example the article by F. Butler in the February 1965 issue of *Wireless World*, correspondence in the following month's issue and also the Mullard booklet on their 95 BFY f.e.t. Two useful books are mentioned in the references[10, 11]).

The Thin-film F.E.T.

Conventional transistors, and those field-effect types so far described, depend for their successful action upon mechanisms within single crystals that have been suitably doped, polycrystalline material being clearly unsuitable. However, a new type of amplifying device, that may loosely be called a transistor, has been described by Weimer[6]. A microcrystalline layer of semiconductor has been used as a channel, and it is claimed that when low resistance contacts are made to the film, thus forming source and drain

137

electrodes, a device is obtained that has a voltage amplification factor greater than 100, an input impedance of greater than 10^6 ohms shunted by 50 pF, gain-bandwidth products in excess of 10 MHz and switching speeds of less than 0·1 μsec. So far as the writer is aware these units are not available from British manufacturers, but if the claims made for the device are realized in units that can be easily reproduced on a commercial scale, then a potentially cheap and popular transistor will be added to the range already available. Development is being pursued feverishly and already Weimer, Shallcross and Borkan[7], with an improved electrode arrangement, have raised the input resistance to 10^{10} ohms and extended the gain-bandwidth product to 25 MHz.

Cadmium sulphide was chosen by these workers for the semiconducting film, presumably because a good deal is known about the solid-state physics of this material as well as the technology associated with its deposition in thin films. *Figure 5.29* shows diagrammatically the coplanar electrode form of a

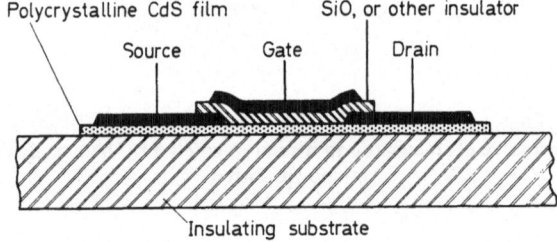

Figure 5.29. A coplanar electrode arrangement of a thin-film transistor

thin-film transistor (t.f.t.). A polycrystalline *n*-type CdS layer, a fraction of a micron thick, is deposited on an insulating substrate and evaporated aluminium contacts are made to form the source and drain. The length of these electrodes is about 2 to 5 mm, and they are spaced about 10 microns apart. An insulated gate is then formed in the usual way, the insulator being about 500 Å thick. Insulating materials found to be satisfactory are silicon monoxide and calcium fluoride. As in the f.e.t. described earlier, the presence of the insulating layer permits positive biasing of the gate without that electrode drawing current of any great magnitude. Typical drain characteristics exhibit the pentode-like characteristics of the f.e.t. It must be emphasized that this type of f.e.t. is not yet in production. Difficulties in producing uniform characteristics between units of the same nominal type make it extremely speculative whether or not the thin-film device will ever become commercially available.

REFERENCES

1. Bardeen, J. and Brattain, W. H. 'Transistor, a semiconductor triode.' *Phys. Rev.* 1948, **74**, 230.
2. Bardeen, J. and Brattain, W. H. 'Physical principles involved in transistor action'. *Phys. Rev.* 1949, **75**, 1208.
   Also
   Shockley, W. 'The theory of pn junctions in semiconductors and pn junction transistors'. *Bell Syst. Tech. J.* 1949, **28**, 435.
3. Olsen, G. H. 'Field effect devices'. *Wireless World*, 1965, **71**, No. 6, 260, June.
4. Application Note No. 22 'Field Effect Transistors and Applications'. Ferranti Ltd., Oldham, Lancs.
5. Hofstein, S. R. and Heiman, F. P. 'The silicon insulated-gate-field-effect transistor'. *Proc. I.E.E.E.* 1963, **51**, 1190, Sept. 1963.
6. Weimer, P. K. 'The T.F.T.—a new thin film transistor'. *Proc. I.R.E.* 1962, **50**, 1462, June.
7. Weimer, P. K., Shallcross, F. V., and Borkan, H. 'Coplanar-electrode insulated-gate thin-film transistors'. *R.C.A. Rev.* 1963, **24**, No. 4, Dec.
8. Zener, C. 'Theory of electrical breakdown of solid dielectrics' *Proc. Roy. Soc.* 1934, 145, 523.
9. Esaki, L. 'New Phenomenon in Narrow Germanium p-n Junctions'. *Phys. Rev.* 1958, **109**, 603, Jan. 15.
10. Gosling, W. *Field Effect Transistor Applications*. Heywood, 1964.
11. Sevin, L. J. *Field Effect Transistors*. McGraw-Hill, 1965.

Suggestions for Further Reading

'Zener Diodes' by J. M. Waddell and D. R. Coleman, *Wireless World*, 1960, **66**, No. 1, Jan.
Introduction to Semiconductor Devices by M. J. Morant, Engineering Science Monographs, Harrap, 1964.
Semiconductors by H. Teichmann, Butterworths, 1964.

6

INDICATING INSTRUMENTS

The solutions of many problems involving circuit quantities such as voltage, current and resistance, can be found only by making appropriate measurements. For the purpose of measuring electrical quantities there are many useful instruments, among which the most common are the ammeter, the voltmeter and the ohmmeter. The incorporation of indicating instruments into electronic circuits to form valve-volmeters, pH meters, potentiometric recorders and the like is discussed in Chapter 13; the present chapter deals mainly with the principles, construction and application of instruments commonly used in the measurement of voltage, current and resistance. By far the most common and important meter movement relevant to these measurements is the moving-coil meter.

'Seeing is believing' is a saying that may often lead to trouble in an electronics laboratory and never more so than in the case of pointer instruments. Quite often an unjustified faith is placed in the accuracy of a given indication. It is therefore necessary to discuss the precautions that must be observed when using pointer instruments.

MOVING-COIL METERS

The indication of any instrument is determined by the combined effect of the deflecting torque and the control torque. The value of the deflecting torque must be a function of the electrical effect to be measured, e.g. voltage or current. The control torque acts in an opposite sense to the deflecting torque. The pointer will rest in its final indicating position when equilibrium is achieved between the two torques.

Moving-coil meters depend upon the fact that a wire carrying a current experiences a mechanical force when immersed in a magnetic field. This force is directly proportional to the current and to the strength of the magnetic field. In practice, the wire is wound into a coil of rectangular cross-section. The coil is pivoted so that it can rotate back and forth within a magnetic field set up by a permanent magnet. The pole pieces of the permanent magnet are curved in shape. Located within the pole pieces is a fixed cylindrical iron core, the function of which is to ensure that the magnetic field is uniform in strength and radial in direction. The effective portion of the coil thus travels always at right angles to the magnetic field. The main constructional features are shown in *Figure 6.1*. Under these circumstances the deflecting torque, T_d, is proportional to the current in the coil, I, the number of turns N and the area of the coil, A.

$$T_d = k\,I\,N\,A$$

To prevent uncontrolled rotation, control torque must be provided. Such provision is made by two phosphor-bronze coiled hair-springs mounted as shown in the diagram. (These springs also serve to lead current into and out of the coil.) For a given meter, it will be seen that the movement of the

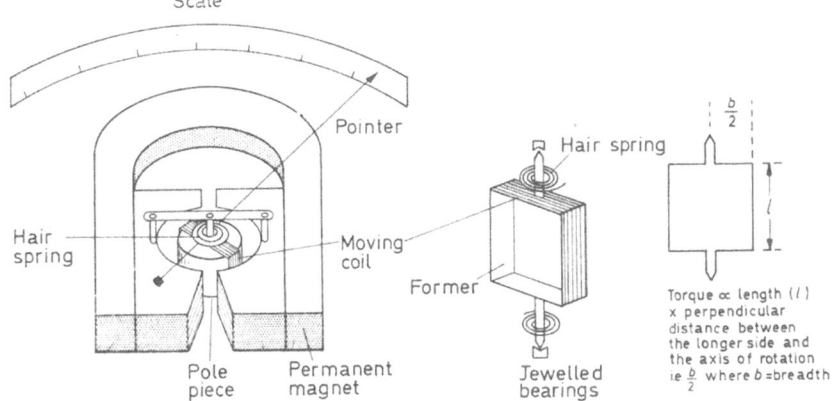

Figure 6.1. Main constructional details of a moving-coil meter

pointer across the scale is directly proportional to the current. The scale is thus a linear one, which is a valuable feature in the type of indicating devices being discussed here. The permanent magnet, coil, indicating pointer and graduated scale form a complete current measuring instrument.

Some form of damping device is essential since without it the pointer would oscillate about the final position before coming to rest; such oscillations are irritating to the observer. In the types of meter movement used in the electronics laboratory, the necessary damping is achieved by winding the coil onto a light aluminium former or frame. The frame constitutes a closed circuit, so when it rotates in the magnetic field eddy currents are induced. The presence of eddy currents in the frame produce a magnetic field which opposes that of the permanent magnet. A braking or damping action is therefore imposed upon the movement and the pointer quickly comes to rest at the final indicated reading.

The inertia of the moving system is reduced to a minimum by the use of lightweight material. Both the coil former and the pointer are usually made from very thin aluminium. To reduce friction to a minimum the shaft that carries the coil and former is made from very hard metal and is set in highly polished jewel bearings.

Meter sensitivity is expressed in terms of the current required to produce full-scale deflection. The smaller the current required, the more sensitive is the meter. For measuring currents in electronic equipment ammeters are used that give full-scale deflections with currents of 1 mA or less. More sensitive multimeters often use movements that give full-scale deflections with currents as low as 10, 50 or 100 μA. The advantages of using sensitive movements is explained in the section on voltmeters.

Ammeter Shunts

To extend the current-measuring capabilities of the meter movement it is necessary to connect a resistor in parallel with the coil. This causes the current that is to be measured to divide, a small part flowing through the coil and the remainder passing through the shunt resistor. The wire from which the coil is made is nearly always copper; unfortunately the resistance of a copper coil varies appreciably with temperature. Whilst the variation is of no consequence in voltmeters (because the resistance in series with the coil is always very much greater than that of the coil) in ammeters where shunts are employed it is important that the effective resistance of the meter should remain constant. A shunt resistor is then used whose resistance is practically constant over the operating temperature range; this guarantees a constant ratio between the coil resistance and the shunt resistance. Manganin is the usual alloy used for shunts because this material has a negligible temperature coefficient at the temperatures involved. There is also a negligible thermoelectric effect when manganin is used with copper. Although shunting a copper coil with a copper shunt may seem to overcome the temperature difficulty it cannot be guaranteed that changes in temperature will be the same in both the coil and the shunt, especially when the latter is carrying large currents. It is therefore impossible to guarantee a constant ratio between the meter and shunt resistances. To achieve a meter resistance that is practically independent of temperature, manufacturers place a resistance in series with the coil. This resistor is also constructed with manganin or some other material having a low temperature coefficient. The series resistance is a good deal higher than that of the coil and thus 'swamps out' the coil resistance. The addition of a swamp resistance increases the voltage drop across the instrument and so lessens its value as an ammeter. Ideally, an ammeter should introduce no voltage drop when placed in the circuit. However, this effect is of less importance than the constancy of resistance.

D.C. MULTIMETERS

In d.c. multimeters, the various shunts are selected by suitable switches as shown in *Figure 6.2*. The obvious, but highly unsatisfactory, arrangement shown in *Figure 6.2a* should never be used. If the switch has poor contacts, it introduces appreciable resistance and the instrument will read high or may even be burned out. The uncertainty of the switch resistance will lead to erratic and inconstant indications. The preferred arrangement, shown in *Figure 6.2b*, shows a ganged switch.

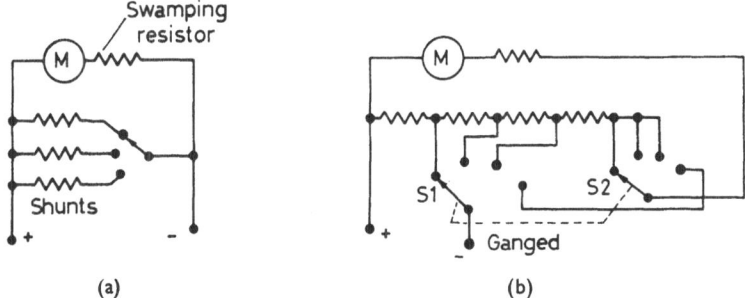

Figure 6.2. Two possible ways of switching shunts to increase the current range of the meter. That shown in (*a*) is unsatisfactory because of uncertainties associated with the switch resistance. The method shown in (*b*) is the preferred arrangement (see text)

Uncertainties in switch S1 merely add a small extra resistance in series with the external circuit, whilst a similar fault in S2 introduces a small additional resistance into the meter circuit. This additional resistance is negligible compared with the swamping resistance. It is important to understand the difference between the two arrangements since many amateurs attempt the shunt circuit of *Figure 6.2a* and proceed to make measurements of dubious value and in a false sense of security. It is the author's opinion that no one should attempt to incorporate shunts in a meter unless he is a trained instrument mechanic.

Precautions in using ammeters

(*a*) The ammeter must always be connected in series with the element through which the current is to be measured.

(*b*) With the switch selectors adjusted to measure current, the instrument must never be connected across a source of voltage, such as a battery, generator or resistance carrying a current. It should be remembered that the resistance of an ammeter, particularly on the higher ranges, is extremely low, and that any voltage, even as low as one volt, may result in a very high current through the meter; damage is often caused by such high currents.

(*c*) Before measuring a current, some idea should be formed of its magnitude. An appropriate switch position can then be selected to produce deflections that are less than full-scale. If there is any doubt about the magnitude of the current to be measured, the highest range should be chosen and then the appropriate scale may be selected by working down the ranges.

Since not all meters are provided with automatic cut-outs to prevent overloads, it should be remembered that the wire from which the coil is made is very fine and is easily destroyed by excessive currents.

(*d*) The 'accuracy' of a test instrument of the kind under discussion is measured in terms of the difference between the indicated reading and the true value of the quantity being measured. Many users of meters entertain an optimistic view of the accuracy of pointer instruments. The specifications followed by many reputable instrument manufacturers are laid down in B.S. 89 : 1954, which deals with portable industrial instruments (e.g. the Avo 8 multimeter). On the d.c. current ranges, the Avo 8 meter will give an indicated reading that is within 1 per cent of the full-scale value over the effective range, i.e. from 0·1 of the scale range to full-scale value. It will be seen therefore that no reliance can be placed upon indications less than one-tenth of the full-scale reading. Even within the effective range the inaccuracy of the indicated reading varies. For example, if the full-scale deflection is 100 mA, any indicated reading may be in error by 1 mA. If the indicated reading is say 95 mA, the percentage error will be approximately 1 per cent but if the indicated reading is 10 mA, the percentage error cannot be guaranteed to be better than 10 per cent. It is always desirable therefore that a meter range should be selected so that the current to be measured produces deflection in the upper portion of the scale. Very often the meter must be used in a horizontal position to produce the highest accuracy, because it is in this position that the meter has been calibrated. The zero setting should be checked before using the instrument. Further restrictions apply when measuring voltage and alternating quantities; these are discussed later in the appropriate sections.

(*e*) It is easy to be careless and connect the meter incorrectly from the polarity point of view. Current must flow through the coil in the correct direction move the indicator needle 'up-scale'. An incorrectly connected meter may lead to no more than a bent needle, which is inconvenient enough, but it is always possible to strain the mechanism severely and jeopardize the future accuracy of the instrument.

The Voltmeter

Moving-coil meters can be used to measure voltage if the validity of Ohm's law is assumed. Although a moving-coil meter is basically a current measuring device, the addition of a linear, stable and accurately known resistance enables the scale to be engraved in volts. Various voltage ranges may be obtained by adding various resistors, called multipliers, in series with the coil

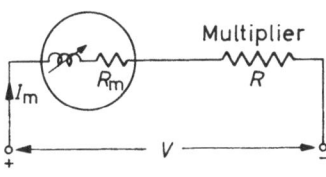

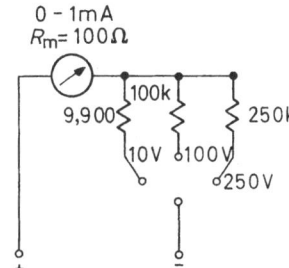

(a) The basic voltmeter circuit. R_m represents the sum of the swamping resistance and the resistance of the moving coil. R is the multiplier resistance

(b) A circuit for measuring voltages up to 250 V in three stages. At the higher voltages the multipliers are large and R_m can be neglected

Figure 6.3. The d.c. voltmeter

and swamping resistor. Taking the case of a 50 μA meter with a total resistance of 1,000 ohms, the addition of a series resistor of 199 k would mean that a voltage of 10 V across the combination would result in a current of 50 μA through the meter. The meter scale may thus be marked accordingly. In general, referring to *Figure 6.3a*

$$V = I_m(R_m + R)$$

No precautions need be taken with the switching arrangement, as in the case of ammeters, since small uncertainties of switch contact resistance are of no consequence; the contact resistance is in series with a very much larger multiplier resistance. The multiplier resistors used by manufacturers of multirange meters are usually wire-wound resistors since these are stable and have a low temperature coefficient; carbon resistors are of no use in this application. Even the high-stability cracked carbon resistor is frowned upon by makers of high-quality instruments, but the new metal oxide resistors are proving very satisfactory because of their good stability and reliability characteristics.

Sensitivity of a Voltmeter

The sensitivity of a voltmeter is measured in ohms per volt and is obtained by dividing the resistance of the meter (including the multiplier) by the full-scale reading in volts. This is just another way of stating the current that produces a full-scale deflection. A voltmeter should have a very high resistance so that it will draw very little current and affect the circuit under examination as little as possible during voltage measurements. Sensitivity therefore is an indication of the measuring quality of a voltmeter. Generally, for measurements in electronics, a meter should not be used unless it has a sensitivity of at least 1,000 ohms per volt. Advances in manufacturing techniques and materials make voltmeters with sensitivities of 20,000 ohms per volt fairly common.

Precautions in using a voltmeter

(a) A voltmeter must always be connected in parallel across that portion of the circuit in which voltage is to be measured. It cannot be too strongly emphasized that, unless the resistance of the circuit is approximately known, there is no value whatever in the indicated voltage. The voltmeter resistance must be considerably higher (about 100 times greater) than that of the portion of the circuit across which the voltage is to be measured. *Figure 6.4* illustrates the point.

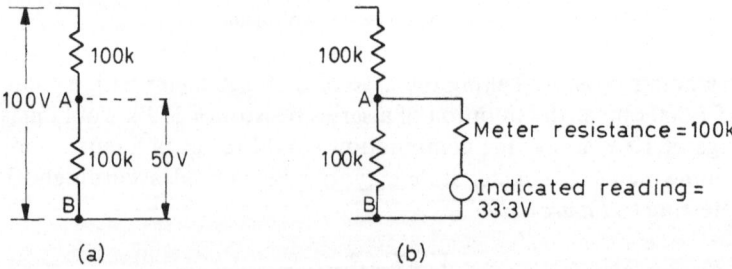

(a)   (b)

Figure 6.4. If the voltage between points A and B of the circuit in (*a*) is to be measured, then placing a voltmeter across the points, as in (*b*), will give a reading which will be very inaccurate if the meter resistance is approximately equal to the resistance across which the measurement is to be made. For the figures chosen as an example, the effective parallel resistance between A and B is only 50 k. Whilst the meter is in circuit the voltage between A and B will be only 33·3 V. When the meter is removed the voltage is 50 V. A much more sensitive meter (with a higher multiplier resistance) is therefore required

146

(*b*) To avoid damaging the meter, careful selection of a suitable position for the range switch must be made. The same precautions should be observed as for ammeters. Care should be taken to observe the correct polarity when connecting the meter across the circuit.

(*c*) The manufacturer's specification must be consulted when considering accuracy. A common specification for an industrial type multimeter is that between full-scale and half-scale deflections the indication will be accurate to within 2 per cent of the indicated value. Below half-scale deflection the indicated value may err by up to 1 per cent of the full-scale value. No reliance should be placed on indications less than 0·1 of the full-scale reading. The zero setting should be checked before using the instrument.

THE OHMMETER

Although the resistance of a circuit element can be determined by measuring the current through the element and the voltage drop across it, and then applying Ohm's law, it is much more convenient to use an ohmmeter with which the resistance can be read directly from a scale. Ohmmeters possess a number of features not found in ammeters and voltmeters. An ohmmeter must supply its own power. Usually this is obtained from a low-voltage battery, although for the measurement of high resistance values, provision is often made for the connection of external power supplies. The basic ohmmeter circuit is shown in *Figure 6.5a* whilst *Figure 6.5b* shows a multirange ohmmeter.

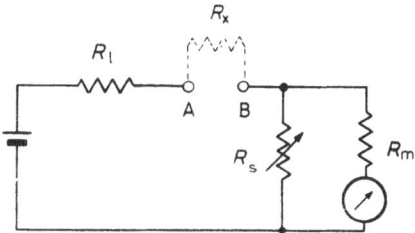

Figure 6.5a. The basic ohmmeter circuit. R_m is the meter resistance, R_s the shunt resistance and R_l a current limiting resistance. Variations of R_l or R_s set the meter to its full-scale deflection position (corresponding to zero ohms) when the terminals AB are shorted. R_s compensates for variations in battery voltage. Varying R_l enables a single scale to be used for various ranges of resistance measurement

In addition to measuring resistance, an ohmmeter is a very useful instrument for checking continuity in a circuit. Often, when locating faults or wiring a circuit, it is not possible to make visual inspections of all parts of the current path. It is not always apparent therefore whether a circuit is complete or not.

The best method of checking a circuit under these conditions is to send a current through it. The ohmmeter is a suitable instrument for this purpose since it supplies the necessary power and the meter to indicate whether or not a current is flowing.

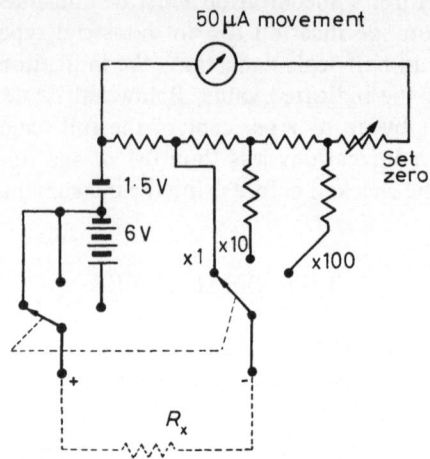

Figure 6.5b. One form of multirange ohmmeter

Using an ohmmeter

(*a*) On resistance ranges, it is first necessary to select the correct switch settings. The meter must not merely start from its normal instrument zero, but must have a reading of zero resistance corresponding to the full-scale deflection. This is achieved by joining the test leads together and adjusting the zero set. On some instruments, e.g. the Avo 8, there is a separate zero set control for each resistance range setting. If on joining the leads it is impossible to obtain a zero setting, or if the pointer position is not constant, but falls steadily, the internal battery must be replaced. The batteries should be examined from time to time to ensure that the electrolyte is not oozing out and damaging the instrument. This is a common fault with Leclanché type 'dry' cells that have reached the end of their useful life.

(*b*) The unknown resistance may now be connected between the test leads and its resistance read from the scale. No attempt should ever be made to measure resistance in a circuit in which there is a current. At least one end of the circuit element must be disconnected; this not only stops the current, but also avoids confusion arising from the resistance of parallel paths.

(*c*) Owing to the non-linear nature of the scale, it is not easy to define accuracy in the measurement of resistance. Generally the minimum error occurs around the centre of the scale increasing to as much as 10 per cent or so near the extremities of the effective range.

The complete d.c. multimeter consists of combining the equivalent of *Figures 6.2b*, *6.3b* and *6.5b* into a single case. Suitable shunts and multipliers are chosen so that only two or three scales are required to cover the whole range of measurements within the capabilities of the instrument.

PROTECTION OF METERS AGAINST OVERLOADS

It is easy to be careless and overload a meter especially when it is only one small part of an experimental arrangement. An attractive feature of some commercial multimeters is the automatic overload protection that is built into the instrument. In the case of the Avometer the cut-out takes the form of a mechanical switch, which is brought into operation by the moving-coil or pointer coming into contact with a trigger just beyond the full-scale position. The circuit to the meter is thus broken and current is prevented from flowing through the coil. A similar trigger is provided at the zero end, thus protecting the meter from overloads in the reverse direction. The cut-out is easily reset by pressing the overload button, thus resetting the trigger mechanism.

An alternative method of protecting the meter is to use a pair of diodes as shown in *Figure 6.6*. When the current is small the resistance of the diodes is

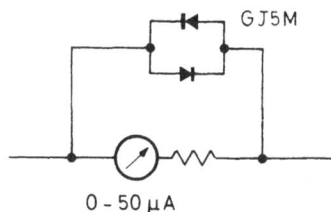

GJ5M

0 - 50 µA

Figure 6.6. The use of diodes to protect the meter from accidental overload. The value of the resistance is 1·6k

very large, as an examination of the characteristics shows. However, as the current increases in a given direction, the diode resistance drops to a low value thus shunting the meter and preventing excessive current from flowing through the coil. It is sometimes necessary to increase the value of the swamping resistance in order to bring the rectifier on to a suitable range of its characteristic. If, however, the meter is a sensitive one, e.g. 50 µA for full-scale deflection, the internal resistance is often large enough for the terminal voltage, at full scale deflection, to bring one or other of the diodes into incipient conduction. *Figure 6.6* is given as an example only; the diodes must be selected for any particular case. It is important to estimate the maximum

149

overload current that is likely to be encountered and to select a diode with an adequate current rating. Some of the silicon *pn* junction rectifiers are very suitable, since they can pass very large currents without damage. Their resistance at normal operating meter currents is very high and thus almost the full sensitivity of the meter is preserved. Some modification of the scale markings is inevitable if the meter is used for measuring purposes. However, as a null detector in a bridge or other circuit arrangement this protection arrangement is ideal. Near the null point the meter is being operated at its maximum sensitivity, whereas off balance excess current is carried by the diodes.

MIRROR GALVANOMETERS

There is a limit to the maximum sensitivity that can be obtained with conventional moving-coil meters of the type already described. Such meters must be able to withstand a reasonable amount of vibration since they are frequently incorporated in laboratory equipment that must be portable or at least capable of being moved from one position to another. The necessary robustness of the suspension and moving parts usually limits the maximum sensitivities to one or two tens of microamps for full scale deflection. When greater sensitivities are required the mirror galvanometer is often used. The usual form of mirror galvanometer operates on the same principles as that of the conventional moving-coil meter in that a coil is suitably suspended in a uniform magnetic field. To increase the sensitivity however the jewelled pivots are replaced by very thin, but strong, wire suspensions. The metallic needle is replaced by a beam of light. The light from a lamp is made to strike a small mirror attached to the filamentary suspension and from the mirror the beam is reflected onto a translucent screen. An optical arrangement brings the beam to a focus on the screen casting an image in the form of a disc of light traversed by a thin black line. Since the screen can be at a considerable distance from the suspension a small rotation of the coil and mirror gives rise to a large horizontal displacement of the image. Laboratory galvanometers often have their screens at a distance of 1 metre from the suspension, and at this distance deflections of 1 mm can correspond to current values of only 0.005 μA. *Figure 6.7* shows the main constructional features of the Pye Scalamp mirror galvanometer which uses a smaller length of light beam. The instrument, although less sensitive than 0.005 μA/mm, is self-contained and is an excellent example of a robust portable laboratory galvanometer.

Damping of mirror galvanometers is achieved either by winding the coil on a metal former (as in a conventional moving-coil meter) or alternatively by using a former made from non-conducting material and adjusting the resistance in the external circuit to a value that will ensure critical damping. If the

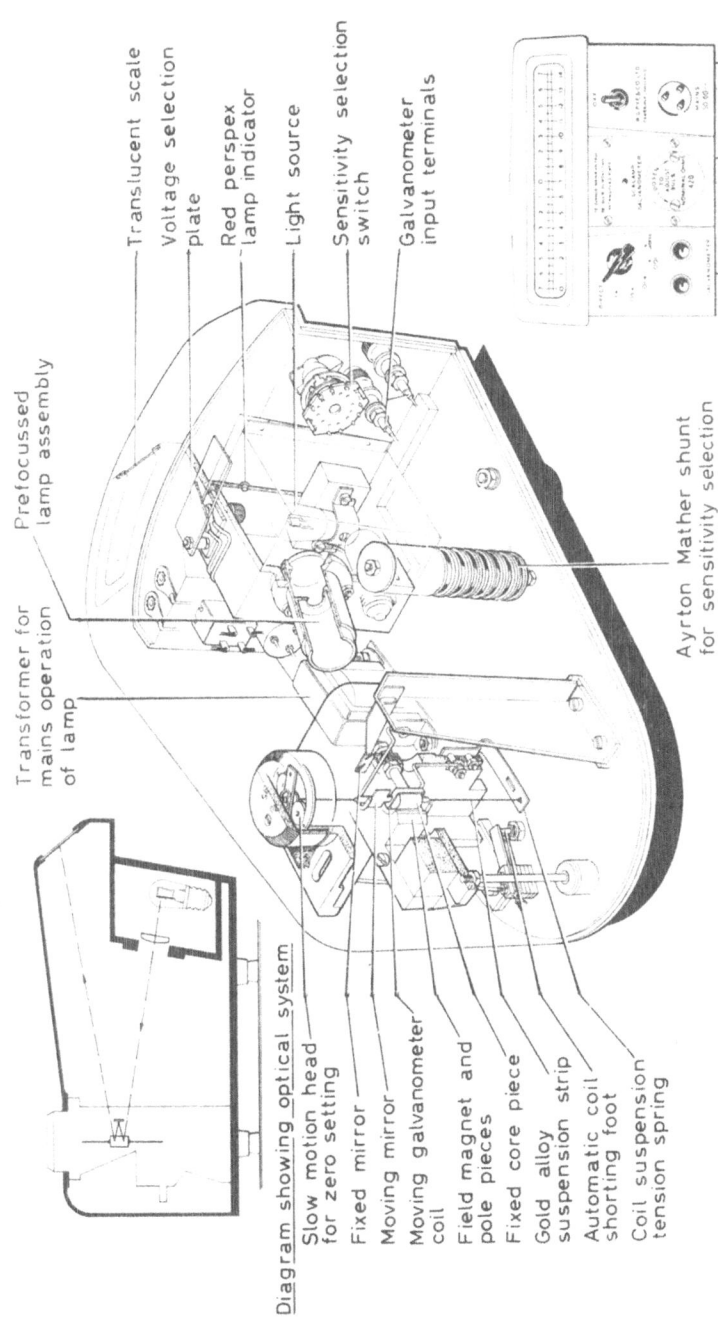

Translucent scale
Voltage selection plate
Red perspex lamp indicator
Light source
Sensitivity selection switch
Galvanometer input terminals

Prefocussed lamp assembly

Transformer for mains operation of lamp

Ayrton Mather shunt for sensitivity selection

Diagram showing optical system
Slow motion head for zero setting
Fixed mirror
Moving mirror
Moving galvanometer coil
Field magnet and pole pieces
Fixed core piece
Gold alloy suspension strip
Automatic coil shorting foot
Coil suspension tension spring

Figure 6.7. The main constructional features of a self-contained mirror galvanometer. This galvanometer is known as the Scalamp Galvanometer and the diagram is reproduced by courtesy of W. G. Pye and Co. Ltd.

151

external resistance is too large oscillations about the desired reading will take place; too small an external resistance causes the final reading to be approached slowly.

A.C. INSTRUMENTS

The moving-coil type of meter movement so far described is not suitable for measuring alternating quantities having frequencies greater than about 5 Hz unless it is modified to include a rectifying device for changing the periodically reversing current into unidirectional current. It will be recalled that the current through the coil must be in a definite direction to produce needle deflections 'up-scale'. Although other forms of meter may be used for measuring alternating current (and these will be described later) they absorb a fair amount of power and usually have non-linear scales. Since moving-coil movements can be made so sensitive and robust, and because their scales are basically linear, it is common practice to use them together with a suitable rectifier, and tolerate the minor disadvantages that the use of a rectifier entails.

Three forms of rectifier are in general use. They are the copper oxide rectifier, the selenium rectifier and the crystal rectifier. The popularity of the copper oxide rectifier is due to the low forward resistance as compared with other types; this form of rectifier therefore causes minimum error on low alternating current and voltage ranges. Unfortunately the copper oxide rectifier introduces comparatively large errors as the frequency of operation increases. Several manufacturers are therefore incorporating crystal rectifiers into their instruments.

· The reader will recall the main features of the characteristics of a *pn* junction rectifier or metal-to-semiconductor rectifier. They are reproduced for convenience in *Figure 6.8*. The half-wave rectifier circuit shown delivers uni-

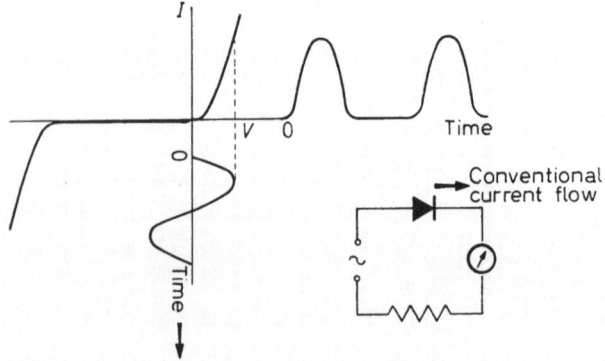

Figure 6.8. The half-wave rectifier circuit feeding a moving-coil meter

directional current pulses to the moving-coil meter. Each pulse of current delivers an impulse to the needle and coil mechanism. During the interval when no current flows the hairsprings attempt to return the coil and needle to the zero position. Because of the inertia of the system the needle is unable to follow the rapid changes in current; it therefore settles in a position which indicates the mean or average current. For a sinusoidal waveform the average current after half-wave rectification is a little less than 32 per cent of the maximum value, I_{max}. (Actually the average is I_{max}/π.) Since most measurements of alternating current or voltage must yield a knowledge of the effective value, i.e. the r.m.s. value, the manufacturer engraves his scales with the appropriate r.m.s. markings. It is emphasized however that since the meter is really measuring average values the marking of the scale readings is valid only for sinusoidal input waveforms The calibration is thus only accurate for this type of waveform; if waveforms of different shapes are to be examined a different type of instrument must be used.

The simple half-wave rectifier circuit shown in *Figure 6.8* is not a practical arrangement. The addition of a series resistor converts the arrangement to a voltmeter, but when used as such it is possible to destroy the rectifying layer of a copper-oxide or selenium rectifier. This is because during the half-cycles when the voltage is reversed the resistance of the rectifier is high. As there is little or no current under this condition, the rectifier is subject to the peak reverse voltage; this high potential can destroy the rectifier.

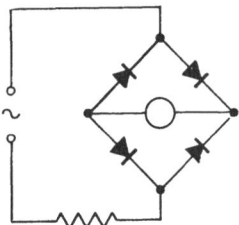

Figure 6.9. The preferred rectifying arrangement (see text)

Although the difficulty may be avoided by placing another rectifier in parallel with the meter, the usual practical arrangement is to employ the bridge rectifying circuit of *Figure 6.9*. Full-wave rectification takes place, and no damage can be sustained by the rectifiers because high reverse voltages are absent. Consideration of the current paths shows that the electrons always flow in the same direction through the meter. The average reading will be about 64 per cent ($2I_{max}/\pi$) of the peak value.

153

A.C. Voltmeters

An a.c. voltmeter with multiple ranges is, in principle, similar to a d.c. voltmeter. Suitable series resistors are used for scale multiplication, but a complication arises because of the rectifier. In the d.c. case the multiplier resistance is calculated from $V = I(R + R_m)$ where R is the multiplier resistance and R_m the fixed meter resistance. In the a.c. case however the expression $V = I(R + R'_m)$ must be used where R'_m represents the resistance combination of the meter and rectifier. Unfortunately R'_m is not constant. Examination of the characteristic of *Figure 6.8* shows that at low voltage values the resistance becomes very high. Using a.c. voltmeters at the lower end of the scale is not therefore recommended. The effective range for a.c. measurements is less than that for d.c. measurements, being from 0·25 of the scale range to full-scale value (B.S. 89 : 1954). The difficulties associated with the rectifier adversely affect the sensitivity. A d.c. meter designed for use as a voltmeter may well have a sensitivity of 20 k per volt. Using the same meter/rectifier combination to measure alternating voltages however requires the consumption of relatively large currents if a linear scale is to be preserved. This is to ensure operation on the linear part of the rectifier characteristic. The sensitivity may well be as low as 1 k per volt on alternating voltage ranges. If the lowest alternating voltage range is from zero to only one or two volts some manufacturers engrave a special additional scale for this range alone.

The range of operating frequencies is limited being commonly 25 Hz to 10 kHz. The newer instruments extend the guaranteed range to 20 kHz.

A.C. Ammeters

The measurement of alternating current and the extension of the meter ranges cannot be carried out as in the d.c. case. In any current meter the voltage drop across the meter should be as small as possible (ideally zero). Operation of the rectifier at very low voltages is not satisfactory because of the high resistance and non-linearity of the characteristic in this region. Ordinary resistive shunts cannot therefore be used as current range multipliers. The difficulty is overcome by using a current transformer. The current to be measured is passed through the primary of the transformer, and the secondary winding, having many more turns than the primary, presents a voltage to the meter rectifier combination that is sufficiently high to ensure that operation is over the linear portion of the characteristic. Various tappings on the current transformer are used to extend the current range of the meter.

ALTERNATIVE INDICATORS FOR
A.C. MEASUREMENTS

Figure 6.10 shows diagrammatically other forms of instruments used for measuring alternating quantities. The moving-iron meter consists of two iron rods A and B of which A is fixed along the length of a short coil and B is

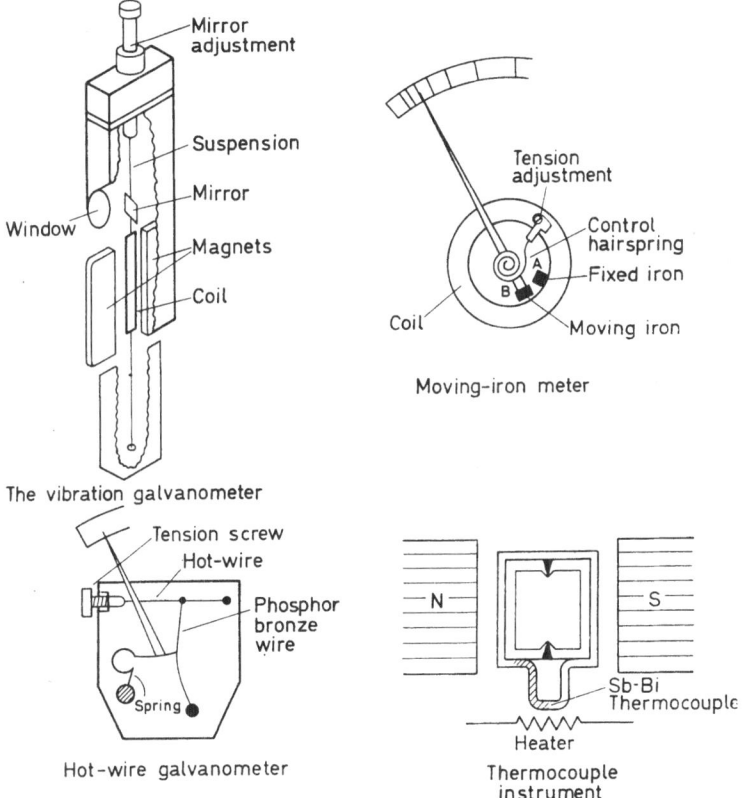

Figure 6.10. Various alternative forms of a.c. meters

movable. When current is passed through the coil both rods become magnetized in the same sense and consequently a force of repulsion exists between them. The movable rod moves about a central pivot carrying with it a pointer that traverses a scale. The displacement of the pointer end is proportional to the square of the current, so the scale is not linear and is very cramped at the lower end.

155

The vibration galvanometer is similar in construction to its d.c. counterpart, but relies on mechanical resonance for the measurement of alternating quantities. The coil, mirror and suspension assembly have a mechanical resonant frequency that depends upon the mass and form of construction. By using small, light parts the resonant frequency can be quite high by mechanical standards. Vibration galvanometers for use with 50 Hz signals are common. Electronically however the frequency range is so limited (up to only tens of cycles per second) that this type of meter is rarely used in an electronics laboratory.

The hot-wire galvanometer is useful for measuring large alternating currents. The meter depends upon the heating effect in wire. On the passage of the current the rise in temperature is accompanied by an elongation of the wire. A suitable mechanical amplifier is then used to move the indicating needle. The instrument is a true r.m.s. reading device, the indication being independent of waveform.

The thermocouple instrument is used primarily at radio frequencies. At such frequencies the conventional meter is useless. The r.f. current is passed through a heating element. The resulting rise in temperature causes an increase in temperature of the antimony/bismuth thermocouple junction. A current is produced, which operates a moving-coil in the normal way. Over a reasonable operating range of current the meter has a linear scale.

SUGGESTIONS FOR FURTHER READING

A.C.D.C. Test Meters by W. H. Cazaly and T. Roddam. Pitman and Sons Ltd., 1951.
Basic Electronic Test Instruments by R. P. Turner. Holt, Rinehart and Winston Inc., 1963.

7

POWER SUPPLIES

All electronic equipment needs to be energized by means of a power supply. In the great majority of cases the power is delivered to the electronic circuit at a steady or fixed voltage. In the early days of radio and electronics the necessary power was derived from batteries, but the large currents and voltages required for thermionic valves made this source inconvenient. Leclanché type batteries were bulky and expensive, and the lead-acid accumulator required periodic attention. Power supplies were therefore invented that took the necessary power from the mains. The invention of the transistor has, however, brought the battery back into favour. Since transistor apparatus usually requires low currents at low voltages (say up to 24 V) the advantages of small size, cheapness and portability can be realized when batteries are used.

BATTERIES

Little need be said here about the lead-acid accumulator or Leclanché type cells since they have been extensively described in many textbooks. Most portable transistorized equipment uses a primary battery of which the Leclanché or so-called 'dry' battery is the best known *(Figure 7.1)*.

These batteries are available in a wide range of sizes and consist of the appropriate number of cells in series. The common cylindrical cell may be used where comparatively high current discharge rates are required, but for h.t. supplies to hearing aids, apparatus using 'battery' valves, and certain types of transistorized equipment, layer type cells are used. When new, each cell unit has an e.m.f. of about 1·5 volts. Such batteries are rated at a definite maximum current discharge rate so that the depolarizing effect will have a chance to keep pace with the hydrogen liberation. For batteries of reasonably large volume, the discharge rate may be about 100-250 mA when they are discharged for 100 hours at a rate of about 5 hours in every 24. Under similar conditions, the smaller cells give less current; for example, the familiar U2 cell gives about 30–40 mA for about 100 hours when discharged for 4 hours in every 24. The end point is reached when the voltage drops to 1·1 V.

Dry cells are intended for intermittent service; they are thus given a chance to recuperate during the rest periods by the action of the depolarizer. This type of cell deteriorates when not in use, the smaller sizes having a shorter shelf-life than the larger. Testing a cell with a voltmeter is of no value when the cell is not delivering current, for even a unit that is almost entirely dis-

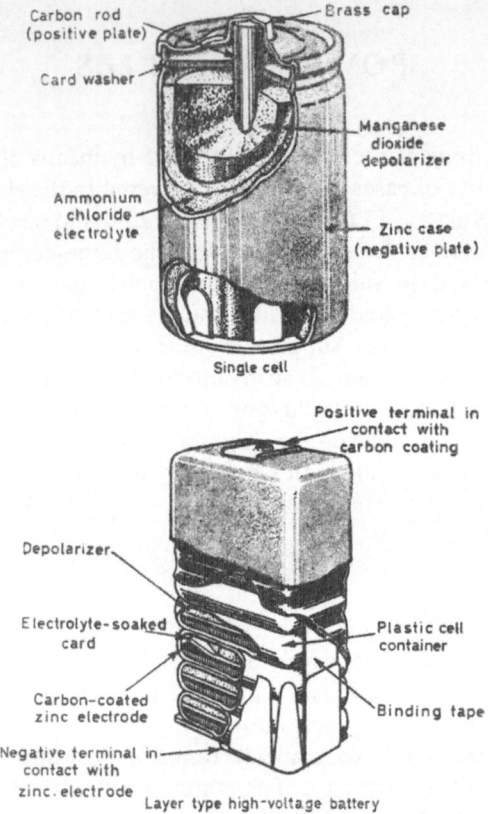

Figure 7.1. Modern Leclanché cells. Reproduced by courtesy of Mallory Batteries Ltd. and *Electrical Manufacture*

charged gives a test reading close to 1·5 V on open circuit. When delivering the maximum rated current, the voltage should exceed 1 V; for some applications 1·1 V is the minimum allowed.

Mercury Cells

The mercury type primary cell (e.g. those made by Mallory Ltd., *Figure 7.2*) was originally developed during the Second World War for use in portable equipment where maximum energy within minimum volume was the

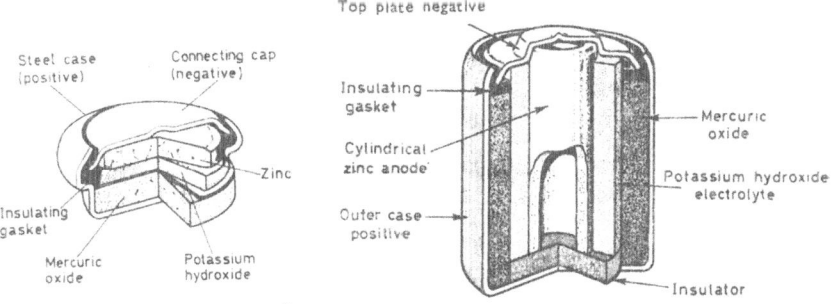

Basic Ruben-Mallory mercury flat cell

Basic Ruben-Mallory cylindrical cell

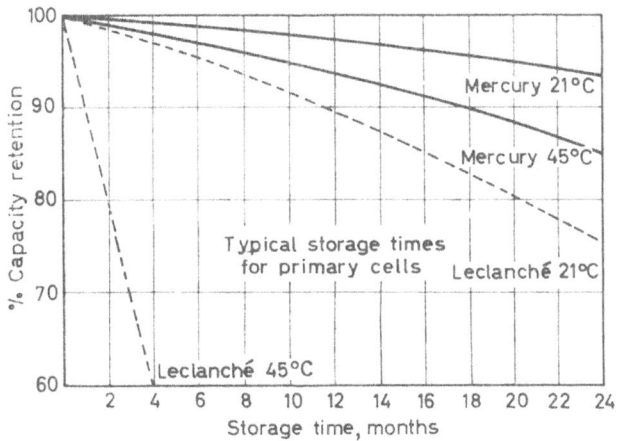

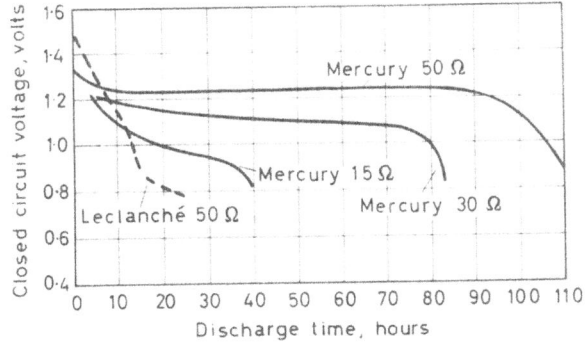

Figure 7.2. The basic Ruben-Mallory cell with comparative storage times and discharge curves. Reproduced by courtesy of Mallory Batteries Ltd. and *Electrical Manufacture*

prime aim. The original type of cell has been much improved as a result of research in the last twenty years, and many of the disadvantages associated with the Leclanché cell have been overcome.

The most attractive feature of the mercury cell is the provision of a steady voltage over nearly all of the cell's useful discharge period. Over long periods of operational use, or after some thirty months of storage, a voltage regulation within 1 per cent of the initial voltage is still maintained. (The data are supplied by Mallory Batteries Ltd.; the diagrams and graphs are reproduced from *Electrical Manufacture*[1].) Greater degrees of stability and regulation may be obtained over shorter operational periods.

The self-depolarizing design of this type of cell when discharged at current drains within the cell specification eliminates the need for 'rest' periods. For continuous operation of commercial and scientific equipment, transistorized devices, medical apparatus and the like, this proves a distinct advantage over the Leclanché type cell. For emergency alarm devices, rescue radio transceivers etc., mercury batteries are ideal since they have a long shelf-life. They can be stored for periods of two years or more in dry conditions and at temperatures between 10 and 20°C without any appreciable loss of capacity.

Nickel-Cadmium Cells

Rechargeable nickel-cadmium cells are useful for use in electronic equipment since they can be sealed, thus avoiding the effect of corrosive fumes (which are given off by lead-acid accumulators). The sealed type of cell has a long life (approaching 15 years with care). They can be completely discharged without ill-effects, and can withstand moderate overcharging. The cell's nominal voltage is 1·2 V. Their physical size depends upon their electrical capacity. Button Ni-Cd cells made by DEAC (Great Britain) Ltd. have diameters of 4·3 cm and 5·03 cm with corresponding thicknesses of 0·76 cm and 1 cm for capacities of 450 mAh and 1,000 mAh respectively, to mention only two of their range. This firm also distributes a rectangular cell with a capacity of 7·5 Ah has dimensions of 4·35×5·05×10·8 cm.

SOLAR CELLS

The silicon photocell is a photovoltaic device that converts light directly into electrical energy. The more familiar selenium photocell discussed in Chapter 12 also makes this direct conversion, but the efficiency of a selenium cell is too low to allow it to be used as a solar battery. The silicon cell has an efficiency approaching 14 per cent in its present state of development, which is about twenty-five times greater than that of a selenium cell. (Efficiency in this context is radiant solar energy falling on the cell divided by the electrical energy available from the device.)

Silicon cells are made by melting purified intrinsic silicon in quartz containers and adding minute traces of a Group V element, such as arsenic or phosphorus. The *n*-type silicon that results solidifies and is cut into slices. These slices after grinding and lapping are then passed into a diffusion chamber and boron is diffused into the *n*-type crystal from boron trichloride vapour. A *pn* junction results. The *p*- and *n*-type surfaces are then plated and terminal wires added.

It will be recalled from Chapter 5 that a barrier layer, in which very few charge carriers exist, is created between the *p*- and *n*-sides of the crystal forming a *pn* junction. When discussing the *pn* junction as a rectifier we saw that the application of a reverse bias voltage increases the potential hill and prevents large numbers of electrons from flowing. The small leakage current that does result is attributed to the production of electron-hole pairs in the barrier layer, the energy coming from thermal sources. In the solar cell there is, of course, no reverse bias voltage, but nevertheless a potential hill exists across the junction. The incidence of radiant energy from the sun creates electron-hole pairs by rupturing the covalent bonds between atoms in the barrier layer. The holes are swept to the *p* side and the electrons are swept to the *n* side *(Figure 7.3)*. If an external circuit exists, electrons flow round from the *n* side to the *p* side dissipating energy in any load that is present. The source

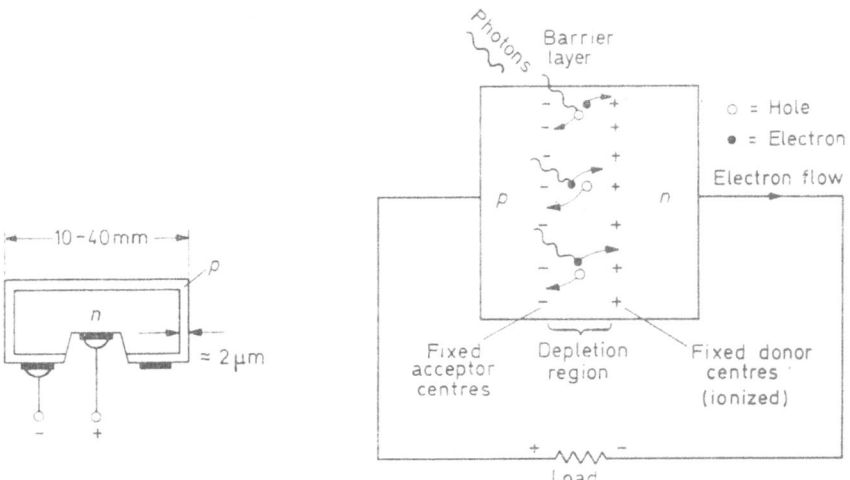

Figure 7.3. Principles of operation of a solar cell. The incidence of photons creates electron-hole pairs, the charge carriers are swept out by the field across the barrier layer. An electric current around the external circuit is then obtained

of the energy is the incident radiation which consists of photons of energy E, where $E = h\nu$, ν being the frequency of the radiation and h Planck's constant. The travelling of electrons to a negatively charged region may seem strange, but it must be remembered that the incidence of photons on the atoms in the barrier layer reduces the potential hill. The Fermi levels associated with junctions in the circuit are disturbed, giving a resultant e.m.f. of the polarity shown in *Figure 7.3*. In rather loose terms, there is an attempt to restore the potential hill to its former value.

Solar cells are the source of power for energizing the transmitting and other electronic equipment in unmanned satellites. At a less spectacular level, they are used as readout devices in computing machines, and in general photovoltaic work. Wherever a source of light is available, solar cells can be used to energize low-powered transistor equipment instead of the more conventional

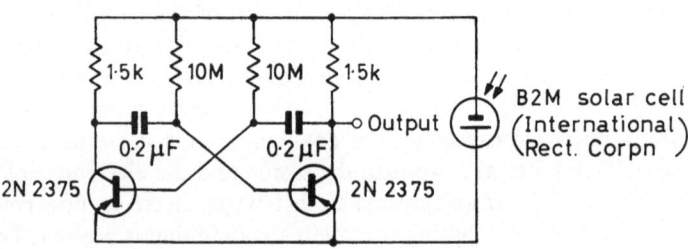

Figure 7.4. A multivibrator circuit powered from a solar cell. This type of relaxation oscillator gives an output voltage that consists of square waves

batteries. *Figure 7.4*, for example, shows a multivibrator being powered by an International Rectifier Corporation solar cell, type B2M. Oscillations are possible with the circuit shown provided the cell is energized with an illumination of at least 10^3 lux ($\approx$ 100 foot-candles).

POWER FROM THE MAINS

When comparatively large amounts of power are needed, the source of supply is nearly always the alternating current supply mains. For powers not exceeding about ten kilowatts, the single-phase supply is usually used. When powers in excess of ten kilowatts are required (e.g. for a transmitter or large industrial equipment) three- or multi-phase systems are used, together with large mercury or gas-filled rectifiers or some of the heavy power-handling semiconductor devices, such as the silicon-controlled rectifier. These heavy current systems are outside the scope of this book.

The power requirements for thermionic valve equipment are:

(*a*) a low-voltage supply to heat the cathode to the correct operating temperatures;

(*b*) a steady high-voltage supply to feed the anode and screen-grid circuits; and

(*c*) steady voltages for special applications, e.g. fixed-bias voltage supplies and reference voltages.

The supply voltages for anode and screen-grid circuits must be free from ripple or noise voltages, and must remain reasonably constant under varying load conditions. Under such circumstances, it is often necessary to stabilize the supply by means of additional apparatus in order to guarantee constancy of output voltage.

In order to meet these requirements using the a.c. mains, it is necessary to construct what is known as a 'power pack'. Frequently, this power pack is assembled on the same chassis as the rest of the electronic apparatus. The power pack consists of a transformer with several secondary windings delivering suitable voltages for the heater and h.t. lines, a rectifier for converting the h.t. alternating voltages into unidirectional ones, smoothing components to eliminate the fluctuations and so deliver a steady voltage, and, if necessary, some additional apparatus for stabilizing the h.t. supply.

Mains transformers

For reasons of efficiency the bulk of the electrical power needed for the country is generated in a few power stations and distributed via a grid network of cables known as the mains. Since enormous powers are involved, it would not be possible to effect the distribution at the voltages considered safe in domestic, laboratory and other locations (i.e. 200–250 volts). This would mean that the corresponding currents would be too large to be carried by cables of practical dimensions. The power is therefore generated as alternating current so that it is possible to transform to a large distribution voltage (e.g. 115 kV). For a given power the current is reduced by the same ratio as the voltage is increased. The cables carrying the current can therefore be comparatively thin and cheap. At the consumer's end of the grid, sub-stations are provided to transform the voltage to a safe value before distribution to domestic and industrial establishments. For industrial locations using large amounts of power, the three-phase 440 V supply is used, but for the type of equipment we are discussing, the supply voltage is from about 220 to 250 volts (r.m.s.) single phase. (A single phase supply is a simple twin-line supply where the voltage on the live line varies sinusoidally with time about a mean earth potential. The neutral line is held at about earth potential.)

The efficient use of this source of electrical energy, like the efficient use of mechanical energy, requires the introduction of some means of converting the form of the energy at the source to a form that can be used by the load. In the case of a motor-car a gear-box is needed between the engine and the road wheels. The gear-box is designed so that the road wheels can turn slowly with great force or alternatively with much greater speeds at less force depending upon the prevailing conditions. In a similar manner, it is necessary to adjust electrical circuits so that the power available may appear at the load as one of various combinations of voltage and current. The electrical device that corresponds to the gear-box is the transformer. It should be noted that neither the gear-box nor the transformer alter the amount of power available if both are 100 per cent efficient. (In practice some loss of power is experienced because of friction in the gear-box or analogous causes in the transformer.) However, the power is suitably adapted to the particular work to be done.

The operation of a transformer depends upon the principle of electromagnetic induction. Fundamentally, a transformer consists of two coils that are electrically isolated, but so placed physically that a changing magnetic field set up by an alternating current flowing in one of the coils induces an alternating e.m.f. in the second coil. Thus, mutual inductance is said to exist between

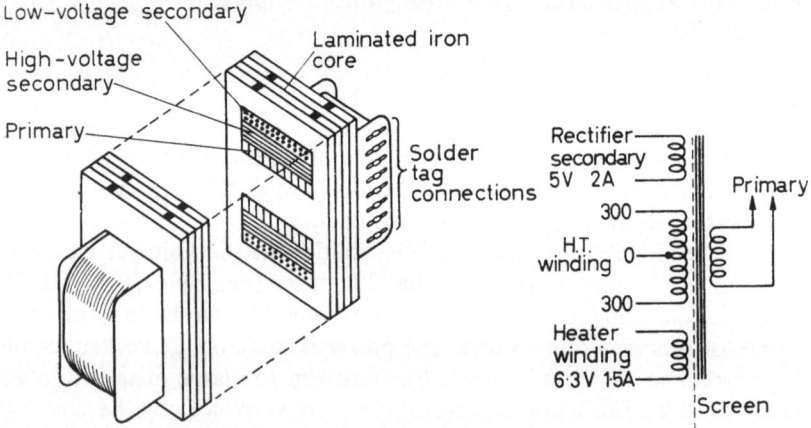

Figure 7.5. Construction of a small mains transformer together with the circuit representation of a typical unit used in simple power supplies

the two coils, and the two circuits are said to be inductively coupled. *Figure 7.5* shows the main physical details together with the diagrammatic representation. The coil connected to the source of power is called the primary winding, and the coil connected to the load is the secondary winding. Several secondary windings may be linked with a single primary winding in order to accommodate several different loading conditions simultaneously. The power

164

delivered by the generator passes through the transformer and is delivered to the load, although no electrical connection exists. The connection between the primary and secondary windings is the flux linkage between the coils. For maximum power transfer all of the lines of flux set up by the primary winding must link the secondaries. To this end, at power frequencies (i.e. 50 Hz in this country) the coils are wound on a suitable former and adequately insulated from each other. Laminations of magnetically soft iron, or suitably ferromagnetic alloy, are then inserted to form the core as in the power choke or inductor described in Chapter 2.

When no power is taken from the secondary, the supply current and supply voltage are 90° out of phase. The power, P, is given by $P = VI \cos \phi$ where V is the supply r.m.s. voltage, I the r.m.s. current and $\cos \phi$ is the power factor. When ϕ, the phase angle, is 90°, $\cos \phi = 0$. No power is therefore taken from the mains even though the primary is connected to the supply.

In an ideal transformer when power is being consumed by the load

$$\frac{i_p}{i_s} = \frac{N_s}{N_p} = \frac{e_s}{e_p} \tag{7.1}$$

where i_p is the primary current, i_s the secondary current, N_s and N_p the number of secondary and primary turns respectively, e_s and e_p being the secondary and primary voltages respectively. The output power ($i_s e_s$) is evidently equal to the input power ($i_p e_p$). (Here we are assuming a purely resistive load, i.e. one in which the current and voltage are in phase.)

Practical transformers depart from the ideal in several respects. Not all of the flux induced by the primary is linked with the secondaries; copper losses, due to wire resistance, and iron losses, due to hysteresis and eddy currents give rise to heat. Copper losses are made good by increasing the number of secondary turns over the value N_s in equation 7.1. The open-circuit e.m.f. of the secondary is therefore higher than the output voltage under load. Eddy-current losses are reduced by making up the core from laminations, each lamination being coated on one side by a thin layer of insulating material. Hysteresis losses and flux leakages are reduced by careful selection of the core material and by taking care with the transformer geometry. The core size depends upon the area of the core, A, and also upon the volume to be occupied by the wire of the coils and its associated insulation. Generous core sizes must be employed if undue rises in temperature are to be avoided; if VI is the volt-amp output the cross-sectional area for small power transformers may be calculated from the empirical relation

$$A = \frac{VI}{5 \cdot 58}$$

where V is in volts, I in amperes and A in square inches.

A transformer used in the power supplies for electronic equipment has several secondaries. The number of secondary coils required, and their respective outputs, depend upon the power consumption and nature of the apparatus to be operated. However, since the basic principles underlying the supply of power to electronic equipment are the same for most installations, it will be necessary to study only a few typical examples.

Since modern low-powered thermionic valves use indirectly heated cathodes, the heaters can be supplied with alternating current. The maximum voltage between the heater and the cathode for trouble-free operation must not exceed that laid down by the manufacturer, which is usually about 50 volts. The majority of valves in a given piece of equipment operate with cathode potentials of not more than 50 volts above chassis potential. It is customary therefore to wire their heaters in parallel, and to have one secondary winding reserved to supply the necessary current; the current is delivered in most cases at 6·3 volts r.m.s. It is usual to have a centre-tap on this secondary winding which should be connected to the chassis or h.t. negative line. (A centre-tap is a wire connected to the electrical centre of the secondary coil.) When this is done the voltage between the heater and the cathode can never be large; additionally, 'hum' voltages are reduced. 'Hum' is the term given to mains-frequency alternating voltage or current components injected into the signal thus forming an unwanted spurious component at the output indicator or transducer. Th eterm 'hum' is used because of its audible effect in loudspeaker equipment.

For reasons that will be obvious later in the chapter, the cathode of a thermionic diode used as a rectifier is at a high potential relative to the h.t. negative supply line. It is not possible therefore to use the secondary winding already mentioned as a supply for the heater of the rectifier because the heater-cathode voltage would then exceed the maximum allowed. A separate winding must thus be provided for this heater. Current for the rectifier heater is also delivered at a low voltage commonly 5 or 6·3 volts r.m.s. One end of the secondary winding is usually connected to the cathode of the rectifier.

In certain circumstances it may be necessary to operate the cathode of a particular valve at some potential substantially different from that of the supply lines (e.g. a valve in an oscilloscope time-base, or the cathode-ray tube itself). In these cases separate additional low-voltage heater windings are required. Special attention may have to be paid to the insulation of these secondary windings from other coils on the transformer.

Many transformers have an electrostatic screen wound between the primary coil and the secondary winding supplying the h.t. current. The screen consists of a layer of copper foil extending over the primary coil. The overlapping ends of the foil must be insulated from each other to prevent currents from being induced in what is effectively a one-turn secondary winding. Such

166

currents would give rise to excessive temperatures and consequent damage to the transformer. The purpose of the screen is to prevent mains-borne interference from reaching the secondary circuits; the interference is caused by electric motors, switches, faulty fluorescent lighting equipment, etc. Since such interference is electrostatic in nature the screen must be earthed or connected to the chassis to be effective. Magnetic fields penetrate the screen without attenuation and thus transformer action is not affected.

The type of winding used for the high-voltage supply lines will depend on the system of rectification adopted. The specifications for this secondary winding are dealt with in the section on rectifiers.

Variable output voltage transformers (often called variacs, although this is the registered trade name of General Radio) are most useful when it is desired to vary the supply to a piece of apparatus. The device is in effect an autotransformer, i.e. a transformer having a single winding; the required voltage is tapped off in a manner reminiscent of the way in which a potential divider or volume control works. By moving a control knob, any voltage may be selected from zero up to about 5 per cent in excess of the mains. Unlike the potential divider no power is consumed since the element is almost wholly inductive. It must be emphasized by way of warning that no isolation from the mains is possible, as is the case with a double-wound transformer having an entirely separate secondary winding.

RECTIFIERS

Because the output voltage of the transformer is alternating, it is necessary to convert it to a direct or steady voltage before it can be used in electronic equipment. The first stage is to convert the a.c. into unidirectional pulses and then these pulses must be passed through a filter whereupon direct current at a steady voltage is supplied to the load. The process of converting the alternating current to unidirectional current is known as rectification, and the filtering after rectification is termed 'smoothing'.

Regardless of the type of rectifier used, the function of all rectifiers is the same, that is, they allow electron flow in only one direction. The thermionic diode has been described in an earlier chapter; and it will be remembered that electrons can flow only from the cathode to the anode. The diode thus acts as a rectifier. Thermionic diodes have been, and still are, popular as rectifiers for low-powered equipment, but they are being supplanted by the devices emerging from recent advances in solid-state technology. Equipment requiring up to 300 mA at voltages up to about 500 V can be supplied by rectifiers having indirectly heated cathodes. Usually two diodes are mounted in one glass envelope and the tube is then known as a double- or duplex-diode.

For domestic or light-current industrial apparatus, each item of equipment can be supplied with its own power pack, and thus there is seldom any need to construct power supplies having huge outputs. If larger outputs than 300 mA are required it is usual to use a directly heated cathode or select a semiconductor device.

The use of indirectly heated cathodes in rectifiers confers important advantages. The other valves in the apparatus have indirectly heated cathodes and therefore several seconds must elapse before these cathodes attain their correct working temperatures. It is desirable from the point of view of valve life that the h.t. voltage should not be applied until these temperatures have been reached. This is automatically obtained if indirectly heated cathodes are used in rectifiers. Some time must elapse before the rectifier cathodes achieve their correct working temperature; during this time the full h.t. voltage cannot be developed. Additionally high-voltage surges across the smoothing and reservoir capacitors are avoided. If the h.t. voltage were immediately available, no power would be drawn until the equipment cathodes had heated sufficiently; the no-load h.t. voltage is higher than that on full load, so the maximum permitted voltage across the capacitors may be exceeded. The indirectly heated cathode of the rectifier prevents this from happening, and there is thus no need for the uneconomic use of capacitors with ratings higher than the operating voltages.

Metal Rectifiers

Metal rectifiers depend for their action upon the special properties existing at a metal-semiconductor boundary. Two common forms are shown in *Figure 7.6*. The copper/copper oxide rectifier is made by treating copper discs so that a layer of cuprous oxide is formed on one side of the disc. Rectification takes place at the barrier layer between the metal and the semiconducting oxide. The lead disc merely makes good electrical contact between the oxide and the cooling fin, and plays no part in the rectification. The selenium rectifier is made by depositing selenium on an aluminium, or nickel-plated steel, disc and applying heat treatment. The selenium is then covered by an alloy of cadmium and tin. The barrier layer is formed by diffusion at the selenium/alloy interface.

Both selenium and cuprous oxide are *p*-type semiconductors. Upon formation of the rectifier cell electrons diffuse into the semiconductor thus filling the acceptor centres. In the copper oxide case the electrons come from the copper, which in consequence becomes positively charged. The oxide becomes negatively charged. When the semiconductor is made positive with respect to the copper the potential hill across the barrier layer is reduced and

168

electrons flow freely from the copper to the semiconductor. The rectifier is therefore forward-biased. Upon reversing the polarity the rectifier is reverse-biased and very few electrons can flow. The reverse resistance is about 1,000 times the forward resistance.

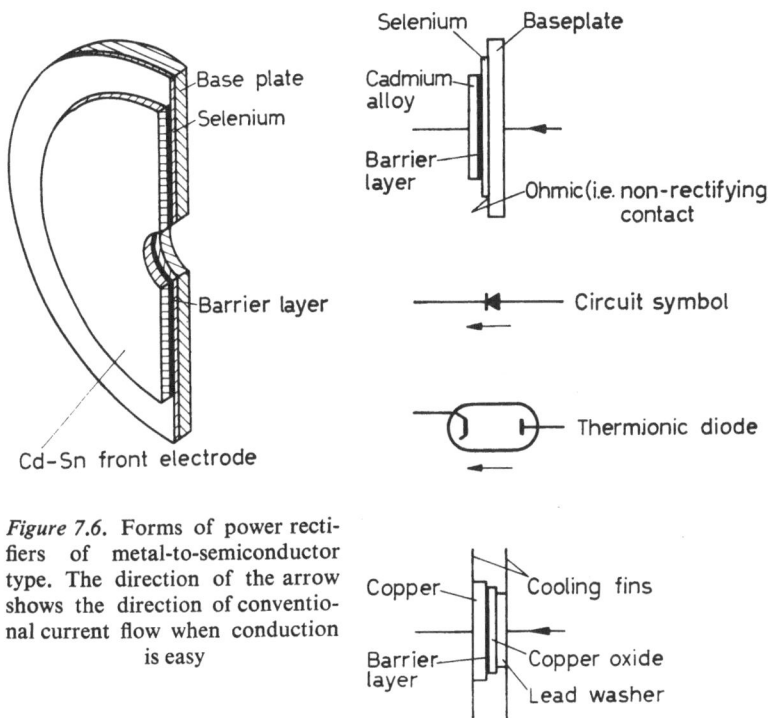

Figure 7.6. Forms of power rectifiers of metal-to-semiconductor type. The direction of the arrow shows the direction of conventional current flow when conduction is easy

In selenium cells it is thought that an *n*-type cadmium oxide and/or cadmium selenide layer is formed on contact with the selenium. When the *p*-type selenium is made positive with respect to the cadmium alloy disc, the rectifier is forward-biased. When the conduction is easy the direction of the conventional current (which is opposite to electron flow) is from the baseplate, via a non-rectifying junction to the selenium, and through to the cadmium layer. The reverse-to-forward resistance ratio is about 5,000.

Since the deposits on the metal plates are very thin, the maximum voltage that each type of rectifier can sustain without rupture is quite low. The peak inverse voltage for selenium rectifiers is about 60 V and that for copper oxide types only about 6 V. To overcome this limitation when higher voltages are to be rectified several discs are combined in series. Thus, the total number of discs determines the peak voltage that can be sustained whilst the effective rectifying area of any one disc determines the maximum current carrying capacity. The heat, which is inevitably generated in the rectifier, must be conducted away from the rectifying discs as quickly as possible. It is usual therefore to interleave each disc with a large cooling fin. All the discs and fins have holes through their centres so that they can be placed in line on an insulating tube. Within this tube is a threaded metal rod. Insulating washers are placed at each end of the stack and nuts at the end of the rod are screwed up holding all the rectifier elements firmly along the tube. The copper oxide rectifying discs are held tightly on the insulating tube, but the barrier layer of a selenium rectifier must not be compressed unduly otherwise the resistance in the reverse direction is reduced.

Since copper oxide rectifiers are easily and irreparably damaged by overloads, their use has usually been confined to supplying low current h.t. lines. The selenium rectifier is more tolerant towards overloading, and, when connected as a bridge rectifier, it is often used in battery chargers and other high-current equipment.

Semiconductor Rectifiers

Because of their small size, efficiency and durability, crystal diodes of germanium, or more usually silicon, are now widely used in rectifying circuits. Although the copper-oxide rectifier is still frequently used in moving-coil meters, this and the selenium rectifier are rapidly being superseded by the semiconductor rectifiers. Since the general theory and construction of crystal diodes have been discussed in Chapter 5, we shall limit our discussion here to their practical applications in power rectifying circuits.

Both germanium and silicon rectifiers show marked advantages over the more conventional types. The germanium reverse-to-forward resistance ratio is approximately 4×10^5 and current carrying capacity is about 50 amperes per cm^2 compared with about 0·25 A per cm^2 for the selenium rectifiers. The peak inverse voltage for germanium rectifiers is often as high as 100 volts. The silicon power diode is an improvement on the germanium type. Not only is the maximum operating temperature raised from about 65°C for the germanium type to 150°C for the silicon, but the peak inverse voltage that modern silicon rectifiers will withstand can be as high as 1,000 V per single *pn* junction. Their reverse-to-forward resistance ratio is about 10^6.

RECTIFYING CIRCUITS

The half-wave rectifier is shown diagrammatically in *Figure 7.7a*. During one half-cycle, the applied voltage has a certain polarity, and during the succeeding half-cycle the polarity is reversed. Since the rectifier conducts for only one direction of applied voltage, electrons can flow only during half a period. It is for this reason that the arrangement of *Figure 7.7a* is known as a half-wave rectifier. The symbol represents a metal rectifier, but obviously a crystal or thermionic diode may equally well be used. The easy-flow direction refers to conventional positive current, i.e. the flow direction is opposite

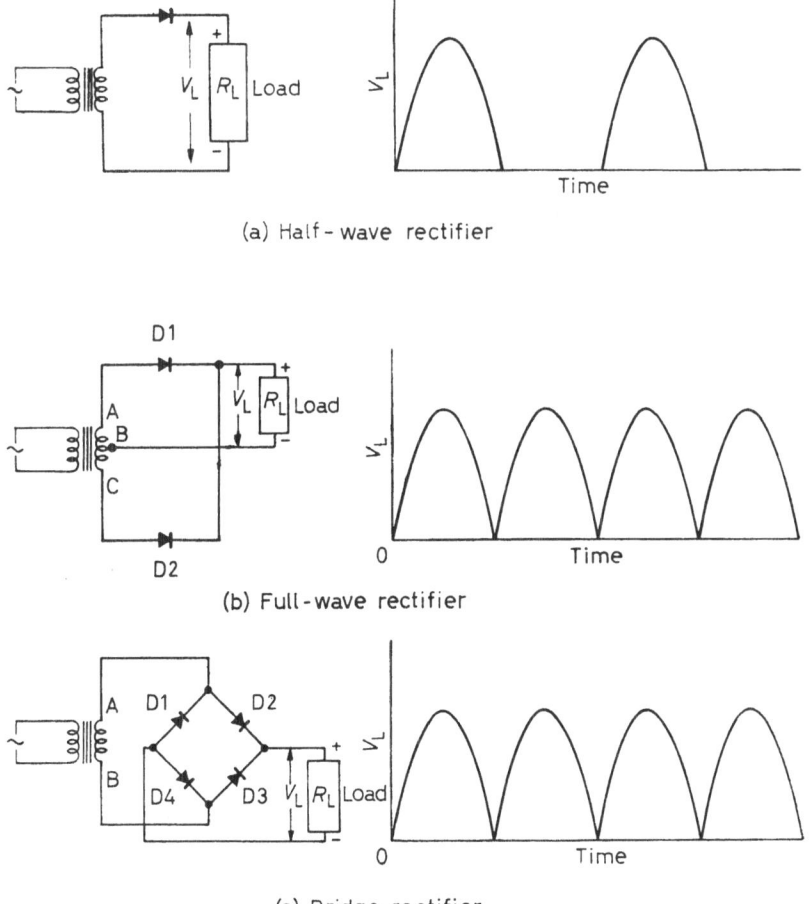

(a) Half-wave rectifier

(b) Full-wave rectifier

(c) Bridge rectifier

Figure 7.7. Rectifying circuits for power supplies

171

to that of the actual electron current. The positive sign is usually omitted; on a crystal or metal rectifier the corresponding lead is indicated by a red spot or band on the body of the component next to the appropriate lead. When connected to an alternating voltage, the rectifier polarizes the resistor representing the load (i.e. the equipment being operated) in the way shown.

The Full-Wave Rectifier

By the addition of a second diode it is possible to have conduction of electrons in the load throughout the whole cycle. The arrangement, shown in *Figure 7.7b*, is then known as a full-wave rectifier. If at an instant of time the polarity of A is positive with respect to the centre-tap B, the polarity of C is negative relative to B and diode D1 will conduct since it is biased in the forward or conducting direction. The direction of the current is therefore from A through D1 and R_L and to the centre tap. Half a period later C is positive relative to B, and D2 is now biased in the forward direction. The direction of the current is then from C, through D2, through R_L in the same direction as before, and back to the centre-tap. The voltage waveform across R_L is therefore as shown.

The Bridge Rectifier

The disadvantage in using the circuit of *Figure 7.7b* is that the transformer secondary must produce twice the voltage of that used in the half-wave rectifying circuit because only half of the winding is used at any one time. This difficulty can be overcome by using four diodes in what is termed a bridge rectifier. The circuit arrangement is shown in *Figure 7.7c*. When A is positive with respect to B, diodes D2 and D4 conduct. On a reversal of polarity between A and B, diodes D1 and D3 will conduct. The resultant waveform across R_L is then as shown in the figure. The arrangement is in effect a full-wave rectifier, and is the popular choice for battery chargers. Since small efficient semiconductor power rectifiers are readily available at an economic cost, many manufacturers use the bridge rectifier circuit as the standard rectifying arrangement in their power packs.

FILTER CIRCUITS FOR POWER SUPPLIES

It is evident that the rectifier circuits, as they stand, do not supply current at the steady or uniform voltage required by electronic apparatus. An examination of the waveform shown in *Figure 7.7* reveals a pulsating voltage.

172

The rectified output must now be modified to level out these pulses and produce current at a steady voltage. Such modification is achieved by using filter circuits.

Filtering is accomplished by using capacitors and either inductors or resistors in the circuit. When used for this purpose, inductors are often referred to as 'chokes' since they choke off any variations of current and allow the easy conduction of only direct current. Filter circuits may take on various forms, but the three basic arrangements are shown in *Figure 7.8*.

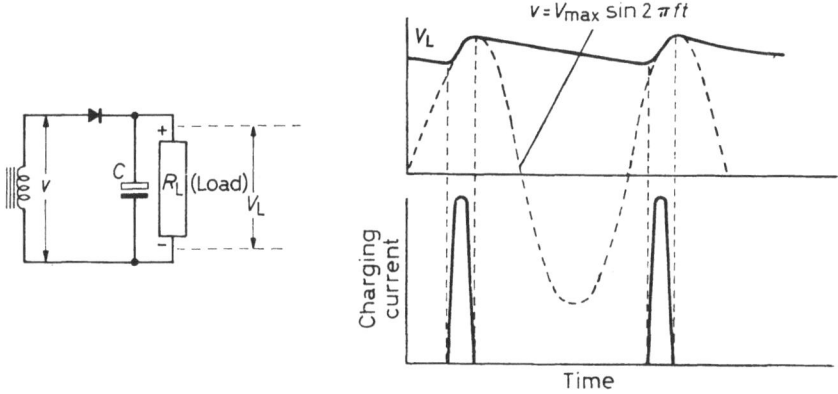

Figure 7.8a. Approximate waveforms for the half-wave rectifier

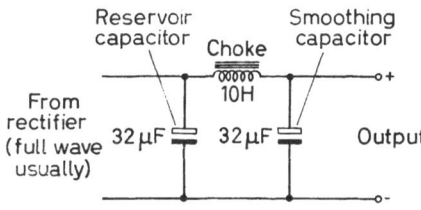

Figure 7.8b. A capacitor input filter with typical values of choke and capacitors. The choke presents a very high impedance to current variations and little opposition to d.c.

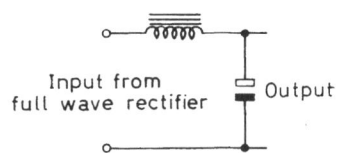

Figure 7.8c. A choke input filter. The smoothing capacitor as in *Figure 7.7b* provides a very low impedance to alternating voltages, but a very high impedance to direct voltages

The simplest filter arrangement, shown in *Figure 7.8a*, consists merely of placing a capacitor in parallel with the load. The value of capacitance must be large in order to present as small a reactance as possible to the pulsating rectified output, and to store sufficient charge so that current may be maintained in the load during the period that the rectifier is not conducting. The

173

reactance of the capacitor should be much less than the resistance of the load. For the kind of loads usually encountered, capacitors of 16–50 μF are commonly used. The rectified pulses charge the capacitor to a voltage close to the peak value delivered by the rectifier. Because of the large value of C, the time constant CR_L is large compared with the periodic time (0·02 sec) of the applied voltage. The voltage across R_L does not therefore fall sinusoidally, but decays exponentially according to the equation $v = V_{max} \exp(-t/CR_L)$. The fall of voltage may be reduced for a given load by increasing the value of C. There is a limit however to the value of capacitance used. It can be seen that the slower the rate of fall of voltage across R_L, the smaller is the time available to recharge the capacitor. The current pulse delivered by the rectifier must therefore have a greater peak value to deliver a given energy. All rectifiers have peak current ratings and these ratings can be exceeded if the value of C is too large, causing damage to the rectifier. For any given rectifier and associated circuit, the maximum value of C that can safely be used is specified by the manufacturer in his published data sheets.

The designer of rectifying circuits must observe an additional precaution. This precaution is concerned with the maximum peak inverse voltage that the rectifier can tolerate. During the time that the rectifier is not conducting, we see from *Figure 7.8a* that the peak or maximum inverse voltage applied to the diode is the sum of the voltage across the capacitor and that across the transformer secondary. The peak inverse voltage (p.i.v.) is thus approximately twice the peak voltage across the transformer secondary. The manufacturer's published data gives the maximum p.i.v. that may safely be applied.

The output from the rectifier when a single capacitor is used as in *Figure 7.8a* is not smooth enough for the high-voltage supply to electronic equipment. The variations would give rise to unwanted components in the signal. For half-wave rectifiers the unwanted component is a 50 Hz one. In full-wave rectification, since two pulses are delivered during each cycle of the mains, the unwanted component has a frequency of 100 Hz. To reduce the ripple to a minimum, it is necessary to employ a further filter consisting of a choke and smoothing capacitor as in *Figure 7.8b*. The first capacitor is termed the reservoir capacitor, and the whole filter arrangement is known as a capacitor input filter.

Filter inductors have already been briefly mentioned in Chapter 2. There, it was pointed out that in order to achieve the necessary high inductance, insulated iron laminations are used as the core material. If the flux density in the core due to the d.c. greatly exceeds that due to the a.c., as is commonly the case, the unwanted d.c. magnetization is reduced by inserting an air gap in the core. This results in a useful increase in the inductance. There is an optimum length of air gap giving the maximum effective permeability for any particular value of d.c. magnetizing current and a.c. flux density. When this optimum gap is used the inductance is almost constant over the range of currents encountered.

For a wide range of load variation, it would be very uneconomical to design and use a choke having a constant inductance at all load currents. To meet this case, inductors are designed to have a specified inductance at some large value of d.c. and a rapidly rising inductance as the d.c. is reduced. Such an inductor is known as a swinging choke. To achieve the desired variation of inductance the air gap is reduced, or dispensed with, the core being continuous. In this latter case, the laminations are inserted in an interleaved arrangement as in the mains transformer.

Capacitor-input filters cause the load current to be delivered by the rectifier in large pulses of short duration. Two disadvantages ensue: firstly, the pulses may be too great for the rectifier; secondly, the losses in a thermionic valve rectifier are considerably higher for heavy pulses than for steady currents of the same mean value. The first disadvantage may be overcome by including limiter resistors in the rectifier lines (in the anode circuit of thermionic rectifiers); power is inevitably lost in the resistors. Because of the second disadvantage the power rating of the rectifiers is reduced. Capacitor-input filters are therefore used only in low-power equipment.

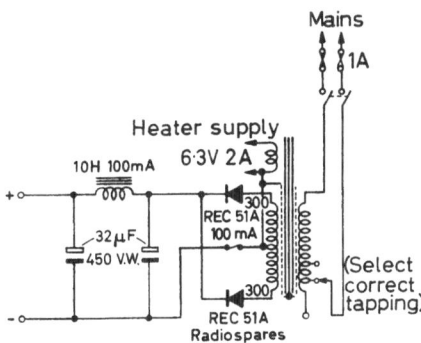

Figure 7.9. A simple power supply to give about 300 V at 80 mA. The values are given as an example. For different outputs a different mains transformer specification will be required (see Appendix 3)

Figure 7.9 shows a complete power pack intended to deliver about 80 mA at 300 V (see also Appendix III). Practical precautions such as the fuses and the double-pole switch in the primary line (to give complete isolation from the mains when the switch is in the 'off' position) should not be overlooked.

175

Choke-input Filters

When high powers are required the choke-input filter of *Figure 7.8c* is used, and then there is a continuous current, each diode conducting in turn throughout half a cycle. (In practice fullwave rectifiers are used.) The maximum current is little greater than the mean current. For continuous current it can be shown that the inductance of the choke must be greater than $R_L/6\pi f$ where R_L is the load resistance and f the supply frequency.

REGULATION CURVES

Regulation curves are graphs showing the output voltage of a rectifier circuit as a function of the load current. Typical examples are shown in *Figure 7.10*. The capacitor-input filter case shows considerable variation of

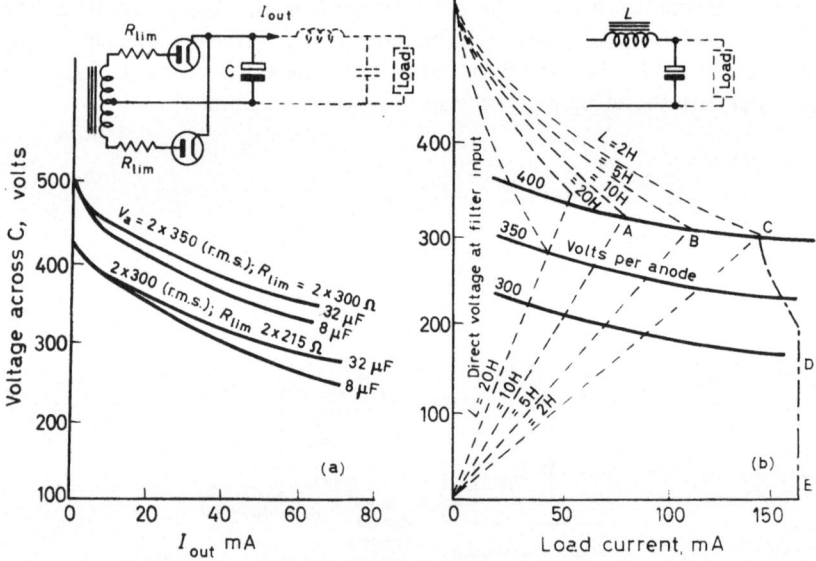

Figure 7.10. Regulation curves (*a*) using a capacitor input and (*b*) a choke input filter

output voltage with load current. The regulation is not so good as when a choke-input filter is used. In fact, if the choke in the latter case had an infinite inductance the regulation would be perfect, but practical chokes depart from this ideal. Manufacturers of rectifiers publish graphs such as these in order to assist with design. Their curves take into consideration rectifier and average transformer losses, and it should be noted that the results given for any particular rectifier are only correct for the set of conditions for which

176

the curves were derived. *Figure 7.10a* is self-explanatory, but *Figure 7.10b* may require some explanation. When the value of the r.m.s. voltage to the rectifier anodes is known a selection of the appropriate solid line curve can be made. The maximum current that can be drawn with safety is that indicated by the intersection of the solid line with the limiting boundary CDE. The minimum current that is likely to be drawn dictates which of the broken lines radiating from the origin must be selected. This determines the minimum inductance that must be used. For example, when the r.m.s. voltage to each anode is 350 V, the maximum current that can be drawn is 150 mA. If it is known that the minimum current likely to be drawn is 60 mA, then the intersection of the solid line representing 350 V and the ordinate for 60 mA is near to the intersection of the solid line and the broken line from the origin representing 10 H. This value of inductance is the minimum that can be used. In practice it is advisable to select a choke with an inductance somewhat higher than the minimum required.

VOLTAGE AND CURRENT STABILIZATION

There are numerous occasions when it is necessary to use a power supply which gives a steady output voltage that is not affected by mains fluctuations or load current variations.

The simplest form of voltage stabilizer is the cold-cathode gas-filled discharge tube. Such a tube consists of a cathode (of relatively large area) and anode structure in an envelope containing gases at low pressure. A gas frequently used is neon at a pressure of the order of millimetres of mercury. About 0·1 per cent of argon is added to reduce the striking voltage. The characteristic of such a tube is shown in *Figure 7.11*. As the applied voltage is raised the current slowly increases.

Although discharge is taking place the current is very small, and there is no visible effect. Eventually the striking voltage is reached and, at this potential difference, the ions and electrons already present gain sufficient energy to cause a breakdown of the gas. When a gas breaks down it glows. Many of its molecules must therefore be excited by collision with electrons, and in addition many others are ionized. The voltage necessary to sustain the discharge is less than that required to initiate it. Prior to breakdown the current is of the order of 10^{-12} A, but when the striking voltage is reached the current suddenly jumps to about 10^{-6} A. In the normal glow region there is a range of current over which the anode voltage is almost constant. For low values of tube current only a portion of the cathode is covered by the glow, but as the current increases the area of this portion increases proportionately, and thus the current density near to the cathode is constant. Eventually the whole of the cathode is covered by the glow; thereafter any further increase in the current causes the tube to be operated in the abnormal glow region. This region is of no concern in voltage stabilization, since precautions are taken to avoid it.

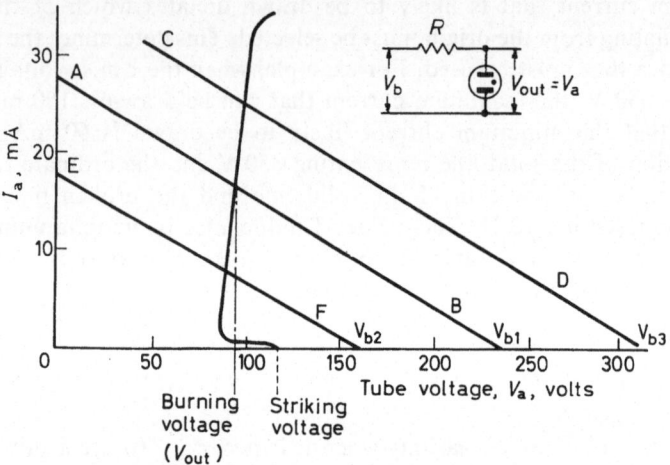

Figure 7.11. Characteristics of a gas-discharge stabilizing tube

Load lines have been drawn in the usual way to illustrate the stabilizing action. The slope of the line is $-1/R$. For a given supply voltage, V_{b1}, a suitable resistance is chosen. This fixes the load line AB. The voltage across the tube is $V_a (= V_{out})$. If the supply voltage now rises to V_{b3} or falls to V_{b2} the load line shifts to CD or EF respectively. The current through R changes but V_{out} remains almost constant.

The position is somewhat altered when the stabilized output is not merely a voltage reference source, but is required to feed a load. The current through R is now $I_a + I_L$ where I_L is the load current in a load R_L. Since V_{out} remains almost constant, any change in the load current will be offset by an opposite change in the stabilizing tube, i.e. $(I_a + I_L)$ remains constant for a given V_b. The value for R (which incidentally includes the internal resistance of the unstabilized supply) is less than that required when only a reference output voltage is required. Also the range of the supply variation over which stabilization is effective is somewhat less than that for the case shown in *Figure 7.11*.

In practice the output voltage, V_{out}, and the supply voltage, V_b, are not usually variable at will; their values are fixed by the prevailing circumstances. The problem then is to decide on the optimum value of R to be used. A compromise must be reached between the optimum value to stabilize

178

against variations of supply voltage and the optimum value for load current variations. Additionally, for a permanently connected load it must be ensured that sufficient voltage is available at the tube electrodes to guarantee striking.

If the required output voltage is higher than the maintaining or burning voltages of the tubes available, two or more may be used in series. To assist striking, shunt resistors should be connected as shown in *Figure 7.12a*. For lower voltages the cathode follower circuit of *Figure 7.12b* may be used.

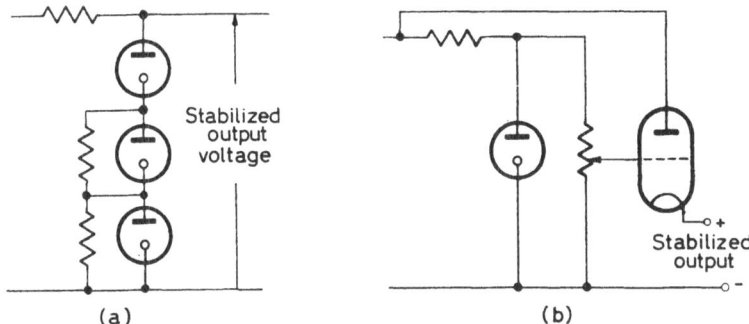

(a)   (b)

Figure 7.12. (*a*) shows a method for stabilizing a higher voltage than any tube will handle individually, while (*b*) shows how to obtain a lower stabilized output voltage from a given tube without vitiating the performance

The variations of burning voltage in any given tube may be as high as five or ten per cent depending upon the tube and the working current range. Hysteresis[2] and step effects[2], drift and temperature effects[2] combine to make the simple glow-discharge tube stabilizer unsuitable for applications requiring a high degree of stability. More complicated electronic circuitry, as described below, is then required.

Zener Diode Stabilizers

The solid-state equivalent of the gas-discharge tube stabilizer is the zener diode. The main characteristics and operation of the device have already been discussed in Chapter 5. By applying sufficient reverse bias voltage, an avalanche condition is reached at which point there is a sudden and substantial increase in the reverse current. Thereafter very small increases in reverse bias voltage cause large increases of current. In effect large current variations are possible, the voltage across the device remaining almost constant. In typical silicon diodes the current may vary over a range of 5–200 mA for a voltage variation of one or two hundred millivolts.

12*   179

Zener diodes are manufactured in a range of preferred voltages (e.g. 4·7 V, 6·8 V, 18 V, 39 V, 47 V, etc.), although intermediate values are also available. Three quantities need to be known about the device, namely, the wattage rating, the 'breakdown' or zener voltage and the tolerance. For close tolerance work a knowledge of the temperature coefficient may also be required. 10%, 5% and 1% are typical tolerance figures; the zener voltage may vary, in silicon devices, over a range of a few tens of millivolts for temperature variations from −50°C to about 100°C. Wattage ratings may be from a fraction of a watt to a few tens of watts.

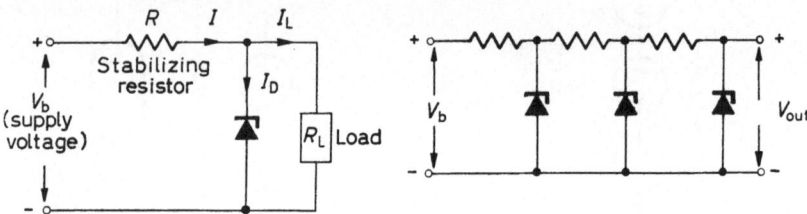

Figure 7.13. Stabilization using zener diodes. The left-hand diagram shows the basic shunt regulator while the right-hand diagram shows how improved regulation is obtained

A simple voltage regulator is shown in *Figure 7.13.* The sum of the load and regulator currents is constant, so that increases of load current are accompanied by corresponding decreases of regulator current. A given zener diode may have an operating range of 5 mA to 200 mA, the upper limit being determined by the wattage dissipation of the device. This is also the range over which the load current may vary. To calculate the value of the stabilizing resistor we need to know *(a)* the supply voltage *(b)* the desired stabilized voltage and *(c)* the load current with its likely current variations. As an example the load current may be taken to be 150 mA, and the maximum variation to be +30 mA i.e. from 150 mA to 180 mA. After selecting a diode with a suitable zener voltage and adequate power dissipation, it is noted that the minimum zener diode current for stable operation is 5 mA. The diode working current is 35 mA (5+30 mA variation in the load). The total current is therefore 150+35 i.e. 185 mA. If the supply voltage is 30 volts and the zener voltage 11·5 V then the voltage drop across the load resistor must be 18·5 V. The value of the stabilizing resistor is therefore $18·5/185 \times 10^{-3} = 100$ ohms. The wattage rating of the resistor is $18·5 \times \times 185 \times 10^{-3} = 3·4$ W. In practice a 5 W resistor would be used to give a margin of safety.

HIGH-VOLTAGE STABILIZED SUPPLIES

When a higher degree of high-voltage stabilization is required than is possible using a gas-discharge tube, it is necessary to use more complicated electronic circuitry. The principle of a regulator giving a very stable output voltage is illustrated in *Figure 7.14.*

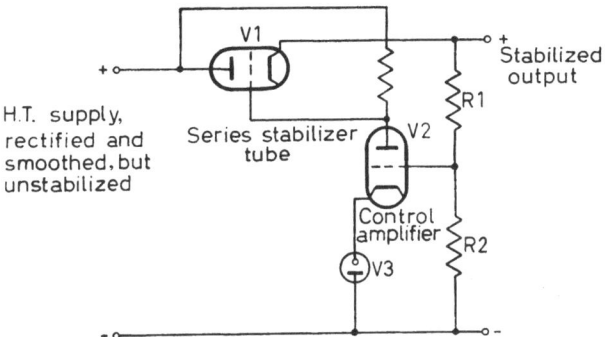

Figure 7.14. The principles of obtaining a stabilized high-voltage supply

The rectified and smoothed d.c. is fed to the anode of the series tube V1, and the output is taken from the cathode. Any change in output voltage is fed to the input of a directly coupled control amplifier, the output of which is used to alter the impedance of the series tube by changing its grid bias voltage. The phases of the amplifier input and output voltages are such that any drop in the stabilizer output voltage is compensated by a reduction in impedance of the series tube. This in turn reduces the voltage drop across the tube and returns the output voltage almost to its original value. A rise in output voltage is compensated in a similar manner.

The amplifier input is taken from a potential divider R1 and R2. These resistors are required because some h.t. must be available for the control valve, and consequently the grid and cathode have to be a good deal negative with respect to the positive line. The cathode voltage is fixed by the reference tube V3. This tube is a cold-cathode glow-discharge tube of the type already discussed. Special tubes have been developed for the V3 position; they are constructed in a way that minimizes the difficulties previously mentioned. The role of V3 is to keep the cathode voltage fixed irrespective of variations of output voltage. Let us suppose that the tube V3 were replaced by a resistor. Changes of output voltage lead to changes of current in the control valve V2. This in turn leads to an undesirable change in the voltage across the cathode resistor replacing V3. For example, if the output voltage increased, the voltage on the grid of V2 would increase, but so would the cathode voltage because V2 would draw greater current. The control valve would therefore be less effective than it would be when a reference tube is used.

A practical circuit based on *Figure 7.14* is shown in *Figure 7.15*. The circuit is capable of providing a high degree of stabilization over the range 150–300 V at 60 mA. The output current may be increased to 120 mA by using a matched pair of EL37's in parallel. The output current is limited at

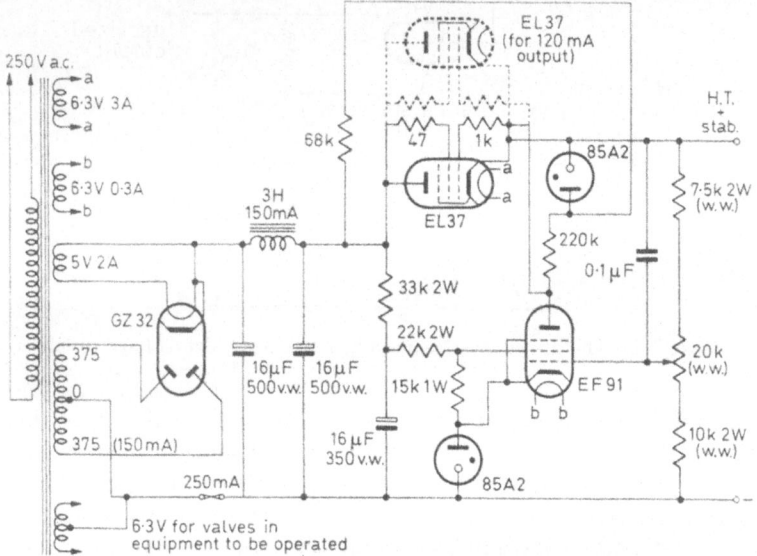

Figure 7.15. Voltage-regulated power supply to give 150–300 V at 60 mA. By using two EL37's the current output may be increased to 120 mA

low voltage outputs by the maximum anode dissipation of the series valve. When the output voltage approaches 300 V the limit is set by the onset of grid current in the EL37 series valve. Stability from minimum to maximum load current, at an output voltage of 250 V, is better than 0·5 V and, under these conditions, the hum output is less than 20 mV. The inclusion of the 0·1 μF capacitor between the control valve grid and the h.t. positive line ensures that no hum voltage is fed to the amplifier. The lead to the control valve grid may be screened by using a coaxial cable.

The circuit of *Figure 7.15* cannot give a variable voltage down to zero volts. Inspection of the circuit will show that insufficient voltage would be available for the control amplifier at low output voltages and control would therefore be lost. When the voltage must vary from zero to some hundreds of volts the

182

circuit of *Figure 7.16* may be used. A separate negative voltage supply, stabilized by gas-discharge tubes, is available for the control amplifier valve. The function of V7, also a gas-discharge tube, is to ensure maximum gain in the control amplifier. The anode load of the control valve should be as large

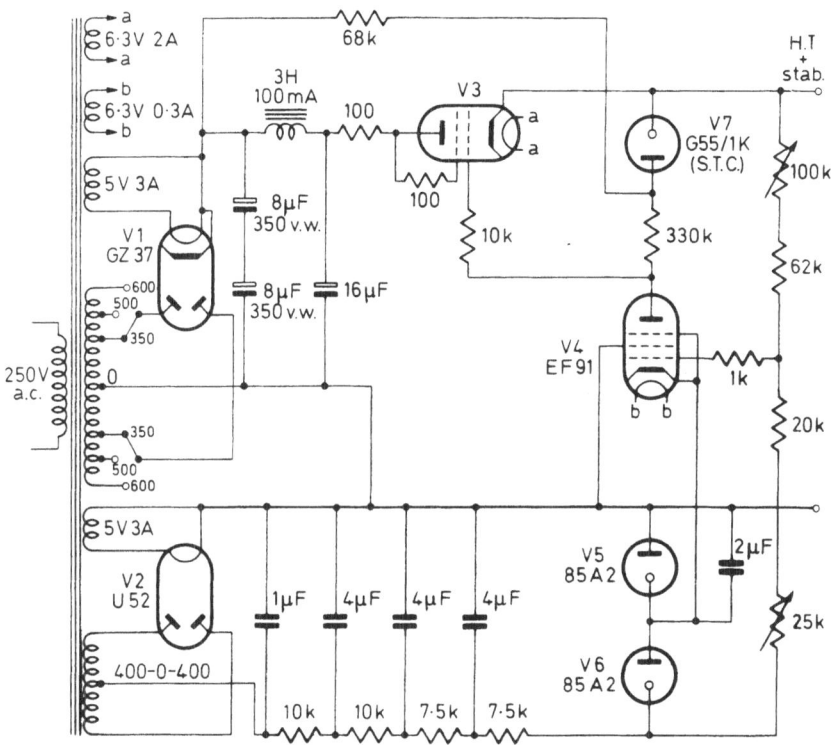

Figure 7.16. A variable stabilized supply to give 0–400 V at 100 mA in three stages; 0–100 V, 100–250 V and 250–400 V

as possible for maximum gain; a limit is reached when the anode-cathode voltage of the control valve becomes too small. The method shown of using a gas-discharge stabilizing tube enables the anode voltage of the control valve to remain high even when the output voltage is zero.

Unstabilized Variable High-voltage Supply

When it is necessary to have an h.t. supply with a variable output voltage, it is not necessary in many cases to use an expensive stabilized supply. Often workers use such a supply because it is the only one available with a variable output. The simple circuit of *Figure 7.17* is worth building so that expensive stabilized supplies may be reserved for work requiring stable output voltages.

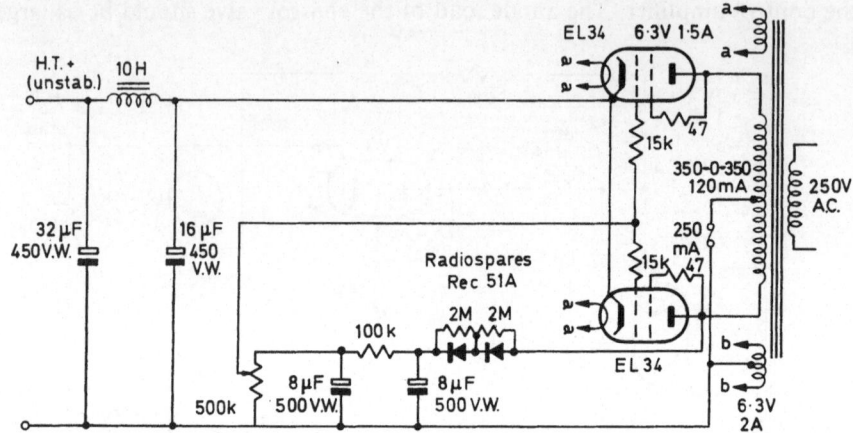

Figure 7.17. A power pack with a variable output voltage from about 20 to 350 V. Both limits of voltage vary depending upon the current drawn by the load. This current should be limited to about 100 mA. The 2 M resistors across the silicon rectifiers equalize reverse-voltage sharing

The two EL34's are being used as rectifiers, but also behave as cathode followers. The cathode voltage cannot rise above the voltage of the grid, and this voltage is set by the 500 k potential divider.

LOW-VOLTAGE SUPPLIES

The ever-increasing use of transistorized equipment has meant the development of low-voltage supplies operated from the mains. The currents required may be only a few milliamps for some equipment, but many items of transistor apparatus need currents of the order of amperes. Such currents are not conveniently obtained from battery sources.

The principles of obtaining unregulated supplies are the same as those used for thermionic valves. The usual arrangement is to supply a bridge rectifier from a low-voltage secondary winding on a mains transformer. The smoothing follows conventional circuitry except that the capacitors have values of several thousands of microfarads. *Figure 7.18* shows a straightforward unregulated power supply with an emitter follower to obtain a variable voltage output.

When stabilized output voltages are required the simple zener diode shunt regulator already described may be used. An improved circuit is needed

184

when the load current varies over several amperes. One version is shown in *Figure 7.19*. It is similar in principle to *Figure 7.18* except that the fixed resistor in the chain feeding the base is replaced by zener diodes. The output voltage between the transistor emitter and the positive line depends upon the voltage drop across the transistor and hence the voltage between the

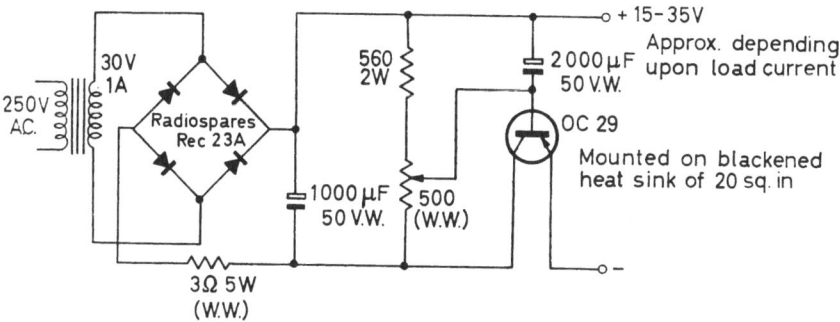

Figure 7.18. Simple low-voltage supply with variable output voltage

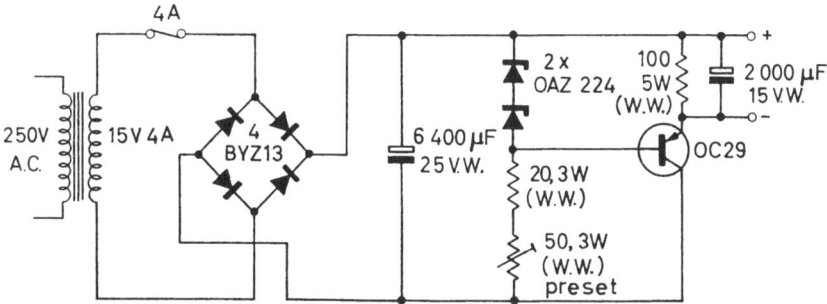

Figure 7.19. Stabilized transistor power supply giving an output current of 0–3 A at a nominal 13 V. At zero output current the voltage is 13·5 V and at 3 A output current the voltage is 12·5. The transistor must be mounted on a blackened heat sink of 22 sq. in.

base and emitter leads. The base-emitter voltage is the difference between the output voltage and the zener voltage.

When a higher degree of stabilization is required the circuit of *Figure 7.20* may be used. Readers will see that the principle of stabilization is the same as that illustrated in *Figure 7.14*. An additional advantage of the circuit is the automatic protection provided if the output should be accidentally short-circuited. When the output falls to zero the zener diode and OC200 become inoperative. The OC35 is starved of base current and thus the current in the short-circuit is limited to a safe value.

185

A precision transistor voltage regulator with inherent short-circuit protection is described by Marshall[4]. The output is 14 V at 300 mA. The output

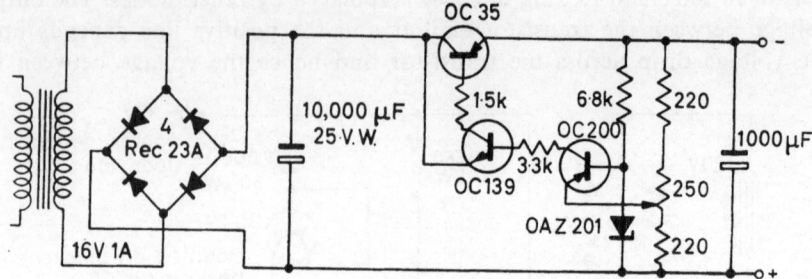

Figure 7.20. A series stabilizer circuit with automatic protection against accidental short-circuit (Pallett[3])

changes by less than 0·2 per cent for the ambient temperature range 20 to 70°C and by less than 0·01 per cent with an input voltage change of 18 to 36 V.

EXTREMELY HIGH-VOLTAGE (E.H.T.) SUPPLIES

When it is required to have a higher direct voltage supply from a given a.c. source than is possible from the types of circuits already discussed, and where, for reasons of weight and economy, it is undesirable to use a large transformer, voltage multiplying circuits may be used. An example of a voltage doubler is shown in *Figure 7.21a*. After steady-state conditions have

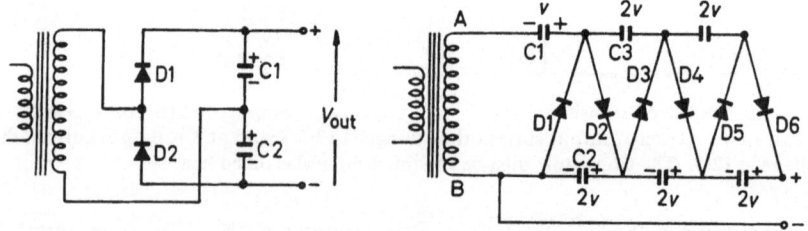

(a) The voltage doubler circuit   (b) The Cockroft-Walton type voltage multiplier

Figure 7.21

been achieved the action proceeds as follows: during one half cycle the diode D1 conducts and C1 charges almost to the peak value of the applied voltage if the load current is low; during the second half cycle D2 conducts and C2 is charged. The polarities are as shown with the result that V_{out} is nearly twice the peak value of the applied voltage. The regulation of such a system

186

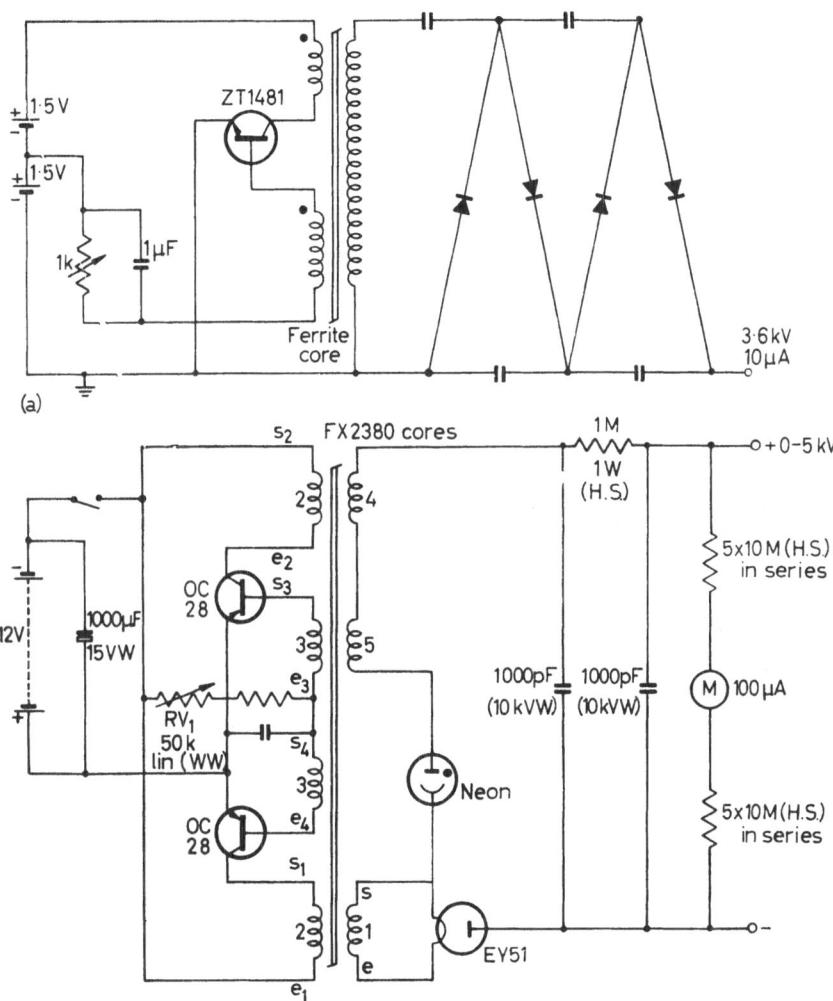

Figure 7.22. Two forms of e.h.t. supply that use high frequency oscillations. The transformer details for (*a*) are as follows:

Core	Ferroxcube FX2277	
Primary	150 turns	28 s.w.g. enamelled cotton wire
Feedback winding	20 turns	28 s.w.g. enamelled cotton wire
Secondary	6200 turns	45 s.w.g. enamelled cotton wire

The primary is wound in three layers with the feedback winding over it. The secondary is then wound on in 70 layers, each layer being interleaved with 0·002 in. thick paper

187

is poor. For higher voltages the circuit of *Figure 7.21b* may be used. With this circuit output voltages of several kilovolts are obtainable. The transformer and diodes 'pump' the charges up the capacitor chain, the final output voltage being the sum of the voltages across the capacitors in any one line.

When B is positive with respect to A, D1 conducts and C1 is charged to approximately the peak secondary voltage, v. During the next half cycle D2 conducts and C2 is charged by a voltage v from the transformer assisted by a voltage v from C1. After a few cycles C2 charges to a voltage $2v$. Whenever B is positive with respect to A, D1 and D3 are forward biased. The charge on C2 is shared with C3. Eventually C3 acquires a potential of $2v$. This occurs right up the chain, the odd numbered diodes conducting during one half cycle and the even numbered diodes conducting during the next half-cycle. All capacitors, therefore, acquire a potential difference of $2v$ across their terminals except C1, which is charged to v. The Westinghouse Brake and Signal Co. Ltd. are a firm who will supply suitable rectifiers as a kit.

For laboratory work e.h.t. supplies can be dangerous; the danger may be averted if the e.h.t. is developed from a radiofrequency oscillator. When overloads are placed on the supply, oscillations cease and the voltage collapses. The most likely use for a supply of this kind is to operate cathode-ray tubes, Geiger tubes and photomultipliers. Flyback systems have been used for several years in developing the e.h.t. for the picture-tube in television receivers. Here the line output oscillator is used as a source of high-frequency power. For laboratory use however an oscillator must be made for the purpose. *Figures 7.22a* and *7.22b* show circuits for high frequency e.h.t. supplies. In *Figure 7.22a* a transistor oscillator is used to energize the transformer. The secondary voltage is multiplied by the Cockcroft-Walton principle. The circuit is given and discussed in Ferranti Application Note No. 20 ('Silicon Rectifiers in E.H.T. Supplies').

The circuit of *Figure 7.22b* provides a fairly safe e.h.t. voltage variable from 0 to 5 kV. The short circuit output current is less than 2 mA. Two power transistors are connected in a push-pull converter arrangement operating from a 12 V supply at an average current of 500 mA. The transformer, operating at about 2 kHz, steps up the voltage which is then half-wave rectified and smoothed in a conventional manner. The neon bulb fires as soon as the oscillator comes into operation and is thus a warning that e.h.t. voltages are being developed. A discussion of the circuit is given in the Mullard publication ('A Variable E.H.T. Supply Unit' — Educational Electronic Experiments No. 4.).

REFERENCES

1. Dalfonso, J. L. 'Miniaturisation and the primary cell'. *Electrical Manufacture.* March, 1962.
2. Benson, F. A. *Voltage Stabilized Supplies.* Chapter 3. Macdonald, London, 1957.
3. Pallett, J. E. Letter to Editor of *Electronic Engineering*, 1963, **35**, No. 419, 46.
4. Marshall, R. C. A transistor voltage regulator with inherent short circuit protection. *Electronic Engineering*, 1963, **35**, No. 420, 106 (February).

Suggestions for Further Reading

Patchett, G. N. *Automatic Voltage Rectifiers and Stabilizers.* Pitman, 1954. See also *Electronic Engineering*, September–December, 1950.

Dzierzynski, O. E. 'Stabilization of a.c. supplies'. *Wireless World*, 1957, **63** No. 10, 491 (October).

Scroggie, M. G. 'Stabilized power supplies'. *Wireless World*, 1948, **54** (October, November and December).

THYRATRONS AND SILICON CONTROLLED RECTIFIERS (SCRs)

When larger amounts of power are required than can conveniently be drawn from the supplies discussed in the previous chapter, the rectifiers described there can be replaced by gas-filled diodes. Such diodes have a robust electrode structure and are filled with an inert gas such as argon, helium or xenon; alternatively, the tube may have mercury added, and thus becomes filled with mercury vapour when the cathode has reached the correct operating temperature. During operation, once the anode voltage has exceeded the ionization potential of the gas, very large amounts of current can be conducted by the tube. The current consists of the primary electrons from the cathode together with the considerable number of secondary electrons and positive ions present in the ionized gas.

In principle, the operation of the rectifying circuits is the same as those previously discussed. The only precaution to be observed is to allow sufficient time for the cathode and gas contents to reach the correct operating temperature. Failure to do so adversely affects the life of the cathode. This is because insufficient ionization in the initial warming-up period leads to excessive anode-to-cathode voltages. The positive ions that are present are then accelerated to the cathode and strike the surface with destructive force.

Frequently it is necessary to control the large amounts of power involved, for example, when heating a furnace or bath of electrolyte to some specified temperature. The gas-filled diode is then replaced by a tube called a thyratron.

THYRATRONS

These are gas-filled discharge tubes that contain a specially constructed cathode, one or more control electrodes and an anode. The cathode is a robust electrode with a large surface area to provide the considerable currents involved. Only in the smallest low-power thyratrons is the cathode of the type used in ordinary valves. Quite often, in the larger tubes, radiation shields are fitted to maintain the cathode temperature by restricting radiation losses. The control electrodes are tubes, or metal rings, as shown in *Figure 8.1*. These electrodes are known as grids since they perform a controlling function. The term is borrowed from high-vacuum amplifying valves, but, as we shall see, the control characteristic is very different from that of the wire grid in an amplifying tube. The presence of the grid modifies the characteristics considerably from those of the corresponding gas-filled diode.

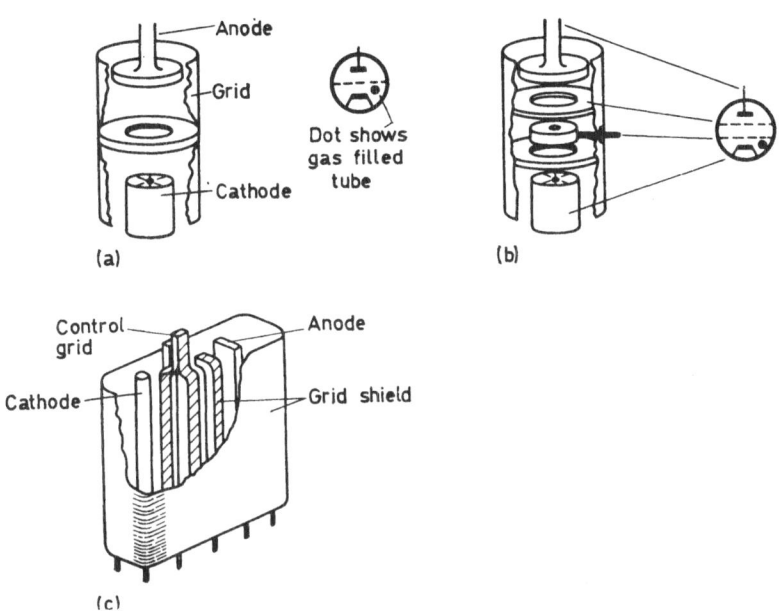

Figure 8.1. Constructional features of triode and tetrode thyratrons of medium power. (*a*) Electrode structure of triode thyratron. The grid is cut away to show the inner structure. (*b*) Electrode structure of a tetrode. (*c*) Structure of a small low-power thyratron of the EN91 or 2D21 type

Since the grid of a thyratron almost completely shields the cathode from the anode, the application to the grid of a small negative voltage is effective in inhibiting ionization. The small negative grid voltage nullifies the effect of a relatively large positive anode voltage. The voltage gradient near the cathode is thus too small to accelerate the electrons to a velocity sufficient to cause ionization when collision with a gas molecule occurs. In this way the grid is able to determine the anode voltage at which breakdown occurs. As soon as conduction does occur, a sheath or plasma of positive ions smothers the grid. The grid then no longer has any control whatever over the anode current, which is determined solely by the resistance of the external anode circuit and the anode supply voltage. The only way in which the tube can be extinguished is by reducing the supply voltage to zero or breaking the anode circuit by means of a switch. Because it takes about 10 milliseconds for the ions in the gas of a mercury filled tube to recombine (i.e. the deioniz- ation time is about 10 msec) the anode voltage must remain below the ioniza- tion potential for at least this period.

The characteristics of a thyratron are usually measured by observing, for several different values of negative grid voltage, the positive anode voltage required to produce ionization and subsequent conduction. A plot of the critical anode voltage against the corresponding grid voltage is shown in *Figure 8.2*. The slope of the graph is known as the control ratio. To explain

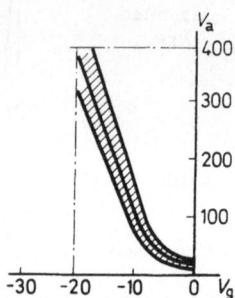

Figure 8.2. Control characteristic for a thyratron with a control ratio of 30. The shaded area represents the region over which the characteristic varies with varying ambient tem perature, varying heater current and over the life of the tube

the term 'control ratio' it may be assumed by way of an example that in a tube with a control ratio of 30, firing will not commence for a grid voltage of − 15 V until the anode voltage reaches 200 V. Had the grid voltage been − 16 V, firing would not have commenced until the anode voltage had reached a value of 230 V.

The introduction of a second grid results in what is known as a tetrode thyratron. The second grid reduces the current to the control grid. Since the latter has a large resistance in the grid circuit, the reduction of current prevents a large voltage drop across the grid resistor. Such a drop in a triode can be troublesome because it may lead to a change in the operating condi- tions or even accidental firing. By adjusting the voltage on the second grid it is possible to control to a limited degree the position of the control charac- teristic.

THYRATRON CONTROL

In many practical applications, thyratrons are used as control or rectifying elements and fed with an alternating voltage. The anode voltage is conse- quently sinusoidal or nearly so. *Figure 8.3* shows the sinusoidal applied

192

voltage. The firing may be delayed by the application of a negative voltage to the control grid. Conduction therefore occurs over a period of time which is less than half the periodic time of the supply voltage. Since the power delivered to the load depends upon the time during which conduction occurs,

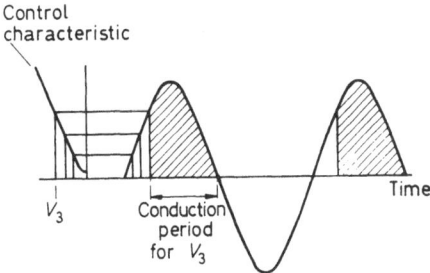

Figure 8.3. Controlling the conduction period (and hence power) by the application of varying grid voltages

it can be seen that it is possible to control the power merely by altering the grid voltage. It is inconvenient to draw the control characteristic in the way shown in *Figure 8.3*. A preferable arrangement is the diagram of *Figure 8.4* where the dotted line represents the value of negative grid voltage required to hold off firing at any corresponding anode voltage.

From *Figure 8.4* it can be seen that the application of a steady negative grid voltage has its limitations. Firing can be delayed only up to 90°. Any further increases of negative grid voltage keeps the thyratron permanently extinguished. To avoid this, two other forms of control have been devised, namely, horizontal control and vertical control.

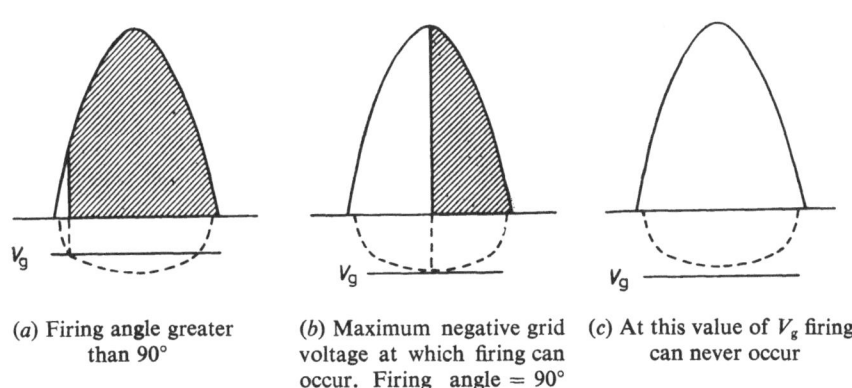

(a) Firing angle greater than 90°

(b) Maximum negative grid voltage at which firing can occur. Firing angle = 90°

(c) At this value of V_g firing can never occur

Figure 8.4. The delay in firing for various values of steady negative grid voltage. The maximum firing delay possible is 90°

Horizontal control (*Figure 8.5*) is achieved by applying to the grid a sinusoidal voltage of the same frequency and about the same amplitude as the supply voltage. The grid voltage is then shifted in phase relative to the supply voltage. The amount of phase-shift determines the firing angle and hence the time during which conduction can occur. *Figure 8.5* shows the phase-shifting of a sinusoidal grid voltage. By using this circuit it is possible to

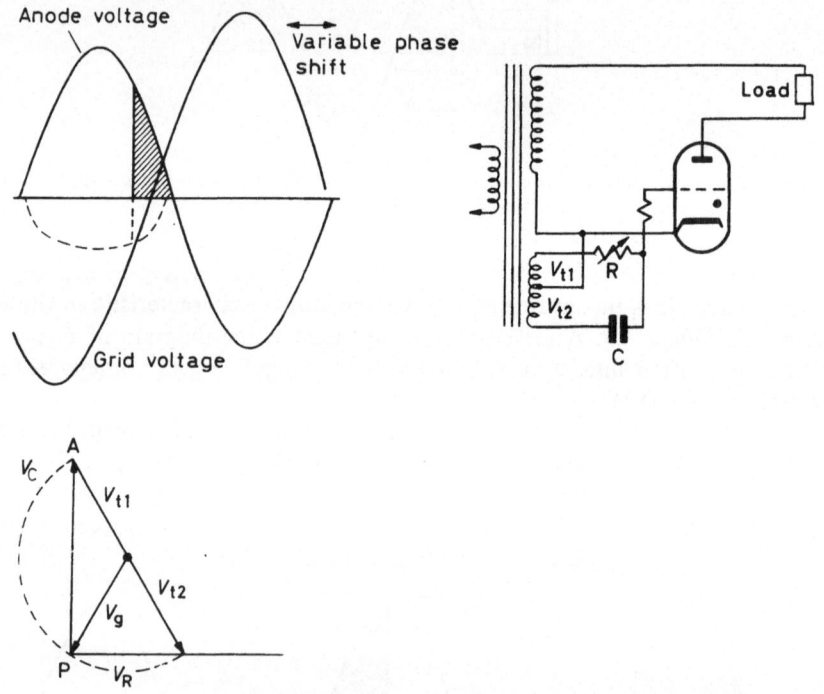

Figure 8.5. Principles of horizontal control

obtain a phase-shift of almost 180°. The phasor diagram shows how this is done. The voltages across R (V_R) and C (V_C) must always be 90° out of phase. The phasor sum of these voltages is the voltage across the whole of the secondary winding feeding the phase-shift network. When R is equal to zero $V_R = 0$ and the grid voltage is in phase with V_{t2}. Upon increasing R to a very high value $V_R \gg V_C$. The point P, therefore, moves round the semicircle towards A. As it does so V_g is shifted in phase until it is almost in phase with V_{t1} i.e. nearly 180° out of phase with V_{t2}.

194

Vertical control (*Figure 8.6*) is achieved by having a control voltage that lags the supply voltage by 90°. This control voltage is then shifted vertically by the application of various bias voltages.

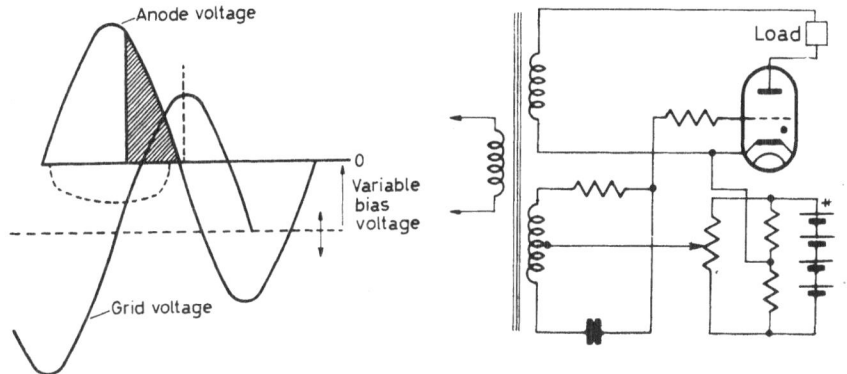

Figure 8.6. Principles of vertical control

THYRATRON APPLICATIONS

A power supply is a basic need for industrial electronic systems, heating and lighting installations, electric motors, etc. Whenever large amounts of power have to be controlled, it is usual to find thyratrons in use. (In recent years the silicon controlled rectifier, S.C.R., has displaced the thyratron, although many installations still use the thermionic device; S.C.R.s are the subject of the next section.) The basic control circuit is shown in *Figure 8.7a*. Two practical examples taken from Kretzmann's book *Industrial Electronics Handbook* (Philips Technical Library) are given. This book gives many other useful thyratron circuits.

When thyratrons are used to energize relays they may be fired, not from a phase-shifting network, but from some transducer such as a photoelectric cell. Smoke detectors, burglar and fire alarms, liquid level warning devices, and counting circuits are typical of the possible combinations of photocells and thyratrons. Some of these applications are described in Chapter 12.

THE SILICON CONTROLLED RECTIFIER (S.C.R.)

This crystal device, known also as a thyristor, is the semiconductor equivalent of the thyratron. It bears much the same relationship to the thyratron as does a transistor to a thermionic triode. Like the thyratron, the S.C.R. is an

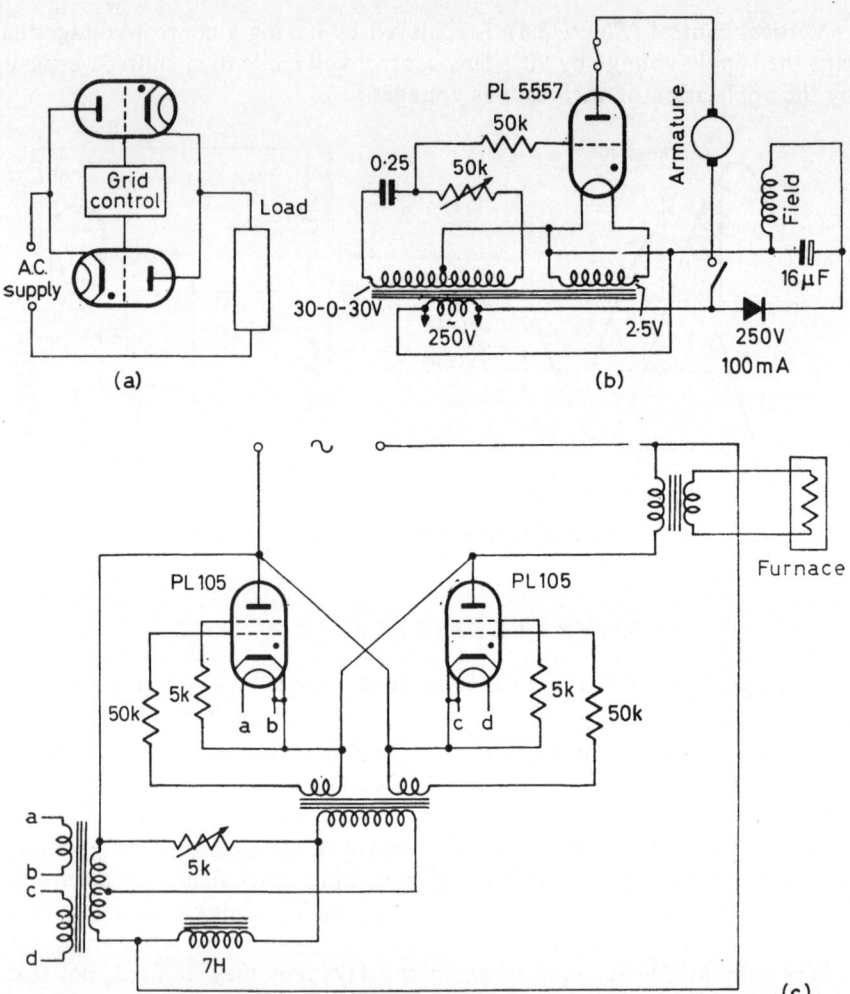

Figure 8.7. (*a*) Basic control circuit. (*b*) Simple single-phase motor control. (*c*) Simple hand-operated temperature control for a furnace consuming a few kilowatts. (*b*) and (*c*) are reproduced from Kretzmann's *Industrial Electronics Handbook* by courtesy of Philips Technical Library

'on', 'off' device. Once in the 'on' state, an S.C.R. remains in this condition so long as there is a substantial current. The control electrode, known as the gate, can only initiate firing; it cannot stop conduction because it loses all control once the main current is started. There are some differences from a thyratron however. The S.C.R. is a high-current, low-voltage device that is triggered by a small current from a signal source. The thyratron is a compara-

196

tively low-current, high-voltage device that is triggered by a small voltage from a signal source. S.C.R.s are more reliable and efficient than thyratrons. The resistance of an S.C.R. in the forward or conducting state is only a very small fraction of an ohm; the voltage drop across the device is therefore much lower than can be achieved in thyratrons. Like the transistor, the S.C.R. is immediately ready for action; it requires no warming-up period. Physically the device is much smaller and more robust than its thermionic counterpart. Powers of tens of kilowatts can easily be controlled by a comparatively small S.C.R.

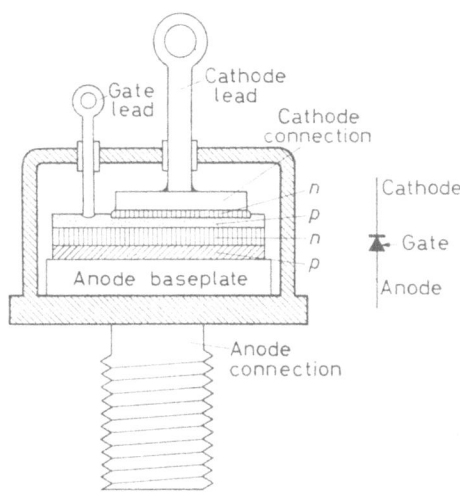

Figure 8.8. Constructional features and circuit symbol of a silicon controlled rectifier

The main constructional features of a controlled rectifier are shown in *Figure 8.8*. The S.C.R. is a three-terminal device consisting of four layers of semiconducting material in a *pnpn* sandwich. There is crystal continuity throughout the structure, and therefore three *pn* junctions exist between the four layers.

When connected in series with a load resistance and d.c. source, current is prevented from flowing in the direction which makes the anode negative. Upon reversing the polarity of the supply voltage, so that the anode is now positive, the device remains non-conducting unless, at the same time, the gate electrode is driven sufficiently positive to produce a gate current that exceeds a certain critical value. As soon as the main current is established the gate loses control, and the anode current is then determined solely by the resistance of the external circuit and the value of the anode supply voltage. As in the case of the thyratron, the current can be cut off only by switching

197

off or reversing the anode voltage, or by introducing so much resistance into the circuit that the anode current falls below a certain maintaining threshold value.

To understand the mechanism, let us consider *Figure 8.9*. When the applied voltage is such that the anode is negative and the cathode positive both the anode junction and cathode junction are reverse-biased. The S.C.R. is then in the blocked condition and its resistance is very high (about 0·2 M). When the controlled rectifier is subject to a forward voltage (i.e. anode positive) then, in the absence of any gate voltage, the gate junction is reverse-biased and conduction through the device is still prevented. In this condition the majority carriers are withdrawn from the cathode and anode junctions and thus all the applied voltage appears across the gate junction. At this junction the field across the depletion layer is high because the voltages that in practice

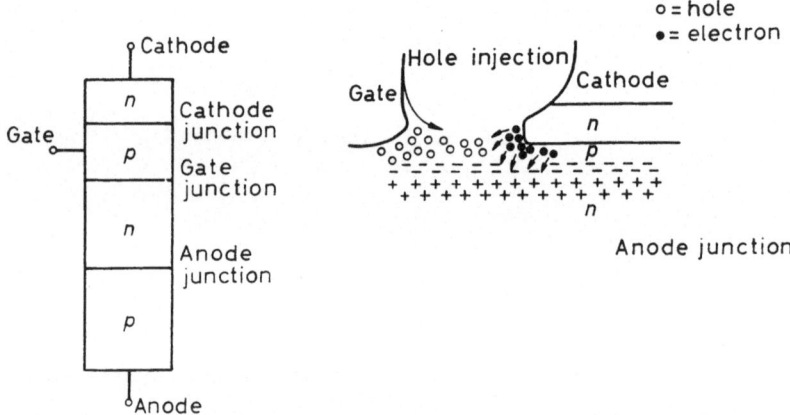

Figure 8.9. 'Firing' of a controlled rectifier

are applied to the anode are large (e.g. 20 or 30 V up to several hundred volts).

To 'fire' the rectifier a pulse of positive current must be injected into the gate. The holes injected by the gate electrode neutralize the negative space charge on the p-side of the gate junction. Because of the presence of a positive space charge on the n-side of the gate junction, the injected holes migrate along the junction. Although the gate current pulse is relatively small (10 mA of gate current may be all that is needed to switch 20 A of main current) the density of holes between the gate and cathode is high enough not only to neutralize the local space charge, but also to induce the injection of a large number of compensating electrons from the cathode. These electrons are attracted to the positive space charge on the n-side of the gate junction.

198

They gain sufficient energy to cause avalanche breakdown (see Zener Diodes; Chapter 5). This avalanche breakdown spreads right along the gate junction. The controlled rectifier now carries a large forward current. Clearly the gate can now no longer influence the conduction process. The rectifier can return to its non-conducting state only when the main carrier density is reduced below the critical value necessary for avalanche multiplication.

Operating Conditions and Firing Circuits for Controlled Rectifiers

Every manufacturer of S.C.R.s publishes, or will supply, a complete set of ratings for his devices. The potential user must ensure that these ratings are not exceeded. It is usual to compute the circuit design so that reliable

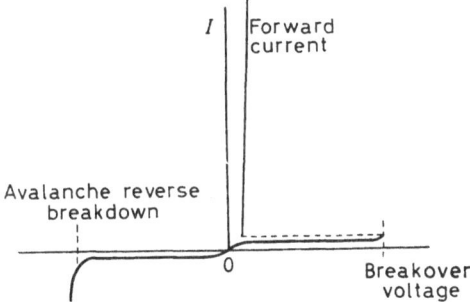

Figure 8.10. Characteristic of a controlled rectifier with zero gate signal, i.e. gate open-circuit

operation can be ensured for all rectifiers in a compatible series. (All rectifiers in a given series are rated for the same load current, but individual members of the series differ in respect of the maximum peak inverse voltages they can tolerate, and in their gate firing requirements.) Factors to bear in mind are variations in supply voltages, component tolerances, maximum forward and reverse voltages, maximum currents to be passed by the gate and main electrodes, and the operating temperatures.

A controlled rectifier that is not in a state of forward conduction has at least one *pn* junction reverse-biased no matter what the polarity of the supply voltage is. When the anode is negative with respect to the cathode the anode and cathode junctions are reverse-biased, and the leakage current is very small since a silicon device is involved. If the reverse voltage is high enough however, an avalanche condition obtains, as in normal junction diodes, leading to unwanted and uncontrolled current conduction. The characteristic in the forward direction (*Figure 8.10*) with the gate open-circuited, differs from that in the reverse direction because only one junction, namely the control junction, is reverse-biased. Increases in rectifier forward voltage mean increases in reverse voltage for the control junction. If the peak applied forward voltage is large enough an avalanche condition is induced in the control junction. Once the avalanche current is established the rectifier suddenly commences to conduct just as though a current pulse had been injected into the gate electrode. The peak forward and reverse voltages for any given rectifier must not therefore be exceeded. From a knowledge of the maximum peak forward and reverse supply voltages likely to be encountered a rectifier with adequate ratings must be chosen; this insures against unwanted or accidental conduction.

Silicon rectifiers can usually be operated in the temperature range −50 to +125°C. To prevent excessive temperature rise of the junction material, it is often necessary to mount the rectifier on a metal plate or moulding, called a heat sink. By ensuring good thermal contact with the heat sink, heat is conducted away from the rectifier, and the junctions remain within the specified temperature range. The size of the heat sink can be calculated from the published data. The sink may be no more than a small metal plate of 100 cm² area; on the other hand, heat sinks for use with devices handling load currents up to 100 A need large surface areas, and generally incorporate several cooling fins.

It is in the design of the gating circuits that difficulties may be encountered. Controlled rectifiers are liable to damage if the maximum safe gate power is exceeded. In particular, the application of large negative voltages to the gate easily leads to failure of the rectifier. More is said about this in the section dealing with the protective circuitry that is necessary when the supply voltage is alternating. The tolerance limits of voltage, current and temperature are neatly summarized in what is called a 'firing diagram' or a 'set of gate control characteristics'. The gate characteristic is, of course, the normal junction characteristic between the gate and cathode leads (*Figure 8.11a*). For any given series of controlled rectifiers, the position of the characteristic is determined by the temperature and the manufacturing spreads for different rectifiers in the group. *Figure 8.11b* shows an upper and lower limit for the position of all possible characteristics for S.C.R.s within a given series and operated within the allowed temperature range. Between the two limits, and near to the origin, there is a shaded area within which firing is possible, but not certain. The shaded area, shown enlarged in *Figure 8.11c*, is bounded by the limiting characteristics and by upper voltage and current levels. The upper voltage level is determined by the low-temperature end of the scale. Even at the lowest temperature all rectifiers in the series will fire if the applied gate voltage exceeds this upper voltage level. The upper current level is set by the highest operating temperature likely to be encountered. For increasing rectifier temperatures, smaller gate currents are required to guarantee firing. A lower voltage level is fixed by the high-temperature end of the scale. Below this level of applied gate voltage no rectifier can fire. To guarantee safe, reliable firing it must be arranged that the firing point lies within the limiting characteristic

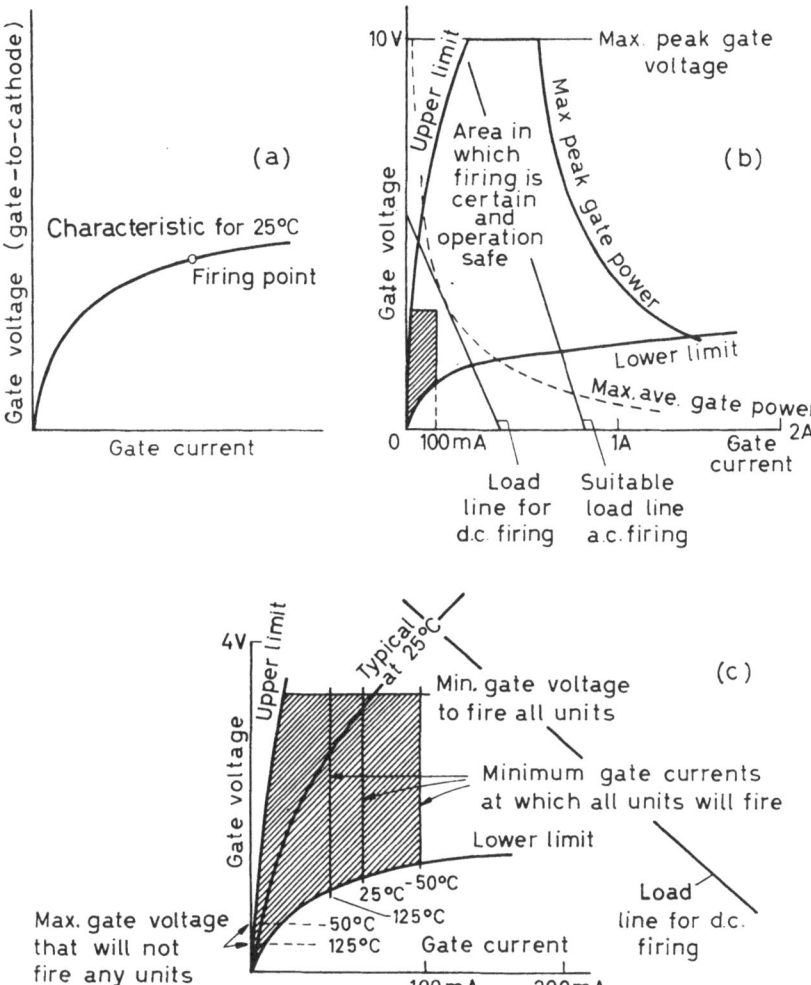

Figure 8.11. Gate characteristics and firing diagrams for S.C.R.s. (*a*) Typical gate-cathode characteristics obtained by applying a positive voltage to the gate. If a positive voltage is applied between anode and cathode and the gate voltage increased, the gate current will increase along the curve. At a certain point the device will fire. (*b*) Example of the limits of the positions of possible characteristics for a certain range of S.C.R.s. (*c*) The shaded area in (*b*) enlarged to show detail

curves and outside the shaded area. The remaining limit is set by the maximum safe gate power that can be dissipated. The position of the power parabola depends upon whether the gate power is dissipated continuously by the application of a steady voltage to the gate, or discontinuously, as when the gate is fed by sinusoidal or pulse signals. The position of the firing point can be made to lie in the area of certain safe firing by choosing a suitable supply voltage and source resistance for the gate. From the diagram we can see that about 100 mA is a suitable gate firing current, a figure that is satisfactory for many commercial rectifiers. From the S.C.R.s available at the present any particular example may need 10–150 mA of gate current to fire; the corresponding gate voltage lies in the range 3–10 V.

A basic circuit for d.c. firing is shown in *Figure 8.12a*. Although the curves for any particular S.C.R. must be consulted, we can take as an example the firing diagram of *Figure 8.11b*. A satisfactory d.c. load line is shown. (For the meaning and construction of load lines the reader should refer to Chapter 4 on amplifier design.) The load line intersects the voltage ordinate at about 6 V, and the current abscissa at about 0·4 A. A series resistor, R_S, is required of 6/0·4, i.e. 15 ohms. The gate supply voltage (V_S) is 6 V. Comparing this with *Figure 8.10c* and taking a typical characteristic, the load line intersects the characteristic at a point with approximate co-ordinates of 4 V, 100 mA. This is the operating point and corresponds to a gate dissipation of $4 \times 100 = = 400$ mW.

Using the circuit of *Figure 8.12b* it can be seen how a controlled rectifier and small switches are used to switch load currents of several tens of amperes. Switching off such a heavy load current cannot conveniently be achieved merely by breaking the supply line to the load. This would imply the use of a specially constructed switch that would be heavy and physically large. In *Figure 8.12b* a second S.C.R. is used to stop the current. By pressing the 'off' switch, S.C.R.2 fires and discharges C. For a short period a reverse voltage is applied to S.C.R.1. This rectifier is consequently extinguished, and reverts to the blocked or non-conducting state. S.C.R.2 extinguishes itself after discharging C because R is made sufficiently large to prevent sufficient current from flowing to keep S.C.R.2 in the conducting state. For all controlled rectifiers there is a threshold current below which the rectifier no longer conducts. This is because the barrier layer at the control junction re-forms at very low currents. The maintaining, or holding current, is usually in the range 10–50 mA depending upon the rectifier type. S.C.R.s used to control heavy currents in the way outlined above are often referred to as 'contactless switches'.

When the main supply voltage consists of a series of rectified sine waves from the unsmoothed output of a half- or full-wave rectifier, it is possible to control the effective power into the load by delaying the firing. Such a firing delay circuit for d.c. (unsmoothed) is shown in *Figure 8.12c*. As the supply voltage increases, the voltage at the gate does not rise immediately since it takes some time for the capacitor to charge to a sufficient voltage to fire the S.C.R. There is a phase delay between the supply and gate voltages determined by the values of C and R. Because of the non-linearity of the resistance of the gate junction, it is not possible to apply simple charging theory to calculate the time delay. The time delay is varied by varying R.

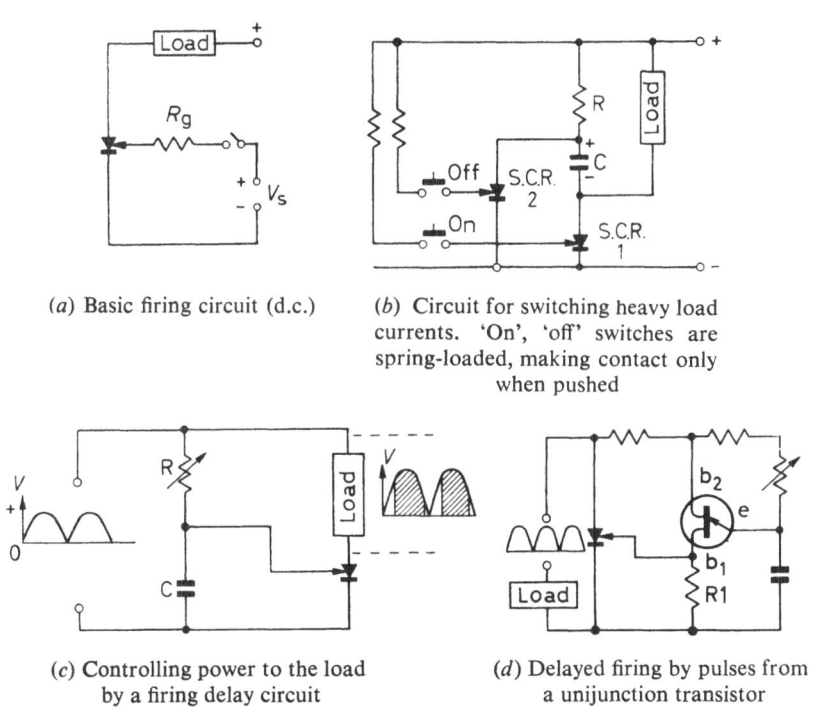

(a) Basic firing circuit (d.c.)

(b) Circuit for switching heavy load currents. 'On', 'off' switches are spring-loaded, making contact only when pushed

(c) Controlling power to the load by a firing delay circuit

(d) Delayed firing by pulses from a unijunction transistor

Figure 8.12. Basic firing circuits for d.c. supplies

In a practical circuit, it is not safe to allow R to approach zero because the current into the gate will be excessive when the supply voltage rises. For example, if the maximum rated gate current is 30 mA and the supply voltage reaches a peak of 300 V then R must be at least 10 k. Such a large resistance would prevent firing when the supply voltage is at some lower value because insufficient gate current is then available. To avoid this difficulty, and give a wider range of delay times, a unijunction transistor may be used to fire the S.C.R. The unijunction transistor (described in Chapter 5) fires the rectifier by delivering to the gate short steep current pulses. The use of pulses is the most satisfactory and reliable method of firing a controlled rectifier. The basic circuit is shown in *Figure 8.12d.* As the supply voltage rises, the voltage to the

203

emitter rises, but with a phase delay determined by R and C. For any desired firing point in the half-cycle it is possible to adjust R to give the right phase displacement and voltage across C so that at the desired time the emitter voltage is β times the voltage across the transistor (i.e. the emitter voltage is then equal to the stand-off voltage). The emitter junction becomes forward-biased and a large current is established in the emitter-base 1 region thus discharging C. As a result, a voltage pulse appears across R1 and fires the S.C.R.

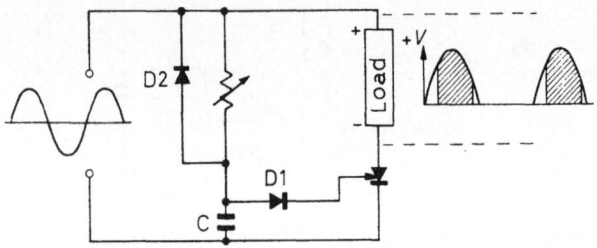

Figure 8.13a. Basic delayed firing circuit for alternating supply voltages. D1 protects gate from excessive reverse voltage when cathode is positive. D2 discharges C during non-conducting half-cycle

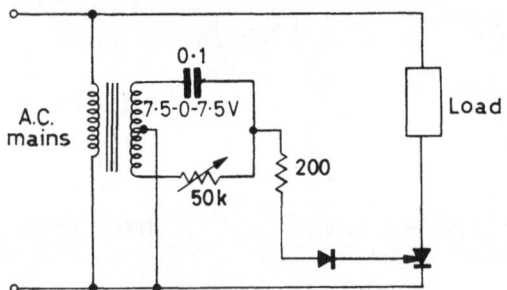

Figure 8.13b. An alternative firing delay circuit in which a phase-shifting network is used. The S.C.R. is chosen to carry the desired load current. The diode is an STC type RS220AF

From the circuit it can be seen that when pulses are absent the gate-to-cathode voltage is never large, since the resistor to B_1 is small (commonly about 50 ohms).

When the supply voltage is alternating, circuitry must be employed to protect the gate junction. A controlled rectifier is easily damaged if a large negative voltage is applied to the gate while the cathode is positive. The simplest way to protect the gate is to connect a diode in the gate lead as shown in *Figure 8.13*. Delay can still be effected with an RC arrangement, but a second diode across R is necessary to dissipate the charge received during the half

204

cycle that the anode is negative. C is thus reset in readiness for the next firing half cycle. Alternatively the phase-shifting network of *Figure 8.13b* may be used.

For full-wave operation the circuit of *Figure 8.14* may be used. Pulsed-gate operation is used to ensure reliability and satisfactory delay. The diodes D1 and D2 serve as routing diodes to ensure alternate firing of the rectifiers. Pulse delay is obtained by the phase-shifting network described in the section on thyratrons. The output from the phase delay circuit is fed into a peaking transformer, T. This has a ferrite core whose hysteresis loop is rectangular in form. When the flux in the core increases, due to the application of a sinusoidal input voltage, the core very quickly becomes saturated. Beyond saturation

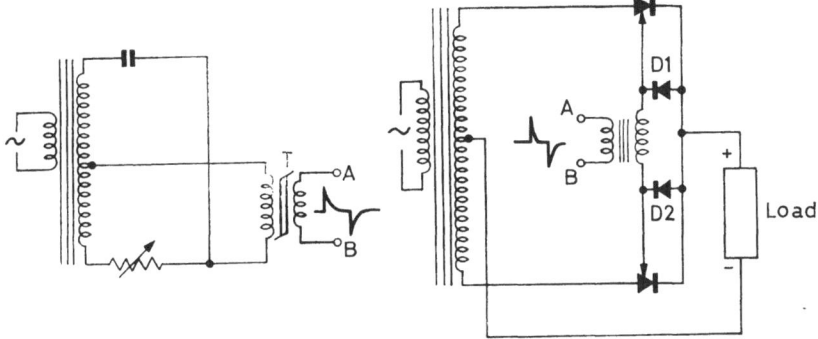

Figure 8.14. Full-wave rectifier with delayed firing facilities. T is a pulsing transformer wound on a ferrite core having square-loop hysteresis properties

there is no rate of change of flux in the core and the output voltage falls to zero. A sinusoidal input voltage thus leads to a series of voltage pulses at the secondary. It is these pulses that are used to fire the S.C.R.s.

PRACTICAL CIRCUITS USING SILICON CONTROLLED RECTIFIERS

A variable voltage power supply based on *Figure 8.13* is shown in *Figure 8.15*. It is a full-wave rectifier circuit in which the circuit of *Figure 8.13* is doubled so that conduction occurs during successive half-cycles. The additional diodes D5 and D6 are effective in isolating the capacitors C1 and C2. These capacitors must have a large value if a reasonable range of voltage variation is required. Since it is convenient on grounds of physical size to use electrolytics, a pair of capacitors connected back-to-back must be used in each position; the combination is then non-polar. Component values in the circuit are given for one pair of S.C.R.s. The circuit can, however, be used to give different outputs provided suitable modifications are made. For a different output from that shown in *Figure 8.15* a suitable pair of rectifiers is chosen that can deliver the required current. In the modified circuit due attention

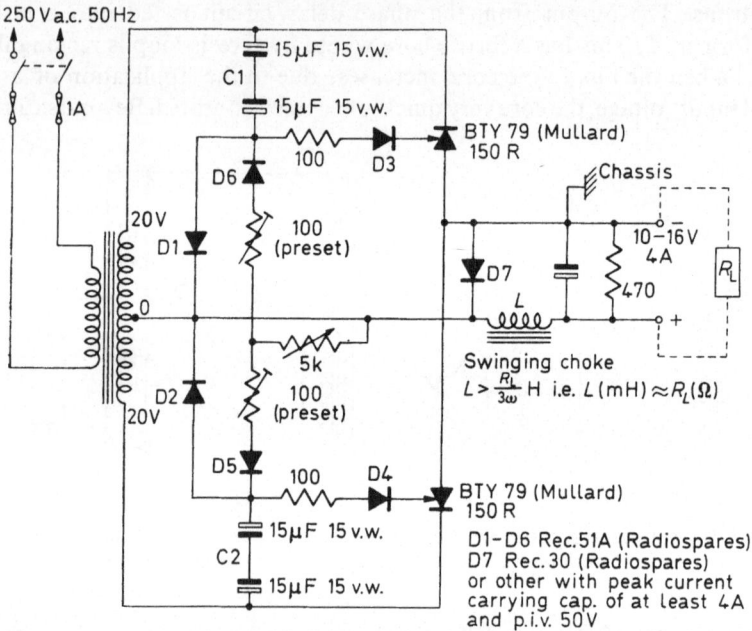

Figure 8.15. Full-wave power supply with voltage variation. The S.R.C.s must be mounted on an adequate heat sink (e.g. the chassis if the negative line may be earthed). D7 maintains the current in the choke when both S.C.R.s are non-conducting. This ensures true choke input conditions throughout the cycle

must be paid to the maximum reverse voltage rating. The gate power must be limited to a value within the manufacturer's stated limit. The circuit shown can safely fire many S.C.R.s in common use. The preset resistors may need adjusting if it is found that one other of the rectifiers fails to fire or fires erratically. Erratic firing may be encountered at low outputs.

Circuits for regulated supplies (i.e. supplies whose output does not vary with load current) are published in several journals. Two circuits for delivering 28 V at 4 A as a regulated output are described by F. Butler (*Wireless World*, Feb. 1963, p. 56).

A circuit published by International Rectifiers is useful as a self-controlled battery charger in laboratories that use rechargeable batteries (*Figure 8.16*). Most of the circuit should be easily understood from previous descriptions. The automatic overload protection is provided by a zener diode. By an appropriate setting of R1, a fraction of the desired final battery voltage (equal to the zener diode voltage) is available at the slider terminal. For all battery volt-

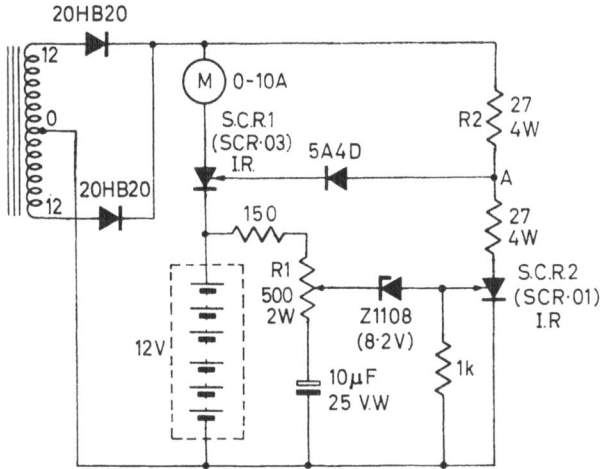

Figure 8.16. A battery charger with automatic overload protection. All the diodes and rectifiers are by International Rectifier Corporation

ages that make the slider potential less than the zener voltage, the zener diodes does not conduct and S.C.R.2 is not fired. As soon as overcharging commences, the battery voltage becomes slightly in excess of the final desired voltage; the zener diode then conducts and S.C.R.2 fires. S.C.R.2 conducts during each successive half-cycle pulse from the rectifier. The current through R2 is then large enough to cause a substantial voltage drop across this resistor, and the potential at A never rises to a sufficiently high value for S.C.R.1 to fire. If the battery voltage falls the zener diode no longer conducts; S.C.R.2 cannot then be fired. The voltage at A is then high enough to allows S.C.R.1 to fire, and charging of the battery recommences.

A circuit which uses an S.C.R. as an overload protection device is shown in *Figure 8.17*. Short-circuit protection is an attractive feature in any form of power supply, especially when it is automatic. With thermionic valves it is possible to use ordinary fuses as protective devices because valves can tolerate overloads for a period long enough to allow the fuse wire to melt. For power supplies in which transistors are used as the controlling elements ordinary fuses do not act quickly enough. The base region of a transistor is damaged by overloads of very short duration. An effective method of protecting the series transistor (Tr1), which is the one that is damaged if the output is accidentally short-circuited, is to use a controlled rectifier between the emitter and base of the series transistor. In *Figure 8.17*, if the output is short-circuited, or if excessive current is drawn from the supply, a large voltage is developed across the 0·5 ohm resistor. A suitable fraction of this voltage is fed to the gate of the S.C.R. and the latter fires. Diode MR1 then conducts. The points A and B

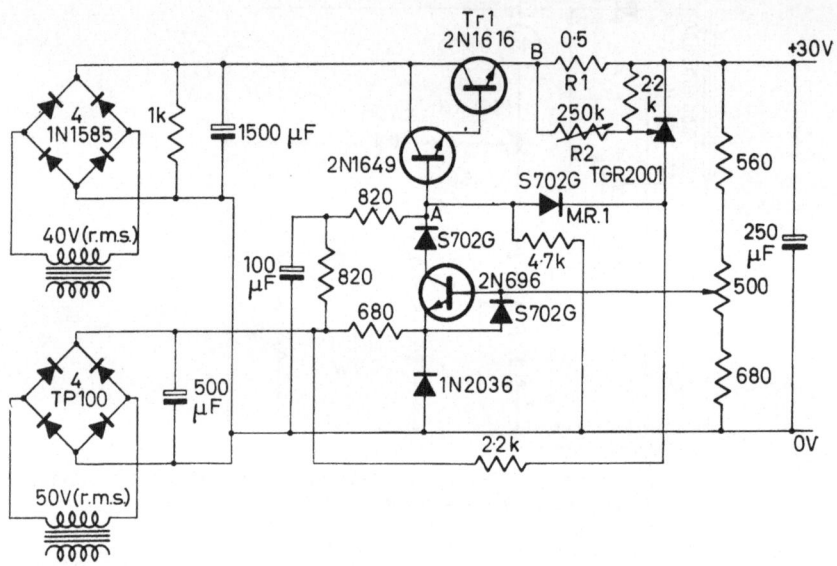

Figure 8.17. Circuit diagram of 30V 2A regulated power supply with short-circuit protection using a controlled rectifier (Ettore Dell'Oro[1])

are then effectively connected together thus protecting Tr1 by shorting the emitter to the base of the Darlington pair. (The Darlington pair arrangement is discussed in Chapter 9.) If the output of the power supply is short-circuited, the short-circuit current passes through the two 820 ohm resistors, MR1 and the S.C.R. Under short-circuit conditions the voltage across this chain is about 50 V. The short-circuit current is therefore limited to about 50/1640 i.e. about 30 mA. The protective circuitry can be made to come into operation before short-circuit conditions are imposed. This is achieved by choosing a suitable value for R1. For his circuit Ettore Dell'Oro has decided that the overload protection should trip at a load current of 2A. This gives a trigger voltage of 1 volt when R1 = 0·5 ohm. R2 controls the gate current and is adjusted to ensure sufficient gate current to fire the S.C.R. reliably at the instant the overload current starts to flow. This technique of overload protection can be used

208

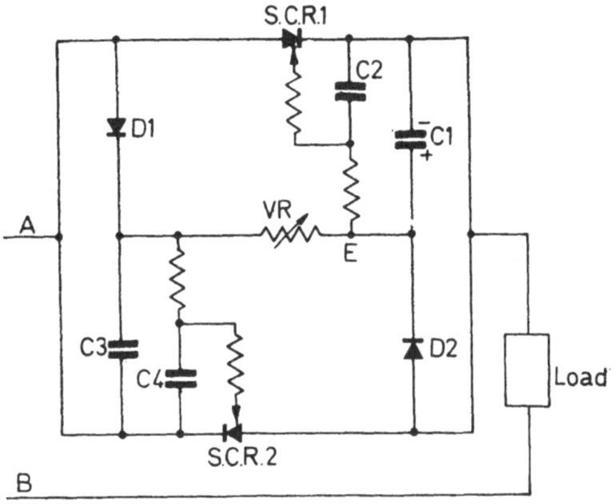

Figure 8.18. Circuit for the smooth control of power to a load. This arrangement can be used in some cases as an economic alternative to the variable transformer especially where high load currents are involved

in any stabilized power supply that uses transistors as a series regulating element.

In *Figure 8.7a* a basic control circuit is shown that uses thyratrons for the control of alternating current. The semiconductor version is shown in *Figure 8.18*, together with suitable gate firing circuitry. This circuit can replace a variable auto-transformer in many applications. Variable auto-transformers are bulky and expensive, especially when large load currents are involved. The circuit of *Figure 8.18* is often an attractive alternative.

The operation of the circuit is as follows. When A is positive with respect to B, D1 conducts and charges C1 via VR. The potential at E rises and C2 charges. Depending upon the setting of VR there is a delay in the firing of S.C.R.1. During the next half-cycle, D2 conducts charging C3 and C4 via VR. At the appropriate time S.C.R.2 fires and carries the load current. The effective power into the load may therefore be controlled by varying VR.

REFERENCES

1. Dell'Oro, E. 'Short-circuit protection of stabilized power supply with s.c.r.s.' Letter to Editor, *Electronic Engineering*, 1962, **34**, No. 408, 118 (Feb.).

Suggestions for Further Reading

Cornick, J. A. F. and Krajewski, Z. A. A. 'The silicon controlled rectifier'. *Power and Works Engineering*, May and July, 1961.
'Firing requirements for silicon controlled rectifiers'. *Mullard Tech. Comm.* Vol. 6, No. 55. March 1962.
'Transistorized circuits for S.C.R. firing'. *Mullard Tech. Comm.* Vol. 7, No. 65. June 1963.
Ross, D. I. and Goodger, B. E. G. 'A 12V to 230V a.c. silicon controlled rectifier invertor'. *Electronic Engineering*, 1965, **37**, No. 451 (Sept.).
Silicon Controlled Rectifier Manual. 4th Edition, General Electric Co., 1967.

9

AMPLIFIERS

It has already been pointed out that the transducers used in a laboratory to convert the quantities under observation into suitable electric signals rarely give sufficient output to actuate an indicating device directly. The electric output from photocells, strain gauges, glass electrodes, thermistors, etc., all need to be increased, i.e. amplified, before the signal can be suitably recorded or indicated. Some intermediate electronic apparatus must therefore be provided between the transducer and indicator to effect the necessary amplification. The design of a suitable amplifier depends upon the transducer and indicator used, upon the power required and upon the frequency of the signal that must be amplified.

Amplifiers may be classified according to the function they perform. If frequency is the classifying criterion then amplifiers are designated z.f., a.f. or l.f., r.f., v.h.f. or u.h.f. The z.f. (zero frequency) types are used to amplify steady voltages or currents. The term 'd.c. amplifier' is often used for this type. Some ambiguity as to the meaning of d.c. arises, but most workers accept the term 'direct coupled'. We shall use the term 'd.c. amplifier' to mean an amplifier that can amplify steady voltages and signals whose frequency range extends to zero frequency. Many of the amplifiers for zero or very low frequency work are directly coupled between their stages. The chopper stabilized types are discussed in a later section in this chapter.

The a.f. (audio frequency) amplifiers are used to amplify signals in the audible range, i.e. 20 Hz to 20 kHz (kilohertz). However, for faithful amplification of the signal waveform the amplifier must have a suitable response in a range of frequencies extending from about 10 Hz to 100 kHz. This is because non-sinusoidal periodic waves of audible frequency have harmonics extending beyond the upper limit of the audio range. The term l.f. (low frequency) amplifier is to be preferred especially when audio work is not involved.

R.F. (radio frequency), v.h.f. (very high frequency) and u.h.f. (ultra-high frequency) amplifiers are used when the signal frequency is much higher than the audio range. The r.f. signals extend over a range that includes the familiar long, medium and short wave communication bands, say from 200 kHz to 30 MHz (megahertz, i.e. millions of cycles per second). The v.h.f. range, used for television and frequency-modulated radio transmissions, goes up to about 300 MHz, whilst the u.h.f. band, used for colour television, for example, extends over many hundreds of megahertz.

The dividing lines between the different bands are not clearly defined. The techniques used in the construction of amplifiers vary enormously depending upon the frequency range to be handled. Since many scientific workers are interested mainly in z.f. and l.f. amplifiers, these are discussed in detail; for other readers however some mention is made of the r.f. types.

A second classification may be made by considering the indicator or other apparatus to be controlled by the amplifier. When the latter is designed to give an undistorted voltage output, it is known as a voltage amplifier. Such an

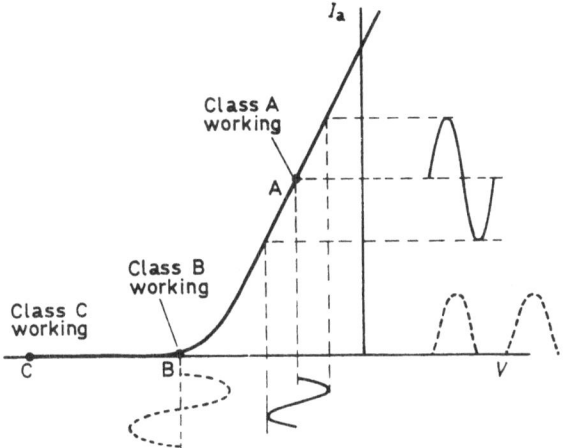

Figure 9.1. The dynamic characteristic of a triode amplifier illustrating the various possible classes of amplifier depending upon the position of the bias point

amplifier may be used to energize the deflector plates of a cathode-ray oscilloscope tube. On the other hand, servo- and other electric motors, loudspeakers, potentiometric recorders, and so on, all require considerable power for their operation. An amplifier designed to supply such power is known as a power amplifier.

A third classification depends upon the position of the bias point, and can be understood with reference to *Figure 9.1*. (A re-reading of the section of Chapter 4 on simple amplifier design may be helpful here.) *Figure 9.1* shows the dynamic characteristic for a single-stage amplifier. If the grid is biased to some negative voltage such that for small variations of grid voltage the waveform of the anode current is the same as that of the input signal waveform, then the amplifier is called a Class A amplifier. Since a resistor is being used as a load the output voltage waveform must also be a faithful, but amplified version of the signal waveform. The bias voltage must be such that the operation of the valve is confined to the straight portion of the characteristic. Anode current is present throughout the whole of the signal cycle, and the input voltage swings are restricted to values that ensure that the grid current is practically zero. The perfect Class A amplifier is an ideal voltage amplifier. The efficiency of such an amplifier is not high; in fact, it can be shown that the maximum efficiency, in terms of output power to total power taken from the h.t. supply, is only 25 per cent. However, power output is of no concern in Class A voltage amplifiers; the aim is to produce an output voltage with the minimum possible distortion.

In a power amplifier efficiency is a consideration; increases in efficiency are achieved by locating the operating point at or very near to the cut-off point (B). Anode current is then present for only half the cycle, and under these circumstances, the amplifier is said to be operating in the Class B mode. The maximum efficiency of a Class B amplifier is 78·5 per cent. Later it is shown that two valves must be operated in a push-pull arrangement with a transformer to restore the complete waveform. The increased saving in power supplied from the h.t. source may be of little consequence in low-powered apparatus, but for transmitters and other high-powered equipment running costs become a major consideration.

For loads consisting of tuned circuits, the efficiency may be raised to over 90 per cent by biasing the valve to a point well beyond cut-off. The valve is then operating as a Class C amplifier. This type of amplifier will rarely be used in laboratories except by those who work with r.f. heaters. Class C amplifiers are not therefore discussed in this book.

When a compromise is being made in a power amplifier between the fidelity of a Class A amplifier and the efficiency of a Class B amplifier, it is possible to run two valves in push-pull, each valve being operated about a point midway between the Class A and Class B operating points. The resulting amplifier is said to be operating in the Class AB mode.

VOLTAGE AMPLIFIERS

Chapter 4 describes how a thermionic triode or pentode can be the basis of a single-stage voltage amplifier. The gain produced by a single stage is often insufficient for most purposes; most amplifiers therefore consist of several stages coupled together. For thermionic valve stages it is reasonable to assume that the performance of the whole amplifier is determined by the combined efforts of each stage, and that each stage performs as it would do were it acting in isolation. The overall amplifier gain is therefore the product of the gains of the individual stages; the total phase-shift is the sum of the phase-shift of each stage. The phase-shift is the phase difference between the output and input signals. Although the phase difference between the anode and grid voltages of a single valve is 180°, other circuit components combine to make the phase-shift of a single stage not always quite 180°. The importance of phase-shift is paramount in feedback amplifiers (treated later in the chapter) but even for straightforward amplifiers a knowledge of the phase-shift is important. This is because an amplifier has often to deal with a complex waveform. If the phase-shift of the amplifier is different at different frequencies then there are relative phase displacements of the harmonics of a complex waveform; the result is that the output waveform differs from the signal waveform, i.e. distortion has been introduced.

In transistor amplifiers the overall performance is more difficult to predict because the individual stages do not act as they would do if they were isolated. This problem is discussed more fully in the section dealing with transistor amplifiers.

Coupling

Since a complete amplifier is made up of stages it is necessary to see how one stage can be coupled to the next. Mere connection of the anode of one stage to the grid of the next is not satisfactory. This is because the mean anode potential is high whereas the grid of the subsequent valve must be negative with respect to its cathode. The steady high voltage on the anode is blocked out by the use of a capacitor, and thus it is possible to preserve the correct bias conditions in the later stage. If a capacitor alone is used, as in *Figure 9.2a*, rises of voltage on the anode side of the capacitor cause some electrons to be attracted out of the electron stream in the second valve and on to the capacitor plate. The capacitor then, in effect, becomes slightly charged. Subsequent cycles of the signal variations increase the charge until eventually a considerable number of electrons accumulate on the plate G of the capacitor. These electrons are unable to return to the main valve current when they are repelled by negative-going changes at A because the grid is not a hot electrode. Eventually the second valve is driven to cut-off and becomes inoperative. To prevent this a grid resistor is connected between the grid and h.t. negative lines *(Figure 9.2b)*. Any accumulation of additional charge is now prevented since this charge can now leak away through the grid resistor R_g. At one time the term 'grid leak' was commonly used for this resistor. The reader may have noticed that the term 'additional charge' has been used. This is because on switching on the amplifier there is an initial charging of C via R_L and R_g. This main charge remains constant however (provided the h.t. voltage remains steady) and plays no part in the transfer of the signal. It has been pointed out (Chapter 3, p. 58) that signal transfer is most efficient when no charging of C *by the signal* takes place. By keeping R_g large very little charging by the signal can occur. This form of coupling one stage to the next is called resistance-capacitance (RC) coupling.

An alternative way of transferring the signal from one stage to the next without upsetting the bias conditions is to use a transformer. The stages are then said to be transformer coupled *(Figure 9.2c)*. Apart from eliminating the d.c. component, it is possible to arrange for some voltage gain in the transformer by having a larger number of turns in the secondary coil than in the

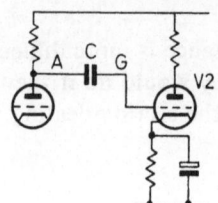

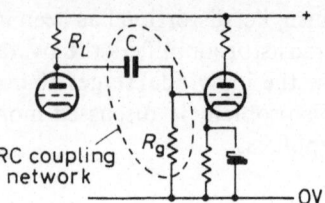

(*a*) Accumulation of negative charge on G side of C, due to positive-going signals at A, upsets the bias conditions on V2 and cut-off may eventually occur

(*b*) By addition of R_g the grid of V2 is maintained at zero volts in the absence of any signal. Correct grid bias conditions are therefore maintained

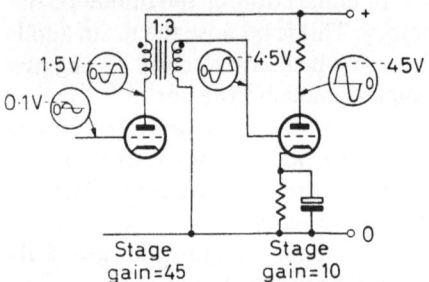

(*c*) Transformer coupling. Note the phase reversals throughout the amplifier. The dots near the ends of the transformer symbols show the ends that have voltages in phase

Figure 9.2. RC and transformer coupling of amplifier stages

primary. Step-up ratios of 3 : 1 and 5 : 1 were once widely used in audio amplifiers when valve technology had not reached its present level. With the production of high-mu triodes and pentodes, the voltage gain of a transformer is not the advantage it used to be. Against the advantage of stepping-up the voltage must be set the inferior performance of the transformer with regard to the range of frequencies it can handle. The distortion of the signal that iron cores inevitably introduce, and the spurious signals (e.g. 50 Hz hum) that are induced into the circuit are additional disadvantages. Wherever possible a transformer should be avoided. The RC coupling gives a superior performance. In spite of this transformers for low-frequency work may still be used to couple a transducer, such as a moving-coil gramophone pick-up or an accelerometer, into the input of an amplifier.

214

The input of the amplifier should always be matched to the type of transducer that is being used. Piezoelectric strain gauges and crystal microphones need to be connected to an amplifier which has an input impedance of at least 0·5 M. With valves this is easily arranged with an RC coupling in which the resistor is 470 k or greater; in transistor amplifiers special circuitry is employed to increase the input impedance of an amplifier. On the other hand tape-recorder heads, moving-coil microphones etc., are often coupled to the input of an amplifier by means of a transformer. As we shall see later, the output stage of a power amplifier needs to be matched to the load for maximum power transfer, and frequently transformers are used as matching components.

Where a single frequency or narrow band of frequencies is involved it is possible to design a satisfactory transformer for coupling purposes. A common example is the coupling between the r.f. stages of a superheterodyne receiver.

Frequency Response

For any individual amplifier, there is a range of frequencies over which the gain of the amplifier is constant. This range is called the mid-frequency range. The actual range depends upon whether the amplifier is designed for l.f., r.f. or v.h.f. operation. All amplifiers are designed to operate satisfactorily over the mid-frequency range, but above and below this range the gain becomes smaller. The performance is often illustrated graphically by what is called a frequency response curve.

Frequency response curves use logarithmic rather than linear scales for the vertical and horizontal axes; logarithmic scales often interpret physical phenomena more satisfactorily than linear scales. Many examples are found in nature in which a quantity, y, grows at a rate proportional to its size, e.g. $y = ke^t$ $\therefore$ $dy/dt = ke^t$. e^t is, of course, an exponential function in which t is the index. Exponent' and 'logarithm' are terms that are equivalent to 'index'. Immediately it can be seen that logarithms are involved. To illustrate the advantage of logarithmic scales when dealing with functions of the type described above, the aural response of two men, A and B can be considered. If the range of frequencies heard by A is from 10 Hz to 10 kHz and that heard by B is 10 kHz to 20 kHz then these physical facts can be interpreted graphically by the use of either a linear frequency scale as in *Figure 9.3a* or a logarithmic scale as in *Figure 9.3b*. Immediately it can be seen that *Figure 9.3b* is to be preferred since A has a frequency range that allows him to hear almost all of the sounds commonly heard by human beings. In fact, many older people have an aural response which extends only to 10 kHz and hear quite well. Although B's response is fictitious, in that no man is known

215

that has this type of response, we know that he would be deaf to all common speech and musical sounds. The graph of *Figure 9.3b* intuitively conveys this to us. From a purely scientific, rather than intuitive, viewpoint, logarithmic frequency scales have two advantages over linear scales. They expand what would otherwise be a cramped representation at the lower frequency end of the scale, and response curves are close approximations to straight lines over extensive frequency ranges.

When comparing the frequency performance of two amplifiers on a single set of axes the vertical scale should be a relative logarithmic one. The logarithmic scale prevents cramping, and, if it is also relative, direct comparisons of performance can be made. The relative logarithmic scale may be divided into a linear scale of units called decibels.

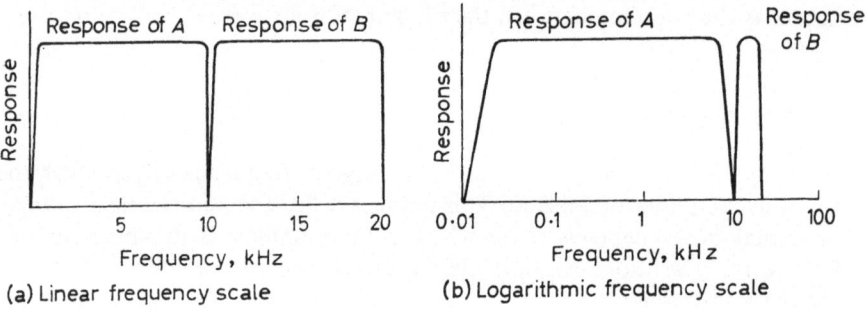

Figure 9.3. Two responses plotted on linear and logarithmic frequency scales

The Decibel Scale

The bel scale was introduced originally by telephone engineers to express the ratio of two powers, and was named in honour of Graham Alexander Bell, the pioneer in telephone work. The bel is, in practice, too large a unit to use and so the unit decibel (dB) was introduced, one decibel being 1/10th of a bel. Hence

$$\text{Power gain (in dB)} = 10 \log_{10} \frac{P_1}{P_2}$$

This is the fundamental and defining equation for decibels. Thus if $P_1 = 2P_2$ then P_1 is $10 \log_{10} 2$ i.e. 3 dB greater than P_2. We often say that P_1 is 3 dB up on P_2. If $P_1 = 0.5 P_2$ then P_1 is $10 \log_{10} 0.5 = -3$ dB compared with P_2. Here we often say that P_1 is 3 dB down on P_2.

To obtain a frequency response curve the output power is observed at different frequencies with the input voltage maintained at a constant r.m.s.

216

value. Since the output power is being developed in the same resistance, R, at different frequencies, it is possible to modify the above equation. If the power at the first frequency is P_{f1} and that at the second frequency P_{f2} then the power gain or loss, N, is given by

$$N = 10 \log_{10} P_{f1}/P_{f2}$$

Now $P_{f1} = V_{f1}^2/R$ where V_{f1} is the output voltage at the first frequency. Similarly $P_{f2} = V_{f2}^2/R$

$$\therefore \ N(\text{dB}) = 10 \log_{10} \frac{V_{f1}^2/R}{V_{f2}^2/R} = 10 \log_{10} \left(\frac{V_{f1}}{V_{f2}}\right)^2$$

$$= 20 \log_{10} \left(\frac{V_{f1}}{V_{f2}}\right)$$

This alternative equation is convenient because it is easier to measure voltages than powers. However, it must be stressed that this alternative equation is only possible because the resistance across which the measurements are taken is the same at both frequencies.

It is often possible to read or hear about the voltage gain of an amplifier being so many decibels. Strictly, it is incorrect to make this kind of statement because the input and output voltages are being developed across different resistances. However, it is so attractive to express the gains in this way that by convention many people use this type of statement. Only very rarely can any ambiguity arise when discussing thermionic valve amplifiers. A straightforward statement such as 'the gain of this amplifier is 60 dB' is taken to mean that $60 = 20 \log_{10} \dfrac{V_{\text{out}}}{V_{\text{in}}}$ i.e. that the output voltage is 1,000 times greater than the input voltage. In this way the overall gain of an amplifier in decibels is easily obtained by adding together the gain in decibels of each stage. In transistor amplifiers care must be taken to distinguish between voltage, current and power gain.

Figure 9.4 shows the frequency response curves for a two-stage amplifier with *(a)* RC coupling and *(b)* transformer coupling. The inferior performance of the transformer coupling due to changes in the impedance of the primary and secondary windings with frequency is obvious. Peaking is due to self-capacitance in the windings.

Whenever specifications for amplifiers are being considered, it is inadvisable to accept a manufacturer's statement that his amplifier has a response out to (say) 100 kHz. From the graph, it will be seen that although some response for the RC coupled amplifier is evident at 100 kHz the amplifier is practically useless at that frequency. All reputable manufacturers state that their amplifiers are flat within so many (2 or 3) dB over a specified frequency range. Alternatively, they may state that the 3 dB points are at two specified frequencies, which means that the gain of the amplifier is constant over the range between the specified frequencies, being no more than 3 dB down at any point along the range. This is a very acceptable specification; 3 dB is often adopted as a figure from which to define the bandwidth of amplifiers, the bandwidth being the range of frequencies between the 3 dB points. For audio work the voltage output is taken at a reference frequency of 1 kHz. For r.f. and other amplifiers any reference frequency in the middle of the

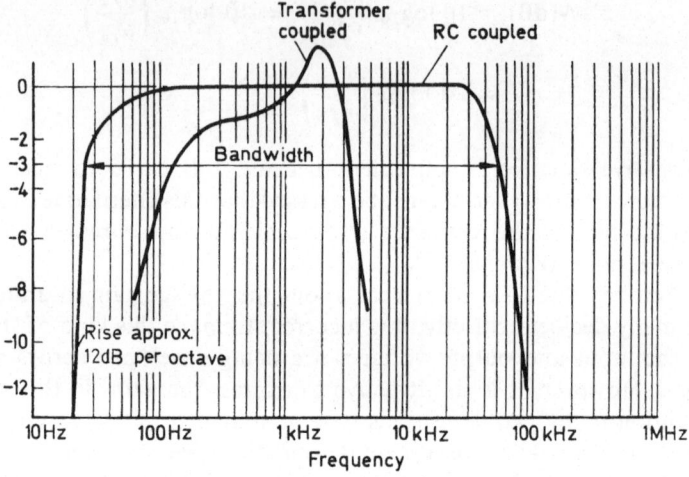

Figure 9.4. Frequency response curves for two two-stage amplifiers, one with RC coupling and the other with transformer coupling

frequency range may be taken. If an a.f. amplifier is 3 dB down at 60 Hz it means that the voltage output at 60 Hz is $1/\sqrt{2}$ times the voltage output at 1 kHz.

Whilst dealing with the subject of frequency response curves it is convenient to explain another common term used to express the rate of change of the response curve. If a curve is rising (or falling) at 6 dB per octave this means that there is a doubling (or halving) of the voltage every time that the frequency is doubled. By using logarithmic scales the rise can be represented by a straight line. A rise of 6 dB per decade means that there is a doubling of voltage every time the frequency is multiplied by 10.

When considering the reasons for the fall in the response of amplifiers at high and low frequencies the components involved must be examined. The time constant of the coupling network is largely responsible for the lack of response at the lower frequencies. The RC combination *(Figure 9.5)* acts as a potential divider for alternating voltages. Efficient signal transfer occurs only

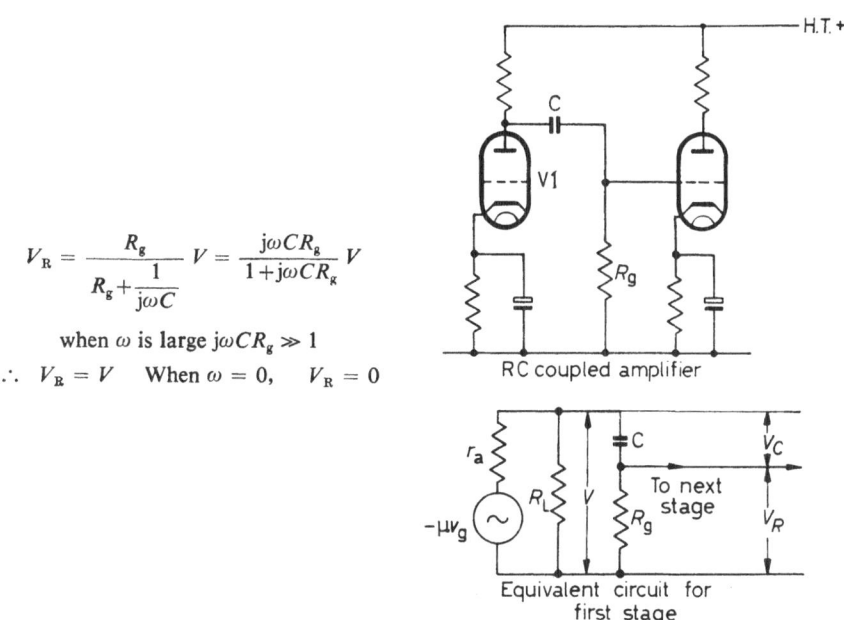

$$V_R = \frac{R_g}{R_g + \dfrac{1}{j\omega C}} V = \frac{j\omega C R_g}{1 + j\omega C R_g} V$$

when ω is large $j\omega C R_g \gg 1$

$\therefore\ \ V_R = V$ When $\omega = 0,\quad V_R = 0$

Figure 9.5. The use of the equivalent circuit to show the fall in amplifier response at low frequencies

when the reactance of the capacitor is negligible compared with the resistance of the grid resistor. As the frequency of operation becomes lower, the reactance of the capacitor rises ($X_C = 1/2\pi f C$); eventually at zero frequency this reactance becomes infinitely large. At very low frequencies therefore, very little of the signal is developed across the grid resistor and transfer of the

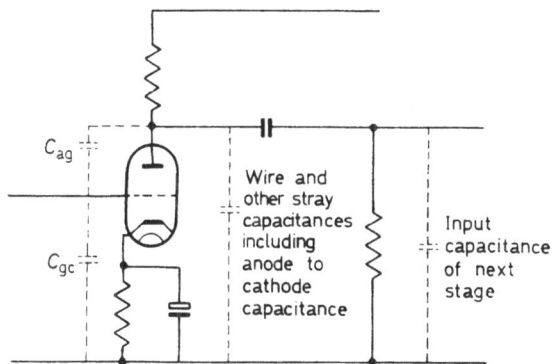

Figure 9.6. The stray capacitances that are responsible for the fall in response at high frequencies

219

signal voltage to the next stage is very poor. A mathematical analysis can be performed with the aid of the equivalent circuit and it is then possible to predict the frequency at which the response is 3 dB down on the 1 kHz level.

At high frequencies, account must be taken of the fact that the circuit diagram shows only the physical components such as valves, resistors and capacitors. The 'hidden' components not shown are the inevitable stray capacitances between various parts of the circuit and the chassis. Connecting wires, printed circuit boards, screening boxes, interelectrode capacitances all play their part in reducing the performance of the amplifier at high frequencies. The stray capacitances can be represented by a diagram as in *Figure 9.6*. Although the capacitances are only a few tens of picofarads, and can be neglected at mid- and low-frequencies, their reactances at high frequencies becomes low. Considerable leakage paths are therefore available for high-frequency signals, and these paths cause a fall in the high-frequency response of the amplifier.

WIDE BAND AMPLIFIERS

In some apparatus (e.g. cathode-ray oscilloscopes, pulse amplifiers and television receivers) the bandwith obtainable with a straightforward amplifier of the type already discussed is not sufficient. Where it is necessary to have amplifiers that operate with almost constant gain over a very wide frequency range, and where signals with a very fast rise time must be amplified with little waveform distortion, it is necessary to employ special circuit techniques.

It is always instructive to apply a signal with a square waveform to the input of an amplifier and observe the output waveform on a high-quality oscilloscope. The square wave is equivalent to a fundamental sinewave signal

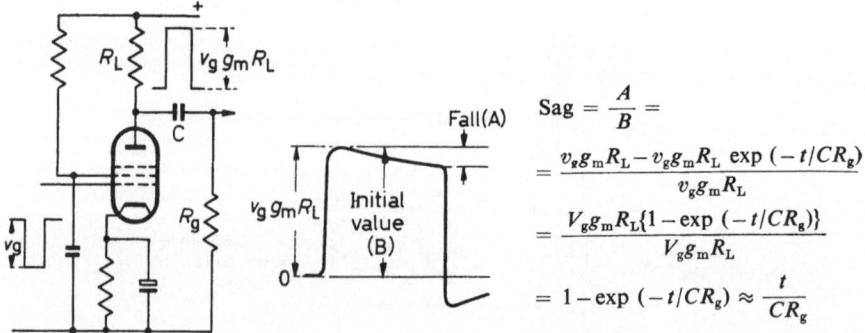

$$Sag = \frac{A}{B} =$$

$$= \frac{v_g g_m R_L - v_g g_m R_L \exp\left(-t/CR_g\right)}{v_g g_m R_L}$$

$$= \frac{V_g g_m R_L\{1 - \exp\left(-t/CR_g\right)\}}{V_g g_m R_L}$$

$$= 1 - \exp\left(-t/CR_g\right) \approx \frac{t}{CR_g}$$

Figure 9.7. By applying a signal with a rectangular waveform, the low frequency response can be judged by the sag in the output waveform. To keep the sag to a minimum all the capacitors in the circuit diagram should be large (see text)

plus all the odd harmonics out to infinity. Assuming that no distortion is introduced into the trace by the oscilloscope itself, departures from the original rectangular waveform are interpreted in the following way. A sag in the waveform *(Figure 9.7)* indicates a falling-off in the response at low frequencies. The major cause of this is the inadequate time constant of the coupling network. Initially when the anode voltage shoots up to a new value the rise in voltage is conveyed without attenuation to the grid of the subsequent valve because no change in the charge of the coupling capacitor could take place in the short time available. If however the anode voltage remains at its new value for a period that is long compared with the time constant of the coupling network then charging by the signal occurs and the grid voltage decays exponentially. The sag after a time, t, can be calculated by noting that the signal voltage across R_L is $v_g g_m R_L$.

The voltage on the grid of the valve of the next stage is given by $v = v_g g_m R_L$ $\exp(-t/CR_g)$ where R_g is the grid resistance of the second valve and C the coupling capacitance. The initial voltage at time $t = 0$ is $v_g g_m R_L$ and hence the fall at any subsequent time is $v_g g_m R_L - v_g g_m R_L \exp(-t/CR_g) = v_g g_m R_L$ $\{1 - \exp(-t/CR_g)\}$. The sag is defined as the fall at any time divided by the initial value, therefore the sag is given by $S = 1 - \exp(-t/CR_g)$. By expanding $\exp(-t/CR_g)$ as a power series and neglecting second and higher order terms it can be shown that $S \approx t/CR_g$. If a minimum value of sag is specified, the time constant that is required in the coupling network can be estimated. For example, suppose a square wave with a repetition frequency of 25 Hz is applied and the sag is required to be not greater than 10 per cent. For this case $t = 0.02$ sec (half the period of the square wave). CR must then be $0.02/0.1$, i.e. 0.2. If the grid resistor R_g is 0.5 M then capacitance must be at least 0.4 μF. In practice a 0.5 μF capacitor would be used. In order to obtain a good response at low frequencies it is evident that an amplifier must have large coupling capacitors. For very low frequencies the coupling capacitor must be very large (say up to 10 μF) and, in practice, this is inconvenient. The coupling capacitor can be dispensed with altogether for a response down to zero frequency, but direct coupling in this way means that special circuitry must be used to preserve the bias conditions in the subsequent stage. Directly coupled amplifiers for working down to zero frequency are discussed in a later section.

Two other components associated with any given stage of an amplifier affect the low-frequency performance. For efficient amplification with a pentode, the potential of the screen grid must be kept constant. This grid is usually fed through a resistor and it is necessary to add a decoupling capacitor between the screen grid and h.t. negative line in order that changes of signal voltage do not cause changes of potential of the screen grid. This latter electrode, being held at a positive potential, acts as an anode to the control grid. If the potential on the screen grid were maintained by only a resistor, this resistor would behave as a load, as it does in the case of a triode. Changes in control grid voltage would therefore bring about changes in screen-grid voltage. To prevent this the decoupling capacitor effectively shorts out the alternating voltage component to the h.t. negative line. Although a value of 0·1 μF is adequate for most amplifiers, wide band amplifiers working at very low frequencies require this capacitor to be increased to 0·5 μF, or even 1 μF.

In both pentode and triode stages inadequate bypass capacitors associated with the bias resistor lead to poor low frequency performance. At mid- and high-frequencies a given bypass capacitor may present so little reactance

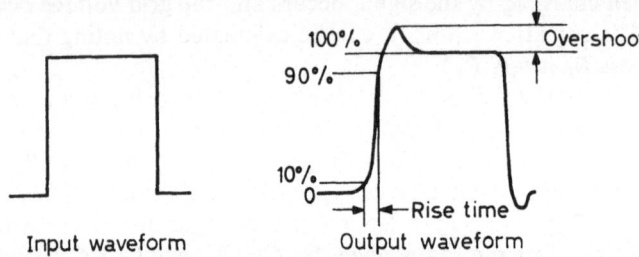

Input waveform   Output waveform

Figure 9.8. The effect of poor high frequency performance is to give a rounding of the leading edge and a consequent increase in the rise time. If poor transient response is also present various degrees of overshoot are evident

that the cathode is effectively at the potential of h.t. negative line so far as alternating voltages are concerned. As the operating frequency is lowered the reactance of the bypass capacitor rises and this component no longer acts satisfactorily as a short-circuit to alternating voltages. Under these circumstances, not all of the signal voltage is applied between the grid and cathode. This is because changes in the signal voltage give rise to changes of current through the valve and hence the bias resistor. Since the bias resistor is not effectively bypassed at low frequencies, there exists across the bias resistor the wanted steady bias voltage together with an unwanted alternating component. The alternating voltage component is subtracted from the signal voltage, the remainder being the effective grid-to-cathode voltage. There is thus a reduction in the gain at low frequencies. If it is not convenient to increase the bias capacitor to the very large value required to prevent the fall in gain, the capacitor can be omitted. This will mean that the alternating component that is being subtracted from the signal voltage is the same magnitude at all frequencies. The fall in response at low frequencies is thus prevented. The price paid however, is a fall in gain at all frequencies, and this may be undesirable. Nevertheless, this fall, due to a process known as negative feedback, is accompanied by certain advantages which are discussed in the section on negative feedback amplifiers.

The main difficulties in producing a wideband amplifier are associated with the high-frequency response, the phase response and the transient response. Poor high-frequency response is evident as a slow rise time of the output waveform *(Figure 9.8)*. It has been found empirically that

$$\text{Rise time (µsec)} = \frac{0.35 \text{ to } 0.45}{\text{Bandwidth (MHz)}}$$

where the bandwidth is the frequency range in megahertz between the 3 dB points. For the expression to be accurate, the overshoot should be zero or small (< 10 per cent). The overshoot is a function of the phase delay through the amplifier at high frequencies. (For the measurement of phase-shifts in amplifiers at different frequencies see Chapter 11 on the cathode-ray oscilloscope.)

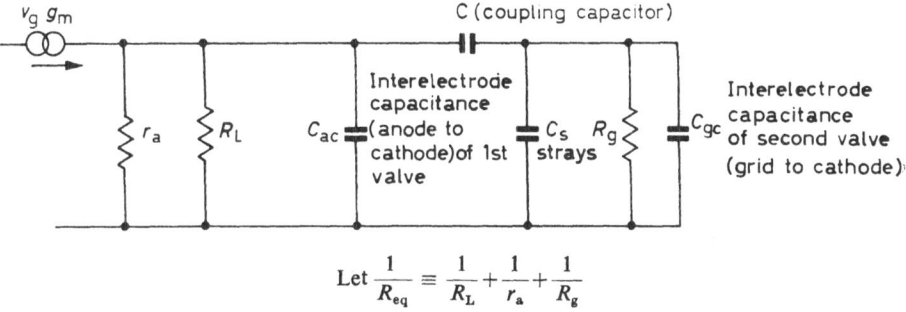

Let $\dfrac{1}{R_{eq}} \equiv \dfrac{1}{R_L} + \dfrac{1}{r_a} + \dfrac{1}{R_g}$

C can be ignored at mid and high frequencies

Let $C_{eq} = C_{ac} + C_s + C_{gc}$

Figure 9.9. Equivalent circuit used for deriving the gain/bandwidth product (see text)

The major source of loss of gain at high frequencies is associated with the stray capacitances illustrated in *Figure 9.6*. If the gain at the mid-frequency range is made high by using large values of anode load resistances then the drop in gain at high frequencies is severe. This is because the shunting effect of the stray capacitances is more noticeable. If the equivalent shunt reactance of all the stray capacitances at a given high frequency is, for example, 5 k, then it is easily seen that the shunting effect across a load of 47 k is serious. If however low values of anode load resistor, say 5 k, were used the shunting effect would not be so noticeable. In wideband amplifiers therefore low values of anode load resistor are used. This solution, like that of dispensing with the bias capacitor, results in a loss of overall gain. Clearly too much gain must not be sacrificed in order to obtain a large bandwidth. An amplifier with a

223

very large bandwidth and very little gain would not be very useful. A figure of merit for an amplifier has been introduced to take account of both bandwidth and gain. The figure of merit is called the gain/bandwidth product and is simply the product of the gain and the bandwidth between the 3dB points. It is desirable to have large gain/bandwidth products.

It can be shown with the aid of the equivalent circuit that for large gain/bandwidth products we should use valves with large mutual conductances and small interelectrode capacitances. For these reasons pentodes are the favoured choice. *Figure 9.9* shows the constant current equivalent circuit together with the strays. In the mid-frequency range the gain is A. The effects of the strays are negligible and the effective load is R_{eq} where

$$\frac{1}{R_{eq}} \equiv \frac{1}{R_L} + \frac{1}{r_a} + \frac{1}{R_g}$$

At high frequencies the reactance of the strays is X_C. The effective load is now R_{eq} in parallel with X_C where $X_C = \dfrac{1}{j\omega(C_S + C_{ag} + C_{gc})} = \dfrac{1}{j\omega C_{eq}}$. Now when $X_C = R_{eq}$ the gain will be reduced by $\sqrt{2}$. (Not 2 because phase must be considered.) The magnitude of $R + jX_C$ is $\sqrt{(R^2 + X_C^2)}$ and if $R = X_C$ this magnitude is $R\sqrt{(1+1)}$ i.e. $R\sqrt{2}$. (See Chapter 2 on a.c. theory.) There is thus a frequency, f, at which $X_C = R_{eq}$ and at that frequency the gain is $1/\sqrt{2}$ times the gain at the mid frequency range, i.e. the fall in response is 3 dB. Since a wideband amplifier is being discussed, f is the bandwidth very nearly. (For example, if the upper 3 dB point is at 5 MHz then $f = 5 \times 10^6$. The lower 3 dB point may be 10 Hz. The bandwidth is then $5 \times 10^6 - 10$ which is very nearly 5×10^6.)

The gain/bandwidth product, P, is Af where A is the gain at the mid frequency range and is equal to $g_m R_{eq}$. Since $X_C = R_{eq}$ at the 3 dB point

$$P = g_m R_{eq} f = g_m X_C f = \frac{g_m f}{2\pi f(C_{ag} + C_S + C_{gc})}$$

$$= \frac{g_m}{2\pi(C_{ag} + C_S + C_{gc})}$$

From this it can be deduced that for wideband amplifiers the ratio between the g_m and the interelectrode capacitances must be large. Pentodes with large mutual conductances are therefore chosen.

There are two ways in which the bandwidth of an amplifier can be extended at the high frequency end. These are *(a)* cathode peaking and *(b)* shunt peaking.

In cathode peaking a small value of cathode bypass capacitor is used. The result is that the gain at mid and low frequencies is reduced by the negative feedback principle already outlined. It is only at the high frequencies that the capacitor is effective in bypassing the bias resistor. At these high frequencies, all of the signal voltage is applied between the grid and cathode; the gain at these high frequencies is therefore enhanced at the expense of the gain at lower frequencies.

In shunt peaking a small inductor is placed in the anode lead. The load for the anode is thus the normal load resistor in series with the small inductor *(Figure 9.10)*. It is usually arranged that the dynamic impedance of the inductor and load resistor combined with the parallel stray capacitances is about half of the load resistance R_L. Thus

$$\frac{L}{C_S R_L} = 0{\cdot}5 R_L \quad \therefore \quad L = \frac{R_L^2 C_S}{2}$$

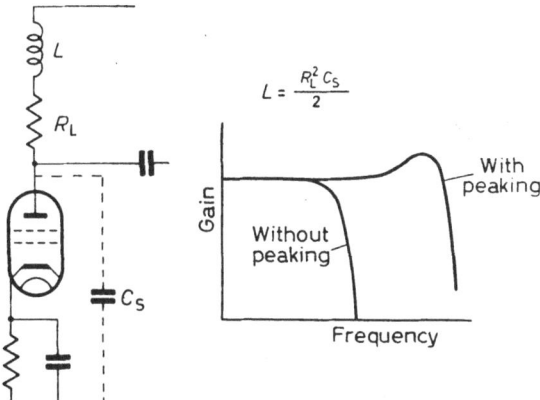

Figure 9.10. Two methods of increasing the gain at high frequencies are (*a*) inclusion of an inductor in the anode load and (*b*) making the bypass capacitor small

(Many of the standard texts on a.c. theory show the dynamic, i.e. effective, a.c. impedance to be L/CR at resonance where L is the inductance of the coil, R the total resistance, usually of the coil alone, and C is the parallel capacitance. See p. 73.)

As an example, let us assume $R_L = 10$ k and $C_S = 30$ pF, both of which are likely values. Then $L = 10^8 \times 3 \times 10^{-11}/2 = 1{\cdot}5$ mH. Some overshoot on fast-rising transients is inevitable with this value. If no overshoot is allowed $L/C_S R_L$ should be made equal to $0{\cdot}25\ R_L$; there is then a reduction in the bandwidth from that obtained when $L/C_S R_L = 0{\cdot}5\ R_L$.

NEGATIVE FEEDBACK AMPLIFIERS

Ordinary amplifiers of the type already discussed suffer from various forms of distortion and their performance is altered by the ageing of components and variations of supply voltages. Straightforward amplifiers are not therefore accurate measuring devices. All modern high performance amplifiers employ negative feedback, and thus are accurate amplifying devices that can be used as the basis of reliable electronic measuring equipment.

Distortion

The output of an amplifier is said to be distorted if a change of waveform occurs between the input and output terminals. The output waveform may contain frequency components not present in the original signal, or, where complex signals are involved, the phase relationship between the various components of the signal may be altered. The relative amplitudes of these components may also be altered.

The actual output of an amplifier is necessarily limited. Although the gain of an amplifier may be, for example, 1,000 this does not imply that any magnitude of input voltage is amplified 1,000 times. An input voltage of 100 V (r.m.s.) does not produce an output voltage of 100 kV in the types of amplifier we are discussing. There is a linear relationship between output and input voltages only over a restricted operating range where overloading of any stage in the amplifier is absent. The relationship between input and output voltages is known as the transfer characteristic *(Figure 9.11)* and this charac-

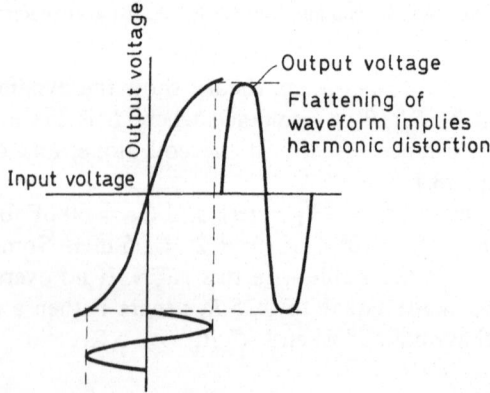

Figure 9.11. Curvature of the transfer characteristic leads to the introduction of harmonic distortion

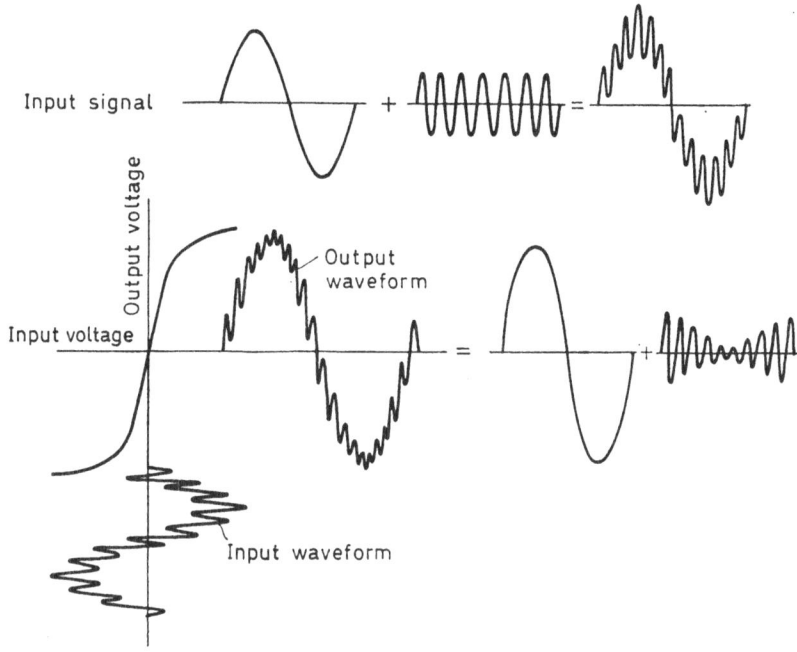

Figure 9.12. When two sinusoidal input voltages are applied simultaneously, the non-linear transfer characteristic reduces the amplitude of the higher frequency signal at times when the lower frequency signal is near to the maximum and minimum voltages. This is intermodulation distortion

teristic is curved at the ends. The gain of the amplifier therefore varies with the instantaneous magnitude of the input signal, and non-linear distortion is said to be present. Curvature of the dynamic characteristics of the valves or transistors contribute to non-linearity of the transfer characteristic. The application of a sinusoidal input voltage results in a periodic output waveform that is non-sinusoidal. Fourier analysis shows that spurious harmonics are present, the result being known as harmonic distortion. Total harmonic distortion, D, is measured as the root of the sum of the squares of the r.m.s. voltages of the individual harmonics, divided by the r.m.s. value of the total signal V i.e.

$$D = \frac{\sqrt{(V_{H2}^2 + V_{H3}^2 + V_{H4}^2 \ldots V_{HN}^2)}}{V} \times 100 \text{ per cent}$$

where V_{H2}, V_{H3}, etc. are the r.m.s. values of the harmonic components.

Intermodulation distortion *(Figure 9.12)* is a form of non-linear distortion whereby the amplification of a signal of one frequency is affected by the amplitude of a simultaneously applied signal of lower frequency. Combination frequencies are produced which have values equal to the sum and difference of the two applied frequencies.

Attenuation distortion is caused by the variation of the gain of an amplifier with frequency. If, for example, a complex waveform has a harmonic of a high frequency and the gain at that frequency is very low clearly this harmonic must be almost absent in the output waveform.

Phase distortion is present when the relative phases of the harmonic components of a signal are not maintained. Such distortion is caused by the presence of reactive and resistive components in the circuit. In cathode-ray oscilloscope amplifiers, television video amplifiers and in radar circuits phase distortion is highly undesirable. It is often said that phase distortion is unimportant in audio amplifiers as the ear is insensitive to moderate changes in phase. Whilst it is true that the ear is insensitive in this respect, it is not true that demands on the audio amplifier can be relaxed. The effect of phase shift is of great importance to the speaker diaphragm from the transient point of view. The quality of a sound depends, among other things, upon the attack and decay times. To obtain similar attack and decay times in the reproduced sound, phase distortion should be reduced to a minimum.

Valve and circuit noise, and 50 Hz components ('hum') are usually classified as distortion when introduced by an amplifier into a signal otherwise free of them.

The Principle of Negative Feedback

Most forms of distortion may be markedly reduced by using negative feedback. Feedback is said to occur in amplifiers when part of the output of the amplifier is added to or subtracted from the input signal. When a fraction of the output is added to the input signal the feedback is said to be positive. The gain of the amplifier rises usually in an uncontrolled way and oscillations occur. This is the subject of the next chapter.

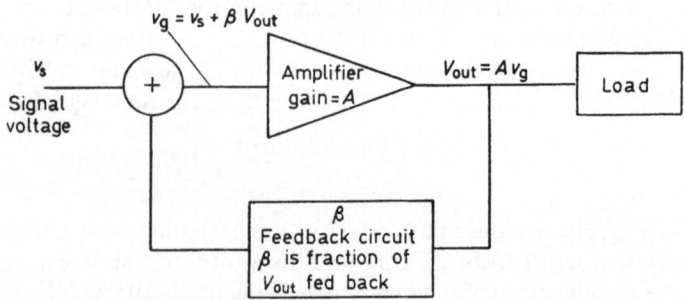

Figure 9.13. Block diagram representation of a feedback amplifier

228

When part of the output is fed back to the input in antiphase (i.e. 180° out of phase) then subtraction occurs and the feedback is said to be negative. *Figure 9.13* is one way of representing a feedback amplifier when we wish to make a quantitative examination.

The main amplifier (of the straightforward type previously discussed) is represented by the triangle and has a gain of A. A fraction of the output voltage, β, is selected by a suitable circuit and fed back to the input. So far as the main amplifier is concerned it 'sees' an input voltage, v_g, which is the signal voltage v_s plus the voltage feedback, βV_{out}, where V_{out} is the output voltage. Therefore

$$V_{out} = Av_g = A(v_s + \beta V_{out})$$
$$V_{out}(1 - \beta A) = Av_s$$
$$\therefore V_{out} = \frac{A}{1 - \beta A} v_s$$

This is the general feedback equation. When the fraction of the output voltage is fed back in antiphase, β is negative. Negative feedback is effected and

$$V_{out} = \frac{A}{1 + \beta A} v_s$$

i.e. the gain with negative feedback, A', is given by

$$A' = \frac{A}{1 + \beta A}$$

From this it can be seen that the gain of the feedback amplifier is lower than that of the main amplifier by a factor $1 + \beta A$. This is the price that must be paid to obtain the advantages described below. The product βA is the gain around the feedback loop and is, therefore, called the loop gain.

It is easy to arrange that the gain of the main amplifier is very high (say 10^5 or 10^6). Then $A\beta$ is much greater than unity with the values of β used in practice. The effective gain of the negative feedback amplifier therefore closely approaches $1/\beta$ i.e.

$$A' \fallingdotseq \frac{A}{A\beta} = \frac{1}{\beta}$$

The gain of the feedback amplifier is thus independent of the gain of the main amplifier provided the latter is large. Variations of supply voltages, ageing of components, and other causes of the variations in gain of the main amplifier are therefore unimportant in a negative feedback amplifier. The gain with feedback depends only on β and this can be made very stable by choosing simple feedback circuits that employ very stable circuit components.

The Effect of Negative Feedback on Gain

Let us suppose that the gain of an amplifier is 10^6 and that 1/100th of the output voltage is fed back in antiphase, i.e. $\beta = 10^{-2}$. The gain of the negative feedback amplifier is then

$$A' = \frac{10^6}{1 \times 10^6 \times 10^{-2}} \fallingdotseq 100$$

If now a serious upset in the main amplifier reduces the gain from 10^6 to 10^4, the gain of the negative feedback amplifier becomes

$$A' = \frac{10^4}{1 \times 10^4 \times 10^{-2}} \fallingdotseq 100$$

which is practically the same as before. The gain of the feedback amplifier has not altered by a large change in the gain of the main amplifier. This independence of gain results from the fact that the input to the main amplifier is the difference between the signal voltage and the voltage fed back. If the gain in the main amplifier falls the difference voltage will increase slightly and so the output remains almost constant.

The Effect of Negative Feedback on the Frequency Response

The upper curve of *Figure 9.14* represents the frequency response of a straightforward amplifier. It has a gain of A_1 at frequency f_1 and a gain of A_2 at f_2. A_2/A_1 is small, resulting in a restricted bandwidth. If now negative feedback is applied the gain at f_1 is $A'_1 = A_1/(1+\beta A_1)$ and at f_2 is $A'_2 = A_2(1+\beta A_2)$

$$\therefore \quad \frac{A'_2}{A'_1} = \frac{A_2}{(1+\beta A_2)} \cdot \frac{(1+\beta A_1)}{A_1}$$

when $\beta A_1 \gg 1$ and $\beta A_2 \gg 1$, $A'_2/A'_1 \fallingdotseq 1$. In other words the gains at the two frequencies are approximately equal. This is shown in the lower curve of *Figure 9.14*. Negative feedback thus increases the bandwidth of the amplifier. One simple practical way of doing this over a single stage is merely to omit the cathode bypass capacitor. As the signal voltage rises the valve current rises. There is an increase in the voltage across the bias resistor resulting in the application of a greater negative bias voltage. This offsets to some extent the rise in the signal voltage, and the grid-to-cathode voltage is not then as great as it otherwise would have been. As the feedback is effective over a very wide frequency range, the benefits of negative feedback are obtained over

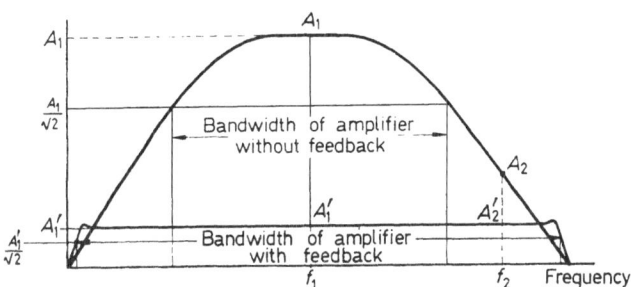

Figure 9.14. The effect of negative feedback on bandwidth. Note also the effect on gain at different frequencies. Without feedback the amplifier has gains of A_1 and A_2 at f_1 and f_2 respectively. A_1 is much greater than A_2. With feedback however the gains at f_1 and f_2 are A_1 and A_2. These gains are equal

all the operating frequency band. There is thus an increase in the bandwidth, greater gain stabilization and a reduction of distortion when compared with an amplifier in which the cathode resistor is bypassed.

The gain of a single stage where the bypass capacitor has been omitted may be calculated in the following way. The total a.c. load for the valve is now $R_L + R$ where R_L is the load in the anode and R the bias resistance. The alternating component of anode current is i_a, therefore the output voltage is $i_a(R_L + R)$. Only $i_a R_L$ is, of course, available to be passed on to the next stage. The voltage effectively fed back and subtracted from the input voltage is $i_a R$. The feedback factor β is thus $i_a R / i_a(R_L + R) = R/(R_L + R)$. Usually R is much smaller than R_L and so β is R/R_L to a good approximation. If a high gain pentode stage is involved then the gain with feedback is approximately $1/\beta$ i.e. R_L/R.

This type of feedback is called current negative feedback because the voltage fed back to the input circuit is proportional to the load current. Voltage negative feedback is the usual form in which a proportion of output voltage is fed back to the input circuit. So far as the input circuit is concerned, it is always a voltage that is fed back to be subtracted from the signal voltage. The ways in which feedback voltages are obtained are discussed later after power and transistor amplifiers have been described.

Distortion in Negative Feedback Amplifiers

One of the major benefits of negative feedback is the reduction of distortion in the output signals. Distortion arises because of non-linearities in the dynamic characteristics of valves and components. The relative independence of negative feedback amplifiers on the characteristics of the amplifier when

231

no feedback is applied, means that smaller distortion components are present in the output signal. This is because a portion of the distortion voltage produced in the main amplifier is fed back in antiphase and is amplified as a signal in the usual way. This amplified distortion neutralizes the original distortion to a large extent.

To obtain a quantitative expression for the reduction suppose that a signal, v, were applied to an amplifier without feedback. An output would be produced equal to $Av+D$ where D is the distortion component. If feedback is now applied and a signal v_s of sufficient magnitude to give the same output as before is used the voltage at the grid of the amplifier does not consist of the pure signal alone. If the distortion with feedback is d then $-\beta d$ is fed back along with the same fraction of the distortionless component of the output voltage ($-\beta d$ because the feedback is negative). Let us consider only the distortion component. So far as the amplifier itself is concerned it 'sees' a grid input voltage of $-\beta d$. This is amplified and distortion is added so the distortion output voltage is $-A\beta d+D$. This is equal to the distortion component d so we have

$$d = -A\beta d+D$$
$$\therefore \ d(1+A\beta) = D$$
$$\text{i.e.} \ \ d = \frac{D}{1+A\beta}$$

The distortion with feedback is therefore reduced by $1+A\beta$ over what it would have been in the absence of feedback.

In making the comparison it is assumed that the outputs are the same with and without feedback. This is necessary because the voltage excursions in the output stage of the amplifier (where nearly all of the distortion is introduced) must be the same in both cases. As the amplifier with feedback has a lower gain than the same amplifier without feedback, the input signal to the feedback amplifier must be raised sufficiently to make the outputs equal. In doing this it is assumed that the driver stage supplying the input at the higher voltage does not contribute to the distortion by being itself overdriven.

The previous paragraphs have shown that negative feedback has the effect of straightening the effective dynamic characteristic. If however any stage of the amplifier is so overloaded that the valves are driven beyond cut-off or are 'bottomed' then feedback is not be able to reduce the resulting distortion. On the contrary, it leads to greater distortion because, having straightened the main part of the characteristic, the discontinuities at the overload points are more severe than they would be were feedback not applied. (A valve or transistor is said 'bottom' when is driven into saturation, i.e. is passing the largest possible current. Under these conditions the anode voltage becomes almost zero and reaches its bottom value.)

Hum and noise when introduced into the amplifier from sources such as the mains transformer, heater lines and the components themselves, are reduced by negative feedback. This does not mean that hum originally in the signal is reduced, because so far as the amplifier is concerned the hum voltage itself constitutes a signal. With negative feedback the ratio of hum to the wanted signal remains the same. Care must therefore be taken that no hum is induced in the input section of the amplifier which may not be within the feedback loop, e.g. the lead to the first valve.

It is not intended in this book to go through all of the mathematical analyses associated with negative feedback amplifiers. It can be shown, for example, that phase distortion is reduced in these amplifiers. For cases where there is voltage negative feedback the input impedance of the amplifier is raised and the output impedance lowered when compared with the same amplifier operated without feedback.

The design of feedback amplifiers is best left to the experts. It is a matter of great difficulty to apply feedback over more than three stages in an amplifier. Because of excessive phase shift at very high and very low frequencies positive feedback is not easy to avoid. When positive feedback occurs there is instability in the amplifier, usually resulting in oscillation. A discussion on stability criteria is highly mathematical and outside the scope of this book.

It is possible, and easier, to construct an amplifier in which negative feedback is applied around each stage individually. The final result, however, is inferior to that obtained by applying negative feedback over the whole amplifier in a single loop.

The Cathode Follower

This circuit, shown in *Figure 9.15*, is a single-stage amplifier in which the load resistor is placed in the cathode lead. The anode is taken directly to the h.t. positive line or is suitably decoupled so that it is short-circuited to the h.t. negative lead so far as alternating signals are concerned. It should be noticed in passing that the h.t. positive and negative lines are at the same potential from the signal point of view. The large smoothing capacitor in the power supply effectively shorts the two lines together so far as a.c. is concerned. The two lines are, of course, at different steady potentials.

The cathode follower is a negative voltage feedback amplifier in which the whole of the output voltage is subtracted from the signal voltage to give the grid-to-cathode voltage. β is therefore equal to unity. The gain of the amplifier is then given by

$$A' = \frac{A}{1+A}$$

233

This is necessarily always less than 1, from which we deduce that the circuit is of no use as a voltage amplifier. However, the cathode follower circuit is intended not as a voltage amplifier, but as a very useful impedance transformer. Its input impedance is so high that it virtually places no load on the circuit supplying the input voltage; this circuit's operating conditions are therefore undisturbed. The cathode follower is consequently very useful as the input stage of a valve voltmeter. (We recall from Chapter 6 that ordinary voltmeters can easily upset the working conditions of a high-impedance circuit.)

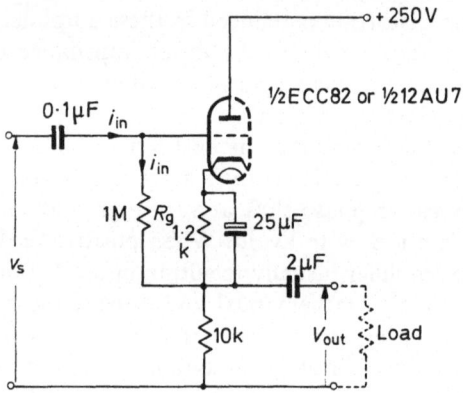

Figure 9.15. The cathode-follower circuit. With the values of components given the gain is about 0·9 and the input impedance 10 M. The output impedance is about 450 ohms (see text)

The output impedance of a cathode follower is so low that considerable power can be supplied to loads of relatively low impedance. The circuit is useful for feeding long transmission lines or signal cables. For audio work it is often unnecessary to screen the cable. The circuit is ideal for coupling a high-impedance microphone to an amplifier via a long lead. The cathode follower has excellent frequency response, and, being a negative feedback amplifier, the distortion of the stage is low.

The expression for the input resistance, R_i, can be obtained by considering *Figure 9.15.*

$$R_i = \frac{v_s}{i_{in}}$$

We have no appreciable grid current when operating a valve with negative grid bias so that the input current flows through the grid resistor R_g. The alternating voltage across the ends of R_g is $v_s - V_{out}$ therefore $i_{in} = (v_s - V_{out})/R_g$.

234

From this we have

$$R_i = \frac{v_s}{(v_s - V_{out})/R_g} = \frac{R_g}{1 - A}$$

where A is the gain equal to V_{out}/v_s. Now the gain of a cathode follower is very close to unity, therefore R_i is very high. For example, if we take typical values of $R_g = 1$ M and $A = 0.9$ then $R_i = 1/(1-0.9) = 10$ M. This is a considerable increase in the input resistance over that of a conventional amplifier. In the latter the input resistance is the grid resistance R_g.

$$C = C_s + C_{gc}$$

where C_s = stray wiring and other capacitances

C_{gc} = grid-to-cathode interelectrode capacitance

C_{ag} is the anode-to-grid interelectrode capacitance

Voltage across $C = v_g$

Voltage across $C_{ag} = v_g + V_{out}$
$$= v_g + Av_g$$
$$= v_g(1 + A)$$

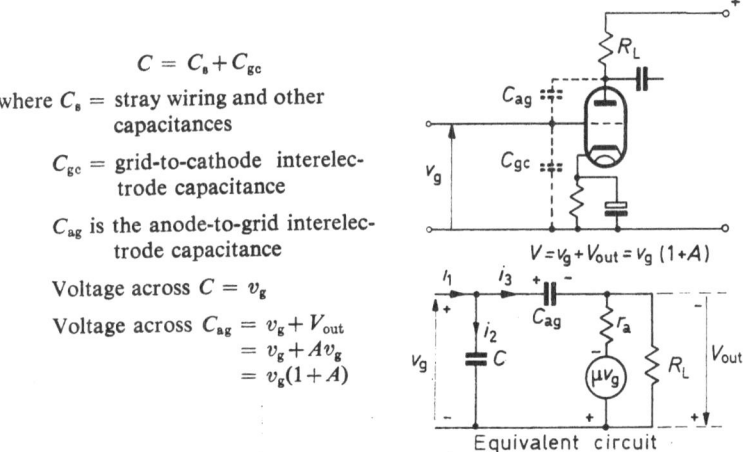

Figure 9.16. Calculation of effective input capacitance due to Miller effect can be made by using the equivalent circuit

When operating at high frequencies the shunting effect of the stray capacitances must be taken into consideration. If the effective input capacitance is say 200 pF then, although the input resistance may be 1 M in an ordinary amplifier, the reactance of the parallel capacitance at 1 MHz is only about 1·6k. *Figure 9.16* represents the position in an ordinary amplifier. Although the stray and interelectrode capacitances themselves may be quite small, their effective value is many times larger due to what is called the Miller effect. This can be shown in the following way. As the grid voltage, v_g, rises the anode voltage, V_{out}, (which is equal to Av_g) falls. Across the anode-to-grid capacitance, C_{ag} there is a voltage equal to the sum of these two voltages, i.e. $v_g(1 + A)$. A current, i_3, therefore passes through this capacitance given by

$$i_3 = \frac{v_g(1 + A)}{X_{C(ag)}} \quad \text{where} \quad X_{C(ag)} = \frac{1}{j\omega C_{ag}}$$

$$i_2 = \frac{v_g}{X_C}$$

235

and $\quad i_3 = i_1 - i_2$

$$\therefore \; i_1 = i_2 + i_3 = \frac{v_g}{X_C} + \frac{v_g(1+A)}{X_{C(\mathrm{ag})}}$$

$$= v_g \left(\frac{1}{X_C} + \frac{1+A}{X_{C(\mathrm{ag})}} \right)$$

The input reactance $X_{\mathrm{in}} = v_g/i_1$

$$\therefore \; \frac{1}{X_{\mathrm{in}}} = \frac{1}{X_C} + \frac{1+A}{X_{C(\mathrm{ag})}}$$

$$\therefore \; j\omega C_{\mathrm{in}} = j\omega C + j\omega C_{\mathrm{ag}}(1+A)$$

$$\therefore \; C_{\mathrm{in}} = C + C_{\mathrm{ag}}(1+A)$$

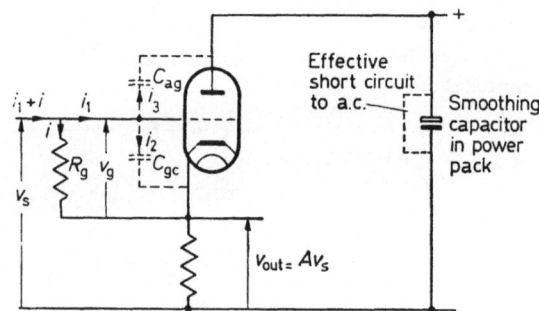

Figure 9.17. Circuit representation used to calculate the input capacitance of a cathode follower

This means that the input capacitance is not merely the input stray capacitance plus the interelectrode capacitance between grid and cathode. The effective input capacitance is increased by $(1+A) C_{\mathrm{ag}}$. If $C_{\mathrm{ag}} = 5$ pF, as it may well do in a triode, and A is 40, for example, $(1+A)C_{\mathrm{ag}}$ exceeds 200 pF.

The position is greatly improved in a cathode follower circuit *(Figure 9.17)*. The voltage across C_{gc} is v_g, and that across $C_{\mathrm{ag}} = v_s$. Now $v_g = v_s - Av_s = v_s(1-A)$, so $i_3 = v_s/X_{C(\mathrm{ag})}$ and $i_2 = v_g/X_{C(\mathrm{gc})}$

$$\therefore \; i_1 = \frac{v_s}{X_{C(\mathrm{ag})}} + \frac{v_s(1-A)}{X_{C(\mathrm{gc})}}$$

The input reactance $X_{\mathrm{in}} = v_s/i_1$

$$\therefore \; \frac{1}{X_{\mathrm{in}}} = \frac{1}{X_{C(\mathrm{ag})}} + \frac{1-A}{X_{C(\mathrm{gc})}}$$

$$\therefore \; C_{\mathrm{in}} = C_{\mathrm{ag}} + C_{\mathrm{gc}}(1-A)$$

236

As A is very nearly unity, the input capacitance is equal to C_{ag} for all practical purposes. This would make $C_{in} = 5$ pF for the previous figures, which is a substantial reduction in input capacitance. A cathode follower therefore has a high input resistance and low input capacitance, i.e. its input impedance is very high.

The gain of a cathode follower can be obtained by considering the equivalent circuits of *Figure 9.18*.

$$V_{out} = \frac{\mu R_L v_g}{r_a + R_L}$$

This is the conventional expression previously derived in Chapter 4. In the case of the cathode follower the input voltage is not v_g but v_s. However the

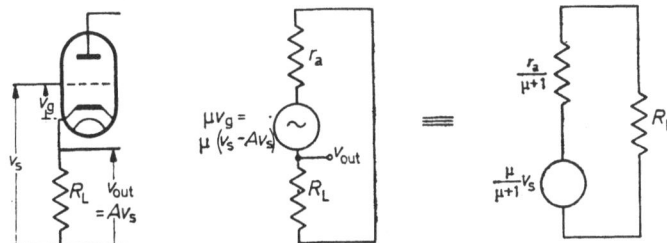

Figure 9.18. The equivalent circuit for finding the gain and output impedance of a cathode follower

grid voltage is given by $v_g = v_s - A v_s = v_s(1 - A)$.

$$\therefore \quad V_{out} = \frac{\mu R_L}{r_a + R_L} v_s(1 - A)$$

$$\therefore \quad \frac{V_{out}}{v_s} = A = \frac{\mu R_L(1 - A)}{r_a + R_L}$$

$$A(r_a + R_L + \mu R_L) = \mu R_L$$

$$\text{and} \quad A = \frac{\mu R_L}{r_a + R_L + \mu R_L}$$

This can be written as

$$A = \frac{\dfrac{\mu}{\mu+1} R_L}{\dfrac{r_a}{\mu+1} + R_L} = \frac{\mu' R_L}{r_a' + R_L}$$

Comparing this with the standard formula it can be seen that the cathode follower behaves like a valve with an amplification factor $\mu' = \mu/(\mu+1)$,

237

and an anode a.c. resistance $r'_a = r_a/(\mu+1)$. The output impedance of a cathode follower is the effective internal impedance of the circuit, i.e. $r_a/(\mu+1)$. As μ is usually much greater than unity this approximates to r_a/μ which is $1/g_m$. In the circuit of *Figure 9.15* the output impedance with half the double triode ECC82 ($g_m = 2 \cdot 2$ mA) is $1/2 \cdot 2 \times 10^{-3}$ i.e. about 455 ohms. Strictly this should be 455 ohms in parallel with the 10 k load resistor. With the component values given the gain is

$$A = \frac{17 \times 10 \, \text{k}}{7 \cdot 7 \, \text{k} + 10 \, \text{k} + 17 \times 10 \, \text{k}} = \frac{170 \, \text{k}}{187 \cdot 7 \, \text{k}} \fallingdotseq 0 \cdot 9$$

(For an ECC82 $\mu = 17$ and $r_a = 7 \cdot 7$ k)

TRANSISTOR AMPLIFIERS

The very large amount of work done on thermionic valve circuits before the invention of the transistor made it inevitable that the new device would be tried in similar circuit configurations. Many successful transistor circuits are therefore the corresponding valve circuits in which transistors replace thermionic valves; suitable adjustments of supply voltages and passive component values are made to accommodate the properties of the semiconductor device.

The simple single-stage amplifier may be designed graphically in a manner similar to that used for the thermionic valve. The most important configuration used is the grounded or common emitter mode. This is roughly equivalent to the standard common cathode mode used with valves. The signal is applied between the base and the emitter and the output is taken from the collector. Since the emitter is at the potential of the bottom supply line so far as a.c. is concerned, the emitter is common to both the input and output circuits—hence the term common emitter mode. *Figure 9.19* shows a typical set of transistor characteristics for the common emitter mode. In the family of curves, each curve shows the relation between collector current and collector voltage for a fixed value of base current; base current is the third variable because the transistor is a current-operated device. (With thermionic valves, grid voltage is the third variable since valves are voltage-operated devices.) Alongside the characteristics is the simplest possible Class A amplifier. The load line is constructed in the usual way from a knowledge of the supply voltage and load resistance. A suitable operating point is chosen and the bias current noted. In a *pnp* device, it is necessary to inject electrons into the base, therefore the simplest way of obtaining the bias current is to connect a resistor from the base to the negative supply lead. In calculating the bias resistor it is noted that the emitter junction is forward biased and so the emitter-base voltage cannot be large. It is approximately 200 mV in germanium transistors and 400 mV in silicon transistors. Compared with the supply voltage (of 12 V in our example) the base-emitter voltage is small and can be neglected. The voltage between the negative supply line and the base can be taken as the supply voltage. It is now an easy matter to calculate the bias resistor from a knowledge of the supply voltage and bias current required. For our example this resistor is $12/(50 \times 10^{-6}) = 240$ k.

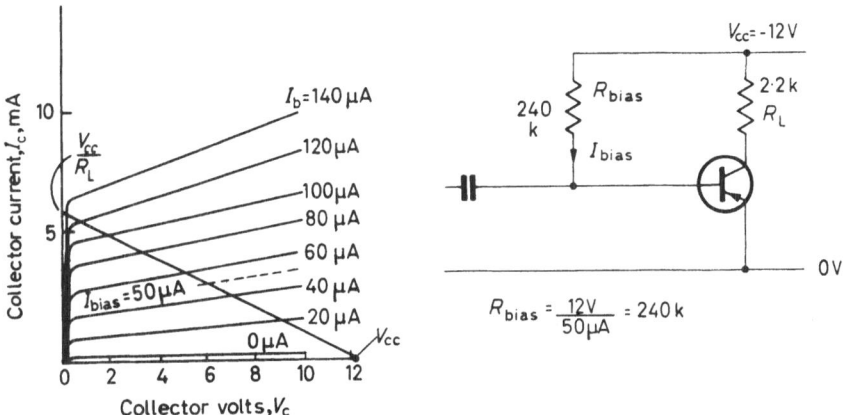

Figure 9.19. The simplest transistor amplifier. The diagram shows how to design the circuit from the load line and collector characteristics

Temperature Effects and Bias Stabilization

The simple amplifier described in the previous section is not satisfactory because its performance is easily and adversely affected by changes in temperature. In normal transistor action the injection of a signal current into the base region effectively increases the number of charge carriers there. In the case of a *pnp* transistor, it is electrons that are injected into the base. Holes are then attracted to the base from the emitter region. In a good transistor relatively little recombination occurs in the base region and the majority of holes are swept into the collector region across the collector-base barrier layer and constitute the collector current, I_C. $I_C = \alpha' I_b$ or βI_b where I_b is the base current and α' (or β) represents the current gain. Unfortunately at room temperatures, germanium, and to a much lesser extent silicon, exhibit intrinsic conduction. It will be remembered from Chapter 5 that such conduction results from the presence of charge carriers (the electron-hole pairs) formed when a covalent bond is ruptured. These charge carriers

239

give rise to leakage currents which, if not properly controlled, can lead to unsatisfactory transistor operation or even to the irreparable damage of the device. The leakage currents, although small in the case of common-base operation, are very much larger for the same transistor in the common emitter configuration. Let us consider first *Figure 9.20*. Here a transistor is

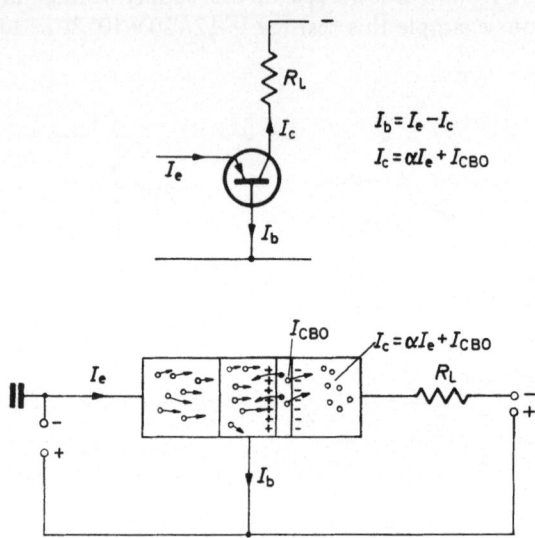

$$I_b = I_e - I_c$$
$$I_c = \alpha I_e + I_{CBO}$$

Figure 9.20. The leakage current in a grounded or common base amplifier

being used in the common-base mode. Most of the emitter current from the bias and signal source diffuses through the base and is swept into the collector region. Very little current passes down the base lead. The collector current is not quite equal to the emitter current. In fact $I_C = \alpha I_e$ where α is defined as the current gain in the common base mode. The gain, α, is less than unity, a usual figure being about 0·98 to 0·99. Power amplification takes place however because the charge carriers constituting the collector current come under the influence of the voltage supply to the collector, which is large compared with the emitter-base voltage. Considerable work can therefore be done by the collector current. It is possible to use a high collector load resistance and hence develop a considerable output voltage. From *Figure 9.20* it can be seen that the collector current consists not only of αI_e but also the leakage current I_{CBO}. (The symbol I_{CBO} is used to represent the leakage current between the collector and the base, the emitter being open-circuit. This is the significance of the subscript letters.) When operated as an amplifier in the common-base mode the same leakage current is produced even though the emitter is connected to the signal source. The collector current is therefore $I_C = \alpha I_e + I_{CBO}$. I_{CBO} is a small fraction of αI_e. Its value at room temperatures is about 2 to 4 μA in a germanium device and about 0·04 μA in a silicon one.

In the common emitter mode the position is very different. *Figure 9.21* shows the position. The leakage current, I_{CBO}, as before is produced by the rupture of covalent bonds in the depletion layer of the collector junction. The presence of the electron from the hole-electron pair in the base region disturbs the position at the emitter junction and holes are induced into the

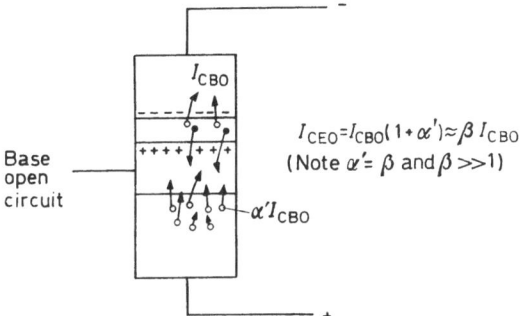

Figure 9.21. The leakage current in a common emitter amplifier, I_{CEO}, is approximately β times that in the common base amplifier

base region just as they are in normal transistor action. No differentiation is made in the base region between electrons from the signal source and those due to leakage current. Normal current amplification occurs and the total leakage current is then I_{CBO} from the covalent bond ruptures, plus $\alpha' I_{CBO}$ due to transistor action. (α' is the current gain in the common emitter mode. This is signified by priming the symbol α; as previously stated β is often used for the current amplification in this mode, i.e. $\alpha' = \beta$.) The total leakage in the common emitter mode, I_{CEO}, is given by

$$I_{CEO} = I_{CBO} + \alpha' I_{CBO} = I_{CBO}(1 + \beta)$$

As β in a transistor may be anything in the approximate range 50 to 250 the leakage current is considerable. For a germanium transistor with $\beta = 100$, the leakage current may approach 500 μA. This very considerable current increases the temperature of the collector junction giving rise to a further increase in I_{CBO}. If suitable circuit arrangements are not made to stabilize the collector current, thermal runaway may result, and the transistor is ruined. Even if this does not happen the presence of such leakage currents shifts the operating point in an undesirable way. The common emitter circuit is therefore unstable from a temperature point of view.

The thermal stability of any common emitter amplifier is considered quantitatively by defining a thermal stability factor F given by

$$F = \frac{dI_{CS}}{dI_{CEO}}$$

where I_{CS} is the stabilized collector current. This stability factor should be as small as possible, i.e. for a given increment in leakage current the increment in the collector current should be be as small as possible (ideally equal to zero).

Biasing Circuits

From the foregoing it can be seen that the problem in biasing a transistor is not so much one of choosing a suitable bias current and operating point as one of stabilizing the position of the operating point with changes of temperature. Without a stable operating point the amplifying stage becomes useless.

Several ways of stabilizing the bias point have been proposed. They all depend upon feedback in some form. Two popular ways are shown in *Figure 9.22*. In *Figure 9.22a* the bias resistor is taken to a tapping along the load. If the leakage current rises the potential at A falls and the bias current is reduced. This, in turn, reduces the collector current thus offsetting the effect of the rise due to increased leakage current. In *Figure 9.22b* feedback is introduced by placing a resistor in the emitter lead. By maintaining the base voltage at suitable constant value any undue rise in leakage current causes the voltage at the emitter to fall. The emitter-base junction then approaches a reverse-biased condition, thus reducing the current through the device. In the absence of a signal the base voltage is held reasonably constant by the potential divider R1 + R2. The lower the values of R1 and R2 the more constant is the base voltage, and the more effective is the emitter resistor in stabilizing the operating point. There is a limit, however, to the reduction in resistance values that can be tolerated. This limit is set by the current drain on the power supply; for battery operation this drain must be reasonably low. Additionally the value of R2 must be high enough to avoid serious shunting of the signal current.

In designing an amplifier stage similar to *Figure 9.22b* let us suppose that the supply voltage is 9 V and the load resistor 3·9k. We are now required to calculate suitable values for R1, R2, and R_e. It is usual to allow the voltage across R_e to be 10 to 15 per cent of the supply voltage. This provides reasonable thermal stability without unduly reducing the voltage available

for the transistor. In our example we may take the emitter voltage to be
$-1\cdot0$ V. The quiescent current is determined as explained below; we shall
choose 1 mA as a suitable value. This gives a value of 1k for R_e. The base
voltage must be 200 mV more negative than the emitter voltage, hence the
voltage of the base is fixed at $-1\cdot2$ V. The bias current is obtained graphically and will be taken to be 20 μA. As an empirical rule the R1R2 chain

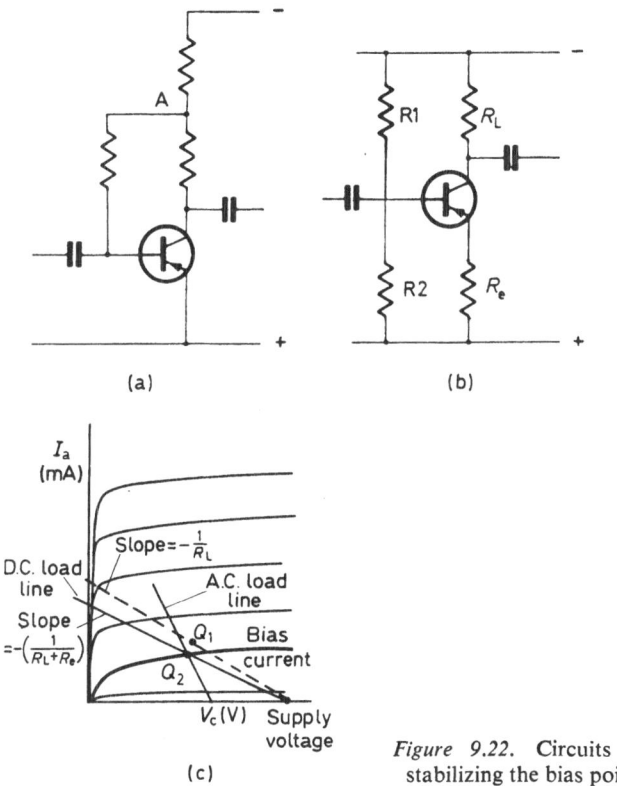

(a)

(b)

(c)

Figure 9.22. Circuits for
stabilizing the bias point

is allowed to draw from 5 to 10 times the bias current. Let us say that the
current in R2 for our amplifier is 6 times the bias current i.e. 120 μA. Since
the voltage across R2 must be $-1\cdot2$ V, R2 must be 10 k. The voltage across
R1 is evidently $9-1\cdot2$ i.e. 7·8 V therefore R2 = 7·8 V/140 μA i.e. 56 k (using
the nearest preferred value). The current through R1 is 140 μA being the
current through R2 plus the bias current to the transistor. We thus arrive
at the circuit of *Figure 9.23.*

It will be realized that the design of a transistor amplifier is more complicated than that of a valve amplifier. We assumed in our example that the

quiescent current was known, whereas in practice this must be estimated from the I_c/V_c characteristics. Plotting the load line on these characteristics is not straightforward because the emitter resistance cannot be ignored as can the cathode resistance in a valve amplifier. The cathode resistance is very small compared with the anode load resistance, whereas in a transistor amplifier the emitter resistance is usually as high as 20 to 30 per cent of the collector resistor. For d.c. purposes, therefore, in estimating the bias conditions and obtaining a suitable working point the effective load consists of

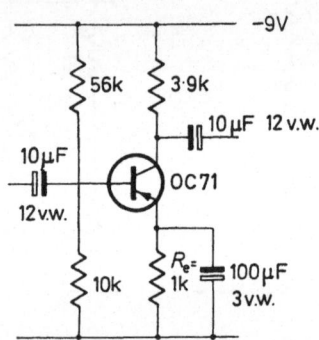

Figure 9.23. A complete single-stage transistor amplifier. The emitter bypass capacitor is necessary to provide a short-circuit path for a.c. components of the emitter current

R_L plus R_e. The resulting load line is known as the d.c. load line and has a slope of $-1/(R_L+R_e)$. Since R_e cannot be calculated without a knowledge of the quiescent current we may proceed by constructing an initial load line for R_L in the usual way. An estimation of the appropriate quiescent current from Q_1 *(Figure 9.22c)* is known to give a high value since the actual d.c. load line has a smaller slope. An intelligent estimate of the position of the correct operating point, Q_2, will give a value for the quiescent current that is correct for all practical purposes. As a guide Q_2 will give a quiescent current about 10 per cent less than that for Q_1. R_e can now be calculated as previously explained and a new load line drawn for R_L+R_e. If this line passes through Q_2 the original estimate was perfect. Provided the d.c. load line is close to Q_2 there is little point in proceeding to a second estimation.

The output voltage available from the stage cannot be predicted from the d.c. load line because the signal output is developed only across R_L. A load line passing through Q_2 with a slope of $-1/R_L$ is known as the a.c. load line and from this the output voltage can be determined. For this we observe the swing of the collector voltage for variations of base current. Variations of the base, i.e. signal, current can be predicted from a knowledge of the input voltage and the input resistance. This latter resistance is obtained from the slope of the graph of base-emitter voltage versus base current. The slope is taken at the point corresponding to the forward bias voltage, i.e. 200 mV in our example. This graph is usually given in the manufacturer's data.

When a single stage is driving a subsequent transistor amplifier the effective load for the initial stage is R_L in parallel with the input impedance, R_{in}, of the following stage. The slope of a.c. load line is then $-1/(1/R_L + 1/R_{in})$. If the second stage is similar to that in *Figure 9.23*, R_{in} for this second stage is about 1·5 k. The first stage will have a current gain of approximately 28, i.e. the current delivered to the second stage divided by the input current is about 28.

Transistor Parameters and Equivalent Circuits

When transistors are incorporated into circuits, it is natural for many electronic engineers to analyse the behaviour of the circuits in order to obtain a better understanding of the design principles involved. For those not wishing to engage in complicated circuit analyses, it is still necessary to know something of the procedures involved so that they can read the literature and understand some of the techniques used.

The analysis of transistor circuits is made easier if the actual transistor can be replaced by an equivalent circuit. The use of an equivalent circuit has already been exemplified when dealing with simple triode amplifiers. With thermionic valves the equivalent circuits are relatively easy to derive, and most workers agree on the constant voltage and constant current equivalent circuits in widespread use. With transistors, however, the position is more difficult because transistors behave in a more complicated way than thermionic triodes. The interaction of input and output circuits is absent in most thermionic devices, whereas with transistors the influence of the output circuit on the input circuit is most marked. Since transistors are current-operated devices and have a comparatively low input impedance, the current drawn by the input cannot be ignored. The more complicated behaviour of transistors is reflected in the large measure of disagreement among electronic engineers and physicists as to the best equivalent circuit to use under a given set of circumstances. A large and bewildering number of parameters have therefore been defined to describe transistor behaviour. Space does not permit a discussion of every parameter system that has been devised.

Some aspects are discussed below of one approach to the subject which is favoured by the author. Not all will agree that it is the best approach. We should maintain an open mind about parameters and not hesitate to use a different system in those circumstances in which it could obviously be better to use the different approach. In many analyses however the system outlined below proves to be satisfactory.

In so far as the transistor is a linear circuit element (i.e. working on the straight portions of the appropriate characteristics) it may be represented as a 'black-box' with two input terminals and two output terminals (*Figure 9.24*). The internal workings of the box are of no concern when considering circuit analyses; the transistor's behaviour from an external point of view is all that needs to be known. Four variables are associated with the box, viz. v_1, i_1, v_2 and i_2, representing input voltage and current and output voltage and current, respectively. Because linear operation is assumed these signal voltages and currents are related by sets of equations. Any two variables can be taken as known and the other two can be calculated from the equations. There are six ways in which two variables from the four can be selected. Not all selections produce useful results. The most fruitful are those about to be described.

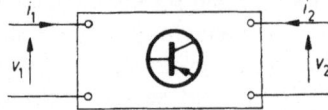

Figure 9.24. 'Black-box' four-terminal network representation of a transistor

Suppose the two currents involved, i_1 and i_2, are known. The voltages, v_1 and v_2 must be related to the currents. The relationships can be expressed by the following equations.

$$v_1 = z_{11}i_1 + z_{12}i_2 \qquad (9.1)$$

$$v_2 = z_{21}i_1 + z_{22}i_2 \qquad (9.2)$$

The z coefficients have numeral subscripts to show their position i.e. z_1's are in the first line, the first being z_{11} and the second z_{12}. In the second line there are z_2's with z_{21} and z_{22} in the first and second position, respectively. Expressed in matrix form the equations become

$$\begin{pmatrix} v_1 \\ v_2 \end{pmatrix} = (Z) \begin{matrix} i_1 \\ i_2 \end{matrix} \quad \text{where} \quad Z = \begin{pmatrix} z_{11} & z_{12} \\ z_{21} & z_{22} \end{pmatrix}$$

The reader need feel no alarm if matrix algebra is a new technique to him. The elements of the subject are very easy to grasp. (They are described in Appendix 5, the description being a precis of two of the author's articles on the subject[1]. We shall avoid the matrix form here.) The z coefficients are called the z parameters of the transistor and are a suitable description of the transistor's external behaviour. The various terms of the defining equations represent the a.c. or signal variations only, and it is assumed that all supply voltages and bias currents are present. The latter play no part in circuit analysis however and can be ignored.

When the transistor is open-circuited to a.c. at its output, i_2 is zero and equation 9.1 becomes $v_1 = z_{11}i_1$. z_{11} is therefore the input impedance, v_1/i_1, when the output is open-circuited. With the output short-circuited to a.c. $v_2 = 0$, therefore $-z_{21}/z_{22} = i_2/i_1$. This ratio is the current gain. If the transistor is in the common-base mode of operation, $-z_{21}/z_{22} = \alpha$. In the common emitter mode primes are used to indicate this mode thus $-z_{21}/z_{22} = \alpha' = \beta$. By opening and shorting the input and output terminals to a.c. and using equations 9.1 and 9.2 it can be seen that z_{11} is the input impedance with output o/c (open-circuited); z_{21} is the forward transfer impedance with the output o/c; z_{22} is $v_2/i_2 =$ output impedance with input o/c and z_{12} is the reverse transfer impedance with the input o/c.

If $z_{12}i_1$ is added to both sides of equation 9.2 and $z_{21}i_1$ is transposed then

$$v_1 = z_{11}i_1 + z_{12}i_2 \tag{9.3}$$

$$v_2 + (z_{12} - z_{21})i_1 = z_{12}i_1 + z_{22}i_2 \tag{9.4}$$

These equations show that the 'black-box' may be replaced by either of the networks of *Figure 9.25*. *Figure 9.25* shows what are called the T-parameters,

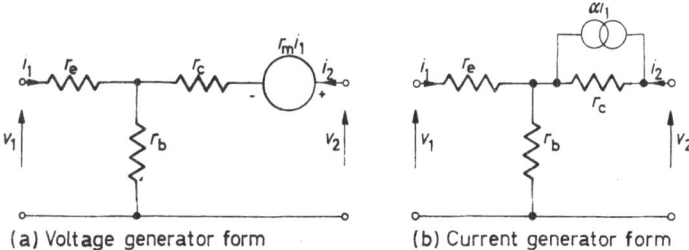

(a) Voltage generator form   (b) Current generator form

Figure 9.25. The equivalent circuits using T-parameters

Figure 9.25a being the voltage-generator form and *Figure 9.25b* being the current-generator form. From *Figure 9.25* it can be seen that

$$v_1 = (r_e + r_b)i_1 + r_b i_2 \tag{9.5}$$

$$v_2 - r_m i_1 = r_b i_1 + (r_b + r_c)i_2 \tag{9.6}$$

By comparing equations 9.3, 9.4, 9.5 and 9.6 and equating coefficients, it is easy to obtain a relationship between the T-parameters and the z-parameters.

The T-parameters are often used as a first approach to transistor equivalent circuits. The resistances r_e, r_b and r_c are respectively the resistances associated with the emitter, base and collector. Typical values for a small a.f. transistor are $r_e = 50$ ohms, $r_b = 500$ ohms and $r_c = 1$ M. r_m, the mutual resistance, is 0·98 M, the corresponding α being 0·98 and $\beta = 50$. The accuracy of the figures is not high. It must be remembered that there are large tolerance spreads in transistors of the same nominal type from the same manufacturer.

From the foregoing, it can be seen how a whole system of parameters may be constructed. The principles for setting up the equations are the same in every case. Two of the four variables are selected and their dependence on the other two, using suitable coefficients, is stated. For example, for certain purposes it may not be found convenient to use the z-parameters. In high frequency work the y-parameters are preferred. They are defined from the equations

$$i_1 = y_{11}v_1 + y_{12}v_2 \tag{9.7}$$

$$i_2 = y_{21}v_1 + y_{22}v_2 \tag{9.8}$$

They are called the y or admittance parameters because each one has the dimensions of an admittance, i.e. a current divided by a voltage.

Both the z- and y-parameters are difficult to measure in the laboratory; a set is therefore defined in which all of the parameters can be easily and accurately determined. Such a set is known as the h-parameters, h standing for hybrid. Many manufacturers now prefer to describe their transistors in terms of h-parameters. The defining equations are

$$v_1 = h_{11}i_1 + h_{12}v_2 \tag{9.9}$$

$$i_2 = h_{21}i_1 + h_{22}v_2 \tag{9.10}$$

Examination shows that h_{11} has the dimensions of an impedance, h_{22} the dimensions of an admittance and h_{12} and h_{21} are pure ratios. This is the reason for calling them hybrid parameters.

In the United States, they prefer not to use the numeral subscripts. Instead they use h_i, h_r, h_f, h_o. A second letter subscript shows the mode of operation, b, e, and c standing for common (or grounded) base, emitter and collector, respectively. Thus

$$h_{ib} = h_{11}$$
$$h_{rb} = h_{12}$$
$$h_{fb} = h_{21}$$
$$h_{ob} = h_{22}$$

In the common emitter mode numeral subscripts are primed thus

$$h_{ie} = h_{11}'$$
$$h_{re} = h_{12}'$$
$$h_{fe} = h_{21}'$$
$$h_{oe} = h_{22}'$$

i, r, f and o stand for input, reverse, forward, and output, respectively. h_{ie} is therefore the input resistance v_1/i_1 of a common emitter transistor with the output short-circuited to a.c. (see equation 9.9). Under the same circumstances $h_{fe}(= h_{21}') = i_2/i_1$. This is the current gain in the common emitter mode, i.e. α' or β. h_{oe} is the output impedance with the input open-circuited ($i_1 = 0$). (When $i_1 = 0$ it is implied that the input current, I, is constant so that the variation or change in input current, i_1, is zero.) h_{re} is the reverse voltage feedback ratio v_1/v_2 when $i_1 = 0$.

Although there is a large, and for the newcomer bewildering, number of parameters, the latter are all logically derived from a simple basic pattern. Once this is appreciated much of the initial confusion is dispelled. A table for the interconversion of z, y and h parameters is given in Appendix 5.

Multistage L.F. Amplifiers

When more gain is required than can be obtained from a single stage, two or more stages are coupled together. The simplest arrangement is to use RC coupling.

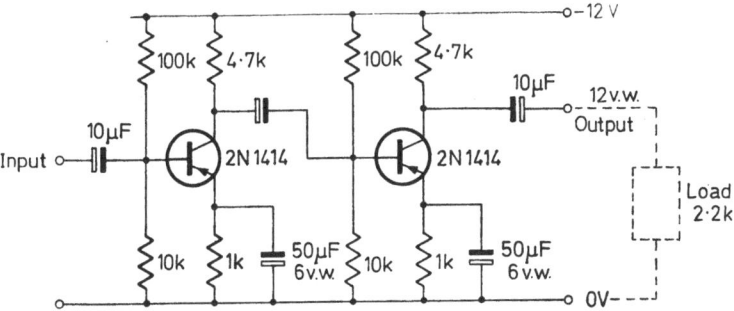

Figure 9.26. A complete two-stage transistor amplifier. The significance of the load is explained in the text

Figure 9.26 shows a simple two-stage amplifier. The term 'gain' when used in connection with a transistor amplifier is ambiguous. The terms 'power gain' 'current gain' or 'voltage gain' must be used. The current gain of the first stage is not the β figure for the transistor. Redrawing *Figure 9.26* in the block diagram form of *Figure 9.27*, it can be seen that if there is an input current of i_1 to the first stage, the first transistor gives an output current of βi_1. This current must be shared between four resistors which are effectively in parallel, the resistor $R_L = 4.7$ k, the 10 k biasing resistor (R_2) in the second stage in parallel with the 100 k resistor (R_1) and a resistance to represent the internal loss in the first transistor. This last resistor is approximately 20 k. If the load resistor is, say, 4·7 k in parallel with $R_2 (= 10$ k) then the resultant effective resistance is about 3 k. (We may ignore R_1.) The input resistance of the second transistor is about 1·5 k and, therefore, only about 67 per cent of βi_1 goes into the second transistor. If this second transistor is loaded with 2·2 k (i.e. about 0·5 of 4·7 k) then the overall current gain of the two stages is approximately $(0·67\beta)^2$ i.e. $4\beta^2/9$. For a 2N1414, β is about 35 and so the current gain of the amplifier of *Figure 9.26* may be expected to be approximately 540; thus for large current gains, the collector resistance should be high compared with the input resistance of the following stage. An alternative way of expressing this is to say that for maximum current transfer from one stage to the next, the input resistance of the latter stage should be low compared with the effective output impedance of the previous stage.

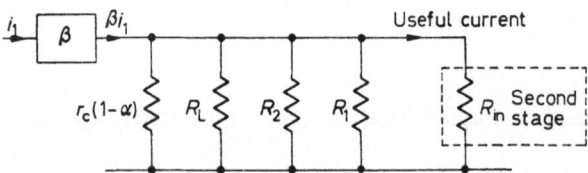

Figure 9.27. Equivalent circuit for the calculation of the current gain of a single stage of amplification. For the significance of R_L, R_1 and R_2, see *Figure 9.22b*

The voltage gain of an isolated single stage may be found approximately in the following way. If the input resistance of the stage is R_{in} and the input voltage is v_1, then the input current to the base must be v_1/R_{in}. The collector current is, therefore, approximately $\beta v_1/R_{in}$ and so the output voltage is $\beta v_i R_L/R_{in}$. Taking $\beta = 35$, $R_{in} = 1$ k and $R_L = 5$ k the voltage gain is approximately 175. When a second stage is added however the voltage gain drops considerably because the effective load for the first transistor is now 5 k in parallel with the input resistance of the second stage. This input resistance may be 1·2 k. The effective load for the first transistor is now approximately 1 k and the voltage gain of the first stage drops to 35. The input resistance of the second stage is therefore of dominant importance when considering voltage amplification.

Within a multistage voltage amplifier not every stage should have the greatest voltage amplification. On the contrary, in the intermediate stages the current gain should be as high as possible so that the signal current driving the final stage produces the largest collector current possible. This means that the greatest output voltage is developed across the collector load. In order to maintain this voltage however the last stage must not lead into a load impedance much smaller than the collector load resistance. It is obviously

250

no solution to reduce the collector load resistance to a value that is small compared with a given load resistance. In practice, the effective output impedance of an amplifier is reduced by using a suitable form of negative feedback. The circuit of *Figure 9.22a* gives a low output impedance with the bias resistor connected to the collector instead of the point A as shown. In addition, the input impedance is also reduced. This circuit is therefore a satisfactory final stage in a voltage amplifier since it has the necessary low output impedance. There is also good current transfer from the previous stage because of the low input impedance.

The first stage of a voltage amplifier should have a high input impedance because of the nature of the signal source. Voltage signal sources have high impedances in contrast to current signal sources which have low impedances. Current amplifiers are relatively easy to design because transistors have an inherent low input impedance. It is when high input impedances are required that difficulties are encountered.

The simplest way of increasing the input impedance of the first stage of an amplifier is to omit the emitter bypass capacitor. This produces voltage feedback and thus, in addition to increasing the input impedance, the benefit of lower distortion and better frequency response is obtained. Omitting the bypass capacitor also has the effect of increasing the output impedance of the stage. By combining this with a following stage that has a low input impedance, there is good current transfer from the first to the second stage. *Figure 9.28a* and *b* shows circuits for the two stages which, when combined, give a two-stage voltage amplifier.

Although omitting the bypass capacitor does increase the input impedance the rise is typically from about 1 k to only a few tens of kilohms. In view of the presence of the resistor, R_2, of the potential divider chain supplying the bias current, the effective input impedance of a conventional stage is not likely to exceed about 10 k. This input impedance is still very low for many purposes. One way of solving the input impedance problem is to use an emitter-follower input stage. *Figure 9.28c* shows the arrangement, which can readily be recognized as the transistor equivalent of the cathode-follower valve circuit. In order to maintain a high input impedance to the amplifier the input impedance of the stage following the emitter follower must not be too low. This is the reason for decoupling the bias resistor of the second transistor. In the absence of this decoupling, feedback would occur as in *Figure 9.28b* and this results in a lowering of the input impedance.

High input impedance means that for a given input voltage the input current must be small. Extremely small input currents can be made to operate the emitter follower if they are first amplified in another transistor; a compound pair (sometimes called a Darlington pair after the name of the investigator of compound pairs) is then formed. In this way the input impedance of the compound emitter follower can be raised to a value of approximately $h_{fe1}h_{fe2}R_L$ where h_{fe1} and h_{fe2} are the current gains of the transistors forming the compound pair and R_L is the emitter load resistor. For $h_{fe1} = h_{fe2} = 50$ and $R_L = 1$ k the input impedance is 2·5 M, a very high value for a transistor circuit. When combined with the 1·5 M bias resistor, the effective input impedance is about 1 M.

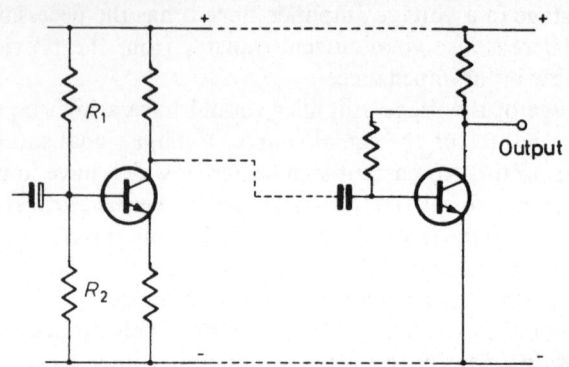

(a) Simple transistor amplifier with bypass capacitor omitted to increase input resistance. Output resistance is also increased. Note change of voltage polarity because an *npn* transistor is being used

(b) The use of feedback from the collector decreases the input and output impedances. Coupled to (a) this circuit forms a two-stage voltage amplifier

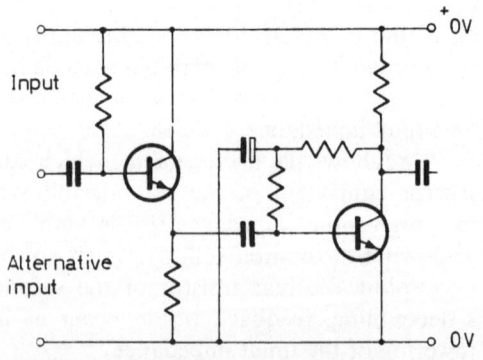

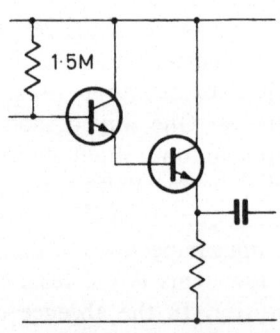

(c) Use of emitter follower stage increases the input impedance of the amplifier

(d) Where the bias for the emitter follower is obtained via another transistor the input impedance can be raised to 1 M approx.

Figure 9.28. The various stages in producing a voltage amplifier

252

Figure 9.28e shows a complete amplifier published by Ferranti Ltd.[2], which combines the features discussed in connection with the circuits of *Figure 9.28a–d*. The last stage returns the feedback resistor R12 to the emitter of Tr3. In doing this the feedback reduces not only the distortion in the last stage but also reduces that introduced by the coupling network C3R9. The frequency response of the amplifier is flat within 5 dB over the range 10 Hz–50 kHz. The nominal input level is 40 mV (r.m.s.) and with this input the voltage output is 3 V into 1k. The total harmonic distortion is said to be less than 0·1%. The input and output impedances are 1 M and 150 ohms respectively taken at 1 kHz.

The Darlington pair arrangement is not confined to emitter followers. Apart from the emitter follower (i.e. common collector) mode the arrangement can also be used in the common emitter and common base modes. In the common emitter mode of *Figure 9.29*, the emitter of Tr1 is connected directly to the base of Tr2, the collectors of both transistors being connected to a common load R_L. The bias resistor R_B is chosen so that both transistors are operating under Class A conditions. Any input signal gives rise to an emitter current that is approximately β_1 times the input current, i_{in}. This emitter current is the base current for Tr2, therefore the load current is approximately $\beta_1 i_1 + \beta_1 \beta_2 i_1$ where β_2 is the current gain of Tr2. Thus the current gain is $\beta_1(1 + \beta_2) \approx \beta_1 \beta_2$. The composite pair therefore has a very high current amplification. The input resistance is higher than for an ordinary common-emitter amplifier.

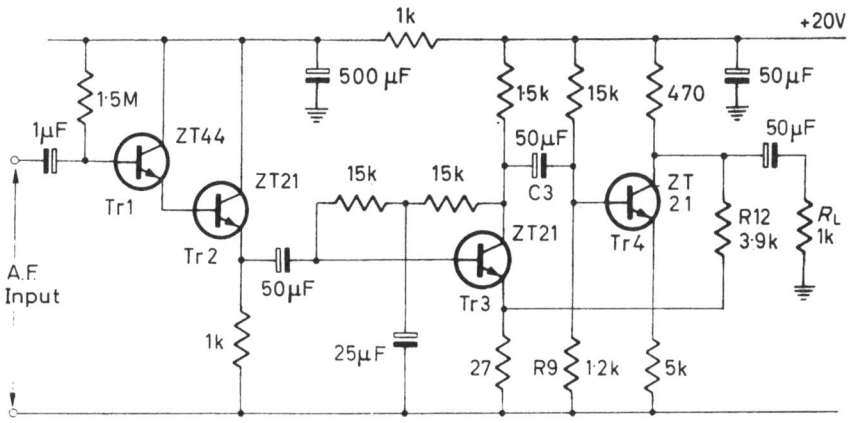

(e) A complete voltage amplifier with an input impedance of 1 M and a frequency response of 10 Hz to 50 kHz within 5 dB. Further details are given in the text. The transistors are Ferranti types and the circuit is reproduced by courtesy of Ferranti Ltd.

with a high input impedance and low output impedance

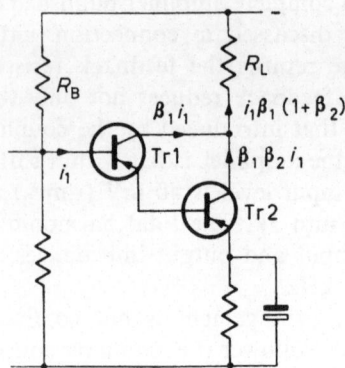

Figure 9.29. The compound connection or Darlington pair

GENERAL PURPOSE AND TUNED AMPLIFIERS

General purpose amplifiers are difficult to design because the specifications vary widely for different applications. No amplifier can therefore be considered ideal for all purposes; however the use of negative feedback can often result in an amplifier that meets wide variety of needs. The basic circuit of *Figure 9.30* due to Texas Instruments Ltd. can be used (*i*) as an amplifier with a fairly wide bandwidth or (*ii*) as a tuned amplifier with a narrow bandwidth. The variable performance is achieved by connecting suitable networks between the terminals O and A.

Basically the circuit uses two Texas silicon *npn* transistors (TI494) with resistance coupling. The collector of Tr1 is connected directly to the base of Tr2. A 27 k resistor between the emitter of Tr2 and the base of Tr1 provides the bias for the first stage. It is, in effect, a d.c. feedback path and produces good d.c. stability. The signals at O and A are 180° out of phase and so networks connected between these points give rise to negative feedback provided there is little or no phase shift in the network itself. By connecting an RC series circuit *(Figure 9.30b)* between O and A, a general purpose feedback amplifier is obtained that has a gain of about 40 dB that is maintained over a temperature range of about −20°C to +100°C. The function of the capacitor is to prevent upsetting the bias conditions in the emitter circuit of Tr1. In effect, it isolates the collector of Tr2 from the emitter circuit of Tr1 so far as direct voltages are concerned. The feedback is effective only for alternating voltages.

If this network is replaced by a frequency-selective one a tuned amplifier results. Let us examine briefly two tuned circuits that would be likely to be

254

used as feedback networks, and note the effect that they have on the frequency response of the amplifier.

For frequencies above 100 kHz the *LC* parallel resonant circuit of *Figure 9.30c* can be used. In Chapter 3 it was explained that when a coil and capacitor are connected in parallel, the impedance of the combination is low at all frequencies except the resonant frequency. At the latter frequency,

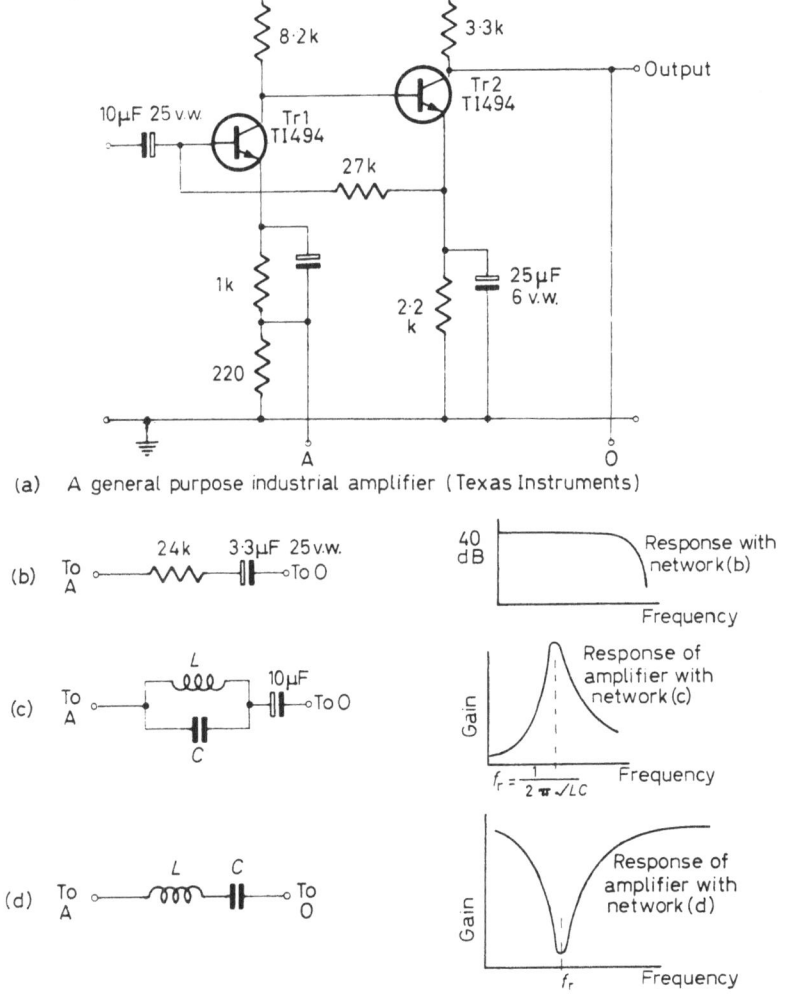

(a)   A general purpose industrial amplifier (Texas Instruments)

Figure 9.30. A feedback amplifier the response of which is variable by connecting different feedback networks

255

the impedance rises to a very high value. The impedance would be infinitely high at resonance if the resistance losses in the coil and the capacitor were absent. When connected between the points O and A there is heavy negative feedback at all frequencies except the resonant frequency. The feedback fraction β is approximately unity and so the gain of the amplifier is also approximately unity. At the resonant frequency however very little feedback occurs because of the very high impedance of the resonant circuit. The gain then rises to a value that it would attain in the absence of any feedback.

The series resonant circuit of *Figure 9.30d* produces an amplifier that has a high gain at all frequencies except the resonant frequency. This is because the impedance of a series resonant circuit is high except at resonance when it reduces to only the resistance of the coil; this can be made low. The result is that at the resonant frequency there is considerable feedback and the gain at this frequency is reduced almost to unity.

For tuned amplifiers where it is undesirable to use a feedback line the normal load resistor can be replaced by a parallel tuned circuit. The dynamic load impedance is given by $Z = L/CR$ *(Figure 9.31a)*. For narrow band-

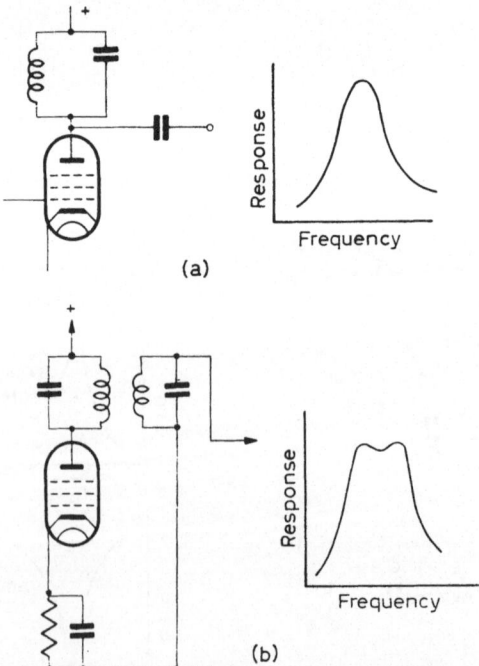

Figure 9.31. Tuned amplifier. (*a*) shows the response with a single resonant circuit for the load. (*b*) shows the response when two circuits tuned to the same frequency are mutually coupled. With close coupling and high-Q coils a 'double-humped' response is obtained

widths the Q of the circuit should be high. The dynamic load impedance is then high at the resonant frequency. At all other frequencies the load impedance and hence gain of the stage is low.

Double-tuned amplifiers are often encountered where an attempt is made to obtain a response over a narrow band of frequencies. The most familiar case is that of the intermediate frequency (i.f.) stages of a superheterodyne radio receiver. The stages are transformer coupled and both the primary and secondary windings are tuned. Provided there is close coupling (i.e. most of the primary flux is associated with the secondary coil) analysis shows that even though both circuits are tuned to the same frequency interaction on close coupling produces the double-humped curve of *Figure 9.31b*.

For frequencies below about 100 kHz, it is still possible to use coils, but high-Q coils for audio frequency work are bulky and expensive. Most workers therefore prefer to use RC networks, the most popular being the parallel-T network of *Figure 9.32*. Unfortunately, the transistor amplifier of *Figure 9.30* cannot be used with this network because the latter requires to be fed from a low impedance source and should lead into a high impedance. Parallel-T networks do not perform satisfactorily if they are heavily loaded.

The principles of using selective negative feedback to produce a tuned amplifier can be employed with any amplifier. For low frequency work a simple solution is to use thermionic valves; this allows the use of the parallel-T network since leading into the grid of a valve means that practically no loading is applied to the network. The parallel twin-T network of *Figure 9.32* has the property of producing infinite attenuation at a frequency f_r given by $(2\pi f_r)^2 = \dfrac{2}{R_1^2 C_1 C_2} = \dfrac{1}{2R_1 R_2 C_2^2}$. This means that if the output of an amplifier is fed into a parallel twin-T network the output from the network is high at all frequencies except f_r. If this network output voltage is fed back to the input in antiphase, there is negative feedback and the overall gain is low. At the resonant frequency however there is no output from the T network and hence no feedback. The full gain of the amplifier is then realized. The amplifier of *Figure 9.32* must be fed from a high impedance source to avoid loading the T network. If this is not possible an isolating stage can be placed before the amplifier and the signal led to the input of the isolating stage.

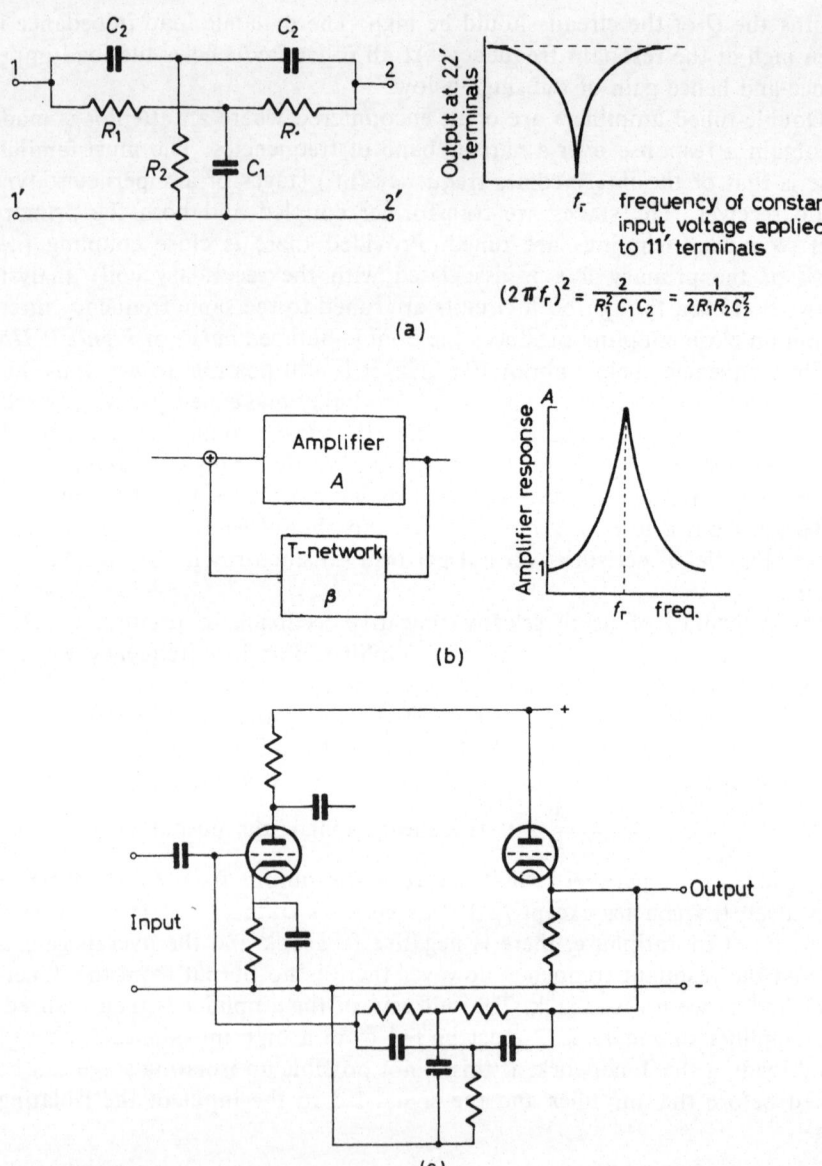

Figure 9.32. A low-frequency tuned amplifier with a parallel twin-T feedback network. (a) shows the response of the network and (b) the principles of the amplifier. (c) is one possible form of circuit in which there must be an odd number of stages before the cathode-follower output. The input must be fed from a high impedance source to avoid loading the T network (see text)

258

INTEGRATED CIRCUITS

Users of electronic equipment who are not interested in building apparatus from the basic components may well need to design a system to meet their requirements. Complete amplifiers, oscillators and other circuits can now be

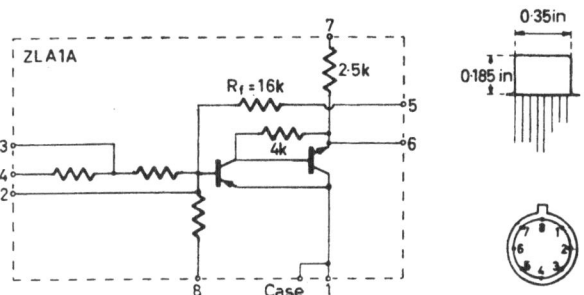

(a) A Ferranti monolithic integrated circuit based on a silicon chip (see text for performance figures)

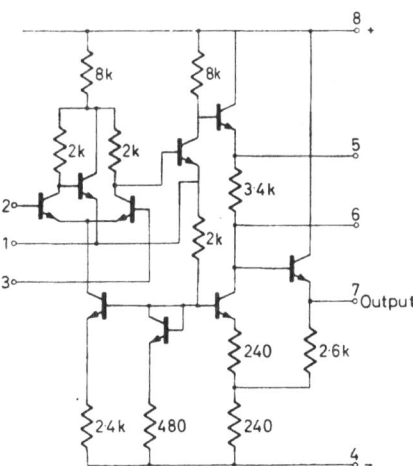

(b) Circuit diagram of a Fairchild wideband differential d.c. amplifier. The complete circuit is available in a TO5 can similar in size and shape to that used for the Ferranti circuit above

Figure 9.33. Examples of integrated silicon monolithic circuits

obtained ready built in microminiature form. Ferranti and Fairchild are two firms who do this kind of work, and the interested reader is advised to write to them for details of their range of circuits. Two have been chosen at random by way of example (Figure 9.33).

These amplifiers are constructed on small silicon chips usually not larger than about 0·1 in. square. Special techniques are used which involve epitaxial processes and masking methods to create within a single chip the necessary transistors, resistors and diodes to fabricate a complete circuit. The Ferranti circuit uses a small TO5 transistor can to house the circuit. Three voltage gains are possible depending upon the connections made at the input and feedback terminals. For their ZLA1A circuit, the voltage gains are 8, 16 and 32. At these amplifications, the frequency response is flat within 3 dB to 1·5, 0·9 and 0·5 MHz, respectively. The corresponding input impedances are 3·2 k, 1·6 k and 0·8 k.

A more complicated amplifier of Fairchild is shown in *Figure 9.33b*. This amplifier is a wideband d.c. amplifier (0–30 MHz) intended for use as an operational amplifier. It can be used, however, as a precision instrumentation amplifier. This amplifier is a monolithic version of what is called a differential or difference amplifier; it is supplied in a small TO5 can or in a 'flat-pack' (a small rectangular encapsulation a little over 6 mm square). Leads are brought out from two opposite edges.

AMPLIFIERS USING FIELD-EFFECT TRANSISTORS

Chapter 5 has already outlined the construction and mode of operation of these devices. Their outstanding characteristic is the high input impedance possible. At the time of writing, the greatest success is enjoyed with reverse-biased junction types. Manufacturers are meeting some difficulty in the production of insulated-gate types, but these difficulties will no doubt be overcome in the very near future.

Figure 9.34 shows a circuit designed by Ferranti Ltd. for their ZFT 12 field-effect transistor. In order to test the claims made for the device the circuit was assembled on a piece of Veroboard (which is the trade name of Vero Electronics Ltd. for their product) which consists of strips of copper bonded to an insulating board as for a printed circuit. Holes are punched out in rows and columns 0·2 in. apart. The gate electrode is taken directly to a polythene insulated terminal. The output of the f.e.t. is taken via a capacitor to the base of a silicon *npn* bipolar (i.e. conventional) transistor. Heavy negative feedback is used to increase the input impedance of the complete amplifier. Ferranti claim for their circuit an input resistance of 500 M, an input capacitance of 4·5 pF and unity gain. The resistors in the test amplifier were ordinary 10% tolerance types. The amplifier was found to have an input resistance of 490 M, an input capacitance of 10 pF (including the very short input lead) and a gain of 0·99. The distortion at 1 kHz was too low to be measured accurately on a Marconi distortion meter when the input voltage was 4 V r.m.s. Clipping of sine waves was not evident on the oscilloscope until the r.m.s. voltage reached 5·1 V. The square wave response was very satisfactory at a repetition frequency of 10 kHz. The claims made for the circuit were therefore fully justified.

The production difficulties associated with insulated-gate f.e.t.s make this device expensive. The cost at present (1966) is from about £5 to £10 per unit. Nevertheless, as increasingly larger numbers are used the cost will come down and it is, therefore worthwhile considering some of the applications.

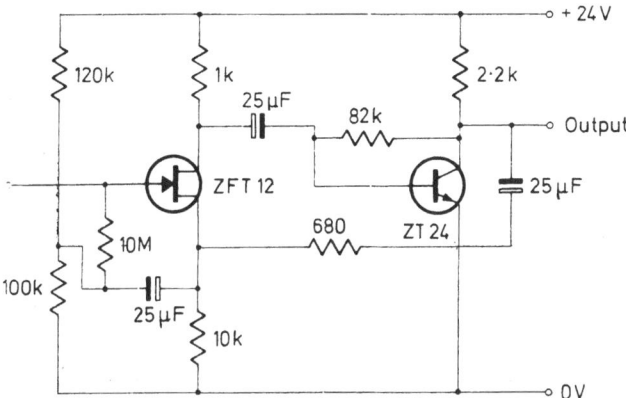

Figure 9.34. High input impedance amplifier of Ferranti Ltd. using a field-effect transistor in the output stage. Input resistance is 500 M in parallel with an input capacitance of 5 pF

Because of the insulated gate, the input resistance of an m.o.s.t. is extremely high. The device combines the desirable input features of a thermionic valve with those of a conventional transistor (e.g. small size, no heater, etc.). The simplest form of amplifier is shown in *Figure 9.35a* and follows very closely the thermionic valve arrangement. The input resistance of the gate itself is about 10^{12} ohms, but the associated input components reduce the effective input impedance of the amplifier. At low frequencies (< 10 kHz) the input impedance can be up to 10 M. An interesting circuit published by C. J. Mills is given in *Figure 9.35b* and was developed for use in circuits requiring electrometer conditions. The input resistance can be as high as 10^{11} ohms. The overall gain of the circuit is 5 and the output resistance is less than 100 ohms. A source follower circuit is used with the f.e.t. to give the high input resistance. (Compare with the cathode-follower.) The output from the source resistance is then taken to a high input impedance compound pair. The transistors are *npn* and *pnp*, so this arrangement is referred to as a complementary pair. This d.c. amplifier is ideal for use with piezoelectric crystal transducers. The range of operating frequencies claimed is 0–1 MHz.

261

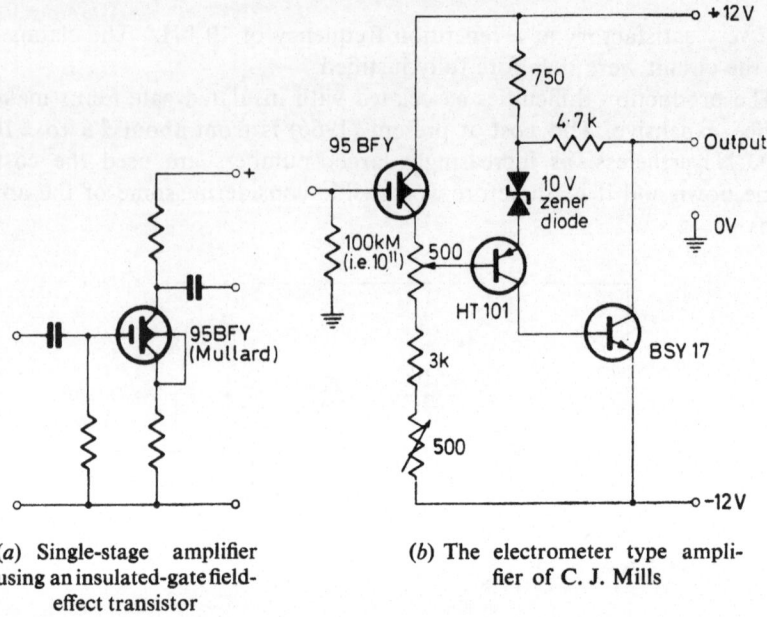

(a) Single-stage amplifier using an insulated-gate field-effect transistor

(b) The electrometer type amplifier of C. J. Mills

Figure 9.35. Two amplifier circuits using insulated-gate field-effect transistors

POWER AMPLIFIERS

The amplifiers considered so far have been designed to give a maximum voltage or current output with minimum distortion. Although they develop power in their anode or collector load circuits, this power is of little importance. The choice of valves or transistors and the associated components is not influenced at all by power considerations except in so far as the components have to be operated within their maximum power ratings. Power amplifiers however are those in which power output is the chief consideration. These are the amplifiers which are designed to operate into loads such as servomotors, potentiometric recorders, moving-coil pen recorders, loudspeakers, meters and other recording devices. The aim is usually to deliver the maximum power into the load, consistent with a reasonably low distortion. The dissipation of power necessarily implies a resistive load since no power can be dissipated in a capacitive or inductive load. There may however be a reactive component associated with the load as, for example, in an electric motor. Here, although the power is dissipated in an equivalent resistance, there is always the inductance of the motor coils to be taken into consideration. This affects the design of the power amplifier in which stable operation must be ensured.

In general, the actual load has a resistive value which is not under control because it is determined by the nature of the indicating or other device connected to the output of the amplifier. In order to obtain maximum power output, the load and the amplifier must be matched with respect to their impedances. If we take as an example the simple case of a triode with a fixed input sinusoidal voltage and a resistive load then, using the equivalent circuit of *Figure 9.36*, it can be seen that the a.c. power in the load resistance

Power in the load $P = i_a^2 R_L$

$i_a = \mu v_g / (r_a + R_L)$

$$P = \frac{\mu^2 v_g^2 R_L}{(r_a + R_L)^2}$$

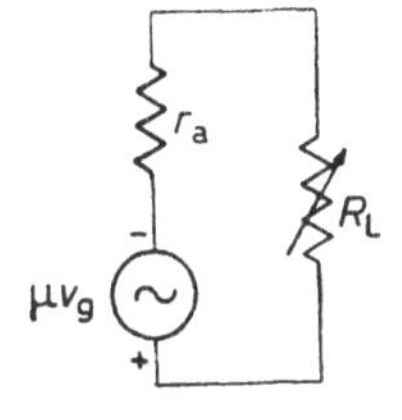

Equivalent circuit

$$\frac{dP}{dR_L} = \frac{(r_a + R_L)^2 \mu^2 v_g^2 - \mu^2 v_a^2 R_L 2(r_a + R_L)}{(r_a + R_L)^4}$$

For maximum power $\dfrac{dP}{dR_L} = 0$

$\therefore (r_a + R_L)^2 = 2R_L(r_a + R_L)$

$\therefore R_L = r_a$

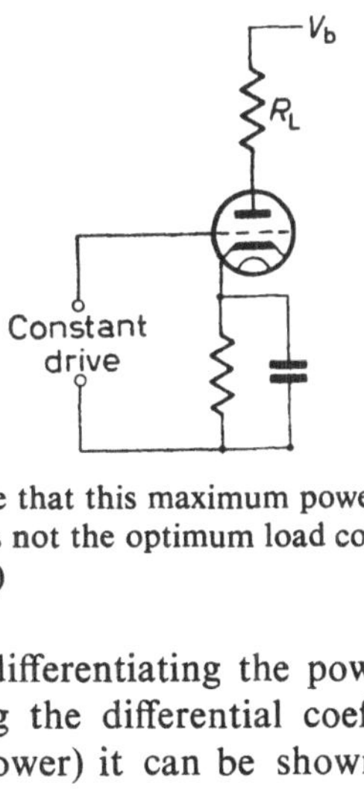

Constant drive

Figure 9.36. The maximum power transfer theorem. Note that this maximum power is for a fixed r.m.s. input voltage. Only R_L is varied. $R_L = r_a$ is not the optimum load condition (see text and *Figure 9.39*)

is $i_a^2 R_L$. By substituting $i_a = \mu v_g (r_a + R_L)$ and differentiating the power expression with respect to the load, on equating the differential coefficient to zero (which is the condition for maximum power) it can be shown that the maximum power is delivered when the load R_L is equal to r_a, the internal a.c. impedance of the valve. Although the valves used for power amplifiers are specially constructed and usually have low amplification factors and anode slope resistances, typical values of r_a are still of the order of a few thousand ohms, whereas the load resistance is often hundreds or perhaps only tens of ohms. This results in a matching problem. The usual solution is to use a transformer. By connecting the primary coil in the anode circuit and the secondary coil to the load, it can be shown the impedance 'reflected' into the primary circuit is $(N_p/N_s)^2 R_L$, where N_p and N_s are the number of turns in

263

primary and secondary coils, respectively, and R_L is the actual load assumed to be resistive only. It is assumed that the output transformer is perfect. By choosing the correct turns ratio, the effective resistance as 'seen' by the valve can be made a suitable value. The transformer thus transforms the actual load impedance to the desired anode load impedance. The most suitable value of anode load impedance is called the optimum load. In spite of the analysis given above, the optimum load is not equal to r_a. In deriving the maximum power theorem it was assumed that only the load resistance R_L varied, the input grid voltage being fixed at some small r.m.s. value. Under these circumstances $R_L = r_a$. In practice, however, the input voltage is not fixed, and the output valve is driven as hard as possible by increasing the grid voltage swing to a suitably large value consistent with a low distortion figure. It is usual to specify that the distortion should not exceed 5 per cent total harmonic (i.e. the square root of the sum of the squares of the r.m.s. values of all the harmonics in the output signal should not exceed 5 per cent of the r.m.s. value of the fundamental. It is assumed that the grid is being driven by a signal with a perfectly sinusoidal waveform for the purposes of this calculation). For maximum power into the load, when driving the grid voltage to the maximum extent without clipping the output waveform, the optimum load is $2r_a$.

When considering the power output, it is often convenient to simplify the analysis by assuming that the anode characteristics are straight parallel lines as shown in *Figure 9.37*. The results of making such an assumption are not

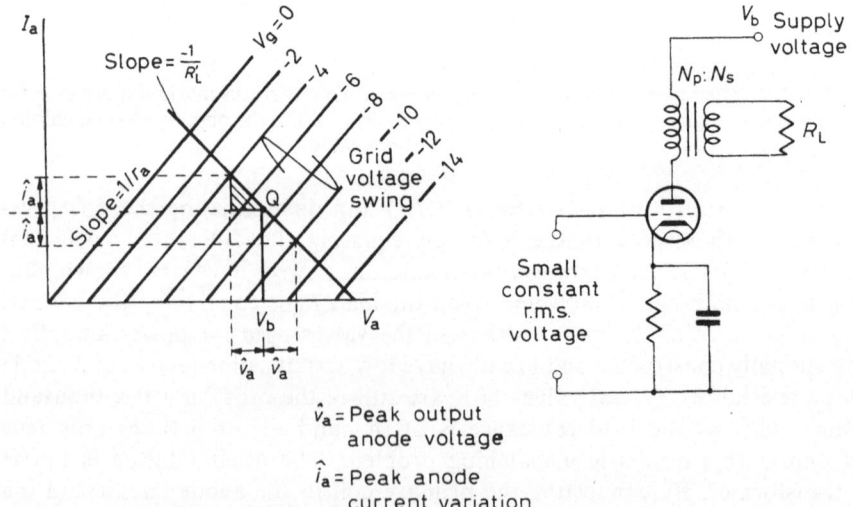

$\hat{v}_a$ = Peak output
 anode voltage

$\hat{i}_a$ = Peak anode
 current variation

Figure 9.37. Idealized triode characteristics with a.c. load line with slope $-1/R'_L$ where
$$R'_L = (N_p/N_s)^2 R_L$$

264

far from those found experimentally. In drawing the load line it must be remembered that the load is transformer-coupled to the valve. The slope of the load line is therefore $1/R'_L$ where $R'_L = (N_p/N_s)^2 R_L$, R_L being the actual load. The anode voltage in the quiescent state is taken as the supply voltage, V_b, because the steady voltage drop across the transformer primary is very small and can be neglected. When a signal is applied to the grid the anode voltage swings about the supply voltage. The anode voltage can rise above the supply voltage because of the self-inductance of the transformer primary. If the input r.m.s. voltage is kept constant the excursions of grid voltage about the bias voltage may be ± 3 V as in *Figure 9.37*. This produces an anode alternating current with a peak value of $\hat{v}_a$ and an anode alternating current with a peak value of $\hat{i}_a$. The output power is, therefore,

$$P_{out} = \frac{\hat{i}_a}{\sqrt{2}} \cdot \frac{\hat{v}_a}{\sqrt{2}} = \frac{\hat{i}_a \hat{v}_a}{2}$$

From the diagram it can be seen that the area of the shaded triangle represents the output power. The region of the triangle is enlarged in *Figure 9.38*.

AQ is a constant for a given input voltage

Area of triangle $\dfrac{\hat{i}_a \hat{v}_a}{2}$

$AB = AQ - \hat{v}_a$

$\therefore$ Area of triangle $= \dfrac{AB}{r_a} \dfrac{\hat{v}_a}{2} = \dfrac{(AQ - \hat{v}_a)\hat{v}_a}{2r_a}$

Differentiating with respect to $\hat{v}_a$ and equating to zero

gives $\hat{v}_a = \dfrac{AQ}{2}$ for maximum area

$\therefore$ $AB = BQ$ $\therefore$ Slope of CA = slope of CQ

$\therefore$ $R'_L = r_a$

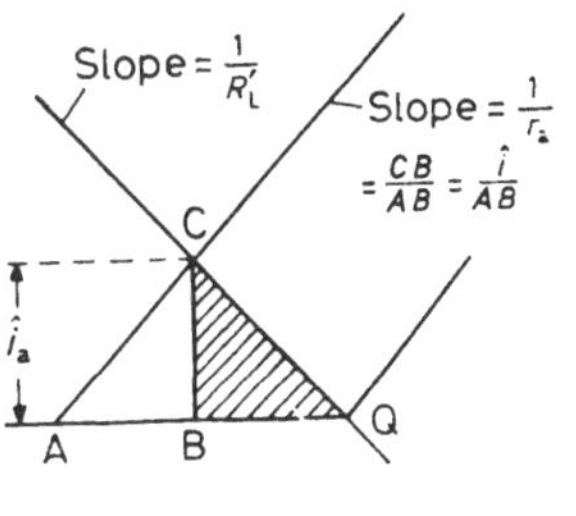

Figure 9.38. Enlarged sketch of the power triangle region

An expression for the area is obtained; this expression is differentiated and equated to zero to obtain the maximum power condition. From this it is found that for a fixed bias voltage (i.e. a fixed operating point, Q) rotations of CQ about this operating point (which is equivalent to changing R'_L) yield the maximum power condition when $R'_L = r_a$. This is in accord with the maximum power transfer theorem.

In power amplifiers this analysis is not relevant because the only concern is to obtain the absolute maximum power into the load, consistent with a satisfactory distortion figure. The absolute maximum power exceeds the maximum power obtained in the previous paragraph if the restriction relating to a fixed r.m.s. value for the input voltage is abandoned. By driving the grid hard (but without clipping the anode current waveform) and adjusting the position of the operating point accordingly the maximum possible output power is obtained when $R'_L = 2r_a$. To understand this let us consider *Figure 9.39*. The two quantities that are fixed on the graph are (*a*) the position of the $V_g = 0$ characteristic, and (*b*) the position of the V_b line. It is assumed that the supply voltage is fixed at the maximum value that the manufacturer allows and that sufficient grid voltage drive is available to give the maximum

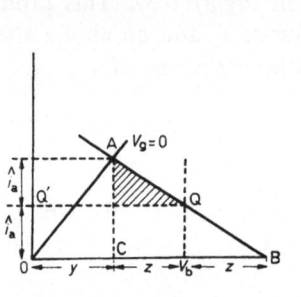

$$\frac{y}{AC} = r_a \quad \therefore \quad y = 2r_a \hat{i}_a$$

$$\hat{v}_a = V_b - y$$

$$\text{Area of triangle} = \frac{\hat{i}_a \hat{v}_a}{2} = \frac{y}{2r_a} \frac{(V_b - y)}{2}$$

Differentiating and equating to zero gives

$$y = \frac{V_b}{2} \quad \therefore \quad y = z$$

$\therefore$ Slope of OA is twice the slope of AB

$\therefore \quad R'_L = 2r_a$

Figure 9.39. When Q can be adjusted with the load R'_L to enable full-drive conditions at the grid the optimum load is $2r_a$. This with full grid drive will give the absolute maximum power to the load for a given V_b

swing about Q up to $V_g = 0$ and down to a value of grid voltage that just produces cut-off. If now various load lines are drawn corresponding to various values of R'_L the position for one of the lines is as shown in *Figure 9.39*. The power for any position of the load line is $\hat{i}_a \hat{v}_a / 2$. By making the substitutions shown in the diagram it is found that the absolute maximum power is developed when $R'_L = 2r_a$. The optimum load for a triode is therefore $2r_a$. The result differs from the former one because the position of Q is allowed to move up and down the dotted V_b line. At each new position of Q the grid voltage is adjusted for maximum drive.

The optimum load for a pentode is governed by the characteristics, which are not so easy to idealize as those of a triode. The anode slope impedance of most output pentodes is of the order of several tens of kilohms. Taking 50 k as a value it can be seen from *Figure 9.40* that an effective load of 100 k (i.e. $2r_a$) would not be realistic. In fact, any load which brings the load line below the knee of the characteristic (e.g. AQ) is unsatisfactory because of the crowding of the characteristics there and subsequent distortion. Positions such as CQ yield output powers that are too small. It is usual, therefore, to

266

define the optimum load as that which will produce a load line in the position of BQ, i.e. just avoiding the 'knee'. The optimum load for a pentode Class A power amplifier is approximately $0.1 r_a$.

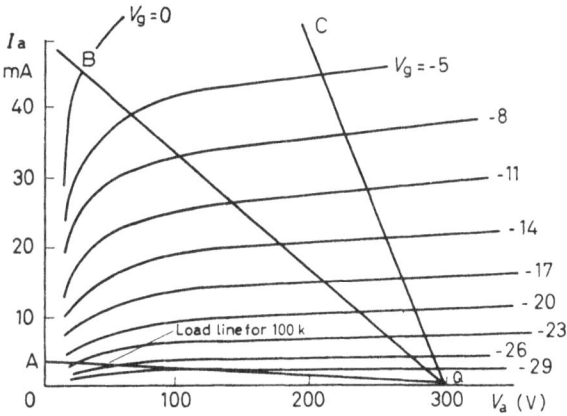

Figure 9.40a. Characteristics for a typical output pentode. Load line AQ represents a load of 100 k and BQ represents 6 k. BQ is about the optimum position

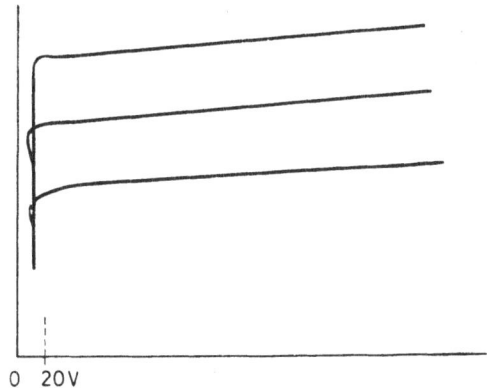

Figure 9.40b. Beam tetrode characteristics showing the knee closer to the ordinate

Efficiency

The efficiency of a power amplifier is defined as the output power divided by the total power supplied by the h.t. line. Under optimum conditions for a Class A triode the area of the power triangle in *Figure 9.39* is compared with

267

the area of the rectangle, $\hat{i}_a V_b$. In this case, when driving fully, $\hat{i}_a$ is equal to the quiescent current and so the total input power is $\hat{i}_a V_b$. Simple geometric considerations show that the area of the triangle is one quarter of the area of rectangle $0V_bQQ'$. The efficiency is, therefore, 25 per cent. Idealized conditions are assumed and therefore 25 per cent is the theoretical maximum efficiency when $R'_L = 2r_a$. In practice, 20 per cent is the efficiency figure more likely to be realized.

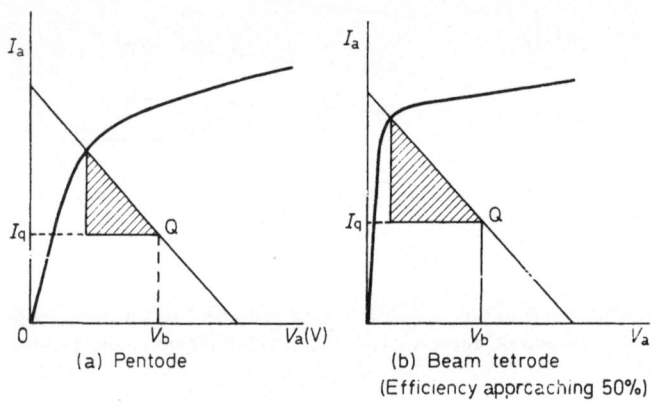

Figure 9.41. Power triangles and efficiency considerations for a pentode and beam tetrode under Class A conditions

In pentodes the efficiency is higher because it is possible to work nearer the current ordinate. In *Figure 9.41a*, for instance, the area of the power triangle is greater than 25 per cent of the area of the input power rectangle $0V_bQ\,I_q$. The beam tetrode can be operated very near to the current ordinate and the efficiency then approaches the theoretical maximum of 50 per cent.

If the characteristic curves for transistors are examined a similarity between them and those of a beam tetrode can be noted. The knee of any transistor characteristic is extremely close to the current ordinate which is indicative of a device that can be used as a very efficient power amplifier.

Efficiencies up to the figures quoted above can be improved upon by operating the valve or transistor under Class B conditions. In this mode, the operating point, Q, is brought down the load line to the cut-off position by the application of suitable bias arrangements. A sinusoidal input signal then produces a series of load current pulses that are half-sine waves *(Figure 9.42)*. The power taken from the h.t. supply is the supply voltage times the mean current, i.e. $V_b \cdot 2I_{max}/\pi$. The power into the load is the r.m.s. value of the current, $I_{max}/\sqrt{2}$, times the r.m.s. value of the voltage across the load

$(V_b - V_{min})/\sqrt{2}$. The efficiency is therefore

$$\frac{I_{max}(V_b - V_{min})}{\sqrt{2}\cdot\sqrt{2}} \times \frac{\pi}{2I_{max}V_b}$$

$$= \frac{\pi}{4}\left(1 - \frac{V_{min}}{V_b}\right)$$

For a pentode, tetrode or transistor V_{min} is small compared with V_b and so the efficiency approaches $\pi/4$, i.e. about 78·5 per cent.

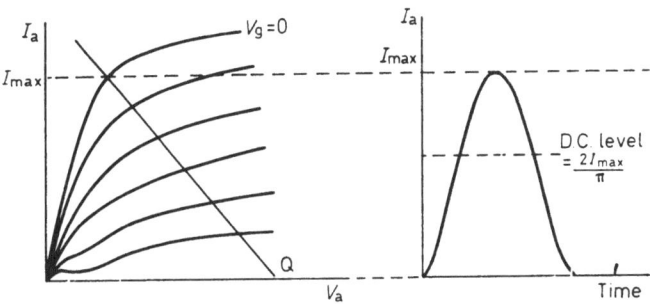

Figure 9.42. Pentode operated under Class B conditions with a sinusoidal grid voltage

Push-pull amplifiers

It is obvious from *Figure 9.42* that it is not possible to use a single valve as a Class B amplifier because half of the waveform is missing. It is necessary to use two valves in what is termed a push-pull arrangement *(Figure 9.43)*. Each output valve or transistor conducts for half a period and the complete waveform is restored in a special output transformer. This transformer has a centre-tapped primary the outer ends of which are connected to the anodes of the output valves, the centre tap being connected to the h.t. positive line. Conduction by each valve for alternate half periods gives rise to signal flux in the transformer core throughout the whole period; the complete waveform is therefore available in the secondary output winding.

There are several advantages in using the push-pull type of output circuit. Compared with a single-valve output stage the same type of valve in a Class B push-pull circuit delivers twice the power with less distortion. Any second or even harmonic distortion components are cancelled in the transformer because the current components associated with these harmonics are fed into the transformer in phase; the fluxes due to these currents in each half of the primary therefore cancel. Only the wanted fundamental (and odd har-

269

monics) are fed in antiphase to the transformer producing the signal current in the load. Triodes, because of their characteristics, produce mostly second harmonic distortion; at one time therefore they were the favoured valve for high-fidelity amplifiers. Pentodes produce mostly odd harmonics which are unfortunately not cancelled in the output transformer. These components are not large however and the usual arrangement nowadays is to take advantage of the higher output powers and efficiencies of modern pentodes and tetrodes, reducing the distortion with negative feedback.

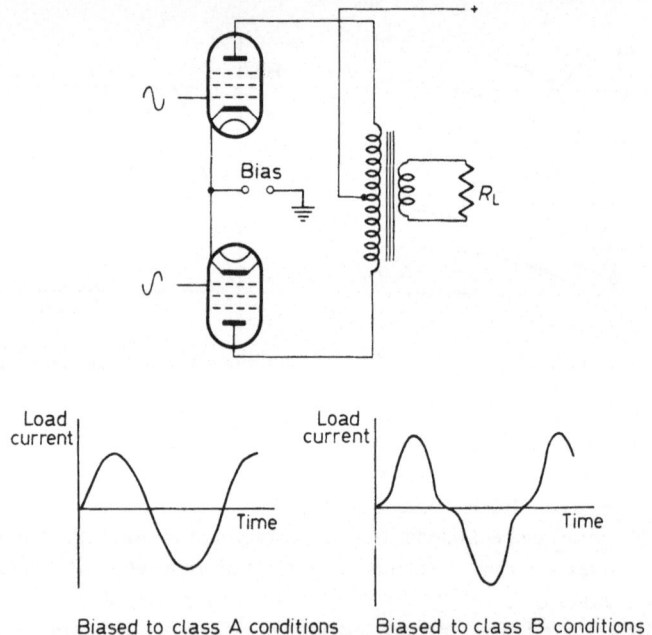

Biased to class A conditions   Biased to class B conditions

Figure 9.43. A push-pull amplifier. In Class A the distortion is low. The distortion with Class B conditions, known as crossover distortion, is due to curvature of the characteristics near to the cut-off point. Many amplifiers are designed for Class AB conditions where a compromise is reached between the high efficiency of the Class B mode and the low distortion of the Class A mode

The push-pull arrangement is not confined to Class B operation, but may also be used with Class A stages. Two additional advantages of the push-pull are then enjoyed. Firstly, the load on the high-tension supply is almost constant in a push-pull Class A output stage. This results in the elimination of voltage fluctuations on the h.t. line and greatly eases the smoothing problem. With single-ended output stages the fluctuations on the h.t. line caused by variations in signal strength are fed back to the earlier stages of the amplifier. If care is not exercised, positive feedback results, causing instability and spurious oscillations. The usual way of avoiding these difficulties is to decouple each stage, which involves interposing resistors in the h.t. positive line. Large value electrolytic capacitors are connected between the h.t. negative line and the ends of the resistors remote from the h.t. positive supply terminal. In effect a separately smoothed supply is available for each stage. Decoupling is not such a serious problem in amplifiers that use Class A push-pull output stages.

The second advantage of the push-pull arrangement when Class A operation is used relates to the quiescent currents. In Class A the anode current necessarily contains a large steady component. In push-pull operation the steady components pass in different directions in the primary of the output transformer; the fluxes due to these components cancel. It is possible therefore to design a transformer that is not large physically, but that has a high primary inductance. This results in an improvement of the low-frequency response of the amplifier.

There is no doubt that the output transformer is one of the major sources of distortion and indifferent performance in cheap power amplifiers. Even in high-quality amplifiers it is difficult to attain faithful reproduction of a square wave signal when transformers are used. A carefully designed output transformer is required which must be physically large to have a good response at low frequencies. A happy solution to this problem is possible now that matched power transistors are available. The output impedance of transistor amplifiers can be made comparable with that of the loads commonly encountered. The use of emitter-follower stages and negative feedback enable transformerless output stages to be designed that have very low output impedances. Transformerless output stages have been designed using thermionic valves, but the elimination of the weight and size of the output transformer is not such a noticeable advantage as it is in transistorized equipment. *Figure 9.44a* shows in principle the circuit details of a transformerless power amplifier designed by Osborne and Tharma[13]. Transistors Tr1 and Tr2 form a composite or Darlington pair (see *Figure 9.28* and the related text). Together they form one of the output 'transistors'; Tr3 and Tr4 together form the other output 'transistor'. The signal is fed to an emitter-coupled driver stage (shown dotted) which produces two versions at the outputs. One version is 180° out of phase with the other. (This and other driver stages are discussed in the next section of this chapter.) Had the output stages been operating in Class B, each section would conduct for only half the time. In fact with this circuit Class AB is chosen to minimize crossover distortion. The load is in the emitter circuit of Tr2 and the driving voltage is applied between the base of Tr1 and the − 52 V line which is effectively a signal earth line. Similar

271

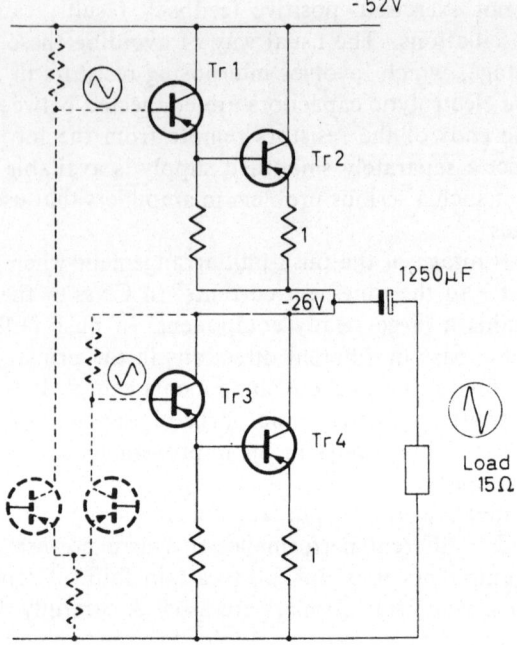

Figure 9.44a. The basic circuit for the Class AB push-pull transformerless power amplifier of Osborne and Tharma[13]

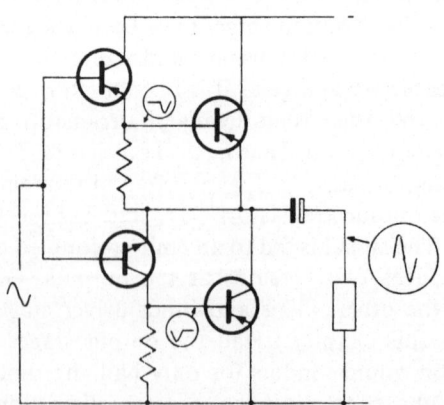

Figure 9.44b. A Class B push-pull amplifier that uses a different driver stage (see also Figures 9.45 and 9.47)

272

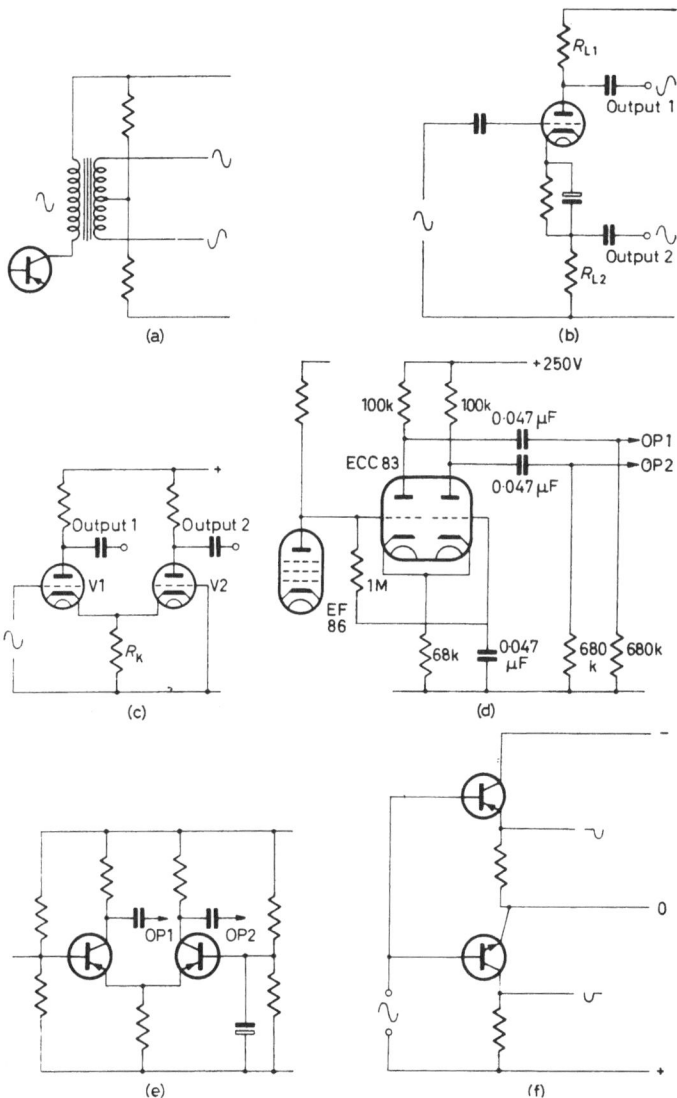

Figure 9.45. Various examples of circuits used for driving push-pull output stages
(*a*) Transformer phase splitter. (*b*) The split-load phase splitter. (*c*) Cathode-coupled phase splitter. (*d*) Practical circuit of the cathode-coupled phase splitter. (*e*) Conventional emitter-coupled phase splitter. (*f*) The use of a complementary pair to provide driving signals for the output stage (biasing arrangements are omitted)

conditions exist for Tr4 as in this case the drive signal is connected between the base of Tr3 and the 'live' end of the load.

A circuit that uses a different driving arrangement is shown in *Figure 9.44b* which represents what is called a single-ended Class B push-pull output stage. (*Figure 9.44a* is a single-ended Class AB push-pull output stage.) Complementary transistors (*pnp/npn*) have been used as output transistors in some versions of this circuit, but because matched complementary power transistors are difficult to manufacture, two *pnp* power transistors are used in the circuit of *Figure 9.44b;* a complementary pair is used in the driving stage, as explained in the following section. The amplifier of Tobey and Dinsdale[4] *(Figure 9.47)* uses this principle.

Push-pull Driver Stages

The output stages of a push-pull amplifier are driven by two signals in antiphase except where complementary output transistors are involved. The manufacturing difficulties already mentioned make this latter kind of amplifier unpopular and we will not discuss it here.

The simplest driver stage is the transformer phase splitter of *Figure 9.45a* which is used in both valve and transistor amplifiers. Transformer coupling is not suitable in high-grade equipment because it introduces phase shift which may lead to instability in feedback amplifiers. Also, as we have seen already, a wider and flatter response can be obtained by using RC elements.

To be satisfactory, a phase splitter should give two outputs of equal amplitude and exactly 180° out of phase. The high-frequency response should be well maintained and, if possible, there should be some useful amplification. Of several possible circuits two are shown in *Figure 9.45b* and *c*. They are superior to other circuits, which accounts for their widespread use. In (*b*) a single valve is used with a load in both the anode and cathode leads. The application of an input signal which is positive-going increases the current through the valve. There is thus a fall in potential at the anode because of the increase in current through the load resistor R_{L1}. Simultaneously, there is a rise of voltage at the cathode end of R_{L2}. This circuit gives two well-balanced signals at all but the highest frequencies, and is economical in the number of components used.

The cathode-coupled circuit of *Figure 9.45c* requires more components, but has advantages over the split-load phase splitter. It is completely self-balancing and the high-frequency response is well maintained. The phase-shifts are small and useful gain is available. The gain is about half of what it would be if the valve were used under similar conditions as a Class A amplifier. The two outputs are obtained in the following way. When the grid of V1 goes more positive than the bias voltage, there is an increase of current in V1. The anode voltage falls, and this constitutes output 1. Since the current through V1 increases there is an increase in voltage across the common cathode resistor R_K. The grid of V2 is held constant for alternating voltages and, therefore, the effective grid-to-cathode voltage of V2 changes. Since the cathode is positive-going, the grid in effect is becoming more negative with respect to the cathode. The current through V2 is therefore reduced. The voltage across the load of V2 falls and the anode voltage of V2 rises. This constitutes output 2. The sum of the currents is practically constant, which helps in maintaining a constant h.t. voltage in the same way as that of a push-pull output stage. A practical circuit is shown in *Figure 9.45d*. It uses an ECC83 valve, which has a high amplification factor ($\mu = 100$). The anode load of the second valve should theoretically be slightly higher than that for the first valve if precise equality of output voltage is required. In practice, with the circuit shown, the use of high-μ valves makes the difference negligibly small. The gain of the circuit from the input to each output is 25. Output voltages as high as 20 V are available with a total harmonic distortion of about 2 per cent. With lower drives and an output voltage of 5 V the distortion is reduced to about 0·5 per cent. Overall negative feedback in the complete amplifier reduces these figures so far as the final output to the load is concerned.

Transistor phase splitters follow similar lines to those of valve circuits. *Figure 9.45e*, for example, shows a conventional emitter-coupled phase splitter. As the action of this phase-splitter is similar to that of the cathode-coupled valve version no more need be said here.

Figure 9.45f is an arrangement that does not have a thermionic valve equivalent. A complementary pair of transistors (one *pnp* and one *npn*) used in this way is very useful in transformerless power amplifiers. The circuit shown is a driver for a single-ended push-pull stage, and is itself operated in Class B. A negative-going signal at the input to the driver stage drives Tr1 'on' whilst Tr2 is non-conducting. With a positive-going input signal Tr1 is non-conducting whilst Tr2 is driven into the 'on', i.e. conducting, condition. The outputs from the complementary pair are then used to drive a pair of *pnp* output transistors as shown in *Figure 9.44b*. Further information on these types of amplifier can be obtained by consulting references 4, 12 and 13.

Two complete power amplifiers are shown in *Figures 9.46* and *9.47*. The valve amplifier of J. R. Ogilvie is a high-grade a.f. power amplifier suitable for laboratory use with resistive, inductive, and capacitive loads. It embodies feedback principles which are the subject of a British patent. His circuit was published in a letter following the appearance of a circuit for a low-frequency laboratory power amplifier designed by A. R. Bailey. The references give details of operation and performance.

The low distortion transistor amplifier published by Tobey and Dinsdale[4] has given the author very good service in the domestic field. Although primarily intended for audio reproduction the amplifier performs well for several laboratory applications. The simplicity and compactness coupled with good

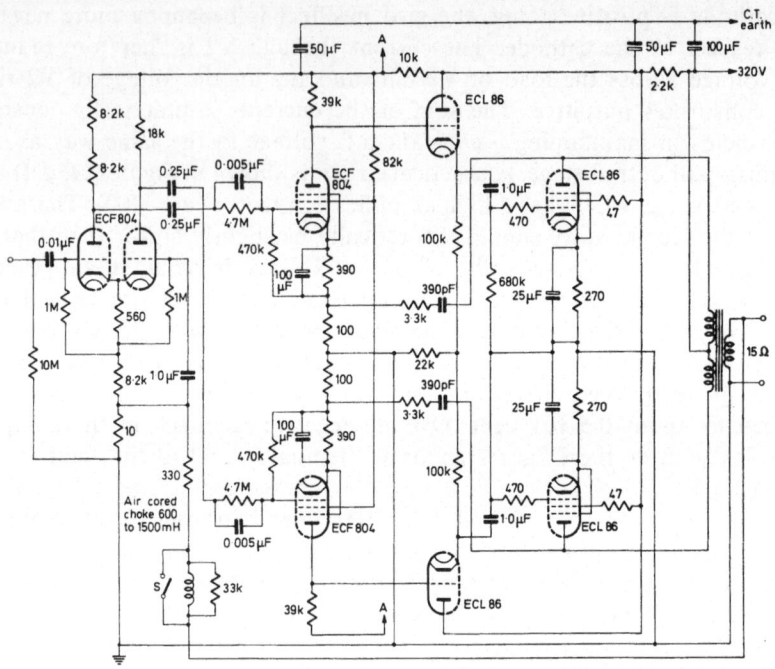

Figure 9.46. Circuit of a 10 W power amplifier for use in the frequency range 20 Hz to 20 kHz. For capacitive loads switch S is opened, whilst for inductive and resistive loads it is closed. The circuit was designed by J. R. Ogilvie[3]

performance make it useful. It operates in Class B mode, thus enabling important economies to be effected in the design of heat sinks and power supplies. The amplifier has a response flat within ±1 dB over the frequency range 40 Hz to 20 kHz, and an output impedance of only 0·25 ohms.

D.C. AND OPERATIONAL AMPLIFIERS

There are numerous occasions in the laboratory where it is desired to amplify a signal of very low frequency or signals going down to zero frequency. Temperature measuring instruments using thermistors or thermocouples,

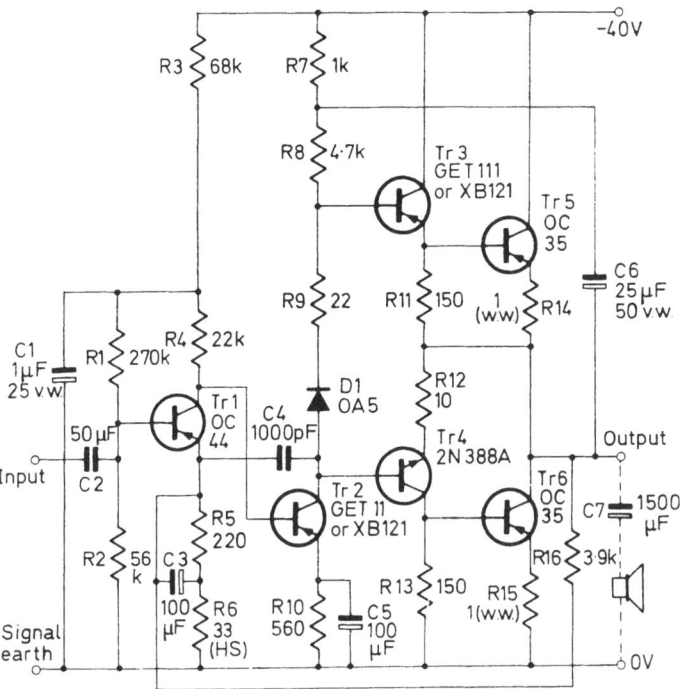

Figure 9.47. The transistor power amplifier of Tobey and Dinsdale[4]. R1 and R9 require adjustment to set the voltage levels. R1 is set to make the collector voltage of Tr6 = 20 V. R9 sets the quiescent current in the output stage (Tr5, Tr6) to 10–20 mA

valve voltmeters, pH meters, photometers, analogue computers and oscilloscopes all require d.c. amplifiers.

The major problem with voltage amplifiers that have to handle steady voltages is zero drift. The difficulty is particularly acute when the drift occurs in the early stages since later stages amplify the drift, and an output voltage is produced even the when input voltage is zero. Later stages are unable to distinguish between true signal variations and slow drift signals. In thermionic valve amplifiers the major causes of drift are variations in heater and supply voltages, and changes in the bias conditions due to changes of component valves with temperature and grid current. With a negative bias on the grid, the grid current is very small. Normally it can be neglected, but in d.c. amplifiers for low level work, i.e. amplifying signals of a few microvolts, the random effect of minute grid currents can be troublesome. In an a.c. amplifier the slow variations are not passed on to subsequent stages because the coupling circuit includes a coupling capacitor. In d.c. amplifiers, however, the coupling is direct.

277

In transistor amplifiers variations of temperature are responsible for most of the drift. The inverse leakage current, previously discussed in the transistor amplifier section, rises exponentially with temperature approximately doubling with every 8 degrees centigrade rise in temperature. The change in β with rise in temperature is an important factor in transistor d.c. amplifiers. For germanium types the β figure may change by as much as 2 per cent per degree centigrade. Silicon transistors are not so sensitive, but changes of about 0·5 per cent per degree centigrade are to be expected. Another factor to consider in low-level stages is the rapid fall of β with collector currents of

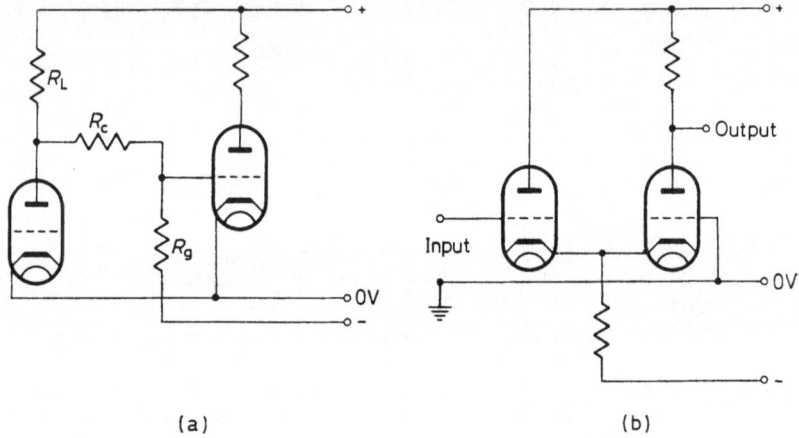

(a)   (b)

Figure 9.48. (*a*) Direct coupling between amplifier stages. (*b*) A directly coupled amplifier in which the first stage acts as a cathode-follower driving an earthed-grid amplifier

less than about 300 μA. When the leakage current rises with temperature the collector current rises and thus there are appreciable changes in the current gain. Fortunately changes in β can be largely nullified by the application of negative feedback.

The most common method of direct coupling is to replace the more familiar capacitor with a resistor as in *Figure 9.48a*. In order to preserve the correct bias conditions in the following stages, it is necessary to have both a positive and negative supply rail. The chain R_L, R_c and R_g must then be designed so that with zero input voltage, the voltage on the grid of the second valve has the desired bias value for Class A operation.

Direct coupling between two stages is effected if cathode coupling is used as in *Figure 9.48b;* instead of using the conventional phase-splitter arrangement, the anode load of the first valve is omitted. The arrangement is, in effect, a cathode follower input stage driving a common (or grounded) grid

278

amplifier. The stage therefore has a high input impedance. Any rise in voltage on the grid of the first valve increases the current through the common cathode resistor. There is a change in the grid-to-cathode voltage of the second valve and this is amplified in the usual way. *Figure 9.48c* shows a simple d.c. amplifier suitable for several purposes. It is an up-to-date version of a circuit published in the *Review of Scientific Instruments*[8], and later used in a modified form for simple analogue computation by Clayton.[7] The circuit uses only a modest number of components; the cathode coupled input stage is combined with conventional direct coupling. High stability (metal oxide) resistors were used

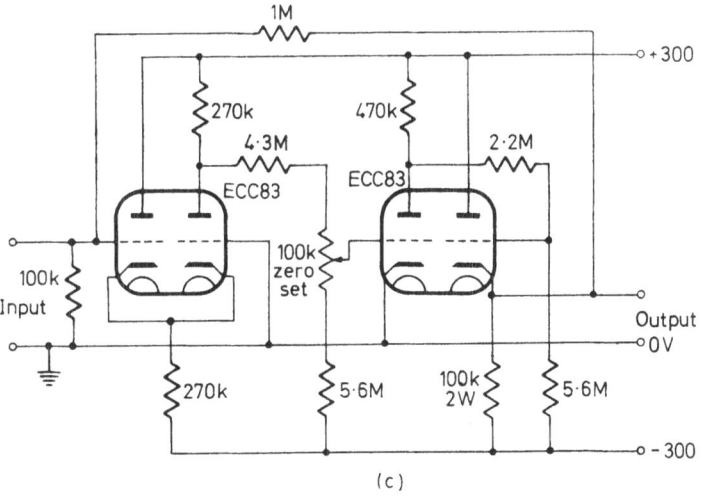

Figure 9.48c. A d.c. amplifier that combines (*a*) and (*b*). Such an amplifier was used by Clayton[7] as an operational amplifier (see text on operational amplifiers later in the chapter). It is essential to use negative feedback with this amplifier by connecting a feedback path as in operational amplifiers

in the author's version. The zero set potential divider may give some initial setting up difficulties. To set the zero, we require the variable resistor to be not too sensitive otherwise very small movements of the control produce swings at the output which are too large; this can be irritating during operation. The 100 k variable is found to give fine enough control, but some adjustment of the 4·3 M and 5·6 M resistors may be necessary. Although the circuit shows single resistors for these components, they are, in practice, made up to the correct value using several components. Initially the 100 k variable resistor may be replaced by a 1 M variable resistor and approximate balance obtained. By measuring the resistance from the slider to each end (with the h.t. supply disconnected), it is easy to design the final chain. The 100 k variable resistor is then returned to the circuit.

In simple transistor amplifiers direct coupling is possible provided it is ensured that the correct base-emitter voltage is maintained. This is achieved in the Mullard circuit[9] of *Figure 9.49* by using zener diodes in the emitter leads of the second and third transistors. The bases can then be connected directly to the collector of the previous stage. Because of the stabilizing action of the zener diodes, changes in current through the second and third transistors do not affect the respective emitter voltages which are held constant at 4·7 V and 6·8 V, respectively. This would not be the case if the initial correct emitter voltages were obtained via emitter resistors since subsequent changes of emitter current would necessarily produce changes of voltage across these

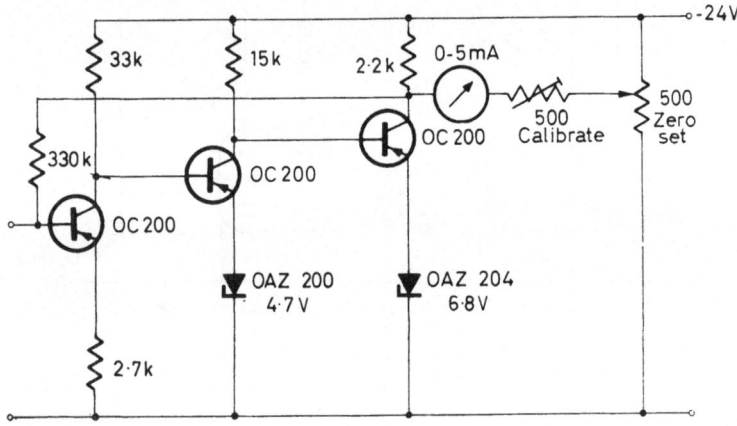

Figure 9.49. Directly coupled amplifier using transistors. The necessary bias conditions are maintained by zener diodes[9]. The source feeding the input to the amplifier must have a resistance of at least 300 k. For low source resistances the circuit of *Figure 9.50* should be used.

resistors. Overall feedback establishes the operating point of the first transistor as well as stabilizing the amplifier against changes in the current gains of the transistors.

The cathode-coupled amplifier mentioned in the section on phase splitters is a useful arrangement that assists in the elimination of drift. In valve circuits, changes of heater and h.t. voltages affect each valve to the same degree. (We deliberately ensure that matched pairs are used for a cathode-coupled pair.) Since it is the difference between the anode voltages that is passed to subsequent stages, changes of anode current due to changes of supply voltages are cancelled out. For example, if a rise in heater voltage is experienced, the anode current of both valves rises. The fall of potential is the same across each anode load, and therefore the difference between the two anode voltages due to this effect is zero. In *Figure 9.50* we have the transistor equivalent, i.e. the emitter-coupled amplifier. Because of the large common-emitter resistor the emitter-coupled amplifier is often called a long-tailed pair. The same term is used for cathode-coupled valve amplifiers. When an input voltage is applied to the input terminals of the amplifier of *Figure 9.50* the base voltage of one transistor becomes more negative with respect to the base of the other transistor of the input pair. The collector current of one transistor rises whilst that

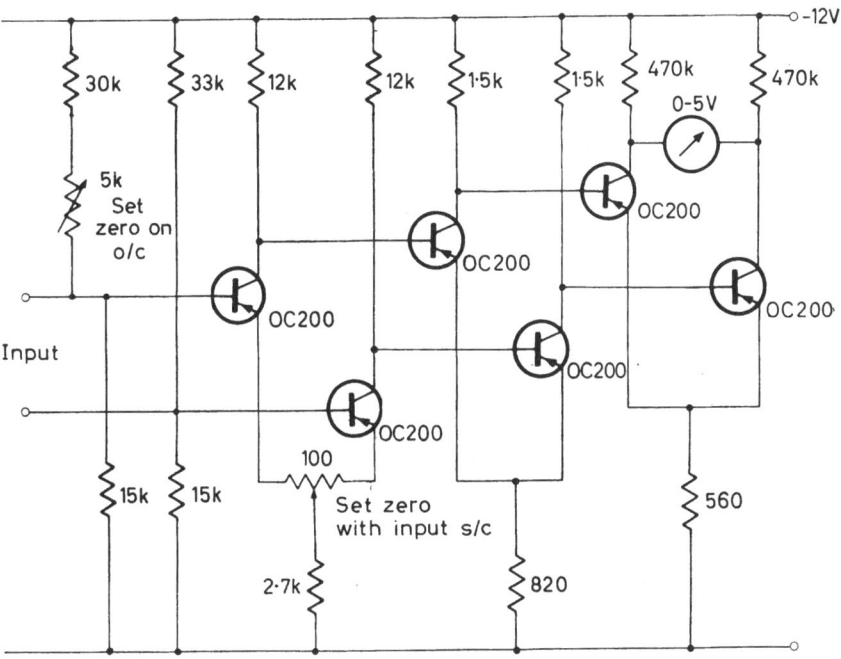

Figure 9.50. A directly coupled transistor amplifier using balanced pairs (long-tailed pairs). The circuit is suitable for any source resistance[9]

of the other transistor falls. The two collector voltages alter and are 180° out of phase. These voltages then form the signal for the subsequent pairs, the final difference voltage being registered by means of a voltmeter. Two zero-set variable resistors are provided. The 5 k alters the bias conditions of the first pair thus setting the d.c. conditions for the amplifier. This resistor is adjusted for zero output when the input is open-circuited. The zero adjustment under short-circuit input conditions is made by altering the variable resistor across the emitters of the first pair. This adjustment corrects for minor differences between transistors in a given pair.

281

Differential Amplifiers

A differential amplifier is one that amplifies the voltage difference that exists at every instant of time between signals applied simultaneously to the two inputs. Implicit in the definition is the fact that rejection of equal amplitude coincident signals takes place, the output being zero under such circumstances. In practice, perfect rejection is not achieved because of the inevitable differences of symmetry of the two portions of the circuit due to small differences in resistor and capacitor values. The active devices also contribute

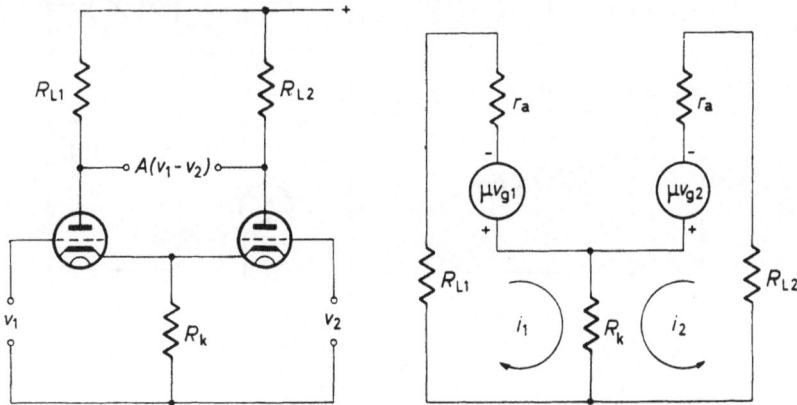

Figure 9.51. A differential (i.e. difference) amplifier

because it is impossible to match them exactly. 'Common mode' is a term used in connection with differential amplifiers that refers to signals that are identical in amplitude and phase; 'common-mode rejection' refers to the ability of a differential amplifier to reject the common mode. The common mode rejection ratio is the ratio between the common-mode amplitude and the amplitude of the small difference signal which it is necessary to apply to the input terminals to produce the same output from the amplifier.

The fundamental difference amplifier is shown in *Figure 9.51*, in which two identical valves are coupled with a large common cathode resistor. The amplifier is therefore a version of a balanced or long-tailed pair. The phase-splitter circuit of *Figure 9.45c* is a special case in which $v_2 = 0$. The circuit is not difficult to analyse if we consider the equivalent circuit. Using Kirchhoff's law that states that the sum of the e.m.f.'s around a closed loop is

282

zero we have

$$i_1(r_a + R_K + R_{L1}) + i_2 R_K = \mu v_{g1} = \mu(v_1 - (i_1 + i_2) R_K)$$

$$i_2(r_a + R_K + R_{L2}) + i_1 R_K = \mu v_{g2} = \mu(v_2 - (i_1 + i_2) R_K)$$

Therefore $i_1[r_a + R_{L1} + (1 + \mu) R_K] + (1 + \mu) R_K i_2 = \mu v_1$

$$i_1 R_K(1 + \mu) + [r_a + R_{L2} + (1 + \mu) R_K] i_2 = \mu v_2$$

The output voltage v_{out} across the two anodes is $-(i_1 - i_2) R_L$ where $R_L = R_{L1} = R_{L2}$ assuming the same value of anode load resistors; the minus

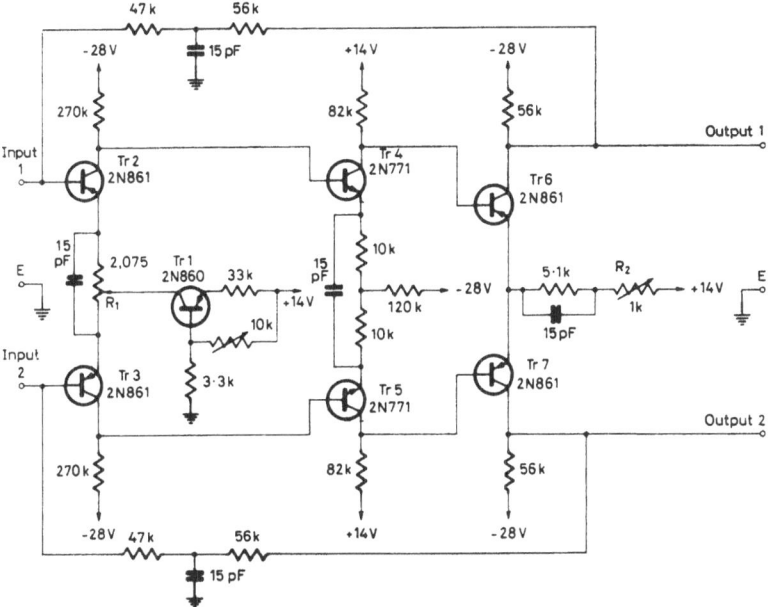

Figure 9.52. A practical d.c. difference amplifier[10]

sign shows the usual phase reversal. We have, therefore,

$$(R_L + r_a)(i_1 - i_2) = \mu(v_1 - v_2)$$

and $$v_{out} = -\frac{\mu R_L}{R_L + r_a}(v_1 - v_2)$$

$$= A(v_1 - v_2)$$

where A is the gain equal to that of a single stage amplifier with the same valve and load resistor. One pair can feed another for greater amplification as we have already seen in the transistor version of *Figure 9.50*. A directly coupled differential amplifier primarily designed for analogue computer work by Philco Ltd.[10] is

283

shown in *Figure 9.52*. The amplifier can, of course, be used as the basis of other instruments that require a differential output, e.g. an oscilloscope amplifier or a voltmeter for use at very low or zero frequencies. Since the output is the difference of two input signals, drift is greatly reduced by using differential stages.

Chopper Stabilization

Amplifiers for alternating voltages do not suffer from drift because of the RC.coupling network. It is an attractive solution to the drift problem to use such an amplifier for steady signals. The steady signals cannot, of course,

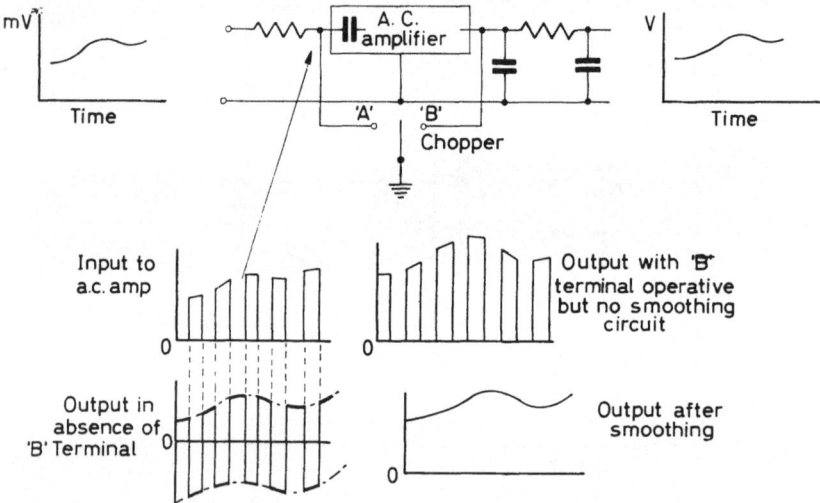

Figure 9.53a. Principles of amplifying a slowly varying signal with an a.c. amplifier and chopper. The chopper polarizes the output by shorting the output during the negative-going excursions

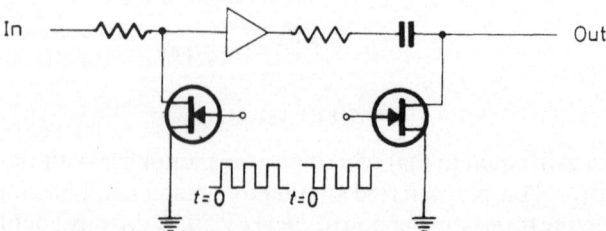

Figure 9.53b. A d.c. amplifier that uses field-effect transistors as choppers

284

be applied direct to the input terminals. The signals must first be chopped by a mechanical switching device (called a chopper) or transistor switches that perform the same function. The principles of such an amplifier are shown in *Figure 9.53a*. By using the chopper the steady or very slowly varying signal is converted into square waves the height of which is proportional to the signal at any time. These square waves can then be amplified in the usual way. The output in the absence of any restoring mechanism would be symmetrical about the 0V line since the d.c. level is lost in an a.c. amplifier. However, by shorting the output periodically as shown the negative-going portions are removed and the d.c. level is restored. A simple smoothing circuit then yields an amplified version of the input signal. Mechanical choppers are limited to a chopping frequency of about 400 Hz which means that the input signal must

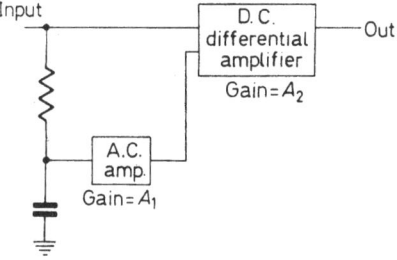

Figure 9.54. Principle of a d.c. amplifier with good high-frequency response. Signals of low or zero frequency are processed by the chopper a.c. amplifier before being amplified by the d.c. section. The gain for these frequencies is $A_1 A_2$. High frequencies that cannot be handled by the chopper are amplified directly by the d.c. section. For these high frequencies the gain is A_2 (see also *Figure 9.57*)

not vary with a frequency greater than about 40 Hz. This frequency limitation can be relaxed somewhat by using transistor choppers. Details of practical circuits can be found in the Mullard handbook of transistor circuits[9] and in several journals. The field-effect transistor is simpler to use than a conventional bipolar transistor. When using bipolar transistors in d.c. amplifiers for low-level work in the medical or biological field for example, the offset voltage is troublesome. When ordinary transistors are used as choppers there is a small output from the chopper amplifier even when the input signal is zero. This output voltage, known as the offset voltage, results from the difference between the unequal voltages that exist across the two *pn* junctions within the transistor when the latter is in the 'on' state. Unfortunately the offset voltage is a function of temperature so that variations of ambient temperature give rise to the introduction of a spurious 'input' voltage. Field-effect devices have negligibly small offset voltages. The principle of a chopper using f.e.t.s is shown in *Figure 9.53b*.

285

Because of the restriction of chopper frequency, good high-frequency performance cannot be expected. It is possible, however, to obtain good high-frequency performance and low drift by combining an ordinary differential d.c. amplifier having good high-frequency response with a chopper amplifier to reduce the drift. *Figure 9.54* shows the principle (see also *Figure 9.57*). It is essential to use negative feedback with this type of amplifier. However, since the usual duty of such an amplifier is for operational work, this is taken care of automatically. The whole arrangement is then known as an operational amplifier.

Operational Amplifiers and the Principles of Analogue Computing

Basically an analogue computer is a machine for solving differential equations. Many problems concerning aerodynamics, ballistics, electron optics, physics, mechanical and electrical engineering are suitable for solving by this machine. All the branches of study mentioned can provide some formidable calculations for human beings; indeed, many research problems have previously remained undeveloped because of the volume of mathematical computation involved.

Computers fall into two main classes, viz. the digital type and the analogue type. The first type performs its calculations in an arithmetic way using a binary system of arithmetic, and possessing a 'memory'. It is suitable for evaluating formulae, calculating salaries, solving traffic problems and dealing with a host of other problems associated with the commercial world. In the field of scientific research the digital computer can handle many problems including those that can be solved by an analogue computer. However, the digital machine is very expensive, so when the problem involves the solving of differential equations, it is usual to use the much cheaper analogue computer.

The analogue computer solves problems presented to it by creating voltages that are proportional to the values of the terms of a differential equation at any time. Means are available in the computer for operating upon these voltages in a way that is analogous to the operations that could be performed by a mathematician using pencil and paper.

The first step in using the computer is to express the problem in mathematical terms. To see the way in which a differential equation may be formed, and rearranged so as to be suitable for the computer, let us consider a simple problem.

Figure 9.55 shows a mass resting on a rough table. The extension of the spring is assumed to be governed by Hooke's law and as a simplifying assumption we shall take it that the frictional force is proportional to velocity. For a step force displacing the end of the spring A by an amount x at

286

time $t = 0$, let the problem be to find the velocity and acceleration at some later time t. The acceleration on the mass, $m\dfrac{d^2y}{dt^2}$, is the sum of two forces, viz. that proportional to the extension of the spring $(x-y$, where y is the displacement of the end B of the spring) and that frictional force, acting in a direction opposite to the velocity, that is proportional to the velocity.

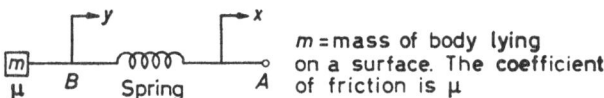

m = mass of body lying on a surface. The coefficient of friction is μ

Figure 9.55

The equation of motion is:

$$m\frac{d^2y}{dt^2} = k(x-y) - \mu\frac{dy}{dt}$$

(where k is Hooke's constant and μ is the dynamic coefficient of friction)

$$\therefore\ m\frac{d^2y}{dt^2} + \mu\frac{dy}{dt} + ky = kx$$

Rearranging this into a form suitable for the computer we have

$$\frac{d^2y}{dt^2} = \frac{kx}{m} - \frac{ky}{m} - \frac{\mu}{m}\frac{dy}{dt} \tag{9.11}$$

If we can now produce voltages proportional to the individual terms of equation 9.11 and then devise a circuit arrangement that operates on these voltages in a manner analogous to that used by a mathematician we are in a position to solve the problem.

Figure 9.56 shows, in block diagram form, an arrangement for solving equation 9.11. Let us assume initially that a voltage proportional to $\dfrac{d^2y}{dt^2}$ exists. (The way in which this voltage arises will be explained shortly.) By feeding this voltage into an integrator a voltage proportional to $\dfrac{dy}{dt}$ is produced. A further integrator produces the corresponding y voltage. By suitable sign reversers and scale multipliers voltages are produced that represent $\dfrac{-k}{m}y$ and $\dfrac{-\mu}{m}\dfrac{dy}{dt}$. If these voltages are now added to a constant voltage representing $\dfrac{k}{m}x$ the output voltage of the summing amplifier is proportional to $\dfrac{d^2y}{dt^2}$. This is the voltage 'we assumed already available. The output

287

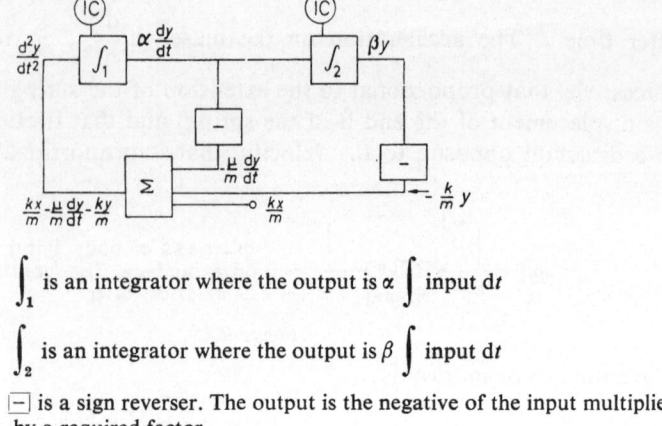

$\displaystyle\int_1$ is an integrator where the output is $\alpha \displaystyle\int$ input dt

$\displaystyle\int_2$ is an integrator where the output is $\beta \displaystyle\int$ input dt

$\boxed{-}$ is a sign reverser. The output is the negative of the input multiplied by a required factor

$\boxed{\Sigma}$ is a summing amplifier

(IC) feeds in the initial conditions

Figure 9.56. Setting up the electronic solution to a second order differential equation

of the summing amplifier is, therefore, fed into the input of the first integrator. From *Figure 9.56* it can be seen that we require electronic apparatus that

(*a*) adds and subtracts

(*b*) multiplies by a constant coefficient

and *(c)* integrates.

Additionally for other problems we require apparatus that can generate voltages analogous to the many types of forcing function encountered (e.g. voltages proportional to x, $\sin x$, x^2, e^x, $\log x$, etc., where $x = f(t)$.

Electronic Operational Amplifiers

The electronic 'boxes' that perform the necessary functions are based on high-gain feedback amplifiers. Since the response is required to go down to zero frequency the amplifiers must be of the d.c. type (*Figure 9.57*). A combination of a high-gain d.c. amplifier together with feedback and input impedances is known as an operational amplifier. It is usual to represent the complete amplifier (with a gain of $-A$) as a triangle; the associated feedback and input impedances are represented by Z_f and Z_1, Z_2, Z_3, etc., as shown in *Figure 9.58*. If the input voltage at the junction (P) of the input and

288

feedback impedances is e, then:

$$\frac{e_1-e}{Z_1}+\frac{e_2-e}{Z_2}+\ldots+\frac{e_n-e}{Z_n}+\frac{e_0-e}{Z_f}=0 \tag{9.12}$$

since there can be no accumulation of charge at this junction.

Now $e_0 = Ae$. Substituting this in equation 9.12 and rearranging we obtain

$$e_0 = -\sum_{k=1}^{n}\frac{e_k}{Z_k}\bigg/\left[\frac{1}{Z_f}-\frac{1}{A}\left(\sum_{k=1}^{n}\frac{1}{Z_k}+\frac{1}{Z_f}\right)\right].$$

If the gain of the amplifier is made very large, i.e., $A \rightarrow \infty$ in the limit:

$$e_0 = -Z_f\sum_{k=1}^{n}\frac{e_k}{Z_k}. \tag{9.13}$$

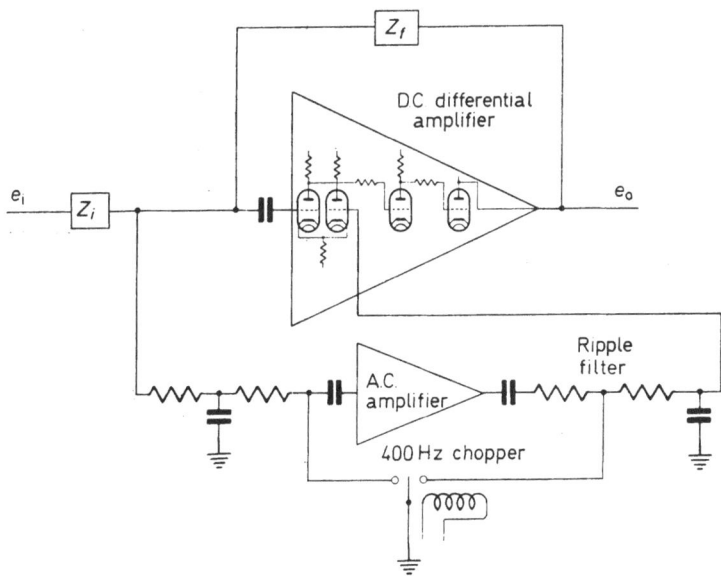

Figure 9.57. D.C. amplifier with chopper stabilization

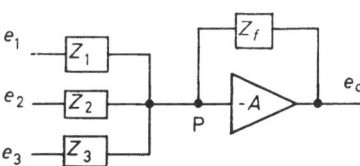

Figure 9.58. Generalized operational amplifier (the triangle represents the combined d.c. and chopper a.c. sections)

Under these conditions enough current is drawn through Z_f to maintain P at almost zero potential. This point is known as a virtual earth. Drift can be reduced to negligible proportions by using the type of operational amplifier that consists of a d.c. amplifier, a chopper system and auxiliary a.c. amplifier. Careful selection of the input valves and operating conditions ensures a sufficiently low grid current.

If we ignore for the time being the effects of drift, grid current, finite bandwidth, finite gain and so on, equation 9.13 may be used to demonstrate the power and usefulness of the operational amplifier.

Consider first the case where $Z_1 = Z_2 \ldots = Z_f = R$. It follows from equation 9.13 that:

$$e_0 = -\sum_{k=1}^{n} e_k$$

i.e. the output voltage is the sum of the input voltages multiplied by minus one.

In general when Z_1, Z_2, Z_f, etc. are all different then

$$e_0 = -R_f \left(\frac{e_1}{R_1} + \frac{e_2}{R_2} + \ldots + \frac{e_n}{R_n} \right).$$

The operational amplifier thus acts as a summing amplifier when the associated impedances are resistances.

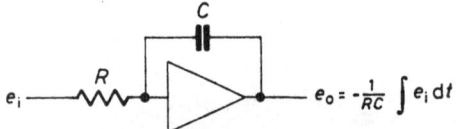

Figure 9.59. The simple integrator

If the feedback impedance is a capacitance then $Z_f = 1/j\omega C$. It is convenient to invoke operational notation in further analyses. We cannot here discuss the use of operational notation from first principles; the reader is directed to such works as *Introduction to Laplace Transforms for Radio and Electronic Engineers* by W. D. Day. We find that algebraic processes can be used in the solution of differential equations, and that useful results are obtained if $j\omega$ is replaced by the operator p. If $f(t)$ is a function of time $pf(t)$ is regarded as the differential coefficient of the expression with time, and $(t)/p$ is regarded as the time integral.

Equation 9.13 then becomes:

$$e_0 = -\frac{1}{Cp} \left(\frac{e_1}{R_1} + \frac{e_2}{R_2} + \ldots + \frac{e_n}{R_n} \right) \qquad (9.14)$$

We regard this as:

$$e_0 = -\frac{1}{CR_1} \int e_1 \, dt \ldots - \frac{1}{CR_n} \int e_n \, dt.$$

The reader should note that:

$$\frac{1}{Cp} \frac{e}{R_1} \quad \text{is not equal to} \quad \frac{1}{CR_1} \int e_1 \, dt$$

as is stated rather loosely in several textbooks. The use of the p operator however allows algebraic processes to yield the same answer as would be obtained by performing the actual differentiation and integration.

Equation 9.14 shows the use of the operational amplifier as a summing integrator with the inevitable sign reversal.

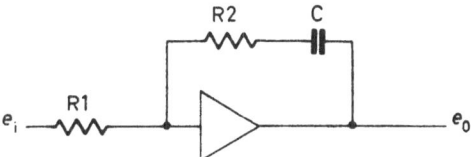

Figure 9.60. Circuit with a more complicated transfer function (e_0/e_i is called the transfer function)

The choice of different networks to represent Z_f and the input impedances gives rise to a wide variety of operational forms. Many of these forms are useful in simulation and servo work. For example, in *Figure 9.60*

$$e_0 = -\frac{R_2}{R_1}\left(1 + \frac{1}{CR_2 p}\right) e_i$$

e_0/e_i is called the transfer function. The example just quoted is well known to control engineers as an 'error + integral of error' circuit. A more complicated example is the quadratic transfer function used by simulation engineers. The circuit of *Figure 9.61* yields the transfer function:

$$-\frac{R_3}{(R_1 + R_2)(1 + \tau_1 p)(1 + \tau_2 p)}$$

where $\tau_1 = C_1 R_1 R_2/(R_1 + R_2)$

and $\tau_2 = C_2 R_3$.

In practical amplifiers the output voltage is not precisely that which may be expected from a knowledge of the transfer function. The departure from an ideal performance is due to imperfections in the amplifier. These imper-

19*

fections are drift, grid current and finite values of gain bandwidth and input impedance.

The effect of having a finite gain in a summing amplifier with infinite input impedance and of a given gain A, is to introduce a percentage error in e_0 of:

$$\frac{100}{A}\left(R_f \sum_{k=1}^{n} \frac{1}{R_n} + 1\right).$$

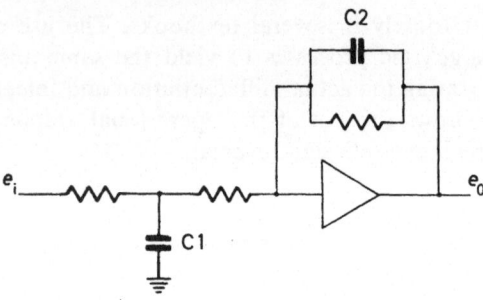

Figure 9.61. Circuit to yield the transfer function often used by simulation engineers

Let us take, for example, an operational amplifier with $R_f = 1$ M and two input resistances $R_1 = R_2 = 0\cdot1$ M; the percentage error would be

$$\frac{100}{A}\left\{1\left(\frac{1}{0\cdot1} + \frac{1}{0\cdot1}\right) + 1\right\} = \frac{2,100}{A}$$

If the error is to be less than $0\cdot1$ per cent it can be seen that the gain must be at least 21,000.

The effects of employing amplifiers with finite bandwidths and gains and the use of 'impure' impedances (e.g. a leaky capacitor or inductive resistor) lead to errors in the solution of differential equations[11]. The roots of the characteristic equations differ from the true roots and additionally the computer solves a higher order equation than the original problem equation. The additional roots so obtained are foreign to the original equation.

To be satisfactory, an operational valve amplifier must satisfy the following design requirements:

(*a*) the gain-bandwidth product must be large;

(*b*) The effects of drift must be minimized by applying continuous correction. For practical purposes the application of the corrections must be automatic; this results in the arrangement outlined in *Figure 9.57*.

(*c*) the grid current must be low ($< 10^{-10}$ A);

(*d*) the gain/frequency characteristic (Nyquist diagram) must be such as

292

will ensure adequate stability. To allow for various loading conditions and satisfactory transient response the stability must be unconditional;

(e) the input impedance must be very high; and

(f) adequate output power must be available. This often results in the use of cathode-follower output stages.

For simple integration and summation, however, the circuit of *Figure 9.49* should prove satisfactory. It is only when several operational amplifiers are combined in a system that the specification of the amplifier is stringent.

Transistor operational amplifiers are replacing valve types rapidly. The circuit of *Figure 9.52* may be used successfully as an operational amplifier.

REFERENCES

1. Olsen, G. H. 'Matrix algebra'. *Wireless World*, 1965, **71**, Nos. 3 and 4, March and April.
2. *Crystal Microphone Amplifier*. Ferranti Application Note No. 7.
3. Ogilvie, J. R. 'An a.f. power amplifier with crossover feedback'. Letter, *Electronic Engineering*, 1966, **38**, No. 456. Feb.
4. Tobey, R. and Dinsdale, J. 'Transistor audio power amplifier'. *Wireless World*, 1961, **67**, No. 11, p. 565. (Nov.)
5. *Field-Effect Transistors and Applications*. Application Note No. 22. Ferranti Ltd.
6. Down, B. 'Using feedback in f.e.t. circuit to reduce input capacitance'. *Electronics*, 1964, Dec. 14.
7. Clayton, A. B. 'Simple analogue computer'. *Wireless World*, 1960, **66**, No. 5. (May.)
8. *Rev. Sci. Instrum.* 1950, **21**, p. 154.
9. *Reference Manual of Transistor Circuits*. Mullard Ltd. 1960.
10. *Application Lab. Report No. 713*. Philco International Ltd.
11. Aspects of these errors are discussed by Macree, A. B. *Proc. I.R.E., N.Y.* 1952, **40**, 3-3, and Marsocci, V. A. *Trans. I.R.E., N.Y. (Electron Computers)* E.C. 5, 1956, 207. Those interested in analogue computation are referred to the books *Analogue Computation* by R. W. Williams, Heywood & Co. Ltd., and *Electronic Analogue Computers* by G. A. Korn and T. M. Korn, McGraw-Hill. 1956.
12. Greiter, O. 'Transistor amplifier output stages'. *Wireless World*, 1963, **69**, Nos. 1-6 (Jan-June.)
13. Osborne, R. and Tharma, P. 'Transistor high quality amplifiers'. *Wireless World*, 1963, **69**, No. 6. (June.)

10

OSCILLATORS

An oscillator is an instrument for producing voltages that vary in a regular fashion; the waveforms of the voltages are repeated exactly in equal successive intervals of time. In many cases the waveform of the output voltage is sinusoidal and the oscillator is then called a sinewave generator or harmonic oscillator. Those instruments that produce repetitive waveforms that are square, triangular or sawtooth in shape are called relaxation oscillators. The term 'relaxation' is used because during the generation of the waveform there is a period of activity in which there is a sharp transition from one state to another. This period is then followed by a relatively quiescent one, after which the cycle is repeated. Several examples of relaxation oscillators occur in nature, the most common one being the heart. During the operation of the heart there is a period of activity in which the blood is pumped through the heart chambers and out into the arteries. This period is followed by one in which the heart muscles relax and prepare for the next burst of activity.

Oscillators can be constructed so as to operate at frequencies as low as one or two cycles an hour or as high as hundreds of megacycles per second. The selection of a suitable frequency or range of frequencies depends upon the function that the oscillator is required to perform. For the testing of equipment, or as a source of power for energizing a.c. bridges, the frequencies are usually in the l.f. or low r.f. range. Conductivity cells and electrolytic tanks are supplied with energy at frequencies of a few hundred cycles per second. Radio-frequency oscillators are widely used in the generation of carrier waves for telecommunication systems and in the construction of non-lethal e.h.t. supplies. Industrial heaters of dielectric materials such as wood, glue and plastics depend upon r.f. oscillators. Physiotherapy departments in hospitals use this type of heater in the treatment of bone and tissue disorders. Where the heating of electrically conducting material, such as metal ingots, is involved, induction coils are fed from power oscillators operating at lower frequencies. The material to be heated is placed within the coil and the eddy currents that are induced within the material cause rises in temperature. Both induction heating and dielectric heating have the advantage that the heating is produced within the bulk of the material. These methods of heating do not therefore rely on conduction from a hot surface layer.

For the applications described above the waveform produced by the oscillator has usually to be sinusoidal or nearly so. In other applications such as cathode-ray oscilloscopes, television receivers, radar equipment, digital computers and automatic industrial controllers, relaxation oscillators are important and necessary sections of the equipment.

SINEWAVE GENERATORS

The most general method of producing sinusoidal oscillations is to use a feedback amplifier in which the feedback is positive at some desired frequency. The feedback circuit must therefore be frequency selective. From the previous chapter we recall that the gain of a feedback amplifier is given by

$$A' = \frac{A}{1 - \beta A}$$

where A is the gain of the amplifier without feedback and β is the fraction of the output voltage fed back to the input. In the cases we have considered so far the feedback has been negative. This results in a fall in gain, but several important advantages are obtained. In the design of negative feedback amplifiers care must be taken to ensure that positive feedback does not occur at any frequency at which the loop gain (βA) is unity. If such care is not taken the amplifier becomes unstable and oscillates at that frequency; the oscillations are uncontrollable and the amplifier cannot perform satisfactorily. Consideration of the gain formula for A' shows that with positive feedback A' is greater than A. (In the preliminary discussion in this paragraph it is implicit that the quantities A and β are both real.) For values of βA less than unity the output voltage for a given input voltage increases as βA increases. When $\beta A = 1$, A' is theoretically infinite and any disturbance in the circuit (such as variations of heater or h.t. supplies) is amplified, fed back in just the right phase to be further amplified, and so on until the amplifier is driven to its full extent. Infinite amplification is impossible in a practical case because of the limits set by the saturation and cut-off points of the valves or transistors. However, even in absence of an input signal any disturbance within the amplifier is able to maintain itself. An amplifier that provides its own input in this way functions as an oscillator. By making the feedback circuit frequency selective we automatically arrange that $\beta A = 1$ at only one selected frequency. Oscillations of only this frequency are, therefore, produced and so the output voltage waveform is sinusoidal. In practice, the non-linearity of the oscillator characteristic causes the output waveform to depart from a true sinewave. In a well designed oscillator, however, the departure is very small, and the harmonic content can easily be reduced below 1 per cent of the fundamental.

Maintenance of Oscillation

The conditions under which oscillations are initiated and maintained require that any losses and attenuation in the feedback line must be made good by the amplifier. In mathematical terms the loop gain βA must be unity. The quantity βA is usually complex and is of the form $X + jY$. For oscillations to be maintained $\beta A = 1 + j0$. The real or ordinary part of βA must be unity and the quadrature component must be zero. This is equivalent to saying that the feedback is positive and that the input and output voltages associated with the amplifier are in phase. The condition βA is $1 + j0$ is known as the Barkhausen criterion for oscillation and is often illustrated diagrammatically by a Nyquist plot[1]. *Figure 10.1* shows two such plots of the ordinary (real) part of βA vs. the quadrature (imaginary) component for all frequencies from zero to infinity. *Figure 10.1b* shows stable conditions and represents the position that must be achieved in a negative feedback amplifier. Since the plot does not enclose the point 1, 0 there is no

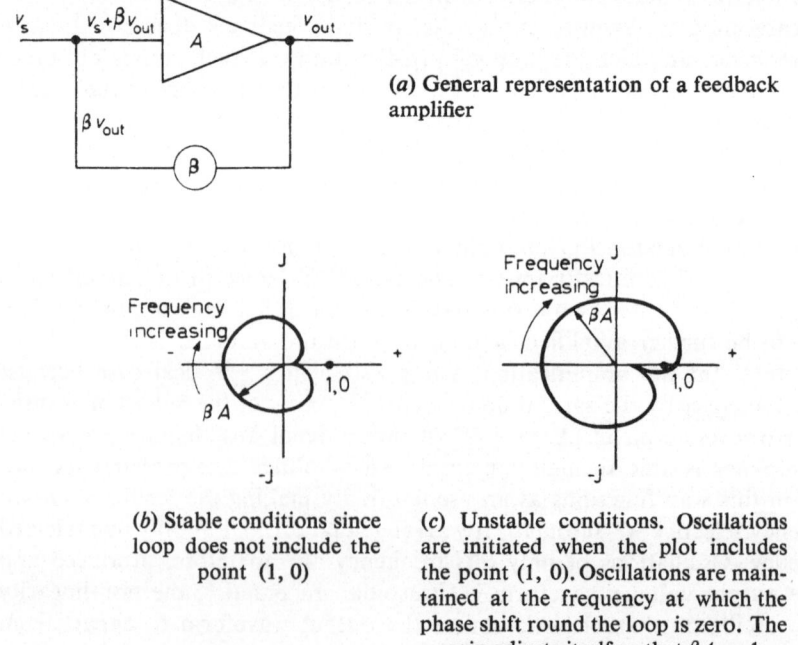

(a) General representation of a feedback amplifier

(b) Stable conditions since loop does not include the point (1, 0)

(c) Unstable conditions. Oscillations are initiated when the plot includes the point (1, 0). Oscillations are maintained at the frequency at which the phase shift round the loop is zero. The gain adjusts itself so that $\beta A = 1$

Figure 10.1. Nyquist plots of the ordinary component of the loop gain βA against the quadrature component for all frequencies from zero to infinity

frequency at which feedback is positive and the loop gain high enough to sustain an oscillatory condition. *Figure 10.1c* shows the plot of an unstable feedback amplifier in which oscillations can be initiated and sustained. When using a feedback amplifier as an oscillator we deliberately ensure that $\beta A = 1 + j0$ at a selected frequency. This is achieved by using a suitable feedback circuit, examples of which are described later in the chapter. Arranging that the input and output voltages of the amplifier are in phase ensures positive feedback. The gain of the amplifier automatically adjusts itself until $|\beta A| = 1$ (i.e. the magnitude of the loop gain — written $|\beta A|$ — is unity). The self-adjustment is achieved by virtue of the non-linearity of the amplifier characteristic near the saturation and cut-off regions, or by the incorporation of a thermistor or lamp as a controlling agency.

The mode of operation of this type of oscillator is thus as follows. On switching on the instrument it may be assumed that a switching surge is present at the output terminals. This surge is fed back in phase to the input circuit via the feedback line and subsequently amplified. The larger output is fed back again and further amplified. Signals of all frequencies but one are heavily attenuated in the feedback circuit. For those frequencies $\beta A \ll 1$ and so the corresponding components in the output voltage die away. One frequency component is not attenuated however, and it is this component that is continually amplified. Eventually the amplifier is driven to its full extent and no further amplification can occur. When the gain settles to a value that makes $\beta A = 1$ a steady state is achieved and oscillations of constant amplitude are sustained.

Oscillators for High Frequencies

When an oscillator is required to operate at frequencies above about 100 kHz the feedback and load circuits associated with the active device usually consist of tuned circuits. A particularly simple high-frequency oscillator is that shown in *Figure 10.2*. The anode load consists of a coil which is inductively coupled to a resonant LC combination in the grid circuit. The coil must be connected into the anode circuit in such a way as to produce positive feedback. The mutual coupling, M, must also be large enough to feed sufficient energy into the resonant circuit to make good losses there. Provided the resonant circuit has a high Q, the coupling need not be too tight and oscillations occur at practically a single frequency. This means that the output waveform is almost purely sinusoidal and little distortion is present.

To find the conditions necessary for maintaining oscillations, we may use the following simplified analysis. The voltage induced in the grid coil is given by $v = j\omega M i_a$ where M is the mutual inductance. Provided the anode coil has small self-inductance it is a good approximation to say that

$$i_a = g_m v_g$$

therefore
$$v = j\omega M g_m v_g$$

To find v_g we note that this is the voltage across the capacitor. This is given by the product of the oscillating current, i, in the grid circuit and the reactance

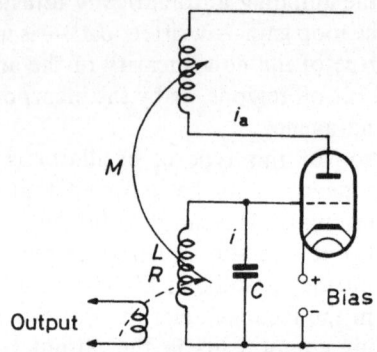

Figure 10.2. The tuned grid or Armstrong oscillator

of the capacitor. Therefore

$$v_g = \frac{i}{j\omega C}$$

and
$$v = j\omega M g_m \cdot \frac{i}{j\omega C} = \frac{M g_m i}{C}.$$

Now
$$i = \frac{v}{Z} \quad \text{where} \quad Z = R + j\omega L + \frac{1}{j\omega C}$$

$$\therefore \ Z = \frac{M g_m}{C} = R + j\omega L + \frac{1}{j\omega C} .$$

At resonance
$$\omega L = \frac{1}{\omega C} \qquad \therefore \ j\left(\omega L - \frac{1}{\omega C}\right) = 0$$

and
$$M = \frac{CR}{g_m}$$

This is the minimum value necessary for the maintenance of oscillations. The frequency of the oscillations is the resonant frequency of the grid circuit namely $1/(2\pi\sqrt{LC})$. A rigid analysis that takes into account the effect of the grid coil on the anode coil and the resistance and inductance of the anode

298

coil produces results that are very close to those given. For the analysis given we have assumed a loosely coupled anode coil of small self-inductance and a high-Q resonant grid circuit; we have also assumed that very little energy is taken from the output coil.

Figure 10.3a shows an alternative arrangement in which the resonant circuit acts as an anode load. *Figure 10.3b* shows a tuned-anode tuned-grid oscillator in which there are resonant circuits in both the anode and grid

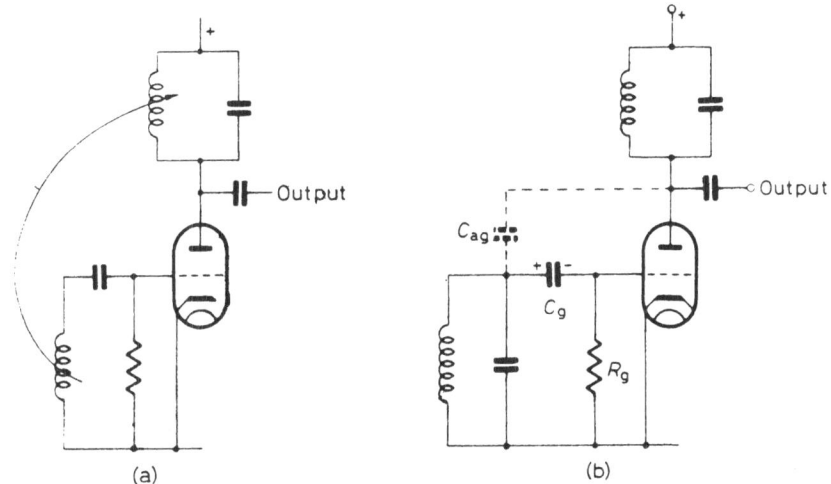

Figure 10.3. (*a*) Tuned-anode oscillator. (*b*) Tuned-anode tuned-grid oscillator

circuits. The two resonant circuits are usually shielded from each other, the necessary feedback being obtained via the anode-to-grid capacitance or the equivalent capacitance in a transistor. Both tuned circuits must be tuned on the inductive side of resonance for oscillations to occur. The circuit with the highest Q determines the frequency of oscillation. As it is usual to obtain the output from the anode circuit the grid circuit is made to have the higher Q and thus controls the resonant frequency. The capacitor C_g and resistor R^g provide automatic bias for the valve. For positive-going voltages at the grid end of the resonant circuit, electrons are drawn from the main stream in the valve and C_g charges with the polarity shown. During the negative-going portion of the cycle the electrons are repelled, but are unable to enter the main anode current stream because the grid is not a hot electrode; the density of electrons in the space charge within the valve assists in preventing electrons from re-entering the stream. To prevent blocking the valve action, a high resistance leakage path, R_g, is provided which allows some of the electrons to leak to the cathode circuit. By suitable choice of R_g and C_g a net voltage is created across C_g and thus the grid is biased to a negative potential with respect to the cathode. The value of the bias voltage depends upon the amplitude of the voltage supplied by the resonant grid circuit.

Two popular and well-tried oscillator circuits, which are basically the same, are due to Hartley and Colpitts and are shown in *Figure 10.4*. When X_1 is a capacitor, and X_2 and X_3 are mutually coupled inductors (frequently a single, tapped coil) then the arrangement is known as a Hartley oscillator. Analysis

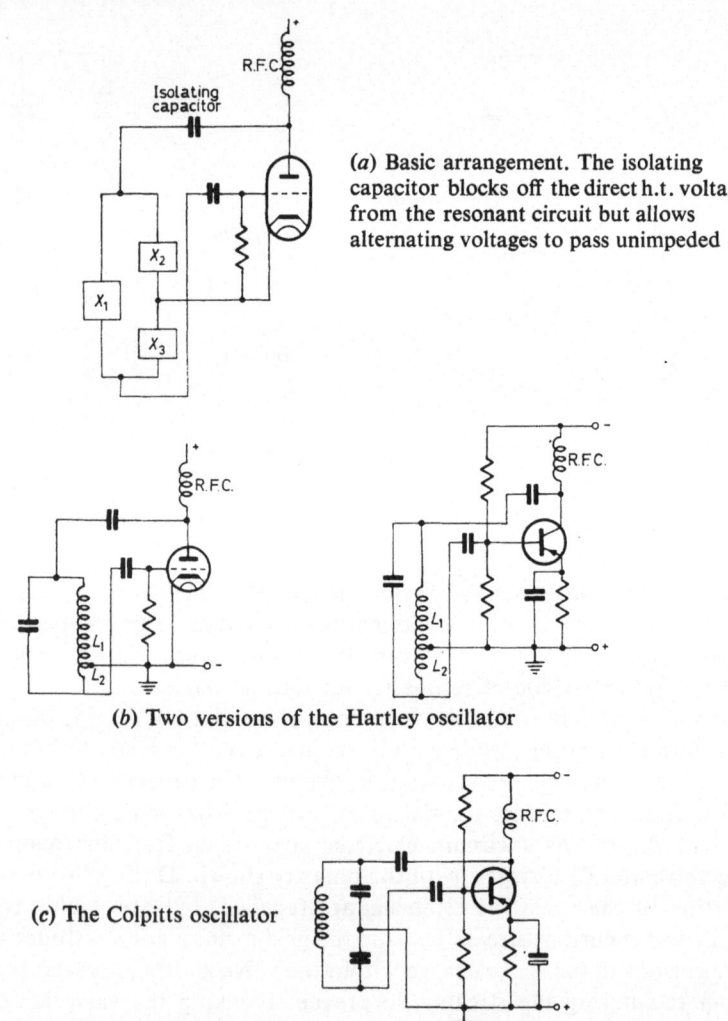

(*a*) Basic arrangement. The isolating capacitor blocks off the direct h.t. voltage from the resonant circuit but allows alternating voltages to pass unimpeded

(*b*) Two versions of the Hartley oscillator

(*c*) The Colpitts oscillator

Figure 10.4. High-frequency oscillators based on the circuits of Hartley and Colpitts

shows the frequency of oscillation to be given very closely by the formula

$$f = \frac{1}{2\pi\sqrt{\{(L_1+L_2+2M)\,C\}}}$$

The operation of both the Hartley and Colpitts oscillators may be understood by considering the basic circuit of *Figure 10.4*. X_1, X_2 and X_3 form a resonant circuit in which quite a large oscillating current is circulating. The small losses that inevitably occur are made up by the supply of alternating power from the active device. The circulating current in the resonant circuit is many times the supply current when the Q is high. For this reason the resonant circuit is often called a tank circuit. A tapping is provided along one of the reactive arms at the junction of X_2 and X_3. The voltages at each end of the tank circuit are necessarily 180° out of phase with each other relative to the tapping. The anode voltage is 180° out of phase with the grid voltage so by connecting the tapping to the cathode and the two ends of the tank circuit to the anode and grid, respectively, positive feedback is introduced into the amplifier and oscillations are sustained. The tapping point is chosen so that a suitable proportion of the voltage between the two ends of the tank circuit is applied between the grid and cathode. As the effective tapping point is moved towards the anode end of the tank circuit the feedback is increased. Several factors such as the Q of the tank circuit, stray capacitances and the damping imposed by the grid circuit make a precise calculation of the tapping point difficult. It is usual to determine the best point experimentally; making $X_2 = 4X_3$ is often a satisfactory starting point.

In the Hartley circuit there is ambiguity about the tapping point. This is because although the tapping point on the coil is precise enough there is also a hidden tapping due to the stray capacitances that exist across L_1 and L_2 *(Figure 10.4b)*. For low radio frequencies, however, the effect of the strays can usually be neglected, but if the operating frequency must be high (say several megacycles per second) then the effect of the strays becomes important and, since the ratio of the stray capacitances is unlikely to be the same as L_1/L_2, uncertain operation can result. At high frequencies, therefore, many workers prefer the Colpitts circuit in which the tapping is effectively along the capacitor. Allowance can then be made for the strays which are merely added to the two capacitors forming the tapping. The frequency of oscillation of the Colpitts oscillator is given approximately by

$$f = \frac{1}{2\pi\sqrt{\left\{L\left(\dfrac{C_1C_2}{C_1+C_2}\right)\right\}}}$$

Where no stringent requirement is placed on the frequency and stability of the oscillator it is usual to obtain the output via a capacitor connected to the anode (or collector) or alternatively to obtain an output directly from the tank circuit by using an additional coil inductively coupled to the coil in the tank circuit. The load on the oscillator necessarily damps the tank circuit and highly stable performance from the oscillator cannot be expected. Where the loading of the tank circuit and oscillator cannot be tolerated, it is necessary to use some sort of buffer amplifier. A cathode-follower circuit is satisfactory. The high input impedance of this circuit ensures that the oscillator is not loaded to any great extent, and the low output of the cathode-follower enables considerable power to be developed in the load. An alternative arrangement is to use electron coupling. An electron-coupled oscillator

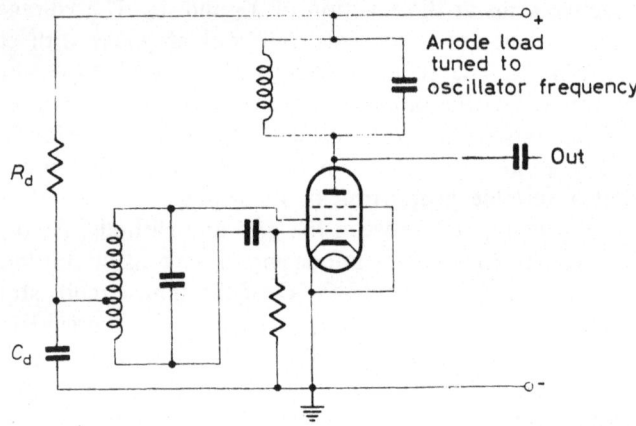

Figure 10.5. Electron-coupled oscillator based on the Hartley circuit

circuit is shown in *Figure 10.5*. The cathode, control grid and screen grid form an internal triode and oscillations are produced in the usual way. Only a small proportion of the alternating valve current is supplied to the tank circuit via the screen grid. Most of the alternating current passes to the anode circuit and hence the load. The suppressor grid is earthed or connected to the cathode in the usual way, and so there is an effective screen between the anode and the screen grid. This results in an effective separation between the oscillating tank circuit and the anode load circuit. The pentode electron-coupled oscillator operates at a frequency that is practically independent of the load. This type of oscillator is sensitive to h.t. supply voltage variations thus it is advisable to use a well-stabilized power supply. R_d and C_d form the screen h.t. supply. C_d must be large enough to decouple the screen satisfactorily. There is then effectively a short circuit a.c. path between the coil tapping and the cathode.

302

Many practical applications of *LC* oscillators are given in the literature. One such application is shown diagrammatically in *Figure 10.6*. It shows a simple oscillator circuit for use in chromatography[2]. Since the supply rails are at the same potential so far as alternating voltages are concerned, the collector load coil is effectively in parallel with the capacitors C_1 and C_2.

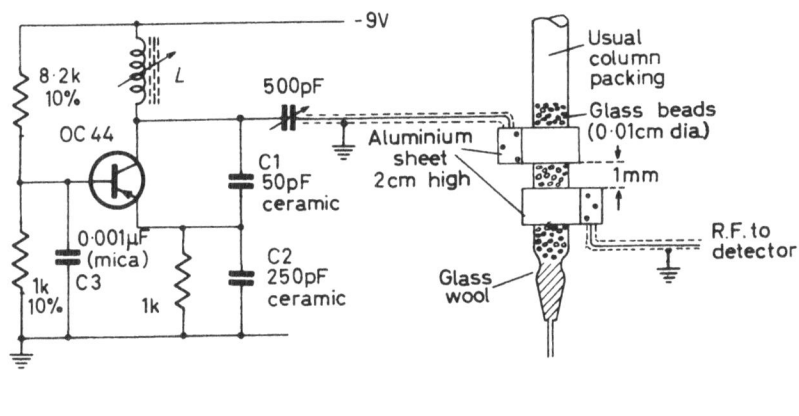

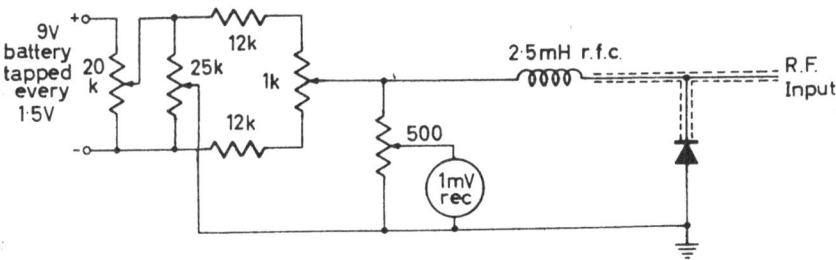

Figure 10.6. A simple high-frequency transistor recorder for chromatography[2]. The 1 mV recorder is a Sunvic high-speed potentiometric recorder (model RSP2). *L* should be adjusted to give the best results with the column used. 1–2 MHz is recommended

The emitter is connected to the junction of C_1 and C_2, the collector to the upper part of the tuned circuit and the base to the lower part of the tuned circuit via C_3; this circuit is therefore a version of the Colpitts oscillator. For convenience, the adjustment of frequency is made by altering the inductance of *L*. This is a ferrite-cored coil and inductance variations are achieved by adjusting the position of the ferrite core relative to the coil. Where frequencies exceeding 10 MHz are required, Jackson recommends the use of an OC171 transistor. This transistor has a better high-frequency performance than the OC44.

303

Crystal Oscillators

There are several applications, notably in telecommunication systems and in laboratory frequency standard equipment, where highly stable frequencies are required. Frequency drift in oscillators is caused by changes in resonant frequency of the tuned circuit, such changes being due mainly to variations of component values with temperature changes. Oscillation at a frequency slightly different from the resonant frequency is caused by the components associated with the tuned circuit, such as leads to valves and transistors, interelectrode capacitances and output loading coils. Changes in the parameters of these components contributes to frequency drift.

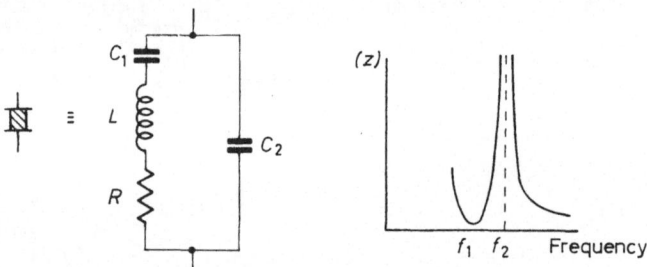

Figure 10.7. The circuit symbol for a crystal together with its equivalent circuit. The graph shows the variation of impedance with frequency in the region of series (f_1) and parallel (f_2) resonant frequencies. The values of L, R, C_1 and C_2 depend upon the individual crystal. One typical sample has values of $L = 5\cdot2$ H, $R = 280$ ohms, $C_1 = 0\cdot01$ pF and $C_2 = 6$ pF. The series resonant frequency is 698 kHz and the Q value 81,400

In nearly all cases the difficulties can be overcome by using tuned circuits with a high Q and constructed from stable components. When such a tuned circuit is available the electron-coupled oscillator previously described gives a very good performance. With ordinary inductors and capacitors however, Q values greater than a few hundred cannot be obtained. Provided the frequency of operation is not required to be variable, very large improvements in frequency stability can be obtained when a quartz crystal is used as the resonating element, in place of the conventional tuned circuit.

 . Quartz crystals exhibit piezoelectric properties[3] that is to say mechanical stresses imposed on the crystal give rise to potential differences across the faces of the crystal, and vice versa. Special cuts are needed relative to the crystallographic axes in order to produce the best performance. Quartz is chosen for oscillator frequency standards because this material is almost perfectly elastic; if mechanical oscillations are initiated it takes a long time for the oscillations to die away. Quartz crystals, therefore, have a very high mechanical Q. So far as the electrical properties are concerned a quartz crystal is equivalent to the LC resonant circuit shown in *Figure 10.7*. The values of L, R, C_1 and C_2 depend upon the physical size of the crystal and the cut used. The crystal itself has conducting electrodes sputtered on to two crystal faces. Connecting leads are then joined to the sputtered electrodes. When the leads are connected to a source of oscillating voltage then mechanical

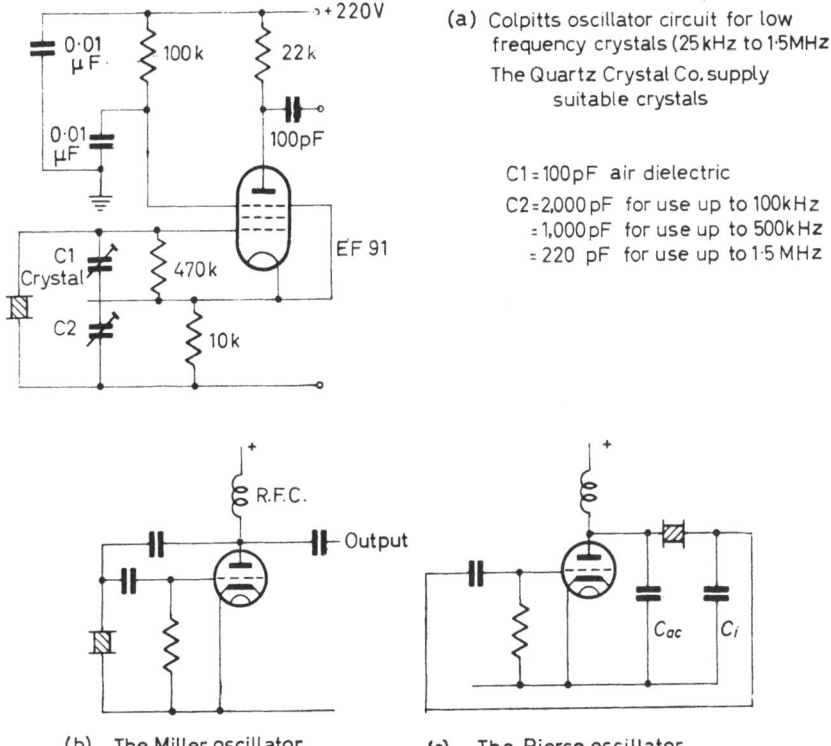

(a) Colpitts oscillator circuit for low frequency crystals (25 kHz to 1·5 MHz)

The Quartz Crystal Co. supply suitable crystals

$C1 = 100\,pF$ air dielectric

$C2 = 2{,}000\,pF$ for use up to 100 kHz
$= 1{,}000\,pF$ for use up to 500 kHz
$= 220\ pF$ for use up to 1·5 MHz

(b) The Miller oscillator

(c) The Pierce oscillator

Figure 10.8. Commonly used crystal oscillator circuits. In the Pierce oscillator C_{ac} represents the anode-to-cathode capacitance and C_i is the input capacitance to the valve

vibrations are established in the crystal plate. Provided the frequency of the oscillating voltage is close to a resonant frequency of the crystal plate, the crystal forces the oscillating voltage to assume a resonant frequency determined by the plate. By using the crystal in place of an LC resonant circuit in an oscillator the frequency is determined almost entirely by the crystal. Q values in excess of 20,000 are easily obtained whilst values up to $0·5 \times 10^6$ can be achieved with care. The frequency stability of a crystal oscillator is therefore very high. The stability depends upon the temperature, but by using crystal cuts that exhibit extremely small temperature coefficients, frequency variations of no more than one part in 10^4 can easily be achieved. Enclosing the crystal in a thermostatically controlled oven improves the frequency stability considerably so that with care the frequency variation can be reduced to one part in 10^8.

The actual oscillator circuits follow much the same lines as those for a conventional oscillator. Some examples are given in *Figure 10.8*.

The Pierce oscillator of *Figure 10.8c* is one of the more popular circuits on account of its simplicity. Apart from the self-biasing components and anode load, the practical arrangement requires no circuit components except the

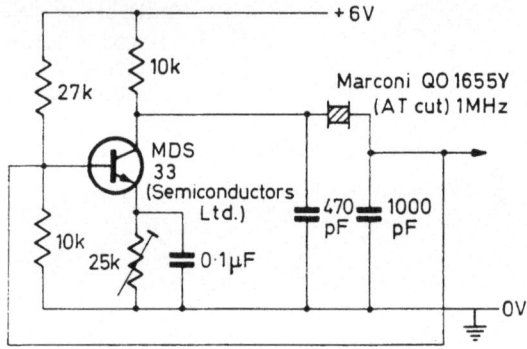

Figure 10.9. One practical version of a Pierce oscillator due to Baxandall[4]

crystal, which is connected between the anode and grid. Improved and predictable performance is obtained if the interelectrode capacitances are swamped by actual components. *Figure 10.9* shows a practical Pierce oscillator based on a micro alloy-diffused high frequency transistor. The crystal is a Marconi type QO1655Y (AT-cut) with an operating frequency of 1 MHz. The circuit was published by Baxandall[4] in an excellent paper describing transistor crystal oscillators. His series-resonant crystal oscillator described in the same paper is reproduced in *Figure 10.10*. The crystal is operated at its series resonant frequency, the amplitude of oscillation being controlled by a symmetrical diode limiter. The power dissipated in the crystal is only 0·4 μW which makes for excellent frequency stability. A 10 per cent change in the supply voltage changes the frequency by only 1·5 parts in 10^8. The ferrite ring on the emitter lead of the first transistor constitutes a one turn inductor, the purpose of which is to suppress parasitic oscillations at frequencies of several megacycles per second. The capacitor C is chosen to form a series resonant circuit with the inductor, the resonant frequency being 1 MHz.

306

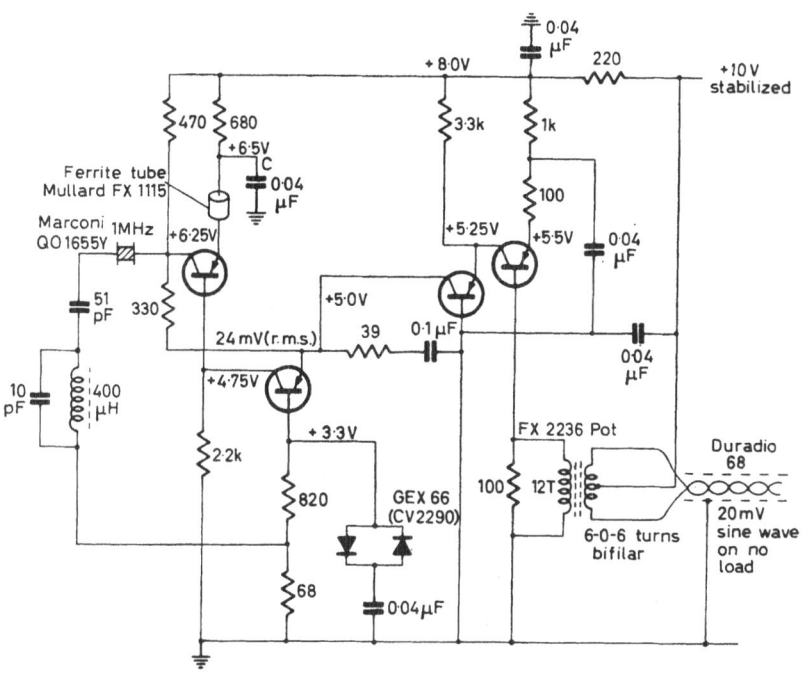

Figure 10.10. The Baxandall crystal oscillator. All the transistors are type MDS 33 (Semiconductors Ltd.)

Negative Resistance Oscillators

When a charged capacitor is discharged through an inductor the voltage across the capacitor does not fall exponentially unless the resistance of the coil is large. In practical inductors used for oscillators it is always ensured that the *Q* factor is large; the resistance of the coil is therefore quite small compared with the reactance of the coil. With such small resistances the current in a parallel LC circuit does not decay exponentially, but oscillates with a decreasing amplitude. The voltage across the capacitor, therefore, has a waveform that corresponds to damped simple harmonic motion i.e. it has the appearance of a sine wave whose amplitude decays exponentially. The cause of the decay is due to energy losses mainly in the resistance of the coil. At high frequencies part of the loss is due to electromagnetic radiation. We have, in previous sections, seen how these losses can be made good by supplying power in the correct phase from an amplifier. Another way of making good the losses is to place in parallel with the LC arrangement a negative resistance. The negative resistance cancels or neutralizes the positive equivalent resistance that represents the losses in the LC arrangement.

With a normal resistor (i.e. one having positive resistance) any rise in applied voltage is accompanied by a rise in current through the resistor.

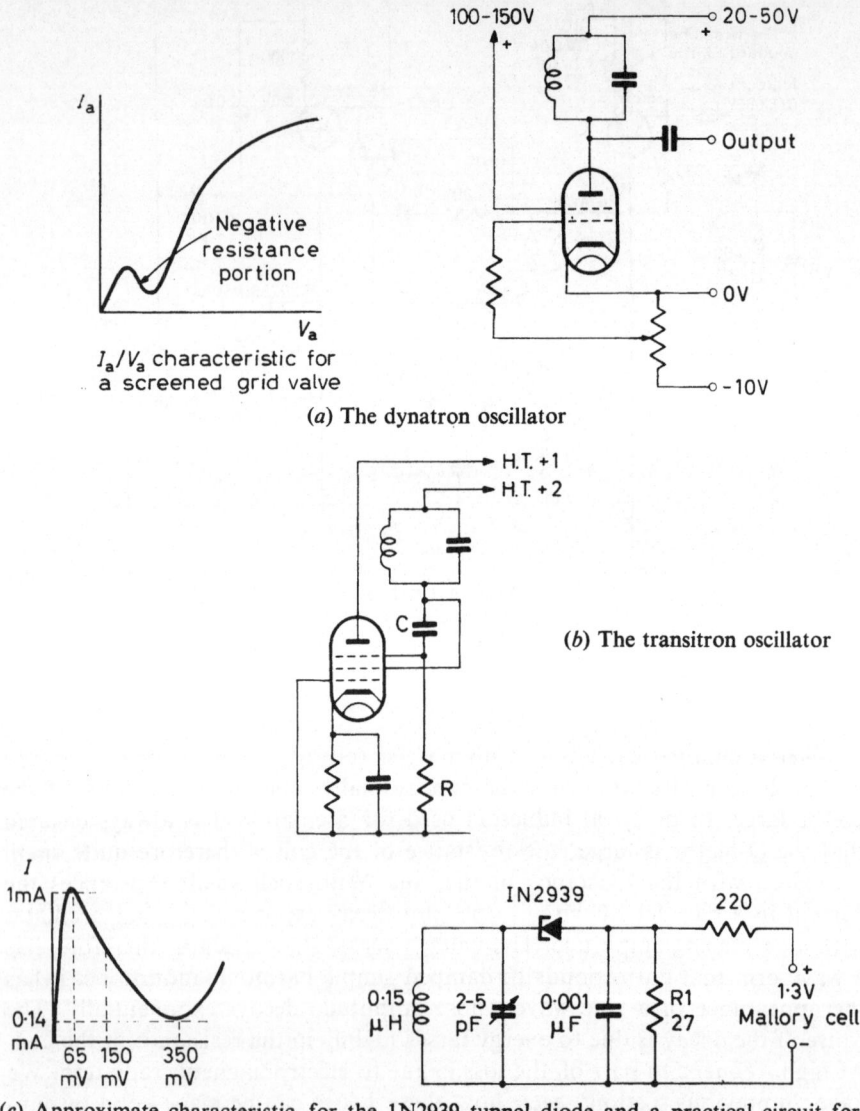

(a) The dynatron oscillator

(b) The transitron oscillator

(c) Approximate characteristic for the 1N2939 tunnel diode and a practical circuit for generating oscillations in the region of 100 MHz

Figure 10.11. Examples of negative resistance oscillators

308

Devices that exhibit negative resistance are characterized by a fall in current as the applied voltage is increased. Conversely falls in applied voltage are accompanied by rises in current. Three devices exhibit this effect over a limited range of applied voltage, namely a thermionic tetrode, a tunnel diode and a unijunction transistor.

The thermionic tetrode or screen-grid valve has already been discussed in Chapter 4. At low anode voltages and for a fixed higher screen voltage, secondary emission accounts for the kink in the characteristic. We may operate over the negative resistance region of the kink and obtain what is called a dynatron oscillator. *Figure 10.11a* shows a typical arrangement. The advantage of this oscillator is that it is essentially a two terminal device with the resonant circuit connected across the two terminals. Screen-grid valves of the type once produced for radio amplifiers were suitable, but are now no longer available. Certain pentodes can be made to act as screen-grid valves suitable for dynatron oscillations by connecting the suppressor grid to the screen grid. It is rarely worth while constructing a dynatron oscillator when so many other circuits are available.

The transitron oscillator is a negative resistance oscillator that makes better use of modern pentodes and is, therefore, superior to the dynatron oscillator. If the suppressor grid is connected for a.c. purposes to the screen grid, the graph of screen current against screen potential has a negative gradient over part of the curve. The valve then behaves as though it had negative resistance between the screen and the h.t. supply to the screen. The connection of an LC resonant circuit between the screen and the supply terminals gives rise to sustained oscillations. The circuit for the transitron oscillator is given in *Figure 10.11b*. The dynamic (or a.c.) connection between the suppressor and screen grids is effected by a capacitor. The capacitance must be such that any variation of screen-grid potential is transferred to the suppressor grid with no time lag and without any appreciable change in amplitude. To do this the reactance of C must be small at the frequency of operation. Any variation of screen-grid potential is then accompanied by the same variation of the suppressor grid potential. In operation the anode and control grid are maintained at constant potentials. We have seen in Chapter 4 that the suppressor grid effectively screens the anode from the cathode and control grid. The total space charge is, therefore, determined almost entirely by the potential on the screen grid when the control grid potential is kept constant. The proportions of the current that goes to the anode and screen grid are determined by the voltages on the suppressor grid and the anode. By keeping the anode voltage constant the suppressor grid voltage determines the relative proportions of the cathode current that pass to the screen grid and anode. When the suppressor grid becomes more negative the field between the suppressor and screen grids is such that the current to the screen grid is increased and the anode current is decreased. If the suppressor and screen grids are dynamically connected rapid changes in screen grid voltage are transferred to the suppressor grid; a condition arises whereby falls in screen-grid voltage are accompanied by falls in suppressor grid voltage which in turn give rise to increases in screen grid current. The I/V characteristic for the screen, therefore, exhibits a negative resistance characteristic. The suppressor conduction current develops a grid-leak bias voltage across R which stabilizes the amplitude of oscillation. This it does by automatically adjusting the steady suppressor bias voltage as the amplitude of the oscillation voltage varies.

The tunnel diode oscillator operates on essentially the same principle as the dynatron. The advantages over its thermionic counterpart are small size, low power consumption and a much higher operating frequency. Typical examples of tunnel diode oscillators operate in the v.h.f. region, i.e. hundreds of megacycles per second. *Figure 10.11c* shows a tunnel diode oscillator that operates at a frequency of about 100 MHz. For correct operation, the diode should be forward biased about 150 mV. The bias supply should be stable, a convenient source being a Mallory type cell. The resistor R1 must be adequately bypassed so that the tunnel diode is connected effectively across the resonant tank circuit. At 100 MHz or thereabouts a capacitance of 0·001 μF is adequate. Connected in this way the negative resistance of the tunnel diode neutralizes the losses associated with the tank circuit.

RC Oscillators

When oscillators are required to operate at frequencies below about 50 kHz, and, especially at audio frequencies, it is inconvenient to use LC circuits. The size of the coil to obtain the necessary inductance is inconveniently large with the result that it is difficult to construct coils with a sufficiently high Q value. Their bulk makes them unsuitable for use in transistorized equipment, and they are prone to pick up 50 Hz hum signals. At high frequencies a variable frequency output is easily obtained by the use of a variable capacitor in the resonant circuit. At low frequencies it is not easy to construct capacitors of sufficiently large value that are also variable. For these reasons oscillators for use at low frequencies are based on combinations of resistance and capacitance.

The principle of operation is the same as that used in LC oscillators; a feedback amplifier is used in which the feedback line consists of a suitable frequency selective network of resistors and capacitors. Two RC networks are in common use. They are the three- or four-section phase-shift network and the Wien bridge network.

Figure 10.12. shows a phase-shift oscillator using three RC sections in the feedback line. In order to have the collector voltage and base voltage in phase, it is necessary to have a 180° phase shift in the RC network. Since there is a 180° phase shift in the transistor, the collector voltage being 180° out of phase with the base voltage, a further shift of 180° in the feedback network brings the overall phase shift to 360°. It is shown in the appendix on matrix algebra that a three-section RC network has a 180° phase shift at only one frequency namely $1/\{2\pi(\sqrt{6})RC\}$ and this, therefore, is the frequency of operation of the oscillator. The attenuation of the network is 29 so the amplifier must have a gain of, at least, this figure. It is quite easy to obtain a gain of 29 in a single stage. The phase advance for a single isolated RC loop is given by $\phi = \tan^{-1}(1/\omega CR)$ and a consideration of the appropriate phasor diagram

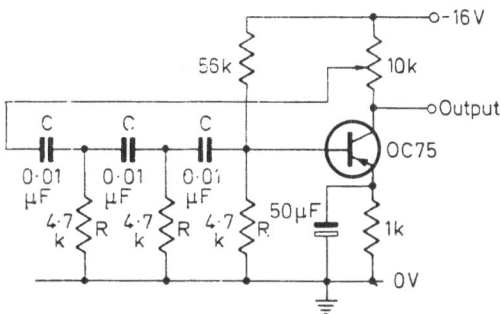

Figure 10.12. A three-section phase-shift oscillator. If the resistors in the phase-shifting network are altered care must be taken to preserve the correct biasing of the transistor by altering the value of the 56 k resistor. The transistor input and output impedances modify the frequency of operation which is not quite $1/\{2\pi(\sqrt{6})CR\}$. The potential divider in the collector circuit allows the voltage fed back to be adjusted for the best waveform

shows that $0° < \phi < 90°$. The output voltage across R would, however, be very small for values near to 90° and figures of approximately 60° are used. When two identical CR networks are cascaded the phase shift is not twice that of a single stage because the first section is loaded by the second. Since it is not possible to obtain a phase shift of 180° with only two sections the minimum, and therefore usual, number of sections to use is three. More sections can be used; for example, if four sections are used the attenuation at a frequency that gives a 180° phase shift is 18·39 so the gain of the amplifier need not be greater than this. The operating frequency becomes $\sqrt{0·7}/(2\pi CR)$. The marginal drop in the attenuation seldom justifies the use of the fourth section.

Where the frequency of the oscillator voltage is to be variable simultaneous adjustment of all the resistors or capacitors in a phase-shift network is not usually convenient. It is possible to use ganged capacitors consisting of three capacitors adjusted by the same spindle; such capacitors are constructed for radio purposes however, and the maximum capacitance of any one section is not usually greater than 500 pF. Where this value is too small it is more convenient to use a Wien bridge oscillator in which the capacitor values are selected by switches and fine frequency variations obtained by using ganged variable resistors. Since only two resistors are involved ganged components can be readily obtained.

The principles of a Wien bridge oscillator are shown in *Figure 10.13a*. The voltages *A* and *B* are in phase at only one frequency given by $f = 1/(2\pi CR)$. If now an amplifier with an even number of stages is used in connection with the bridge, a Wien bridge oscillator results. The gain of the amplifier must make up for the attenuation in the bridge network. (It is shown in the appendix on matrix algebra that the attenuation of this network is 3.)* By using an even number of stages the output voltage is in phase with the input voltage. Connecting the bridge to the input of the amplifier as shown produces positive feedback at one frequency and oscillations at that frequency are sustained when the gain exceeds three.

Figure 10.13b shows a practical Wien bridge oscillator with variable frequency output, designed by Ferranti Ltd.[5] The first stage consists of a Darlington pair. The current gain and input impedance are high which is desirable when operating this type of oscillator. The high input impedance ensures that the lower half of the Wien bridge is not upset by being loaded. To ensure that the voltage amplification is independent of frequency negative feedback is introduced into the amplifier. A high initial gain is needed before the application of the feedback to make up for the fall that occurs when negative feedback is applied. A further advantage of the use of negative feedback is the low output impedance obtained; the loading effect of the Wien bridge is then minimized. The feedback is provided in this case by R1 and R2. A fraction of the output voltage is fed back in series with the emitter of the first stage. By using a thermistor (S.T.C. type R53) for R1, amplitude control is achieved.

---

* Alternatively using the j notation

$$B = \frac{Z_2}{Z_1 + Z_2}\, A \qquad \therefore\ \frac{A}{B} = 1 + \frac{Z_1}{Z_2} \qquad \text{where} \qquad Z_1 = 1 + \frac{1}{j\omega C}$$

$$Z_2 = \frac{R}{1 + j\omega CR}$$

$$\therefore\ \frac{A}{B} = 1 + \frac{\left(R + \dfrac{1}{j\omega C}\right)(1 + j\omega CR)}{R} = 3 + j\left(\omega CR^2 - \frac{1}{\omega C}\right)$$

A and *B* are in phase when the j term is zero, i.e. when $\omega = 1/(CR)$. The frequency of oscillation is given by $f = 1/(2\pi CR)$. The attenuation is then 3. *A* and *B* are in phase because A/B is positive.

Any tendency for the output voltage to change is counteracted in the following way. A rise in output voltage causes an increase in the current through R1 and R2. The resistance of the thermistor, a small glass encapsulated bead type, decreases because of the inevitable rise in temperature due to the increased current. The feedback fraction thus increases which automatically reduces the gain; hence output of the amplifier is reduced to almost its former value.

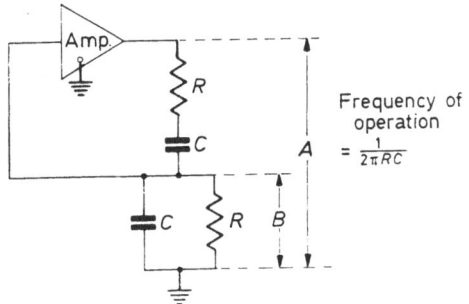

(a) Principles of a Wien bridge oscillator

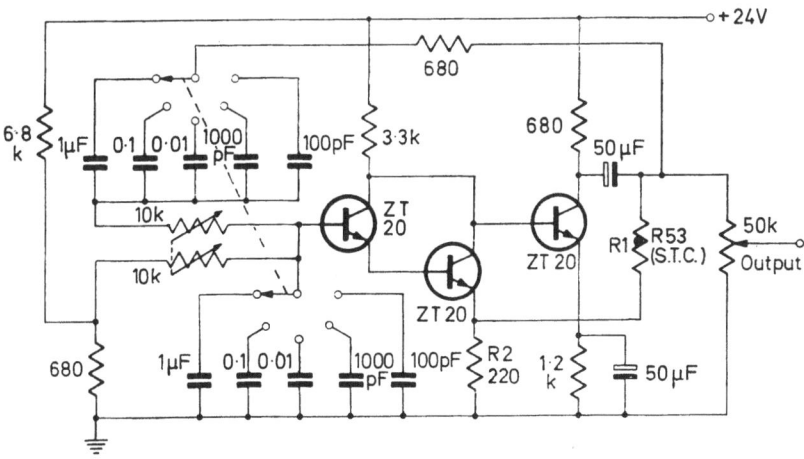

(b) A practical Wien bridge oscillator to operate over the range 15 Hz–2 MHz (Ferranti Ltd.)

Figure 10.13. The Wien bridge oscillator

Decreases in voltage output cause a rise in the resistance of the thermistor and a fall in the fraction of the voltage fed back to the emitter. A rise in gain results, which returns the amplitude of the output voltage almost to its former

313

value. With the capacitor values given the frequency ranges covered are 15–200 Hz, 150 Hz–2 kHz, 1·5 kHz–20 kHz, 15 kHz–200 kHz, 150 kHz–2 MHz. The output voltage is about 1 V (r.m.s.). For a change in supply voltage of 4 V the change in output voltage is less than 1 per cent and the change in frequency less than 2 per cent.

RELAXATION OSCILLATORS

The production of non-sinusoidal voltages is achieved by using relaxation oscillators. The simplest relaxation oscillator consists of a gas discharge tube, capacitor, and resistor connected as shown in *Figure 10.14a*. The capacitor charges via the resistor and when the potential across the plates reaches the striking voltage of the tube, the gas ionizes and a very large current flows, discharging the capacitor rapidly. As the voltage across the capacitor falls towards the extinction voltage the current through the tube falls. The maintaining voltage rises for small tube currents, and since the voltage is not available the tube extinguishes. The capacitor then charges via the resistor again, and the cycle is repeated. The frequency of operation and the amplitude of the waveform are determined by the values of C and R and the striking voltage of the tube. Since the striking voltage of the tube is fixed, it is necessary to change the tube in order to vary the amplitude of the signal. This is obviously inconvenient. A solution to this problem is to use a thyratron, and hold off the striking voltage in the way described in Chapter 8. *Figure 10.14b* shows a thyratron circuit useful for producing sawtooth waveforms. Such waveforms provide the time-base voltages in simple cathode-ray oscilloscopes in which the frequency of operation need not exceed about 120 kHz.

Relaxation oscillations having a sawtooth waveform can be produced by using a unijunction transistor in the circuit of *Figure 10.14c*. Pulses are simultaneously available from this circuit. The voltage across the capacitor rises exponentially until it equals the stand-off voltage. The emitter junction then becomes forward biased and the emitter-to-base 1 resistance falls suddenly. The capacitor must then discharge rapidly via this path. The reasons for the fall in resistance and the operation of a unijunction transistor have already been described in Chapter 5.

All three circuits of *Figure 10.14* suffer from the disadvantage that the rise in voltage across the capacitor is not linear with time. The voltage across a capacitor, v, is directly proportional to the charge, q, on the plates and $v = q/C$. The rate of change of voltage is given by $dv/dt = (dq/dt)/C = iC$. Evidently to obtain a linear rise of voltage (dq/dt = constant) the charging current must be constant. In the simple RC circuits considered so far constant-current charging conditions can be approached by charging via a very

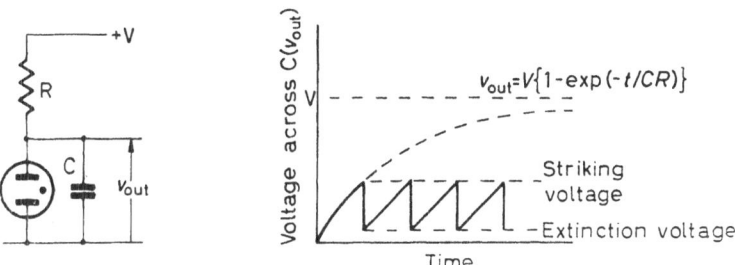

(*a*) Generation of relaxation oscillations using a gas discharge tube

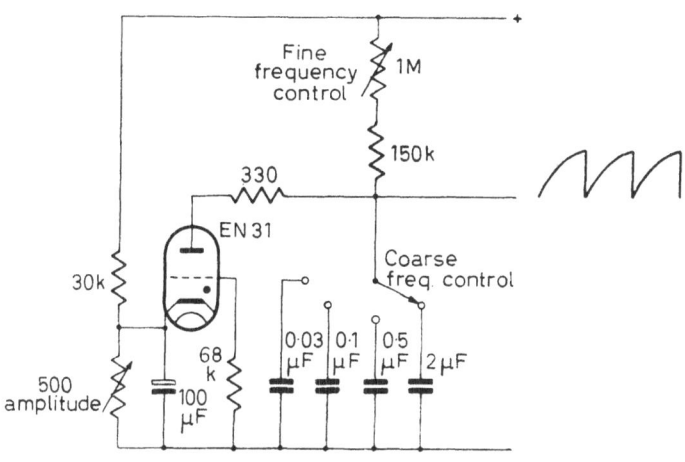

(*b*) Timebase generator using a thyratron

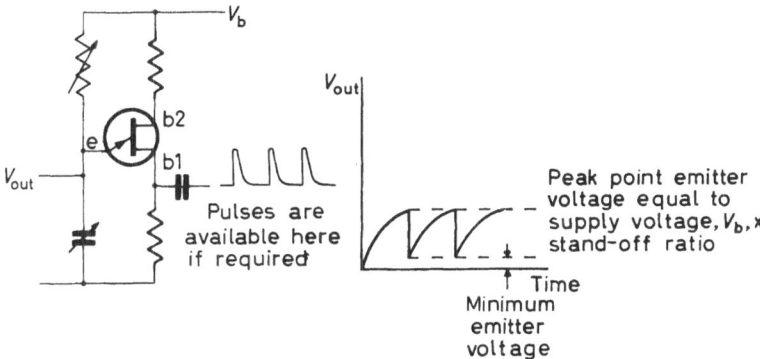

(*c*) Relaxation oscillator based on a unijunction transistor

Figure 10.14. Various forms of relaxation oscillators for producing sawtooth waveforms

315

high resistor (e.g. 1 M) from a voltage source as high as 1,000 V. The voltage across the capacitor is given by $v = V_{max}\{1 - \exp(-t/CR)\}$ which, on expansion to a power series, gives $v = V_{max}\left(1 - 1 + \dfrac{t}{CR} - \dfrac{t^2}{2C^2R^2} \cdots\right)$. For small values of t and large values of CR the second and higher order terms may be neglected giving $v = V_{max}\dfrac{t}{CR}$ where V_{max} is the supply voltage. Where the amplitude of the sawtooth waveform must be a few volts (say as high as 10 or 20 volts) then V_{max} should be at least 1,000 V if reasonable linearity is to be

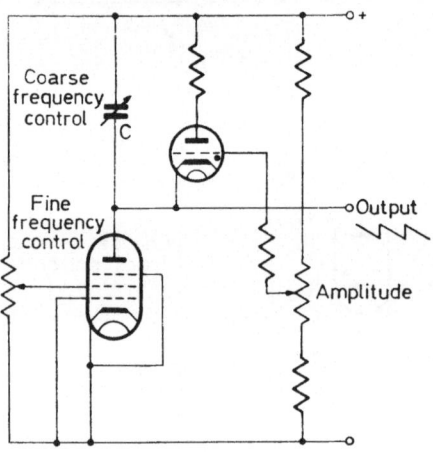

Figure 10.15. Basic circuit for producing sawtooth waveforms. The circuit can be used as a timebase in simple oscilloscopes

preserved. Taking typical figures as an example, let the amplitude of the sawtooth wave be 10 V. To obtain linearity t/CR must be small say 0·01, i.e. the period of the waveform, t, is only one hundredth of the time constant CR. Substituting into $v = V_{max}t/(CR)$ gives $V_{max} = 1,000$ V. It is usually inconvenient to make available a 1,000 V supply, so several circuits have been devised to overcome the difficulty. One of the simplest ways of obtaining a constant charging current with reasonable supply voltages it is to use a pentode in the circuit of *Figure 10.15*. When a pentode is operated above the knee of the characteristic, the anode current is not greatly influenced by changes in the anode voltage. This is shown by the anode characteristics which are almost flat horizontal lines. The pentode is, therefore, a constant current charging device. Since the rate of change of voltage across the capacitor

316

is given by $dv/dt = i/C$ adjustment of both i and C alters the rate of change of voltage. When combined with a suitable electronic switch (e.g. a thyratron or hard valve switch) alterations of i and C alter the frequency of operation. C is altered by selecting various values with a switch selector. The charging current i is varied by altering the screen voltage, and hence current. This provides a fine adjustment of the frequency.

The Miller Transitron

A more sophisticated way of producing linear sawtooth waveforms is to apply a step voltage to a simple single valve amplifier connected as an integrator (see Chapter 9, p. 290). *Figure 10.16a* shows the basis of such a circuit in which the output voltage at the anode is the integral of the input voltage. If the latter voltage is constant the anode voltage falls linearly with time. We may obtain a constant input voltage to R by closing the switch, S. The voltage on the grid rises which causes the anode current to increase. The increase is not sudden, however, because the fall in anode voltage produced by the rise in anode current through R_L is communicated to the grid via C. Provided the gain of the stage is large the anode voltage falls in a linear fashion. Eventually the valve bottoms and no further change occurs in the anode voltage. If repetitive action is required, C must be rapidly discharged and the cycle repeated. An automatic way of doing this is to use the circuit of *Figure 10.16b* in which a single valve is used both as a transitron relaxation oscillator and a Miller integrator. The appropriate waveforms are shown alongside. Starting at the point in the cycle where g_3 has been driven negative the voltage on the suppressor grid is rising exponentially since C1 is charging via R and R_S. Eventually the suppressor grid voltage rises to a point where it can control the division of cathode current between the anode and the screen. The screen current falls causing a rise in screen voltage which is communicated to the suppressor via C1. Very rapidly the screen current is cut off and the screen voltage rises to the supply voltage. There is no sudden rise in anode current, however, because of the voltage drop across R_L, which is communicated to the grid via C2. There then occurs the linear run down in anode voltage as described in the previous paragraph. Meanwhile the suppressor voltage decays exponentially to zero. The increase in cathode current gives rise to a slight increase in screen current and a resultant fall in screen voltage. Although this must be communicated to the suppressor, no rapid change in voltage takes place, since this would result in a rapid change of anode current. The mutual conductance from the control grid to the anode is much greater than that between the screen and suppressor grids, with the result that any rapid cumulative action is suppressed. The linear run down

317

therefore proceeds until bottoming occurs. Once this happens the anode voltage is temporarily constant and the voltage on the control grid is free to rise. The consequent increase in screen current results in a drop of both the screen and suppressor grid voltages. A rapid cumulative action occurs in which the screen current shoots to a high value and the anode current is sharply cut off. The suppressor grid is driven to a negative value. The suppressor grid voltage rises exponentially due to the charging of C1 via R and R_S whereupon the whole cycle is repeated.

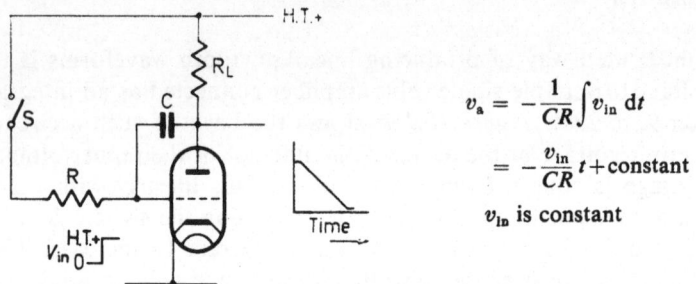

$$v_a = -\frac{1}{CR}\int v_{in}\, dt$$

$$= -\frac{v_{in}}{CR}\, t + \text{constant}$$

v_{in} is constant

(a) The Miller integrator

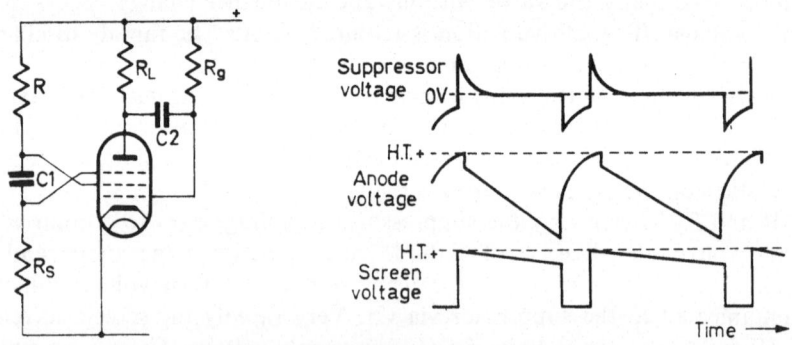

(b) The Miller transitron oscillator (the Phantastron)

Figure 10.16

The Blocking Oscillator

A relaxation oscillator can be made by using the basic circuit of the Armstrong oscillator *(Figure 10.2)*. The inductive coupling between the anode and grid coils is made so tight that M exceeds considerably the value required for oscillation; in consequence an appreciable grid current is estab-

318

lished. *Figure 10.17* shows the circuit of the blocking oscillator. The most useful application of this type of oscillator is in the production of high-current pulses of short duration. The valve is normally cut-off owing to the bias voltage applied via R. On the arrival of a trigger pulse the grid voltage is driven above the cut-off point and anode current starts to flow. The transformer then produces a secondary voltage that drives the grid voltage more

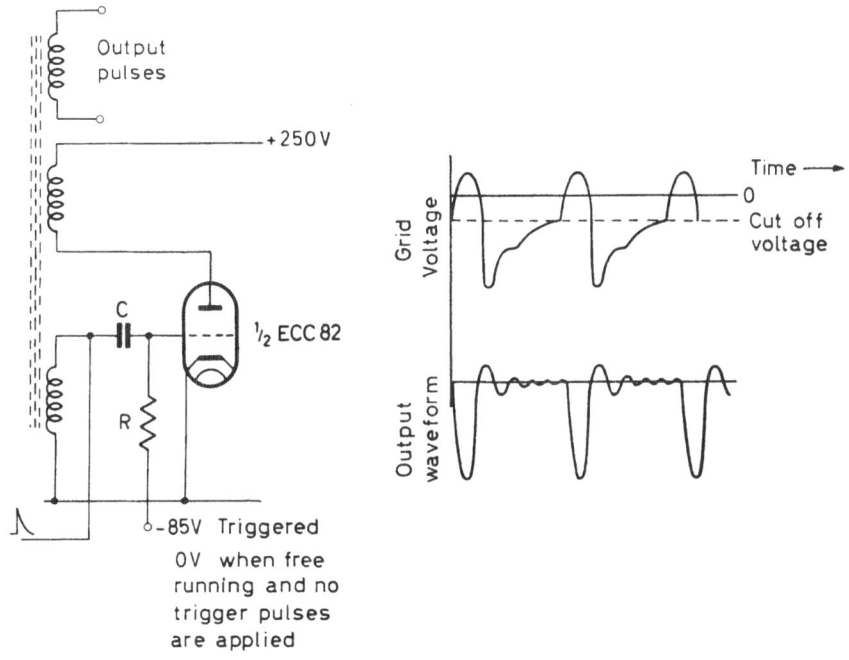

Figure 10.17. The blocking oscillator

positive resulting in a further rise of anode current. A regenerative action takes place until the anode current reaches its maximum value. Thereafter with no further change in anode current the rate of change of flux in the transformer becomes zero and the secondary voltage falls. The grid voltage also starts to fall as the voltage on C falls towards − 85 V. The resulting drop in grid voltage causes the anode current to fall and a further regenerative action takes place in which the anode current is sharply cut-off. The valve is now ready to receive the next trigger pulse and the action is repeated.

A free-running version is produced when the grid resistor R is connected to the cathode instead of − 85 V. Here the initiation of the rise in anode current occurs when the grid voltage is higher than the cut-off value. Once the anode current reaches a maximum and becomes temporarily steady, no secondary voltage exists at the secondary of the transformer. During the swift rise of anode current the grid is driven positive and the capacitor C charges owing to the grid current. Once the secondary voltage of the transformer falls to zero the grid becomes negative and the anode current falls.

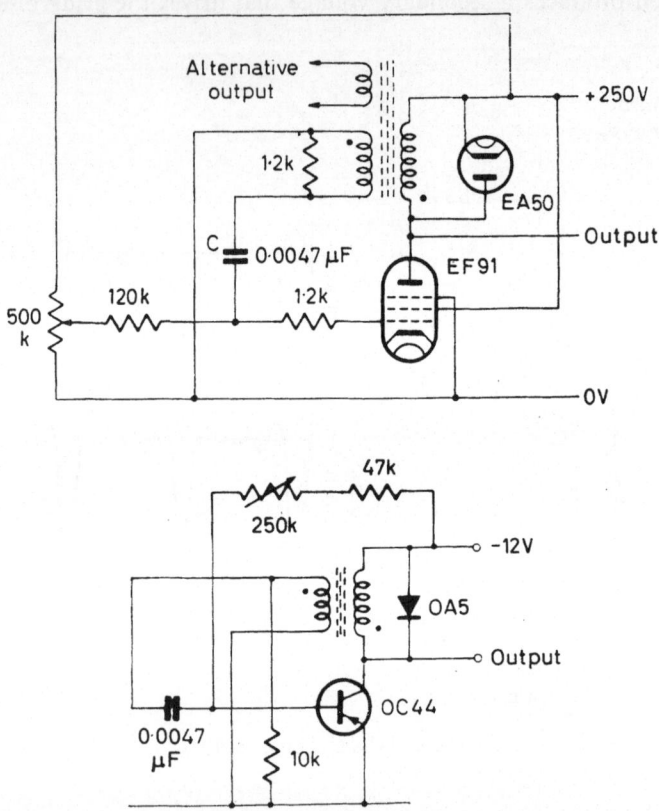

Figure 10.18. Valve and transistor versions of the blocking oscillator

The voltage produced by the transformer secondary together with the charge on C drives the grid voltage to well beyond the cut-off value. The charge on C must leak away via R before the next pulse can be produced. Some control of the pulse repetition frequency can be effected by using the circuit of *Figure 10.18a.* The capacitor C now discharges to a voltage set by the potential divider. In the practical circuit given, a diode across the primary coil and a resistor across the secondary coil effectively damp out the 'ringing' oscillations associated with the transformer. A 1·2 k resistor prevents parasitic

320

oscillations associated with the grid circuit. Used in this way, the 1·2 k resistor is often called a grid stopper. The transformer may be made on a ferrite pot core (e.g. Mullard L23 or similar) having 150 primary turns and 250 secondary turns. If an output voltage is all that is required the output can be taken from the anode. For a current pulse a third transformer winding will be necessary. The number of turns depends upon the current required. *Figure 10.18b* shows a transistor version.

The Generation of Rectangular Waveforms

When we wish to produce oscillating voltages that have a rectangular waveform it is usual to resort to two-state circuits in which there is an abrupt transition from one state to the other. The multivibrator is the most commonly used circuit, which gets its name from the fact that square or rectangular waves are rich in harmonics. The basic circuit is given in *Figure 10.19* in which two RC coupled amplifiers are used, the output of one being connected to the input of the other and vice versa. The valve (or transistor in a corresponding circuit) is alternately switched from the 'on' to the 'off' state.

To assist in understanding the mode of operation of this circuit, the waveforms at different points in the circuit are shows in *Figure 10.19* along with the circuit diagram. Before considering the action in detail it will be helpful to recall the way in which a capacitor transfers a signal from one part of the circuit to another. Provided the charge held by a capacitor remains constant the potential difference across the plates must also remain constant. If, therefore, one plate, A, of a capacitor is connected to a line having a voltage of +200 V and the other plate, B, is connected to a line having a voltage of 0 V then, after charging, the voltage across the capacitor is 200 V. If now the potential of plate A is suddenly dropped to +50 V the voltage on plate B must drop to −150 V provided there is no change of charge in the capacitor. If charging or discharging of the capacitor can take place via resistors then the voltage changes from −150 V to some new value; the time it takes to change to the new value depends upon the values of resistance, capacitance and supply voltage.

After energizing the circuit, oscillations are initiated by a switching transient or some other circuit disturbance, and very quickly steady-state oscillations are established. At one point in the cycle the grid of V1 is driven well beyond cut-off for reasons that will shortly be made clear. V1 is, therefore, cut-off. At this time V2 is heavily conducting, the anode current is high and the corresponding anode voltage is only a few tens of volts above zero. Although V1 is sharply cut-off, the anode voltage does not rise immediately to zero because C1 is charging via R1 and the low resistance grid-to-cathode path in the heavily conducting valve V2. In a short time, however, determined mainly by the time constant C1.R1, the charging is virtually complete and the anode voltage of V1 then assumes the h.t. voltage, V_b. Meanwhile C2,

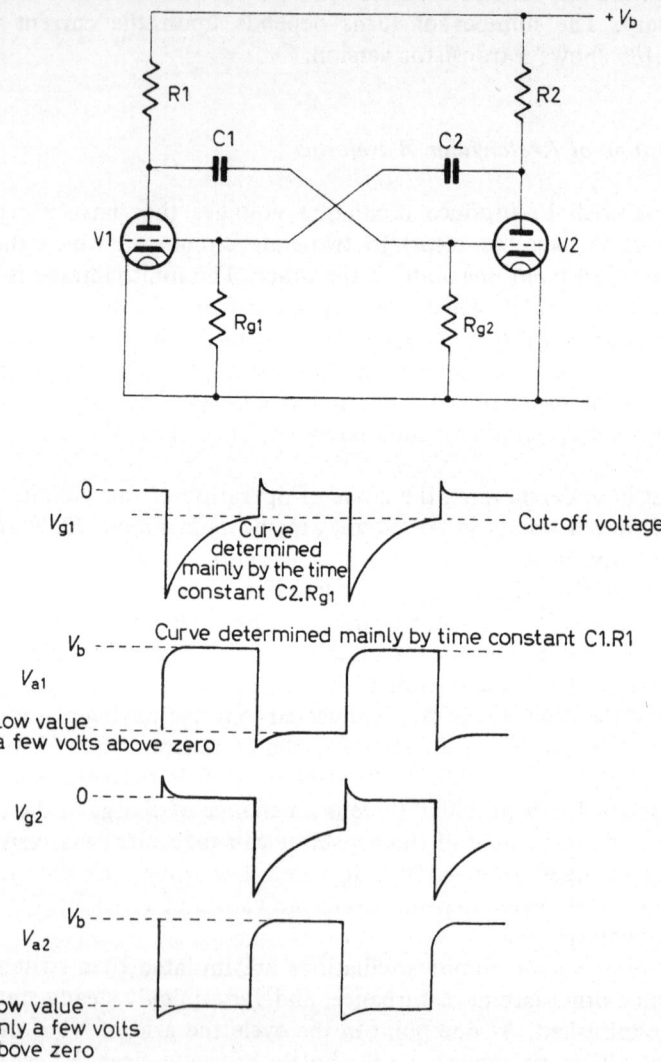

Figure 10.19. Basic multivibrator circuit and waveforms. When R_{g1} and R_{g2} are returned to the h.t. + line the frequency stability is improved (see text). Suggested values for the components are given in the text

322

which had previously been charged by a similar mechanism to that which charged C1, is discharging via R_{g1} and the low resistance path offered by V2. Since R_{g1} is large the time constant $C2.R_{g1}$ is long compared with C1.R1. The grid voltage of V1 therefore rises rather slowly. The time taken to reach the cut-off point can be calculated from a knowledge of $C2.R_{g1}$, the cut-off voltage of V1 and the negative voltage to which the grid of V1 was driven. This latter voltage is almost the supply voltage V_b. Eventually the voltage on the grid of V1 reaches the cut-off voltage. Thereafter further increases of the voltage on the grid of V1 cause this valve to conduct. There is a fall of anode voltage as a consequence of the rise in anode current. This fall of voltage is communicated to the grid of V2 via C1 whereupon the anode voltage of V2 rises because of the fall in anode current in V2. This rise in voltage is communicated to the grid of V1 and the rise in the anode current of V1 is greatly accelerated. A cumulative action develops, and in a very short space of time V1 is driven hard into the conduction condition. The large fall in anode voltage is communicated via C1 to the grid of V2, which is driven to well below the cut-off point. There has thus been a rapid 'flip' in V2 from the non-conducting to the conducting state. A similar action now takes place in connection with V2 whereby C2 charges via R2 and the low resistance path provided by V1, which is now conducting, and C1 discharges via R_{g1} and V1. Once the grid of V2 reaches the cut-off point, V2 starts to conduct again, and a further cumulative action develops. The state of the circuit then rapidly switches to its original condition.

The output from this oscillator may be taken from either anode. The mark-space ratio and repetition frequency are determined by the values of C1, R1, C2 and R2 together with R_{g1}, R_{g2} and the supply voltage. For a mark-space ratio of 1 and a repetition rate of 2 kHz the anode resistors may be 10 k, the coupling capacitors 0·001 μF each and the grid resistors 220 k. A rather low value of anode load is chosen so that the time constant C1.R1 is short. This has the effect of minimizing the rounding of the corners of the waveform. The supply voltage should be 250 V for an ECC82 double triode.

Improved frequency stability can be obtained if the grid resistors R_{g1} and R_{g2} are connected to the h.t. positive line instead of to the earth line. When connected in this way the charging is towards the h.t. positive voltage instead of towards zero volts. To maintain about the same frequency of operation the grid resistors need to be increased to 820 k. Charging towards the h.t. positive voltage causes the slope of the grid voltage waveforms to be steeper as the voltage crosses the cut-off value. Random fluctuations of grid voltage near the cut-off value are then not so important as they would be if the waveform crossed the cut-off voltage line at a shallower angle. The onset of the cumulative action is then more definite. The random grid voltage fluctuations are due to noise, hum and supply voltage variations, and cause the grid to reach the cut-off voltage a little earlier or later than the time determined by the charging rate of the capacitors alone.

Figure 10.20 shows a transistor version of a multivibrator, together with the associated waveforms. The pronounced curvature of the waveform when C1 is charging is due mainly to the lower supply voltages involved. In valve multivibrators the supply voltage is of the order of hundreds of volts, whereas

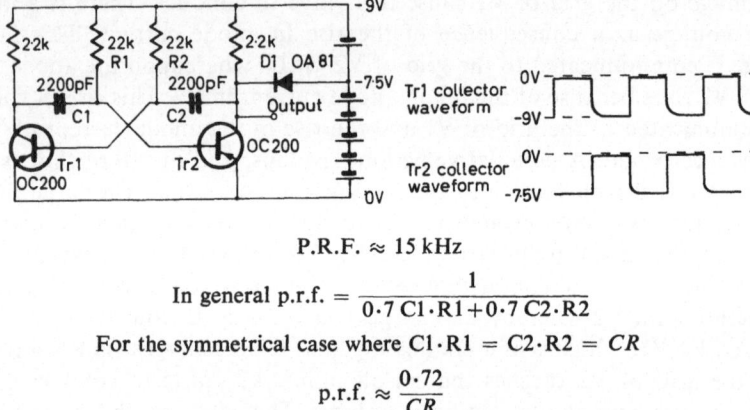

$$\text{P.R.F.} \approx 15\,\text{kHz}$$

$$\text{In general p.r.f.} = \frac{1}{0\cdot7\,C1\cdot R1 + 0\cdot7\,C2\cdot R2}$$

For the symmetrical case where $C1\cdot R1 = C2\cdot R2 = CR$

$$\text{p.r.f.} \approx \frac{0\cdot72}{CR}$$

Figure 10.20. Transistor version of the astable or free-running multivibrator. The clamping diode improves the output waveform

in transistor versions only a few volts are available. The low charging rate difficulty can be overcome by using a clamping diode as in *Figure 10.20*. Diode D1 does not allow the voltage on the collector of Tr2 to go more negative than $-7\cdot5$ V. This results in a chopping off of the more curved portion of the waveform.

The multivibrators described above are astable or free-running. If the frequency of operation of such an oscillator is not stable enough for a given application, improved constancy of frequency can be effected by injecting suitable pulses into the appropriate part of the circuit. The pulses may be derived from an oscillator that operates with a known and highly stable frequency. Such a secondary oscillator may be controlled by a quartz crystal.

An important role for free-running multivibrators that are synchronized in this way is in frequency division. Although an oscillator based on a quartz crystal gives us an excellent frequency standard, we are confined to a single frequency. Several standard frequencies can be obtained if the quartz crystal oscillator is used to synchronize a chain of free-running multivibrators. Very often it is arranged that successive multivibrators in the chain operate at frequencies of one tenth, one hundredth, one thousandth, etc., respectively, of the frequency of the crystal standard. An example is given in *Figure 11.7* of Chapter 11.

Where frequency division is not involved a controlled multivibrator may be constructed so as not to be free-running. Instead of having two unstable states, we arrange to have only one. The circuit is then known as a monostable multivibrator, a univibrator or a 'flip-flop'. *Figure 10.21* shows one

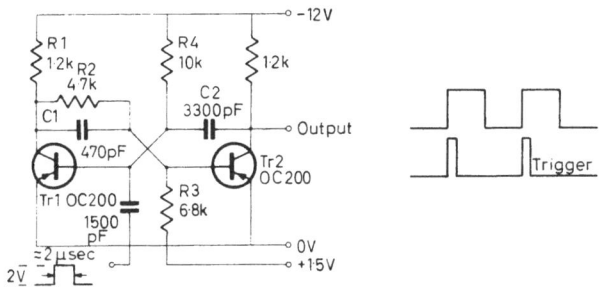

Figure 10.21. Monostable triggered multivibrator (flip-flop)

possible arrangement. In the absence of any triggering pulses the circuit remains in the stable state. This is arranged in the given circuit by returning the base resistor, R3, to a source of positive potential instead of to the earth line. A resistor R2 is placed in parallel across the coupling capacitor from Tr1. The values for R3 and R2 are chosen so that together with the load resistor of Tr1, a potential divider is formed that ensures that Tr2 remains cut-off during quiescent periods. On the arrival of a positive pulse at the base of Tr1, that transistor (previously bottomed) is cut-off whereupon the collector voltage of Tr1 falls. The purpose of C1 is to transfer such a fall quickly to the base of Tr2. The consequence is that Tr2 is swiftly caused to conduct, the usual multivibrator cumulative action occurs, and Tr1 is held in the non-conducting state whilst Tr2 is driven hard into conduction. As now the only current through R1 is that in the chain R1, R2 and R3, the potential at the base of Tr2 is sufficiently negative to keep that transistor conducting. Eventually C2 discharges via R4, and the potential of the base of Tr1 becomes negative. Tr1 now conducts and Tr2 is cut-off, and thus the original position is regained. The effect of the pulse has been to 'flip' the circuit into the unstable state whereupon normal multivibrator action allows the circuit to 'flop' back into the original stable state. The output may be taken from either collector. The output pulse has a duration that is determined mainly by the time constant C2.R4. The trigger pulse must be of sufficient magnitude and duration to allow switching to occur.

325

An alternative way of producing square waves of known repetition frequency is to clip and amplify a sine wave of known frequency. *Figure 10.22* shows the principle. The sine wave has its top and bottom clipped by a pair of biased diodes or more conveniently a pair of zener diodes connected back-to-back as shown. The output from the clipper is then considerably amplified and the output subject to a further clipping stage. The sides of the 'square' wave are then almost vertical. If they are not sufficiently vertical for a given application, further amplification and subsequent clipping are used. One of

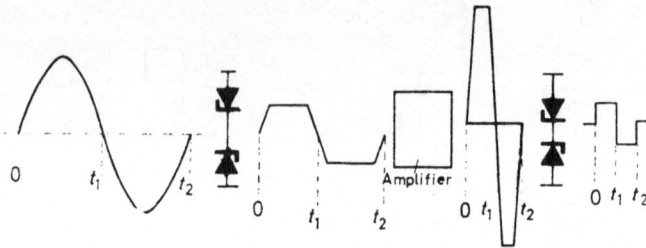

Figure 10.22. Use of zener diodes as clippers to produce square waves from a sinusoidal input

the main advantages of this method is that the repetition frequency of the square wave is easily varied by varying the frequency of the sine wave. The addition of squaring and clipping stages converts any sine wave oscillator into one which produces both sine waves and square waves at will. The zener voltage sets the levels of the signal amplitude throughout the squaring unit. If the output is taken from the zener diode an accurate calibration signal is available of known and almost constant amplitude. This can be useful, e.g. in providing calibration voltages for cathode-ray oscilloscopes. If the square wave amplitude is to be varied then the last pair of zener diodes may be followed by a cathode follower stage with variable output facilities.

REFERENCES

1. Nyquist, H. 'Regeneration theory'. *Bell Syst. Tech. J.* 1932, **11**, 126.
2. Jackson, A. 'A simple high frequency transistor recorder for chromatography'. *J. Chem. Education*, 1965, **42**, p. 447 (Aug.).
3. Heising, R. A. *Quartz Crystals for Electrical Circuits—Their Design and Manufacture.* D. Van Nostrand, 1946.
   See also
   Terman, F. E. *Electronic and Radio Engineering*, p. 506 et seq., McGraw-Hill, 1955.

4. Baxandall, P. J. 'Transistor crystal oscillators and the design of a 1Mc/s oscillator circuit capable of good frequency stability'. *J.I.E.R.E.*, 1965, **29**, No. 4 (April).
5. 'Wide Range Transistor Phase-Shift Oscillator'. *Application Note No. 1.* Ferranti Ltd.

Suggestions for Further Reading

Pulse and Digital Circuits by J. Millman and H. Taub, McGraw-Hill, 1956.
Waveforms by B. Chance, V. Hughes, E. F. MacNichol, D. Sayne, and F. C. Williams, McGraw-Hill, 1956.
'Multivibrator Design' by Foss, R. C. and Sizmur, M. F., *Wireless World*, 1961, **67**, No. 4 (April), p. 221 and No. 5, May, p. 257.

11

THE CATHODE-RAY OSCILLOSCOPE

The modern cathode-ray oscilloscope is designed as a measuring instrument, and is the most useful of all electronic test devices. So great is its versatility that workers in every branch of scientific activity now find the instrument to be almost indispensable. If limited to the purchase of a single item of electronic measuring equipment, the majority of experienced workers would select an oscilloscope.

The instrument delivers its information in the form of a graph or trace on the screen of a cathode-ray tube. By giving an immediate visual display of the amplitude and waveform of the quantity under consideration, a rapid insight is gained into the functioning of electronic and electrical circuits. Where the phenomena to be studied are not electrical, e.g. the mechanical vibrations in beams and machinery, the variations of pH in a solution, temperature fluctuations, patients' heart-beats, etc., then all that is necessary is to find suitable transducers that give output voltages that correspond to the variations of the quantities involved. Strain gauges, piezoelectric crystals and moving-coil microphones, photoelectric cells and thermocouples are examples of commonly used transducers. The electrical output of a transducer is applied to the input terminals of the oscilloscope, and the latter automatically draws out the waveform on the screen. The displaying of waveforms constitutes an important advantage over the usual moving-pointer instruments, since the latter can yield only amplitude information.

The inertia of the moving parts of such electromechanical systems as moving-coil meters, potentiometric and other pen recorders, etc., imposes a severe limitation on the maximum signal frequency to which such instruments can respond. The oscilloscope, on the other hand, is a truly electronic instrument having no moving parts except the beam of electrons, which is, for all practical purposes, inertialess. The cathode-ray tube can thus respond to very rapid alternations of voltage. General-purpose laboratory oscilloscopes having adequate responses up to 30 MHz are now common, and special purpose instruments are available that can respond to signal frequencies as high as 1,000 MHz.

The oscilloscope may be considered as a combination of a cathode-ray tube together with appropriate amplifiers and power supplies. In addition a special relaxation oscillator, known as a timebase, is incorporated.

THE CATHODE-RAY TUBE

This tube is the major device in an oscilloscope. It consists of a source of electrons, focussing electrodes, a deflection system and a fluorescent screen all contained in a suitably shaped glass tube. It is illustrated diagrammatically in *Figure 11.1*. The tube is highly evacuated to pressures of the order of 10^{-6} torr (mm Hg). The aim within the tube is to produce a narrow beam of electrons that comes to a sharp focus at the screen.

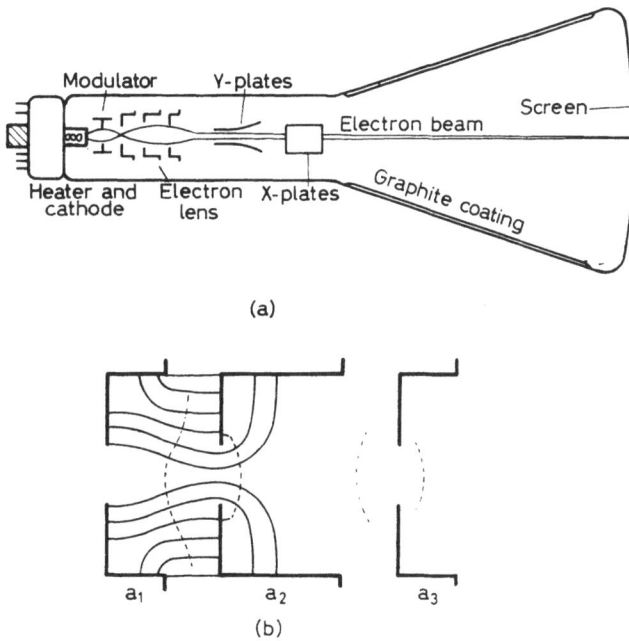

Figure 11.1. (*a*) Diagrammatic representation of a cathode-ray tube showing the main electrode features. (*b*) The electric field within the three-anode electron lens system. The solid lines show the shape of the electric field, and the broken lines show the position of the equipotential surfaces. The similarity between the shape of the equipotential lines and that of converging glass lenses should be noted

The screen is a thin layer of phosphor that glows at the point of bombardment of the electrons. The phosphor, binding chemicals, and activators to increase the luminous efficiency, are deposited on the inside of the face of the tube. The type of phosphor used determines the colour and persistence of the trace. The maximum sensitivity of the light-adapted human eye is at about 5,550 Å and thus for viewing work yellowish-green traces are used. This colour is obtained by using zinc orthosilicate with small amounts of manganese as an activator. Persistence is expressed as short, medium or long,

329

and refers to the length of time taken for the trace to become invisible after the electron beam has been removed. Medium persistence (i.e. where the brightness of the trace is reduced to about 1/1,000th of its original value in approximately 50 milliseconds) is adopted for viewing tubes. Blue, short persistence traces are used for photographic work since the sensitivity of film emulsions used in oscillograph recording is usually highest in the blue region of the spectrum. Orange and yellow long-persistence traces are used in radar equipment and in some large demonstration oscilloscopes.

The electrons are produced by heating a small area of barium and strontium oxides deposited on a nickel surface. The beam current is controlled by the modulator, i.e. a nickel tube that is coaxial with the cathode. The modulator is maintained at some potential which is negative (about -15 V to -100 V) with respect to the cathode. As the potential is made increasingly negative, fewer electrons reach the screen and thus the trace becomes fainter. Eventually, by applying a sufficiently high negative potential, the beam current is reduced to zero and the tube is then said to be 'cut-off'. The knob on the front of the instrument that is used to control the modulator voltage is marked 'brightness' or 'brilliance', since by operating this control the brightness of the trace on the screen may be varied. Because of the similarity between the action of the modulator in a cathode-ray tube and the grid in a thermionic valve, the modulator is frequently called the grid, in spite of its physical shape.

The electrons are accelerated towards the screen and simultaneously brought to a sharp focus by a system of three anodes, known as the electron lens system. These anodes, a_1, a_2 and a_3, consist of a system of cylindrical electrodes, arranged in line coaxially with the modulator and neck of the tube. The anodes a_1 and a_3 are connected together electrically within the tube and are maintained at a high positive potential relative to the cathode. Different tubes require different anode potentials, but values commonly used lie between 1 kV and 5 kV. Between a_1 and a_3 there is a focussing electrode a_2. The voltage on this electrode may be varied by operating the focus control on the front panel of the instrument. This voltage is commonly one sixth to one third of the final anode voltage. The difference in potential between a_2 and the other anodes, a_1 and a_3, creates an electric field through which the electrons must travel. The electric field configuration is such that the electrons are made to converge to a point. The action of the electric fields on the electrons is analogous to the action of glass converging lenses on beams of light, which is why the anode system is referred to as an electron lens. The adjustment of the potential on a_2 by the focus control alters the focal length of the lens system. With the correct adjustment the electron beam is brought to a sharp focus at the screen. The heater, cathode, modulator and anode system are collectively referred to as the 'electron gun'.

330

The Deflection System

In order to make the electron 'pencil' draw on the screen some suitable method of deflecting the electron beam must be employed. In nearly all cases of oscillographic work the deflections are produced by electric fields. The electric fields are created by applying voltages, proportional to the quantities to be represented, to two pairs of deflector plates, arranged mutually at right-angles and between which the beam passes. When one plate of a given pair is made positive with respect to the other, the negative electrons are accelerated towards it. The beam may therefore be deflected in a direction at right-angles to the given pair of plates. Since two pairs of plates are provided, the beam may be made to scan the whole of the useful screen area. The pair nearer the anodes has a greater effect on the beam therefore this pair is reserved for the signal; following the usual convention in graphical work, the deflections produced are in the Y-direction. These deflectors are therefore called Y-plates. The other pair are called X-plates because they produce deflections in the X-direction. The plates diverge at the ends nearest the screen to prevent the beam from being interrupted when large deflections are required.

The deflection produced on the screen is directly proportional to the deflecting voltage, a fortuitous arrangement that yields a linear scale for the axes. In practice, the spot diameter will limit the smallest deflection that can be detected. The resolution of the cathode-ray tube may be defined as the voltage applied to the deflector plates that produces a deflection equal to one spot diameter.

Magnetic Screening of the Tube

Because the beam may be deflected by magnetic fields, precautions must be taken to avoid the influence of stray fields. The mains transformer is the major source of unwanted magnetic fields. Apart from the correct positioning of the transformer (preferably behind the tube) the cathode-ray tube must be magnetically shielded. This is best done by enclosing the tube in a cylinder or sheath of high-permeability metal. Mumetal screens are most often employed for this purpose. Inadequate screening from alternating magnetic fields causes an inclined stroke or an ellipse to appear on the screen instead of a spot. On the application of beam deflection a wave or ribbon appears instead of a thin line trace. Since interference from stray magnetic fields is one of the commonest causes of indifferent focus performance, it is advisable to test for the presence of such fields. For the test the deflector plates must be connected to the final anode. To avoid burning the fluorescent screen the brightness must be considerably reduced below its normal operating level. If now a magnet is rapidly moved near to the neck of the tube the trace will be similar to *Figure 11.2a* provided stray alternating magnetic fields are absent. A trace such as that of *Figure 11.2b* indicates the presence of such unwanted fields; a trace similar to *Figure 11.2c* reveals modulations of the beam current by the presence of spurious voltages on the grid of the tube.

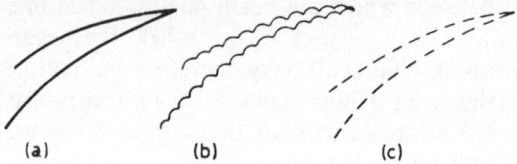

(a)   (b)   (c)

Figure 11.2. Traces obtained when testing for the presence of unwanted stray magnetic fields within the cathode-ray tube. Trace (*a*) is obtained in the absence of any field, whilst trace (*b*) shows the presence of an unwanted magnetic field. The method of obtaining the trace is described in the text. Trace (*c*) reveals the presence of unwanted modulation voltages on the modulator

Double-beam Tubes

It is frequently an advantage to have two quantities displayed simultaneously on the screen as separate graphs. This may be achieved by creating two beams of electrons within the tube. The single beam may be split by a beam-splitting plate or alternatively two electron guns can be used. Each beam passes through its own Y-plates, but both beams pass through a single pair of X-plates. The deflections in the X-direction are therefore common to both traces.

Electronic switching can be employed to give two traces with a single-beam tube, this method being favoured by some manufacturers. An electronic beam-switching circuit consists of a pair of identical amplifiers that are controlled by a square-wave generator. The outputs of the amplifiers are connected together and fed to the input of the single-beam oscilloscope. Each amplifier input receives one of the signals to be examined. The square-wave generator alternately biases the amplifiers so that whilst one amplifier is operating the other is cut-off. When the square-wave switching frequency is much greater than either of the signal frequencies, portions of each signal are alternately presented to the input socket of the oscilloscope. The resulting trace is then composed of portions of each waveform, but because of the high switching frequency an illusion of two separate traces is obtained. Variable bias voltages are applied to the amplifier and square-wave generator so that the two signal traces can be separated for easier examination.

332

Post-deflection Acceleration

Tubes that have been designed to achieve the brightest possible trace on the screen without affecting unduly the deflection sensitivity are known as post-deflection acceleration (p.d.a.) tubes. The deflection sensitivity of an ordinary cathode-ray tube is inversely proportional to the final anode voltage, and thus brightening the trace by the use of greater accelerating voltages reduces the sensitivity. Higher deflection voltages are then required when producing a given deflection with a brighter trace, and this entails more costly amplifiers to operate the deflector plates. However, it has been discovered that if the main acceleration of the electrons in the beam is effected after deflection, the reduction is sensitivity is not very great; p.d.a. tubes therefore have an internal ring or graphite spiral on the glass wall near to the screen; this final accelerating electrode is maintained at a very high voltage (about 10 kV in many cases) relative to the cathode.

When electrons strike the screen the latter would normally acquire an undesirable charge. This is prevented by collecting the secondary electrons emitted by the screen after bombardment by the primary beam electrons. The collection is effected in p.d.a. tubes by the final graphite electrode. In other tubes, a simple graphite coating on the walls near the screen is connected to a_3, and thus has sufficient potential to collect the secondary electrons.

OSCILLOSCOPE COMPONENT PARTS

A complete oscilloscope is shown in block diagram form in *Figure 11.3*. It is necessary to have power suplies in order to operate the cathode-ray

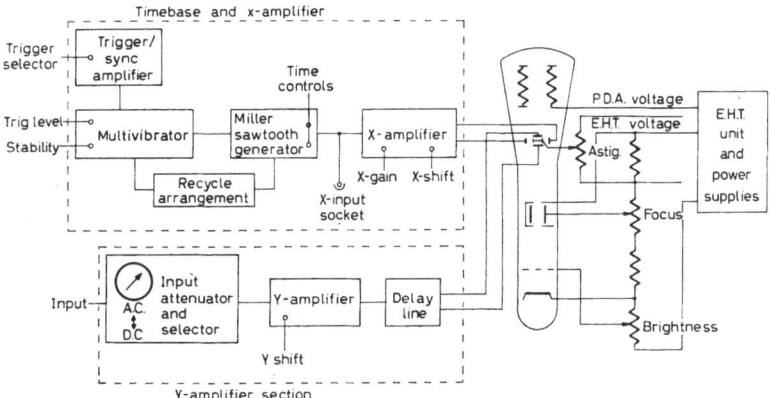

Figure 11.3. Block diagram showing the basic sections of a modern oscilloscope. Double-beam instruments have the Y-section duplicated and use a beam-switching circuit or a double-beam tube

tube. These supplies provide the heater currents and high voltages necessary to operate the tube electrode system and the remaining electronic circuitry. Since signal voltages are rarely large enough to produce satisfactory deflections when applied directly to the deflector plates, intermediate X- and Y-amplifiers must be used. A timebase produces the time axis along the horizontal or X-direction in a way described below.

Power Supplies

The power supplies for a cathode-ray oscilloscope follow conventional lines. The supplies to the timebase and amplifiers are similar to those for any other equipment requiring current at 300–350 volts. In the more expensive instruments stabilized supplies are often incorporated. Considerable currents are needed for the Y-amplifiers if high deflection voltages are required, particularly when high upper frequencies are involved.

The high anode voltage needed for operating the cathode-ray tube may be obtained by simple half-wave rectification of the appropriate alternating voltage. It should be remembered that the cathode is at a high negative potential relative to the chassis since the final anode must be at or near to earth potential in order to avoid distortion of the trace on the screen. The current required from the e.h.t. (extremely high tension) supply is quite small and need not exceed one milliampere. A high-voltage supply developed from a radio-frequency oscillator is an attractive alternative to the conventional arrangement mainly because it is non-lethal. The method has not so far been popular in commercial instruments although at least two manufacturers use it (Marconi and Tektronix). The principles of operation are similar to those used in the rectification of current at mains frequency; however, in the case of the r.f. supply a separate oscillator produces alternating current at frequencies of the order of 100 kHz (see chapter on power supplies).

The Timebase

The majority of phenomena studied with an oscilloscope are periodic, that is to say, the variations are repeated exactly in equal successive intervals of time. All oscilloscopes therefore include a special type of relaxation oscillator and this, together with the X-amplifier constitute what is known as a timebase. Relaxation oscillators for producing timebase waveforms have been described in the previous chapter. The function of the timebase is to deflect the beam in the X-direction so that the spot moves from left to right across the screen with uniform speed. Having travelled from left to right the spot must then be returned to its original position as rapidly as possible (ideally in zero time). The output waveform of the timebase should therefore approximate very closely that shown in *Figure 11.4*. From the diagram it can be deduced that the distance travelled horizontally by the spot must be directly proportional to time so that, in effect the horizontal axis is a time axis. If simultaneously the signal voltage is applied to the Y-plates, the spot delineates the signal waveform. To be effective for viewing purposes the timebase frequency must be that of the signal or a submultiple of it. When this condition obtaines, successive traces coincide and, because of the viewer's persistence of vision and the tube persistence, an illusion is created of a stationary pattern.

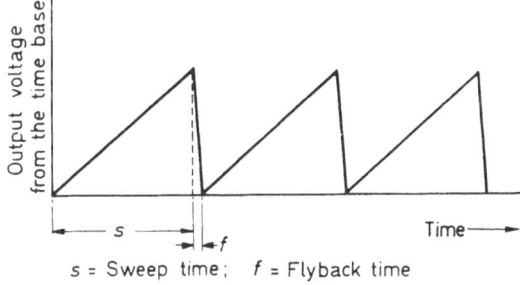

s = Sweep time ;   f = Flyback time

Figure 11.4. The waveform required from an ideal timebase. Strictly, *f* should be zero, but in practice we must accept a small finite value where $f \ll s$

The timebase frequency is varied by adjusting two controls at the front of the instrument. One control selects the approximate frequency and is often called the time control, and the other, the fine frequency control, adjusts the oscillator frequency to the desired value. The stability of modern timebase generators is good enough to enable manufacturers to abandon the former practice of marking the controls in terms of the sweep repetition rate. Instead these controls are marked in time units specifying the sweep rate. The coarse frequency control that selects the size of the capacitor in the main charging circuit of the timebase oscillator is now marked in time units, and the fine frequency control acts as an interpolator between these fixed ranges. The sweep rate can then be read off directly. Thus, if from the controls it is known that the sweep rate is 50 msec/cm, then a phenomenon producing a trace that extends over say 3 cm in a horizontal direction has a duration of 150 msec. The calibration of the timebase in time units is particularly helpful when examining pulse waveforms.

The locking of the timebase frequency to that of the signal is termed 'synchronization'. This is essential since the frequency of the timebase, and often that of the signal, is not quite constant, consequently, if the timebase were free running, successive traces would not coincide and a confused screen pattern would result. When considering the purchase of an oscilloscope, it is important to know what type of synchronization is used. In the older 'recurrent' types still found on some of the cheaper models, synchronization is achieved by feeding a small fraction of the signal voltage directly into the timebase. (This fraction may well come from the Y-plate voltages.) This has the effect of firing the timebase oscillator at the right time to ensure coincidence of successive traces. The frequency of the timebase must be carefully adjusted to be marginally slower than the signal frequency, and positive locking is not always easy to achieve. A much better system found in all good quality oscilloscopes uses a triggered system of synchronization. In this way the input signal exerts a more positive control over the repetition frequency of the timebase. The trigger selector switch often has six positions. The + and − internal positions means that the trigger impulse is obtained from the Y-amplifier, the sign determining whether the sweep starts as the signal waveform is going positive or negative. The external positions have a similar significance except that the trigger signal is obtained from external sources. The alternative positions ensure synchronization with the mains frequency, the trigger voltage usually coming from one of the heater supplies. This is often useful when a determination of signal frequency is to be made, this frequency being not much higher than ten times the mains frequency; triggering from the mains is also useful when studying phenomena associated with machinery that is mains operated. The improvement in synchronization obtained by using a triggering system, rather than older methods, is due to the fact that a precise trigger pulse of the requisite amplitude and short duration is used to fire the timebase at an accurately controlled time. Most triggering systems use two stages, namely a trigger input amplifier and a trigger generator. For normal triggered operation the generator is connected as a Schmitt trigger circuit (see Chapter 13, p. 423). The bias point is adjusted by what is often called a trigger 'level' control. By varying the level control it is possible to select the point on the input waveform at which triggering occurs. With many oscilloscopes the trigger level control can be turned to an 'auto' position. When operating in the auto-trigger mode the Schmitt trigger is converted to a free-running multivibrator running at a frequency of about 40 Hz in the absence of a signal. As soon as the input signal is applied the multivibrator locks to a submultiple of the signal and triggers the timebase automatically. This mode of operation can be used for 90 per cent of all ordinary input signals. Modern oscilloscopes are provided with high-frequency triggering facilities. Selecting the correct high-frequency position of a selector switch causes the multivibrator to run at a frequency of approximately 1 MHz. Input signals with frequencies in the approximate range 1–15 MHz can then be correctly synchronized.

The timebase oscillator depends on the controlled charging and discharging

PLATE I

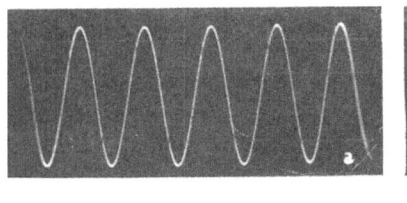
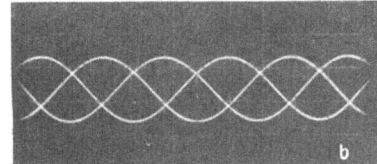
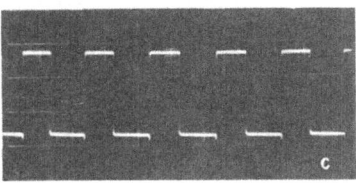

Photographs taken of traces on ordinary film (Kodak Plus-X) with an inexpensive 35 mm camera. All exposures were 1 sec at f.11. The exposure time could be reduced by opening the aperture to f.2·8 since no depth-of-field considerations need to be made. (Trace (a) shows a 1 kHz sinusoidal wave with the oscilloscope correctly adjusted. Trace (b) shows incorrect settings of the timebase controls. Trace (c) shows a square wave with a repetition rate of 1 kHz. The oscilloscope used was a Solartron CD1400

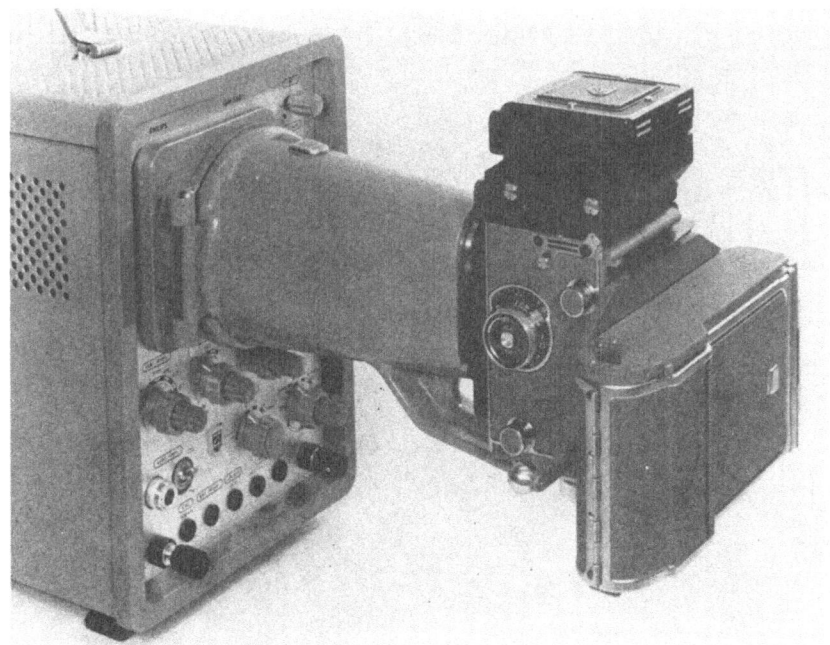

Photographic recording of oscilloscope traces using Philips equipment. A Polaroid-Land attachment is fitted to a Rolleicord Vb yielding prints within 10 seconds of exposure. Reproduced by courtesy of Research & Control Instruments Ltd.

PLATE II

Internal view of the Solartron oscilloscope CD 1014.2. Reproduced by courtesy of Solartron Ltd.

The Marconi dual trace oscilloscope TF 1331. Reproduced by courtesy of Marconi Ltd.

PLATE III

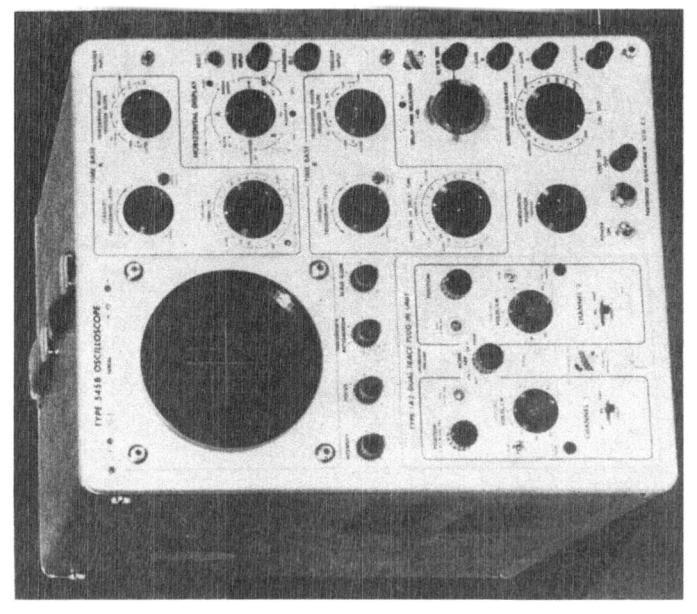

The Tektronix 545B oscilloscope. Reproduced by courtesy of Tektronix Inc.

The Tektronix 310A oscilloscope. Reproduced by courtesy of Tektronix Inc.

PLATE IV

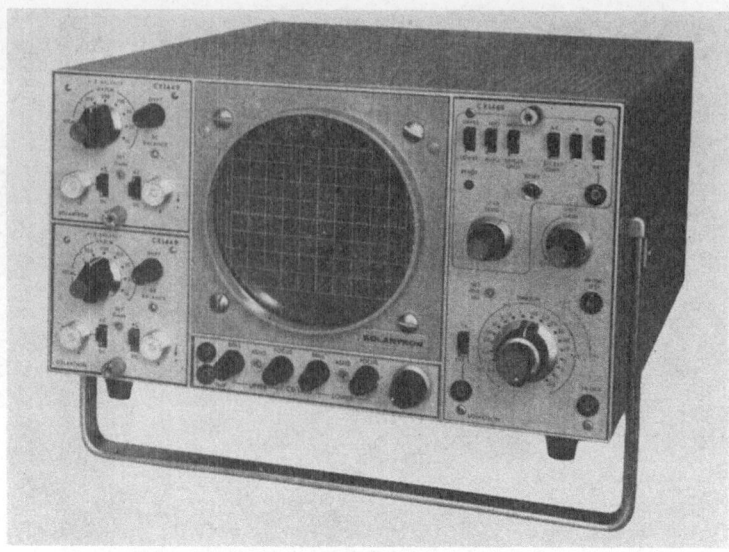

The Solartron CD 1400 oscilloscope

One practical arrangement of the circuit of *Figure 13.22.* The smaller components are conveniently mounted on the tag board and subsequently the tag board assembly is mounted in the chassis; connections are then made to other components in the circuit

Vertical mounting of the components on printed circuit board or Veroboard produces a very compact form of construction

of a capacitor of the appropriate size. Many modern oscilloscopes make use of the excellent linearity of the Miller timebase (Chapter 10, p. 317). Alternatively the linear sweep may be generated by constant current charging of a timing capacitor from a voltage provided by a bootstrap circuit (p. 381). The section that charges the main timing capacitor is preceded by a bistable circuit which 'flips' when a synchronizing pulse is received from the trigger generator. On 'flipping', a pulse is delivered to the charging section and the sweep is initiated. At the termination of the sweep the voltage across the capacitor is high enough to operate, via a diode, a circuit that resets the bistable in readiness for the next trigger pulse. Associated with the bistable circuit is a hold-off arrangement which consists of an appropriately valued hold-off capacitor, selected by a switch operated by the same spindle as selects the main timing capacitor. The hold-off circuit ensures that sufficient time elapses for the charges on relevant capacitors in the timebase to reach their quiescent values before the timebase bistable circuit can be re-triggered. The hold-off capacitor, after receiving an initial charge ensures that an appropriate valve is held in a cut-off condition which prevents the bistable circuit from operating. The hold-off capacitor discharges via a resistor, and after a suitable delay, determined by the time constant of the capacitor-resistor combination, the valve that was heavily cut-off arrives at the condition whereby the bistable can revert to the trigger-sensitive condition. The ultimate potential to which the hold-off capacitor discharges is determined by the setting of the 'stability' control. A system of concentric spindles is usually used so that the 'stability' control knob is physically close to that knob controlling the trigger level. Turning the stability control in a fully anticlockwise direction causes the hold-off capacitor to remain discharged at a voltage that prevents the bistable from operating. No horizontal sweep is possible under these conditions. If the stability control is then rotated, a point is reached where the discharged potential of the hold-off capacitor is such that the appropriate valve is no longer held below the cut-off voltage. The bistable then runs freely as an oscillator and recurrent horizontal sweeps are obtained. For normal operation of the oscilloscope the stability control is adjusted in the absence of a signal so as to be slightly below the free-running condition. When an input signal arrives the trigger-generator is then able to deliver a pulse that just flips the bistable. The timebase frequency must then synchronize perfectly with that of the signal. For applications where the timebase is not required the stability control is turned to the fully anticlockwise position.

The Y-amplifiers

There can be no doubt that a consideration of the specifications of the Y-amplifier is of paramount importance before the purchase of any oscilloscope. The classification of oscilloscopes and their cost depend largely upon the type of Y-amplifier used. The potential user must therefore consider carefully his requirements since failure to do so may lead to the purchase of an instrument inadequate for his needs. Alternatively, money may be wasted in buying facilities that are rarely or never used. If a potential user is uncertain of his future requirements a good solution is to buy an instrument with plug-in units, such as those marketed by Tektronix, Telequipment and Solartron. A basic oscilloscope sufficient for his immediate needs may be bought at reasonable cost, and, if later work demands a higher specification, it will be possible to buy the necessary plug-in units to extend the facilities of his instrument.

The cost of the Y-amplifier section, and hence the oscilloscope, depends largely on the facilities required. Factors to consider are bandwidth, sensitivity, rise time, transient response and delay facilities.

Bandwidth is defined as the range of frequencies over which the amplifier's response does not fall below 0·707 (i.e. $\sqrt{2}$) of the maximum gain at mid-frequencies (0·707 corresponds to a loss of 3 dB—see Chapter 9 in which the decibel scale is described). For fast oscilloscopes that are required for pulse-type waveforms, it is becoming the custom to replace the 3 dB criterion by the rise time and transient response specification. The rise time is the time taken for the spot to travel from 10 to 90 per cent of the vertical distance representing the amplitude of the signal. When rising quickly there is a tendency for the spot to overshoot and oscillate before settling to the final position. This is known as overshoot and ringing, and, together with reflections that may occur in the delay line, these forms of distortion alter the waveform of a pulse or square wave and introduce various 'spikes', 'squiggles' or 'bumps' into the final display. Such distortion may be avoided by having an amplifier with a good transient response. Optimum transient response is obtained when the product of the rise time and frequency response lies between 0·33 and 0·35. For example, a rise time of 0·02 μsec and a bandwidth of 15 MHz gives a product of 0·3. Factors larger than 0·35 probably indicate overshoot in excess of 2 per cent, whilst those larger than 0·4 indicate possible overshoots of 5 per cent or more. The Tektronix 'scope 515A is an example of a good laboratory instrument having a rise time of 23 nsec (0·023 μsec) with a 15 MHz bandwidth.

The sensitivity is expressed as the minimum voltage necessary to produce a deflection of 1 cm. Details of typical specifications are given below.

Delay lines are often incorporated in the more expensive instruments. This valuable feature is a special type of line consisting of sections of inductors

338

and capacitors. When a signal is applied at the input to the line, it takes a finite time for it to appear at the output terminals. This is very useful when we wish to initiate the sweep before applying the signal to the Y-plates. The input signal is made to trigger the timebase in the usual way, but this signal is first sent through the delay line before being fed into the Y-amplifier.

The input impedance of all modern oscilloscopes is high. It is usual to express the impedance as an equivalent resistance in parallel with the input capacitance. The input resistance should not be less than 100 k, but many reputable manufacturers arrange for the value to be up to 10 M. The input capacitance should not exceed 50 pF. Usually this value is small enough not to be troublesome, but its effect when measuring across tuned circuits should be remembered; at high frequencies 50 pF can alter the resonance conditions considerably.

All oscilloscopes have shift controls that enable the operator to move the trace to a convenient position on the screen. In the case of Y-deflections the shift control alters the biasing conditions in the Y-amplifier so as to shift the steady potentials on the Y-plates sufficiently to move the trace to the desired position; X-shifts are produced in a similar fashion.

For simple routine measurements on the bench, a bandwidth of about 100 kHz and a maximum sensitivity of 500 mV/cm will probably prove adequate. The Telequipment Servicescope Minor with its $2\frac{3}{4}$ in. screen, a sensitivity of 200 mV/cm and a bandwidth of 30 kHz has proved satisfactory in many educational and other establishments of modest means. Its cost (1966) is about £25. The Dartronic 'scope model 381 is an example of an inexpensive instrument that has proved satisfactory in the author's laboratories when used for elementary purposes. The 1965 price was just under £40, the bandwidth being d.c. to 9 MHz and the sensitivity 100 mV/cm.

With general purpose laboratory instruments one should expect a bandwidth up to 10 MHz, a sensitivity of 50 mV/cm or greater, and a rise time of about 0·03 sec. The author is having good service from a Marconi TF 1330, a single beam oscilloscope, in which the bandwidth is d.c. to 15 MHz and the rise time is 0·025 sec. The overshoot is less than 1 per cent for an input pulse having a rise time of 0·01 sec. The maximum sensitivity is 50 mV/cm. The cost in 1965 was £300. The Solartron CD 1400 is a less expensive instrument that has proved to be very satisfactory. This instrument uses a cathode-ray tube having a double electron gun to produce the dual trace. With CX 1441 plug-in Y-amplifiers the bandwidth is d.c. to 15 MHz with a sensitivity of 100 mV/cm. For higher sensitivities 10 mV/cm facilities are provided by a switching arrangement. The bandwidth at this sensitivity is d.c. to 0·75 MHz. The CX 1444 timebase can give horizontal sweep times to 0·5 µsec/cm of 2·5 sec/cm. The cost of the basic unit with two Y-amplifiers and the CX 1444 timebase was £186 (1967 price). The Tektronix 310 A is a small, portable single beam that has a 3 in. screen, a bandwidth of d.c. to 4 MHz and a rise time of 90 msec. At £276 (1965) it is rather expensive, but the workmanship is very good. Other Tektronix machines with plug-in units although expensive give excellent service and performance. The type 551, for example, has a bandwidth of d.c. to 25 MHz, a calibrated sweep range of 0·1 sec/cm to 5 sec/cm and a rise time of 14 nsec with a 50 mV/cm sensitivity. The performance can be varied by the use of different plug-in units. By the time a suitable range of units is bought the total cost will approach £1,000, but obviously a first-class instrument is obtained.

In naming certain instruments, it is not implied that other manufacturers produce inferior equipment. On the contrary, Advance, Cossor, Dawe, Roband-Philips and Hewlett-Packard all produce good instruments. These are not, however, used by the author.

The specification of the X-amplifier is usually not nearly so stringent as that for the Y section, although several manufacturers use identical amplifiers for both X- and Y-deflections. A separate X-amplifier with its own input terminal has obvious advantages, especially when it is realized that the horizontal axis need not always represent time. Like the Y-plates, the X-plates are voltage operated, so that if we can produce voltages proportional to frequency, for example, we have a frequency axis along the horizontal direction. In general, if the voltage applied to the X-plates is proportional to a quantity X then an X–Y tracing oscilloscope is obtained.

Special Oscilloscopes

Special 'scopes are produced for many sophisticated purposes. It is not possible here to go into any details, but the reader should be aware that such 'scopes exist. A recent development has been the design of a 'sampling' oscilloscope in which progressive samples of adjacent portions of successive waveforms are taken. Bandwidths approaching 4,000 MHz are possible using special techniques of processing the samples. Interested readers should consult Tektronix or Hewlett-Packard, both of whom produce such instruments. The sampling method provides a means for examining fast-changing signals of low amplitude that cannot be examined in any other way.

Storage or memory oscilloscopes, such as the Dawe-Cawkell Remscope 741 employ special tubes that store the trace on a special grid. Provision is then made for scanning the trace at some later time to make it visible. Such machines are costly and are required only for specialized purposes.

For those who wish to display transistor characteristics (because, for example, an unknown transistor is encountered and no curves are available) then special curve tracers (e.g. The Tektronix 575) are available.

Oscilloscope accessories are available of which the most important are the special leads and probes supplied by most manufacturers, at extra cost, for their instruments. The direct probe is merely a shielded coaxial cable and is used to shield the signal carrying wire from the input terminals to the oscilloscope right up to the point in the circuit at which the measurement is to be made. By using such a probe the induction of unwanted signals by stray electrostatic fields is avoided. The mains is often a troublesome source of unwanted 50 Hz signals. Using a length of unscreened wire as the lead to the oscilloscope often results in large amplitude unwanted signals that prevent a satisfactory observation of the signal to be examined. To reduce the effect of loading on the circuit under test the $\times 10$ probe of *Figure 11.5* may be used. This has the effect of reducing the input capacitance and increasing the input resistance to the oscilloscope.

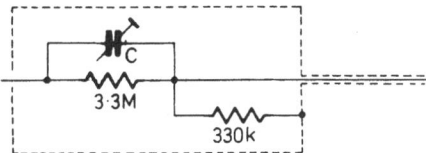

Figure 11.5. Input attenuator probe to give a voltage reduction of about 10 : 1. C should be adjusted so that no overshoot or rounding of the corners occurs with square-wave inputs. A value of approximately 10 pF should prove satisfactory

PRACTICAL CONSIDERATIONS WHEN USING AN OSCILLOSCOPE

In spite of the seemingly complex array of knobs and dials on the front of many laboratory oscilloscopes, no alarm need be experienced in using the instrument. Apart from one exception, no damage can result from incorrect settings of the controls. The handbook for the instrument should be available and consulted before putting the machine into operation. The handbook explains the function of each control and how to make adjustments to obtain satisfactory operation. The exception mentioned above relates to the brightness or brilliance control. In operation, this control must be adjusted to give a trace with the minimum brightness consistent with satisfactory viewing; the risk of burning the screen is then eliminated. Stationary spots within the viewing area should also be avoided to reduce the risk of burning. The timebase should be kept running and the spot drawn out into a line. If this is not possible then a shift control should be used to move the spot to the edge of the screen where screen defects are of relatively minor importance. An increase in contrast can be obtained by operating the oscilloscope in a

341

position which avoids direct light falling onto the screen. Visors or hoods are useful in this respect; coloured filters in front of the screen are also used to increase the useful contrast. For quantitative observations, it is useful to have the filter ruled with a graticule. A convenient and much used arrangement involves the ruling of the main axes, suitably subdivided, together with horizontal lines spaced at intervals of one centimetre.

When an oscillogram is being displayed, it is necessary to adjust the brightness, focus and astigmatism controls until the best picture is obtained. 'Astigmatism' is an optics term which means an inability to focus correctly simultaneously in two directions. In an oscilloscope these direction are the X- and Y-axes. The defect is minimized by ensuring that the mean voltage on the Y-plates is correctly adjusted; this is the function of the astigmatism control. For the best focus, it is necessary to adjust the focus and astigmatism controls together since they are not independent in their action on the trace. For periodic waveforms the timebase should be set so as to give three or four complete cycles of the waveform. The last waveform will not be quite complete due to the finite time it takes for the spot to return to the origin. On most modern 'scopes the return is not shown on the screen because a negative blanking pulse, applied to the modulator during flyback, cuts off the beam.

Failure to obtain a trace on the screen is common with newcomers to the instrument. Apart from ensuring that mains power is being delivered to the instrument (usually by observing the pilot light) users should check that the brightness control is sufficiently advanced to make a trace possible. If the X-shift or Y-shift controls are not correctly adjusted sufficient bias voltage may be being applied to the X- or Y-plates to deflect the spot off the screen altogether. With the X- and Y-shift controls in approximately their mid-positions the continued absence of a trace is then probably due to incorrect settings of the stability or level controls.

It is advisable to check the calibration of the Y-deflections and the timebase at regular intervals throughout the life of the instrument. Minor adjustments are possible from outside the case, but, unless the operator is skilled at repairing electronic instruments, it is best not to delve into the oscilloscope. If, for any reason the outer case is removed, it should be ensured that the mains plug is removed from the mains socket and that the high voltage capacitors have been discharged (preferably with an insulated screwdriver, the operator having one hand in his pocket). Only then is it safe to replace valves or faulty components.

The calibration of the Y-amplifier and the associated deflection on the screen can most conveniently be made by using the calibration voltage available from the instrument itself. The clipping of a sine wave with zener diodes produces a square wave of accurately known amplitude; many manufacturers are now using this method to provide a calibration unit. Where the calibration voltage must be variable the circuit of *Figure 11.6* may be used. The calibration unit, designed by Mullard Ltd., produces a square wave whose amplitude is directly proportional to a direct voltage which is measured by the best available voltmeter. The output waveform can be used for calibration over a range of voltages from 100 mV to 100 V with an inaccuracy of only about ±3 per cent. The square waveform is produced by a free-running multivibrator using V1a and V1b. When the pentode V1b is cut-off the anode voltage is that of the positive h.t. line, which is earthed. The cathode of the

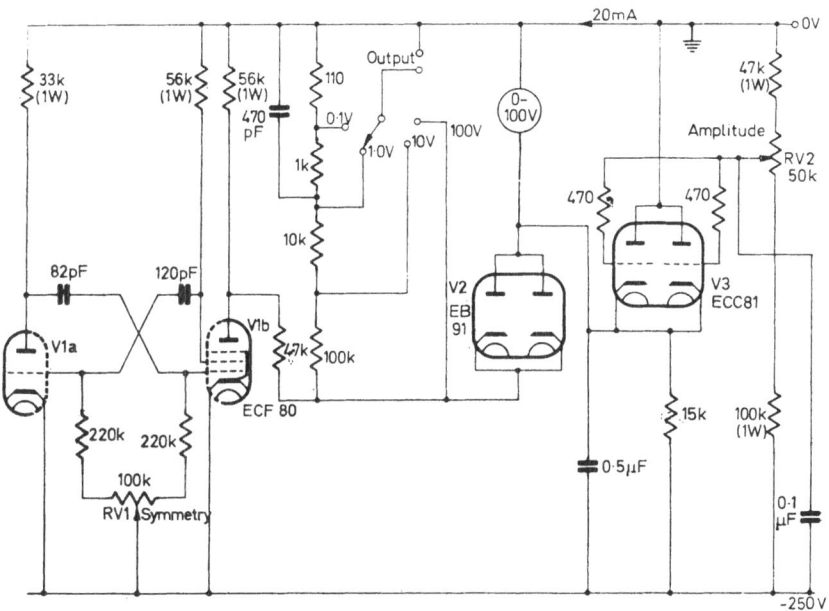

Figure 11.6. Mullard circuit for an oscilloscope calibration unit with variable output voltage facilities

diode is at the same potential. When the pentode conducts the anode voltage falls, but the diode cathode potential can fall only to the potential of the diode anode which is measured by a voltmeter and set to 100 V by means of the amplitude control RV2 and V3. A resistive chain of high-stability, high-accuracy resistors attenuates the square wave amplitude to amplitudes of 0·1 V, 1 V and 10 V. The square waveform for the calibration voltage is the most convenient shape to use since it is easy to align the flat tops and bottoms with horizontal graticule markings.

The timebase can easily be checked by building or buying a crystal-controlled oscillator. A 1 MHz oscillator provides one cycle in every microsecond.

343

By using a chain of multivibrator dividers as in *Figure 11.7* we obtain the submultiple frequencies of 1 MHz. The multivibrators are locked to the crystal oscillator, and hence the calibration frequencies are very accurately known[1].

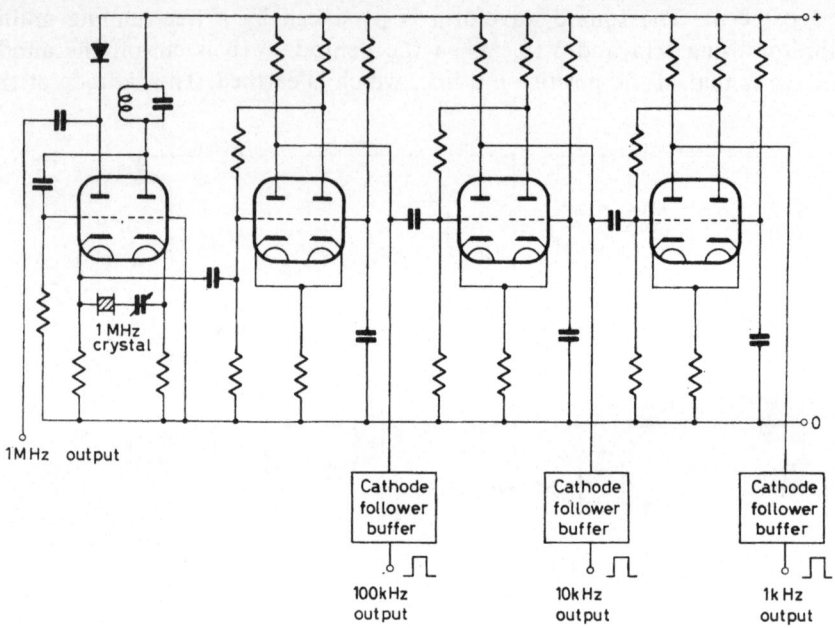

Figure 11.7. Method of obtaining accurately known frequencies from a crystal oscillator drive. Practical details of the circuit for a frequency dividing network that uses a different crystal drive are given by Attew[1]

VOLTAGE MEASUREMENTS

Having calibrated and checked the attenuator settings associated with the Y-deflections the application of the signal produces deflections from which the input signal voltage can be easily determined. Ideally calibrations should be made immediately prior to making a measurement; errors arising from changes in amplification with time due to fluctuations in supply voltages, ageing of components, etc., are thus eliminated. Due regard must be paid to two further errors. The first error arises because of non-linearities of the Y-amplifier and cathode-ray tube. If these are appreciable then strictly a calibration voltage should be of a similar amplitude to that of the signal voltage to be measured. The second error relates to the frequency response

344

of the Y-amplifier. Since this response is never perfect the frequency of the calibration voltage should be similar to the frequency of the voltage to be measured.

Some oscilloscopes have their shift controls calibrated in volts. Thus by shifting the trace relative to a fixed horizontal line it is easy to read off the voltage difference between any two points on the waveform. The accuracies attainable with this method are not usually very high, although at least one firm (Marconi Ltd.) claim that with their TF 1330 oscilloscope the inaccuracy of the indicated voltage is not greater than 2 per cent of the full-scale value.

Time measurement presents a rather more difficult problem. The simplest method is to take advantage of the stability and linearity obtainable from Miller-integrator and other feedback circuits. When using these circuits it is possible to calibrate the timebase controls directly.

If greater accuracy is required it is necessary to have recourse to an external time-marker. Such a time-marker usually consists of an oscillator whose output is suitably distorted and passed through a differentiating circuit. The differentiated output potential then consists of narrow pulses, and, when applied to the modulator of the cathode-ray tube, these pulses produce a brightening of the trace at known time intervals. In order to ensure that the pulses appearing on successive traces shall be superimposed on one another, the oscillator must be quenched after each trace and restarted at the beginning of the sweep. The periodic quenching is achieved by injecting a signal from the timebase into the time-marker generator. Because of the difficulty of starting and restarting a quartz crystal oscillator, this excellent form of frequency standard is not usually considered suitable as a time-marker for ordinary oscillographic work. Ringing, tuned circuits locked to the timebase are the usual choice.

LISSAJOUS FIGURES

Frequency measurements can be made very elegantly and accurately by means of a Lissajous' figure if sinusoidal voltages are involved. The signal voltage of the frequency it is desired to measure is applied to one set of deflector plates, via the appropriate amplifier, and a sinusoidal voltage of accurately known frequency is applied to the other set of deflecting plates, again, via the appropriate amplifier if necessary. The timebase must be switched off by turning the stability control fully anticlockwise. The resulting traces on the screen are known as Lissajous' patterns, so called after Professor Lissajous, who first produced them by combining at right angles the component oscillations of two tuning forks; the combination in his case was effected by optical means. By studying the shape of the pattern an accurate comparison of the frequencies can be made. For any given frequency ratio many patterns are possible depending upon the relative phases of the two voltages. *Figure 11.8* gives some possible Lissajous' patterns.

If the signals involved are v_y, of unknown frequency and v_s, the standard of known frequency, then let

$$v_y = A \sin (2\pi f_y t + \phi)$$

$$\text{and} \quad v_s = B \sin 2\pi f_s t$$

where ϕ is the phase angle between v_y and v_s. If two mutually perpendicular and centrally placed axes are drawn through the Lissajous' figure, the latter crosses the y-axis n times and the x-axis m times. Where a curve crosses itself on an axis, two intersections on that line are counted. The frequencies of the sources producing the horizontal and vertical deflections are in the same ratio as the numbers of times the pattern is intersected by the y- and x-axes, i.e. $f_y/f_s = m/n$.

In practice it is often not possible to obtain an absolutely stationary pattern and counting the intersections on the axes may prove difficult. A somewhat easier method is to count the number of horizontal loops along the top edge of the pattern, h say, and then to count the number of loops along the vertical edge of the pattern, v say. Then $f_y/f_s = h/v$. Where the pattern does not consist of loops alone then each loop may be counted as one unit and a termination of pattern in a sharp line counted as a half unit. Although seemingly complicated when described in words, *Figure 11.8* should clarify

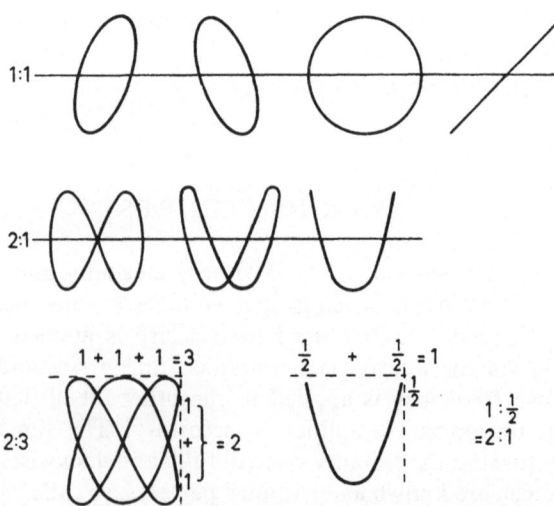

Figure 11.8. Typical Lissajous' patterns. For the significance of the figures and the estimation of the frequency ratios see text

346

this method. In practice, it is an extremely simple way of determining the frequency ratio especially with moving patterns.

Where the ratio between the unknown and standard frequencies is high, the pattern contains too many loops for reliable counting. The method of

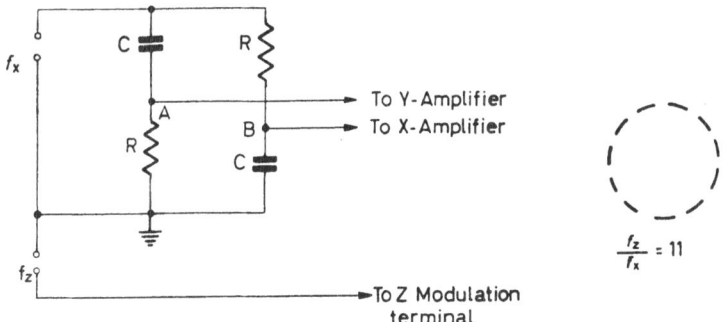

Figure 11.9. Phase-shifting circuit to produce a circular pattern. $R = X_c$ for the voltages at A and B to be 90° out of phase (e.g. 0·1 μF and 32 k when $f_x = 50$ Hz). In practice some adjustments of the gain controls in the oscilloscope are usually necessary to produce circular traces

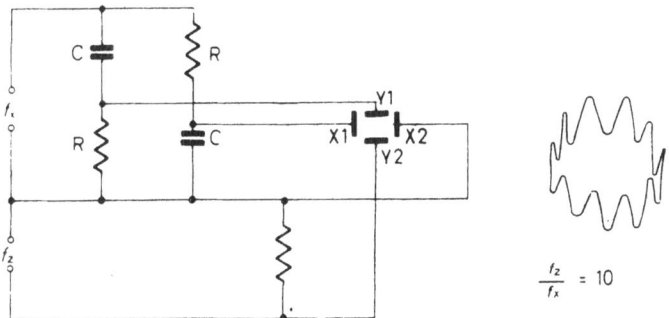

Figure 11.10. Circuit for producing 'coronets'. This system is useful where Z-modulation facilities are not available. Only the 'tops' or 'bottoms' of the 'coronet' must be counted—not both

Figure 11.9 is then more satisfactory. The known frequency source is applied to the Y-plates and also to the X-plates, after being shifted by 90° with the phase-shifting network. Adjustments of R and the gain controls should then produce a circle. The unknown frequency is then made to modulate the beam by applying the signal to the Z-axis input. (The Z-axis is the direction of the beam, which is, of course, perpendicular to both the X- and Y-axes.) A broken ring pattern is then formed from which the unknown frequency may be

347

estimated as shown in *Figure 11.9*. If the user's oscilloscope has not got a Z-axis input facility then the ring may be modulated by using the circuit of *Figure 11.10*. The unknown frequency must be higher than the standard frequency in both cases; in the latter case, with the modulated ring pattern, the amplitude of the unknown signal must be low enough to avoid distortion of the pattern.

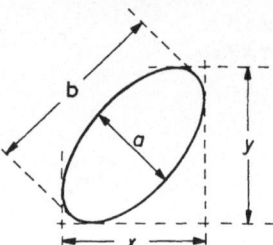

$$\text{Tan } \frac{\phi}{2} = \frac{a}{b}$$

where a and b are the minor and major axes of the ellipse and ϕ is the phase angle

Figure 11.11. When the phase shift in a network or amplifier is required, the input and output voltages can be fed to the X- and Y- inputs to produce the 1 : 1 Lissajous figure. The gain settings must be such that the amplitudes of the voltages presented to the deflector plates produce a pattern in which $x = y$. The phase angle ϕ is then given by the above formula[2]

The phase difference between two sinusoidal waveforms can be measured using the arrangement of *Figure 11.11*. We are in effect using a Lissajous' figure that indicates a 1 : 1 ratio. In applying the formula we must ensure that the amplitudes in both the X and Y directions are adjusted to be identical. The phase-shift in each amplifier channel must also be the same, so that wherever possible an oscilloscope with identical X- and Y-amplifiers should be used.

WAVEFORM ANALYSIS

Perhaps the most frequent use made of the oscilloscope in most laboratories is the general viewing of waveforms present in the various items of electronic equipment being used. By knowing what waveforms are obtained when the apparatus is functioning correctly, an operator with experience can quickly diagnose many faults when viewing the patterns obtained from apparatus that is not functioning correctly. It is advisable to build up from memory and records a set of oscillograms associated with newly bought equipment known to be operating successfully. Such records then supplement the manufacturers' data regarding the fixed voltages to be expected at various points in the circuits of the apparatus. Many of the circuits in this book

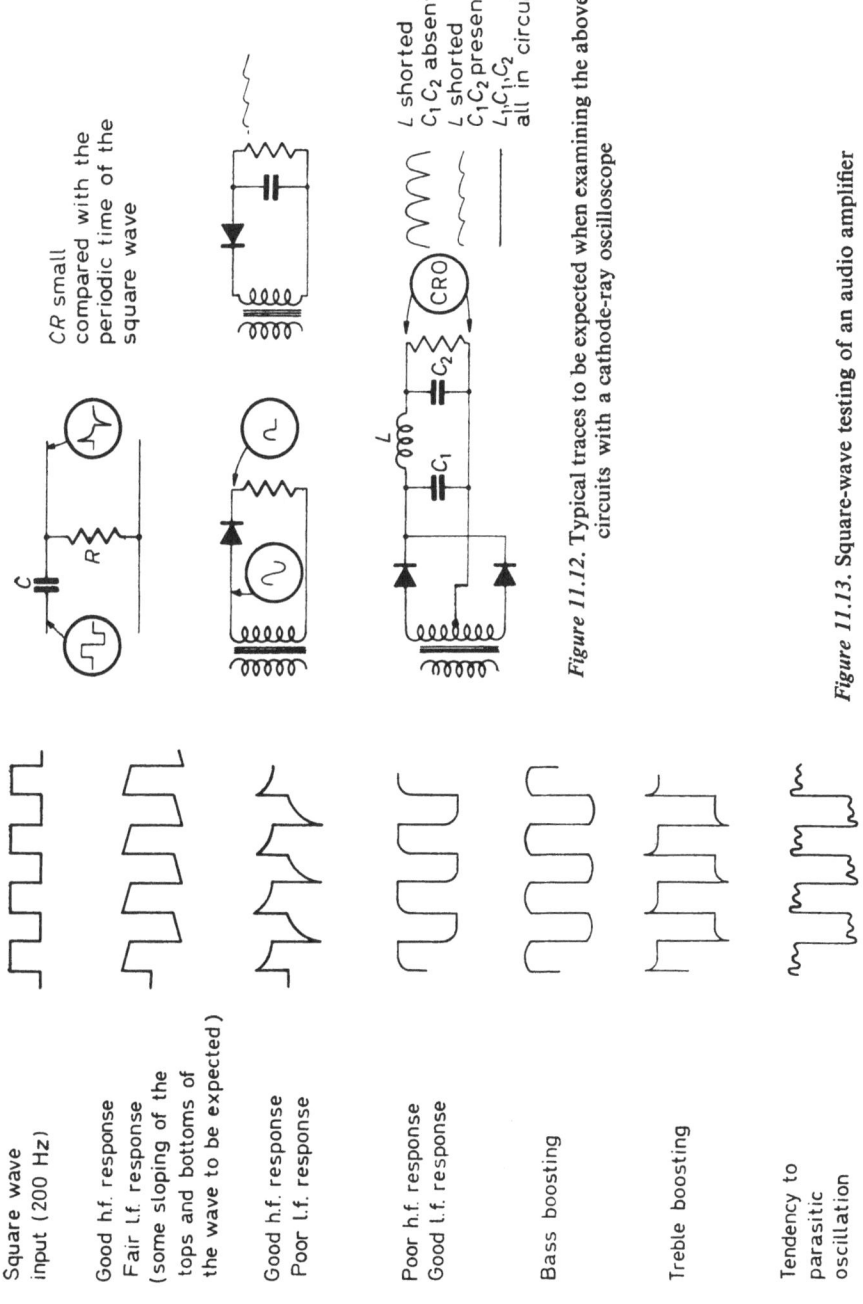

Square wave
input (200 Hz)

Good h.f. response
Fair l.f. response
(some sloping of the
tops and bottoms of
the wave to be expected)

Good h.f. response
Poor l.f. response

Poor h.f. response
Good l.f. response

Bass boosting

Treble boosting

Tendency to
parasitic
oscillation

CR small
compared with the
periodic time of the
square wave

L shorted
$C_1 C_2$ absent
L shorted
$C_1 C_2$ present
L_1, C_1, C_2
all in circuit

Figure 11.12. Typical traces to be expected when examining the above
circuits with a cathode-ray oscilloscope

Figure 11.13. Square-wave testing of an audio amplifier

349

yield waveforms that should be examined by an oscilloscope and recorded for future reference. *Figures 11.12* and *11.13* indicate by way of example, the waveforms to be expected with various circuits.

PHOTOGRAPHING OSCILLOSCOPE TRACES

Apart from making records of oscillograms for diagnostic purposes, useful measurements and studies can be more easily undertaken on photographs of traces, rather than on the traces themselves. Most modern instruments are provided with fixtures that enable suitable cameras to be attached. Two methods of taking photographs of oscilloscope traces are in general use.

In the first method, the film is static throughout the exposure time. For repetitive waveforms the display on the screen must be quite stationary whilst the shutter is open. The exposure time is usually one second at about f.6 when low-voltage tubes are used. The exposure time required depends on the colour of the light emitted from the screen as well as screen brightness. If much photographic recording is anticipated it is advisable to use a screen giving blue traces. In any event, it is a good idea to consult the manufacturer of the particular instrument to be used. For high-voltage tubes Super XX Pan film with an f.2 aperture allows the exposure time to be reduced to as low as 0·02 sec. Too rapid a shutter speed may, however, prevent a complete cycle from being photographed. In cases where transients are to be recorded, synchronization of the start of the sweep with the opening of the camera shutter is essential. Some oscilloscopes make special provision for the initiation of the sweep by contacts fitted to the camera, such contacts being associated with the shutter operating mechanism.

The second method involves leaving the shutter open and drawing the film through the camera at a steady rate. The timebase is switched off for this purpose, and the trace on the tube screen is thus a straight line in the Y-direction. The method is useful when comparatively low-frequency waveforms are to be examined over long periods of time. Briggs[3] has used the method to obtain the frequency response of audio amplifiers. The rate of the film past the screen is related to a change of frequency so that in effect the response of the amplifier at any given frequency can be determined.

When photographing single-shot events by the first method, it is necessary to start the timebase sweep before the transient voltage appears at the Y-plates. In these circumstances, it is usual to allow the transient to initiate the sweep by applying the voltage to the external trigger terminal of the oscilloscope. Simultaneously, the transient voltage is applied to the input side of a delay line, and the output from the delay line is taken to the input of the Y-amplifier. If a delay line is incorporated within the oscilloscope so much the better. So many comparatively inexpensive instruments nowadays incorporate trigger and single shot facilities that most laboratories possess or can acquire an oscilloscope from which transient phenomena may be photographed.

The advantages of recording are several: the permanent records can be enlarged if necessary; it is easy to compare several oscillograms and make quantitative measurements on them; records of very short duration traces can be made, and the recording process, if made automatic, eliminates the need for constant observation.

For most recording where only single pictures are required, the author has had excellent service from a Rolleicord Vb together with a Polaroid-Land attachment. With this system, the prints are available for inspection in 10 seconds. Philips supply a complete kit to fit any oscilloscope; This includes the Rolleicord camera, polaroid attachment and the necessary fitments. The excellent camera can, of course, be used for normal photographic purposes in the laboratory and elsewhere (see Plate I).

For film work from which enlargements can be made, it is best to select a film designed for this type of recording. The Ilford recording film 5G91 has proved successful, although, no doubt, Kodak have equally efficient emulsions. The 5G91 is a very fast orthochromatic film of high contrast. Its antihalation properties are good. The film is recommended by its makers for recording moving spots on cathode-ray screens that give a blue or green trace. Very satisfactory results have been obtained with this film on developing it in Kodak D76 developer for 8 minutes at 20°C. Ilford high speed liquid fixer (Hypam), diluted in the volume ratio of 1 part of fixer to 4 parts of water gives satisfactory fixing after about 4 or 5 minutes. The film should be washed for at least 30 minutes in running water before drying. Prints, especially enlargements, are best made on a hard (i.e. contrasty) paper.

For ordinary recording work, practically any of the usual films and papers may be used with a slight, sometimes negligible, deterioration in the picture. Examples are given in the upper part of Plate I of photographs taken on Kodak Plus-X film with the author's 35 mm camera (a modest Petri 7). A simple close-up lens was used, together with a cardboard tube to exclude ambient light. Normal development and fixing procedure was used.

For specialized work, however, it is always best to refer the problem to Kodak or Ilford. The latter firm produce a very informative book on the recording of cathode-ray tube traces[5].

The property of certain plastics of adhering firmly to a smooth surface can be made use of to provide a cheap and convenient method of recording cathode-ray tube traces. A. W. Gooder has described a method based on such a plastic sheet[4]. By pressing the flexible transparent sheet on to the face of the tube the picture may be traced by means of a chinagraph pencil or ball-point pen. The sheet may then be preserved or used again, if required, by wiping off the trace. A suitable sheet is Vybak flexible sheet DUB 239 clear No. 5, available from Bakelite Ltd., Birmingham.

REFERENCES

1. Attew, J. E. 'Decade multivibrator design'. *Wireless World*, 1952, **58**, No. 2 (March).
2. W.T.C. *Wireless World*, Oct. 1952, p. 432. (The author suspects that W.T.C. = W. T. Cocking, a regular contributor to this journal.)
3. Briggs, G. A., *Sound Reproduction*. Wharfedale Wireless Works, 3rd. Edn, 1953.
4. Gooder, A. W. 'A simple and inexpensive method of tracing cathode-ray tube waveforms'. *J. Sci. Instrum*. 1964, **41** p. 392.
5. Hercock, R. J. *The Photographic Recording of Cathode-ray Tube Traces*. Ilford Technical Monographs, 1947.

Suggestions for Further Reading

The Cathode-ray Oscilloscope by J. Czech, Philips Technical Library, 1957.
The Oscilloscope at Work by A. Haas and R. W. Hallows, Iliffe and Sons Ltd. 1954.

12

PHOTOELECTRIC DEVICES

There are many occasions on which it is convenient to replace the human eye and operator with some sort of automatic device. In gas- or oil-fired furnaces, for example, where flame failure would constitute a serious hazard, it is not economic to employ a man solely for the purpose of detecting flame failure and raising the necessary alarm. Such detection and subsequent action can well be carried out electronically. In applications such as burglar and fire alarms, signalling and warning systems, smoke detection, counting and automatic control, colour comparison, etc., the human operator can now be replaced by reliable electronic apparatus. Such apparatus depends upon photoelectric devices. The devices, often called photoelectric cells, have the ability to convert light signals into corresponding electric signals. The latter can then be processed by conventional electronic equipment to produce desired results.

Photoelectric cells can do more than merely replace the human eye. Because of their ability to measure light intensity accurately, and to respond to radiations in the infra-red and ultra-violet regions, it is possible to build apparatus for specific purposes; infra-red spectroscopy, the estimation of turbidity, fluorimetry studies and photometric work are appropriate examples.

Photoelectric devices may be classified according to the mechanism of their operation. The three main classes of photocell are (a) photoemissive cells (b) photoconductive cells and (c) photovoltaic cells.

PHOTOEMISSIVE CELLS

In 1887, Hertz discovered that when ultra-violet light fell on the polished terminals forming a spark-gap, then the breakdown voltage was very much reduced, i.e. sparks were more readily initiated. The discovery led Hallwachs, Stoletow, Elster and Geitel, Lenard, Einstein, and Millikan to pioneer what is now termed the photoelectric effect. It was found that many metals such as zinc, caesium, sodium, potassium and rubidium emit electrons when subjected to electromagnetic radiations of short wavelength. The alkali metals exhibit the effect with visible light. For monochromatic light the number of electrons emitted is found to be directly proportional to the intensity of the incident radiation. The maximum energy of the emitted electrons was unexpectedly found to be independent of the intensity. This maximum energy is directly proportional to the frequency of the incident radiation provided the frequency is greater then a certain threshold value. The threshold value depends upon the substance used and the nature of the emitting surface.

In seeking an explanation of the photoelectric effect, we must look to the modern theories of quantum mechanics and solid-state physics. Einstein, using the ideas of quantum theory, regarded light as consisting of photons each having a discrete quantum of energy, E. The energy associated with each photon is proportional to the frequency of the electromagnetic radiation, and hence $E = h\nu$ where ν is the frequency, and h is Planck's constant.

Before an electron can be emitted from a material, that electron must acquire sufficient energy to move to the surface and then break through the surface forces that constitute a barrier there. Associated with the surface of every material there is a quantity known as the work function, ϕ. The work function is equal to the work done in removing an electron from the surface to infinity. In practice, a distance of a millimetre or two represents infinity on the atomic scale. When an electron at the surface acquires a quantity of energy, w, which is in excess of the work function, the excess is manifest as kinetic energy, the electron being emitted with a velocity v. Thus $w = e\phi + mv^2/2$ where e is the electronic charge, m its mass and ϕ the work function in volts. In thermionic emission, as we have seen when discussing valves, the energy is acquired from thermal sources. In a photocell the energy is obtained from the incident radiation. For a given emitting surface it can be seen that the incident photons must impart energy at least equal to the work function. If the frequency of the radiation is too low, i.e. the wavelength too long, then no photoemission can occur. Once the critical frequency has been exceeded emission is possible, and the number of electrons emitted is then proportional to the number of incident photons, i.e. to the intensity of the light. It is found empirically that

$$\frac{c}{\nu_0} = \lambda_0 = \frac{12 \cdot 4 \times 10^3}{\phi}$$

where ν_0 is the critical frequency, λ_0 the corresponding threshold wavelength in angstroms, ϕ is the work function in volts and c is the constant representing the velocity of light. For a clean copper surface the work function is $3 \cdot 9$ V. To obtain photoemission the threshold wavelength is about 3150 Å. Visible light, therefore, does not produce photoemission from a copper surface. Since most photocells must operate in the visible or infra-red regions of the spectrum, pure metals are never used because their work functions are too high. Although the alkali metals respond to visible light, emitting surfaces on these metals have efficiencies that are too low for practical purposes. The presence of atomic films of foreign elements can have a very great effect upon the emission. Practical emitters, therefore, use films whose chemical nature enhances emission considerably.

Photoemissive cells are diodes consisting of an anode and cathode sealed into a glass envelope that may be either highly evacuated or gas-filled. The cathode usually consists of a large area of photoemissive material deposited on a metal semicylinder; the anode is a wire placed along the axis of the cylinder. Light is caused to strike the inside of the semicylinder and the electrons emitted from the surface are collected by the wire anode, which is maintained at a positive voltage relative to the cathode. *Figure 12.1* shows a typical construction.

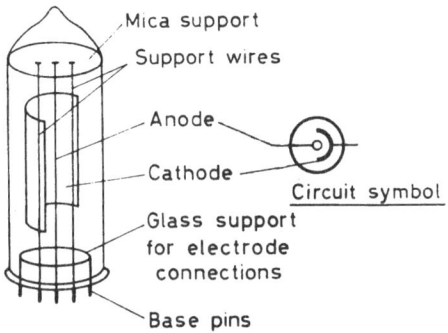

Figure 12.1. Typical construction of a photoemissive cell

The spectral response of the cell is determined by the nature of the cathodic surface. In practice three types of surface are in common use. The silver-oxygen-caesium (Ag-O-Cs) surface is produced by evaporating caesium from a bead onto a silver-plated cathode with caesium oxide present. The threshold wavelength for this cathode is about 12,000 Å (i.e. 1·2 μm). To conform with British Standards the unit of wavelength is now quoted in micrometres (μm) where 1 μm = 10^4 Å. The quantum yield is low (about 0·3 per cent) which means that on average 1,000 photons produce 3 electrons. Photocells using the Ag-O-Cs cathode are used for light containing a preponderance of red light, e.g. tungsten and other industrial light bulbs. For the blue end of the spectrum antimony-caesium (Sb-Cs) cathodes are used. These cathodes have a quantum yield of between 20 per cent and 30 per cent. The surface of the cathodes have a monatomic layer of caesium on an antimony base giving a peak response between 0·38 μm and 0·45 μm. These cells are best suited for use in photometry, colorimetry and daylight signals.

23*   355

No cell has a spectral response identical with that of the human eye, although attempts have been made to produce a surface whose response is very close to that of the eye. The Bi-O-Sb-Cs cathode, especially when used with appropriate filters, has a response that closely approximates that of the eye. *Figure 12.2* shows the response of the Ag-O-Cs and Sb-Cs cathodes along with that of the average human eye.

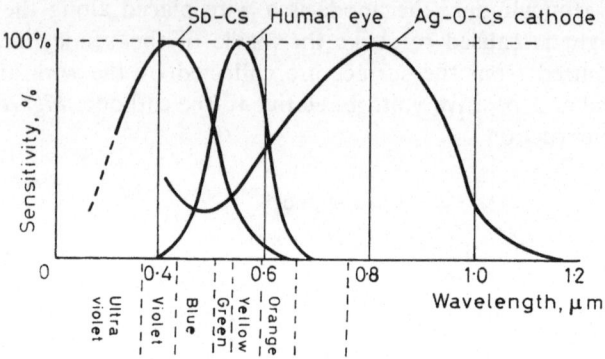

Figure 12.2. The spectral response of two commonly used cathodes compared with that of the human eye. The Sb-Cs response is shown dotted in the u.v. region. Usually this region is cut off by a conventional glass envelope. For response in the ultra-violet a quartz window must be used in the cell

The characteristic curves (i.e. the I_a/V_a graphs) depend upon whether the cell is a vacuum type or is gas-filled. *Figure 12.3* gives the curves obtained with two Mullard type photoemissive vacuum cells in common use; the curves are pentode like in their appearance. For a given light intensity the current rises rapidly for small increases in anode voltage. Further increases in the anode voltage bring about little change in the anode current since operation is now in the saturated region where almost every electron emitted is gathered by the anode. Load lines are constructed in the same way as for thermionic valves. Two points of interest should be noted: the anode loads are high, being of the order of megohms, and the linearity is good, i.e. for a given supply voltage and load, equal increments of light intensity give rise to equal increments of current. The voltage increments across the load are therefore proportional to changes in light intensity. This type of cell is consequently ideal for photometric work. Two additional advantages of the vacuum cell are (*a*) the relative immunity of the photoelectric current to changes in supply voltage (provided such voltage is high enough to produce saturation), and (*b*) the cell's rapid response. For a well-designed cell with low interelectrode capacitance, a response up to several megacycles per second

356

is possible. No other type of photocell has yet been developed that surpasses the vacuum photoemissive cell in speed of response.

The sensitivity of vacuum cells is low, being about 30 μA per lumen. This sensitivity can be increased about tenfold by introducing into the envelope a small quantity of inert gas. A common choice is argon at a pressure of about 1 mm of mercury. When the anode voltage is high enough (greater

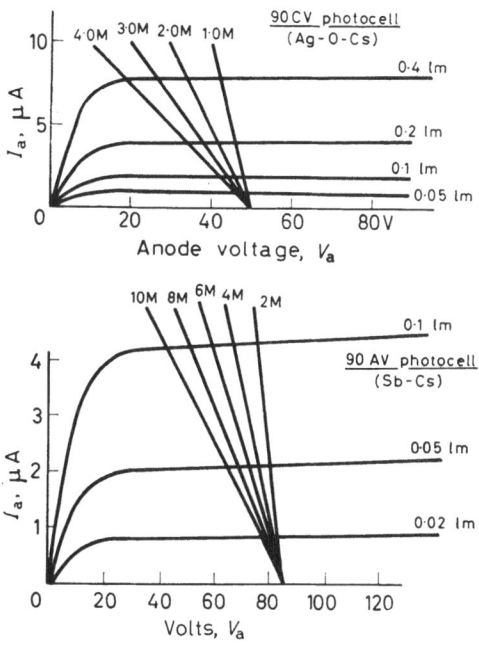

Figure 12.3. Characteristic curves for vacuum photocells (Mullard). The C and A letters refer to cathode types and the letter V indicates a vacuum cell

than about 20 V) a process known as gas multiplication takes place. An electron emitted from the cathode collides with a gas molecule with sufficient energy to ionize the molecule. Two electrons are then released which in turn ionize further molecules. In this way an amplification of the primary photocurrent is obtained. Although an increase in sensitivity is obtained, the I_a/V_a characteristics show a marked non-linearity. *Figure 12.4* shows the characteristics for a 90 CG Mullard gas-filled cell.

The frequency response of a gas-filled cell is poor. Above 1 kHz there is a considerable drop until at 10 kHz the response is almost zero. In spite of this, the high sensitivity has made this cell popular for sound reproduction in cinema film equipment; the high load resistors used with the cell make

357

distortion due to the non-linearity of the characteristics negligible. Gas-filled cells are also useful in industrial applications where high sensitivity is of greater importance than linear response.

The light parameter shown in *Figures 12.3* and *12.4* is measured in lumens and represents the total light flux falling on the cathode, not the illumination level. Some readers may experience confusion in connection with photometric

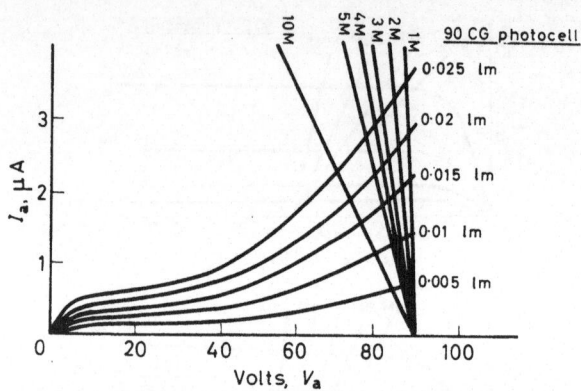

Figure 12.4. Characteristics for a Mullard gas-filled cell (90CG)

units. This is not surprising in view of the large number of units that have been proposed from time to time. Some useful references on photometry are given at the end of the chapter, but it may be helpful to define some commonly used units. The preferred units are laid down in British Standard 1991 : 1954. The intensity of a luminous source is measured in candelas (formerly candles) the defining source being platinum at its solidification temperature under specified conditions. The total luminous flux from a point source of 1 candela (cd) radiating equally in all directions is 4π lumens, that is 1 lumen per unit solid angle. Illumination is the luminous flux falling on a surface. The unit is the lux (lx) (1 lumen per square metre) or, alternatively, the phot (1 lumen per square centimentre). 1 foot-candle, although not a preferred unit, is 1 lumen per square foot.

The maximum life of a photoemissive cell is obtained by operating the device within the manufacturer's specified limits. In the case of gas-filled cells, the maximum anode voltage must not be exceeded, otherwise glow discharge may occur, in which case the tube is rapidly damaged. Maximum stability of the cells is obtained by running them well within their limits. Low values of light intensity and cell current should be used. Continuous operation of a photocell reduces the sensitivity temporarily, but, provided the cell has not been over-run, storage in the dark restores the sensitivity completely or almost so. Photocells should be stored in the dark when not in use.

Owing to the way in which the cathode surfaces are formed, the sensitivity is not uniform over the whole of the surface. It is usual, therefore, to make the beam of light operating the cell diverge so as to illuminate the whole of the cathode area. This is achieved by conventional optical means. When a cell is operating in a circuit that is energized, a small current flows even if the cell is in darkness. This current is known as the dark current and is negligibly small (e.g. a 90 AV cell has a dark current less than 0·1 μA).

Applications

The anode current of a photoemissive cell is very small and is unable to activate relays and other devices directly. The output of the cell is usually obtained as a voltage across a load resistance of several megohms. This voltage is then processed by later electronic circuitry. The output voltage may be predicted by drawing the load lines in the usual way as shown in *Figures 12.3* and *12.4*. In many applications the photocell load is the grid resistor of the valve used to amplify the signal. The high load imposes a restriction on the amplifying valve. Manufacturers state the maximum value of grid resistor permissible with their valves. If grid current is likely to be large then the grid resistance must be small. An electrometer valve is, there fore, the ideal one to use to amplify the output from a photocell. A semi-electrometer valve (such as an under-run EF37A or EF86) will also prove satisfactory. Mullard recommend, with their EF86, an under-run heater obtained by using 5·4 V as the heater supply. $V_a = V_{g2} = 75$ V, $R_a = 220$ k, $R_{g2} = 1$ M, $R_K = 22$ k. The supply voltage should be 93 V. With an I_a of 100 μA, $V_{g1} \approx -2·0$ V.

Although these conditions are ideal, satisfactory operation can often be obtained with conventional amplifiers. The absolute maximum sensitivity is not then available, since somewhat lower values of photocell load resistor must be used.

Figure 12.5 shows a photocell with a straightforward single-stage pentode amplifier.

A simple photoelectric relay circuit using a 90 CG photocell is shown in *Figure 12.6*. In the absence of illumination the grid of the thyratron is held ne ga-tive with respect to the cathode during the half-cycle that the anode is positi ve. Throughout the cycle, therefore, the thyratron is held in the non-conduction state. As soon as the cell is illuminated the photocell current causes a voltage drop across the 1 M resistor and the grid of the thyratron is positive with respect to cathode when the anode is also positive going. The thyratron therefore fires and energizes the relay. The 1 μF capacitor across the relay holds the latter on during the non-conducting half-cycle. The light level at which the circuit operates is determined by the setting of the 5 k variable resistor. *Figure 12.7* shows a similar circuit in which, by a slight modification of the circuit of *Figure 12.6*, the relay is off whilst the cell is illuminated, and on when the cell is in darkness. This circuit uses a vacuum cell (90 CV).

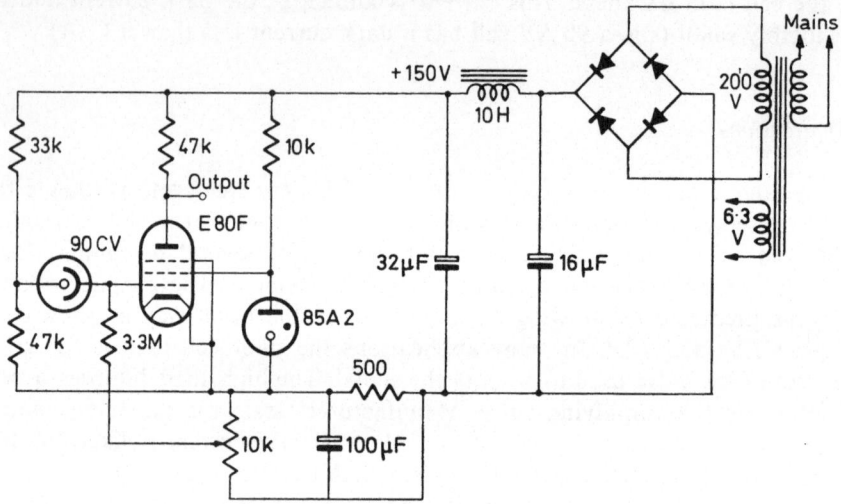

Figure 12.5. A vacuum cell with single-stage amplifier. The screen of the E80F is held at a steady voltage by the stabilizing tube 85A2

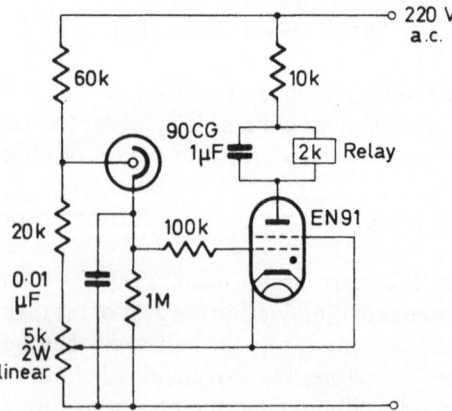

Figure 12.6. A simple photoelectric relay circuit in which the thyratron activates the relay when the cell is illuminated

360

Both circuits are for use on a.c. only. If supplied by steady voltages loss of control of the thyratron would occur once the thyratron had fired.

Photoelectric relays may be designed to incorporate hard valves (i.e. vacuum types) rather than gas-filled types. *Figure 12.8* shows a relay circuit in which a hard valve is used. On illuminating the cell the E80L is biased to beyond the cut-off value when the anode is positive-going. On interrupting the beam the bias is removed and conduction occurs during the half-cyles that the

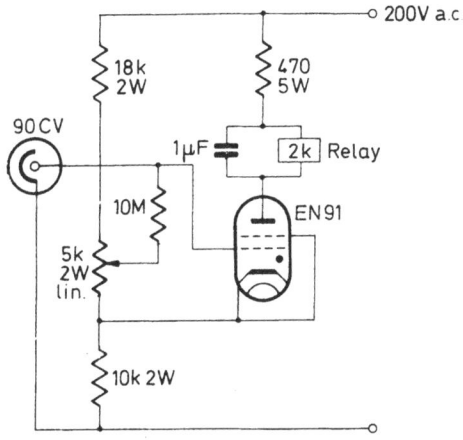

Figure 12.7. Modification to the circuit of *Figure 12.6* to enable the thyratron to be held in non-conduction while the photocell, a vacuum type, is illuminated

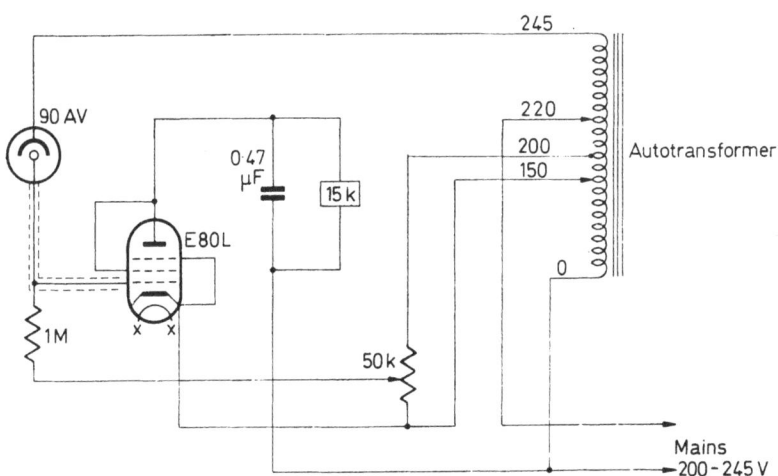

Figure 12.8. A photoelectric relay circuit using a hard valve to operate the relay

361

anode is positive. The relay is then activated. To prevent 'chatter' a capacitor is wired in parallel with the relay coil. ('Chatter' is the vibration caused by feeding the relay coil with current pulses.)

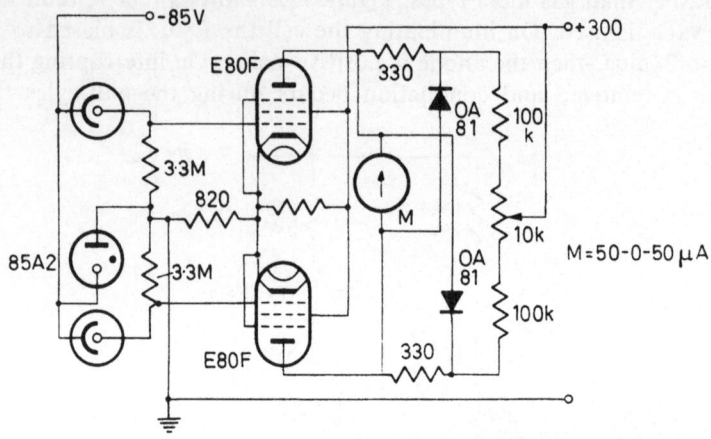

Figure 12.9. Colour and brightness comparator. Cells may be 90 CV or 90 AV depending on whether red or blue sensitivity is needed

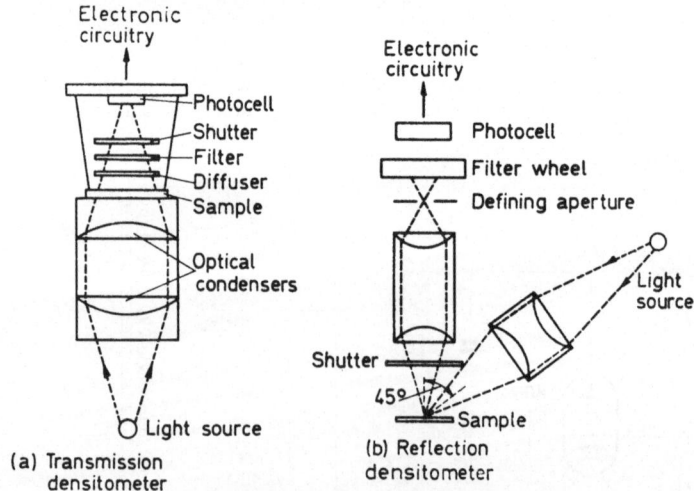

Figure 12.10. Principles of the transmission and reflection densitometers of Baldwin Instrument Co. Ltd.

Figure 12.9 shows two photocells in a balanced pair arrangement. The circuit may be used for colour or brightness comparison, smoke detection, absorptiometers, or any other application requiring a signal when equality of the light beams is upset. Although a meter is shown in the diagram this may well be replaced by an alarm system or a servomechanism. *Figure 12.10* shows the principle of the use of vacuum cells as transmission and reflection densitometers.

Photomultiplier Tubes

Photomultiplier tubes are photoemissive tubes of great sensitivity. Strictly they are misnamed, since multiplication of light does not occur. This type of tube should be called an electron-multiplier tube. However, the term 'photomultiplier' is in common use.

Electron-multipliers have photoemissive cathodes which produce electrons in the usual way. These electrons are then accelarated by a system of anodes, called dynodes. Each dynode along the system is maintained at a progressively higher voltage, the usual voltage between adjacent dynodes being about 100 V. Each dynode is coated so that when a primary electron strikes the electrode secondary electrons are emitted. Between two and three electrons on average are produced by the impact of a single electron. Since the tube may have between 10 and 13 stages it will be realized that considerable electron multiplication occurs. The sensitivity of the tube may be between 500 and 2,000 A per lumen. The maximum current collected by the final anode is, however, quite small, being of the order of a few milliamperes.

To prevent the electrons emitted from the cathode and adjacent dynodes from travelling directly to the final anode, the dynodes are specially shaped. Two common arrangements are shown in *Figures 12.11* and *12.12*. In the

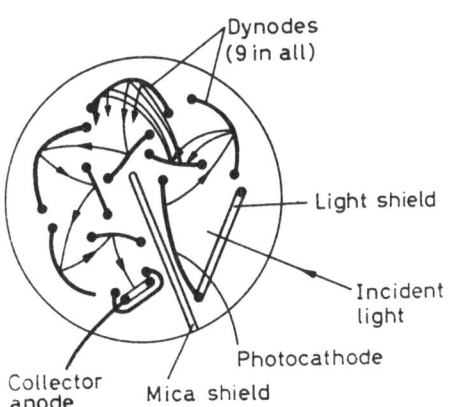

Figure 12.11. Structure of a circular photomultiplier tube

circular type of electron multiplier, the dynodes are specially shaped so as to produce electric fields which direct the electrons to each dynode in turn. In the 'venetian blind' arrangement each dynode consists of a layer of electrodes inclined so as to produce electron paths as shown in *Figure 12.12*.

The advantages of the electron multiplier lie in the very high sensitivities which are obtained without recourse to an associated high-gain amplifier. The signal-to-noise ratio is, therefore, much better than it would be if convention-al amplifiers were used. The linearity, stability and frequency response are

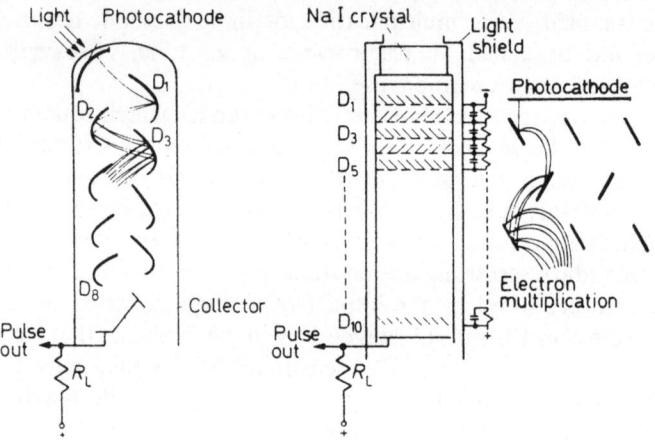

Figure 12.12. Alternative photomultiplier structures. That on the right is frequently used to detect nuclear radiations. The particle, on entering the NaI crystal, causes a flash of light which initiates the release of the primary electron. Multiplication then takes place as the secondary electrons proceed down the dynode arrangements. Each dynode is maintained at the appropriate positive potential by a stabilized power supply

very good. Typical applications are the counting of particles in nuclear radia-tion studies (where the particle gives rise to a scintillation in a suitable crystal e.g. sodium iodide), and the molecular weight determinations of polymers using the very low intensity light scatterings from a solution of the polymer in an appropriate solvent. Polymer Consultant Ltd. of Colchester manufacture an instrument for the latter purpose.

PHOTOCONDUCTIVE CELLS

We have seen in Chapter 5 that one way of interpreting the behaviour of semiconducting materials is to consider energy-level diagrams. If somehow an electron can acquire sufficient energy to overcome the forbidden energy gap between the valence and conduction bands, then that electron is available for conduction; electron-hole pairs are formed and this results in an increase in carrier density manifested as an increase in the conductivity of the material. For photoconductive cells, the source of the required energy is electromagnetic radiation in the visible or infra-red part of the spectrum. The quantum energy, $h\nu$, in this region of the spectrum is sufficient to raise the electrons into the conduction band. Since such energy is smaller than that needed to effect electron emission from the surface, photoconductive cells have a higher quantum yield than any photoemissive cell, and can respond to radiations of much longer wavelength.

Devices that vary their electrical resistance in accordance with the incident radiant energy have been known for a long time. Selenium was frequently used as a standard high resistance in the bridge measurements of the resistances of submarine and other telecommunication cables. Variable results were obtained depending upon whether the standard resistance was in darkness or illuminated. Interest in the photoconductive process led to the investigation of several chemical compounds that exhibit the phenomenon. Because of the high quantum yield and consequent high sensitivity, we find that the output of photoconductive cells is measured in milliamps rather than the microamps produced by photoemissive cells. Infra-red sources can be detected over distances of several miles, a fact that lead to the development of photoconductive cells during the Second World War for aircraft detection at night.

At present the compounds that have been found to be most useful are cadmium sulphide, lead sulphide, lead selenide, lead telluride and indium antimonide. These compounds have peak responses in that order from the visible part of the spectrum to the far end of the near infra-red. Indium antimonide has a useful response out to 8 μm. Other types of cell, using different materials, are under development and no doubt the range covered by photoconductive cells will be improved and extended.

The usual construction of a photoconductive cell is to deposit specially prepared material as a layer a few microns thick onto the electrodes as shown

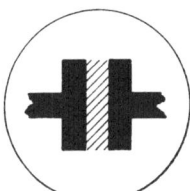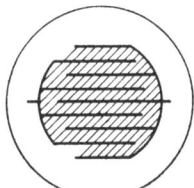

Figure 12.13. End-on views of the electrode structure of two forms of photoconductive cell. The semiconductor material is shown shaded between the metallic contacts. The structure on the left is used for spectroscopic work

in *Figure 12.13*. The arrangement on the left is intended for spectroscopic work, whilst that on the right is for general work. The substrates are usually of glass, and the whole cell is sealed into a protective mount.

Applications of Photoconductive Cells

In industrial locations where red light (from tungsten filament bulbs) or infra-red radiation is used, this type of cell may be employed in flame failure detection, smoke detection, burglar alarms, the detection of overheating in bearings and other machinery parts, etc. For scientific use photoconductive cells are found in apparatus used for astronomy, chemical analysis and measurements. The similarity between the spectral response of cadmium sulphide and that of the eye and certain common photographic emulsions has lead to the use of cadmium sulphide cells as light or exposure meters in photography. Their small size has enabled them to be included in the camera body; their high sensitivity is resulting in the displacement of the photovoltaic type of exposure meter.

Photoconductive cells can operate on low voltages, and, because of their sensitivity, relays may be energized without recourse to intermediate amplifiers. The cells are relatively inexpensive, are shock and vibration proof, and have a long life (> 10,000 hours). They are, however, easily damaged by overheating, and strong light adversely affects their performance.

Photoconductive cells are valuable as the detecting agencies in infra-red spectrometers. Unfortunately, their development is not sufficiently advanced to allow them to replace thermopiles and bolometers. The response of these

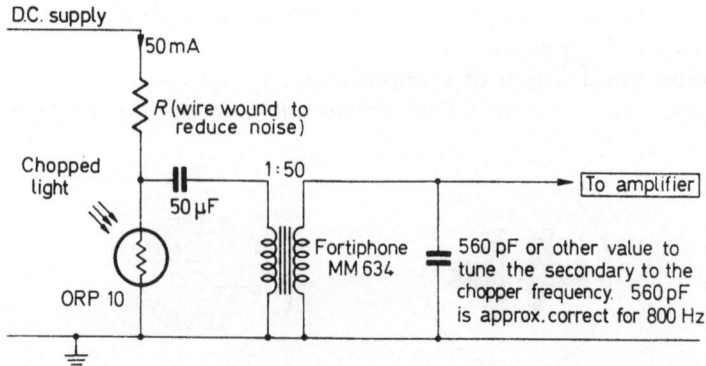

Figure 12.14a

Figure 12.14. Circuit for use with the Mullard ORP10 photoconductive cell in the measurement of radiation out to 8 μm

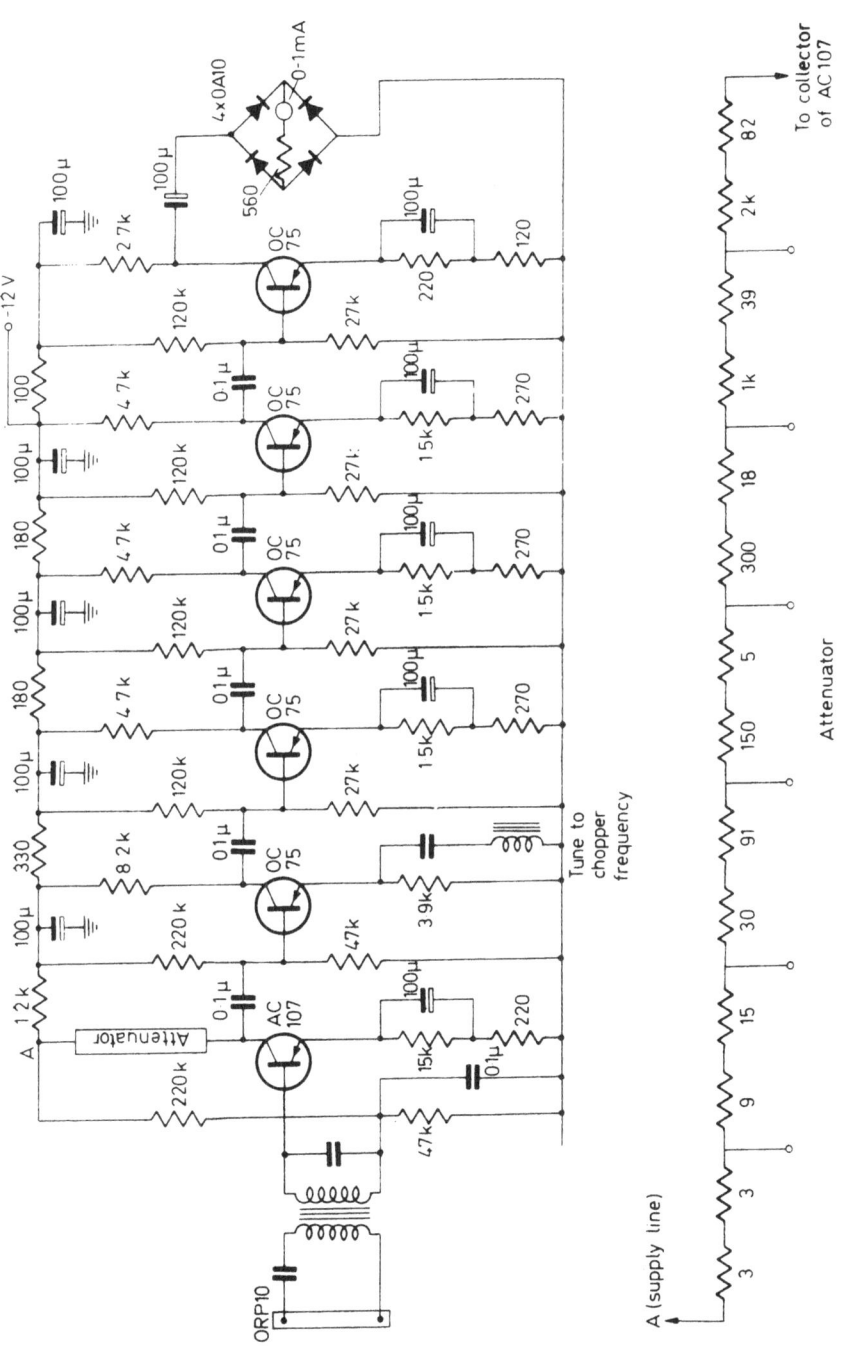

Figure 12.14b

latter devices is satisfactory out to 25 μm, whereas the photoconductive cell can be used in measurements only to about 8 μm. Research is likely to extend this range however. One circuit for use with the Mullard indium antimonide cell type ORP10 is shown in *Figure 12.14*. The low level a.c. amplifier is used to amplify the output from a cell that receives its radiation in the form of pulses from a mechanical chopper. The energy-chopping frequency of 800 Hz is recommended. The maximum sensitivity is 0·5 μV for full-scale deflection and the noise referred to the input is 0·07 μV. The ORP10 responds to incident energy from the visible band out to 8 μm.

PHOTOVOLTAIC CELLS

In this type of cell illumination falling on the device generates an e.m.f. between the cell's terminals. The phenomenon is known as the photovoltaic effect. The cell consists of a metal–semiconductor or semiconductor–semiconductor arrangement; it is the potential barrier which is formed at the respective junctions that provides the mechanism for the generation of the e.m.f. within the cell. In the metal–semiconductor types the semiconductors used are usually selenium or cuprous oxide, both of which are *p*-type. *Figure 12.15* shows the arrangement in a selenium photocell. During the formation

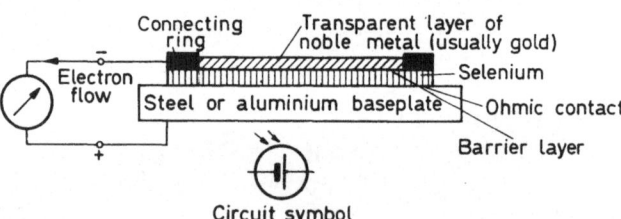

Figure 12.15. Construction of a selenium photovoltaic cell. The contact between the selenium and the baseplate is an ohmic one (i.e. no barrier layer is formed at the interface). The selenium and the baseplate are therefore electrically continuous. Positive and negative signs associated with the terminals predict the direction of current flow in the external circuit. The *p*-type selenium is however negatively charged (see text)

of the cell electrons diffuse from the metal (i.e. gold) into the *p*-type material filling the impurity acceptor centres near to the junction. In this way a barrier layer is created. The height of the potential hill associated with the barrier layer increases until eventually no further electrons can diffuse into the semiconductor. The metal, therefore, becomes positively charged and the semiconductor becomes negatively charged. The incidence of photons on the barrier layer raises some of the valence electrons there to the conduction band

and electron-hole pairs are formed. The electric field that exists across the barrier sweeps the electrons into the metal and the holes into the semiconductor. If an external circuit exists the electrons travel around the external circuit from the metal to the semiconductor. The barrier layer prevents them from travelling from the metal to the semiconductor within the cell. The mechanism for the cuprous oxide cell is similar. Initially, in the absence of illumination, the copper becomes positively charged owing to the diffusion of electrons from the metal to the acceptor impurity centres in the oxide semiconductor. In the presence of illumination electron-hole pairs are formed; the electrons travel into the metal around the external circuit and back to the semiconductor.

Confusion in understanding the mechanism may sometimes arise since the electrons are travelling towards a negatively charged region. We must remember, however, that for a given metal-to-semiconductor arrangement, and at a given temperature, there is always an attempt to preserve the height of the 'potential hill.' Any attempt at lowering the potential brings into play a mechanism to regain the former position. In the case of photovoltaic cells the electron-hole pairs produced by the incident photons reduce the potential hill. Electrons, therefore, travel around the external circuit in an attempt to return the potential difference across the barrier layer to its former value. The energy to maintain this flow is derived from the incident radiation. From the external circuit's point of view, therefore, in a selenium cell, the terminal connected to the selenium is the positive one, and the terminal connected to the gold layer is the negative one. In the case of the copper/copper oxide cell the cuprous oxide is the positive terminal, the metallic copper being the negative terminal. The currents are thus in the opposite directions to those observed when the devices are used as rectifiers.

The details of manufacturing processes are often carefully guarded trade secrets since good operation depends so much on empirical factors. In general, the selenium cell is made by coating the roughened steel or aluminium baseplate with bismuth or some other metal that has been found to form a good ohmic contact with the selenium. A barrier layer is thus not formed at this junction. The selenium is laid down in vacuum, and later annealed so that the original semi-amorphous material is converted into its crystalline allotrope. The counter electrode of an inert metal, usually gold, is then sputtered on; the external connections are then made. This type of cell is often called a front-wall cell since the light, after travelling through the semi-transparent gold layer strikes the barrier layer immediately. The cuprous oxide/copper cell is an example of a back-wall cell in which the light must travel through the semi-conducting material before entering the barrier layer. The latter type of cell is not very responsive to the blue end of the spectrum since the shorter wave-length radiations are absorbed in the semiconductor.

A modification to the selenium cell as outlined above is frequently used today. The counter electrode is formed by the vacuum deposition of cadmium or cadmium oxide. The barrier layer is thought to be formed by the creation of cadmium selenide at the interface. A semiconductor *pn* junction is thus form-ed. The mechanism which provides the photovoltaic current can easily be deduced from the explanations given above. An ultra-thin gold deposit is frequently used to increase the conductivity of the cadmium layer.

The spectral response of the selenium cell is similar to that of the eye, which is why this type of cell is frequently used in colour comparators. For a closer

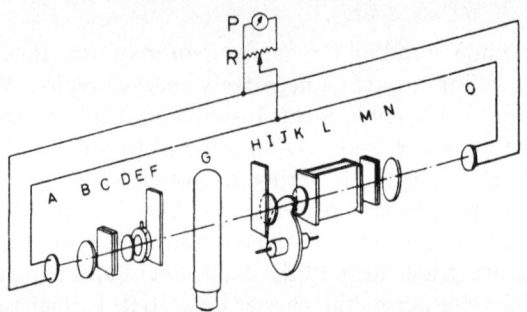

Figure 12.16. The Spekker absorptiometer optical system and photocell circuit. Reproduced by courtesy of Hilger & Watts Ltd.

A and O, photocells	I, window
B, window	J, variable shutter
C, filter	K, lens
D, lens	L, cell for liquids
E, iris diaphragm	M, filter
F, heat absorbing filter	N, lens
G, lamp	P, spot galvanometer
H, heat absorbing filter	R, sensitivity control

approximation to the eye's response an Ilford filter type 827 may be used in front of the cell. The response of a front-walled cell is maintained at the ultra-violet end of the spectrum.

The Spekker absorptiometer and fluorimeter are two items of chemical analysis apparatus that use this type of cell. Two cells are used in conjunction with other equipment as shown diagramatically in *Figure 12.16*. The outputs of the cells are fed, in opposition, to a galvanometer which is shunted to pre-vent overloading. By using two cells, errors due to lamp output variations are reduced. The single-cell absorptiometer is particularly susceptible to this form of error unless apparatus is used to maintain the lamp supply constant. Even then ageing of the lamp is unavoidable. Light from the lamp passes in one direction through a heat absorbing filter, an iris diaphragm and a colour

370

filter, which has a maximum transmission near the absorption peak of the solution. The light is then focussed on to the compensating photocell, A. In the other direction the light, after passing through a heat absorbing filter and iris diaphragm, passes a calibrated measuring disc and then through the absorbing liquid. A similar colour filter and optical system are used. Equality of output from the cells is arranged by setting the measuring disc, J, fully open indicating zero density. The cell containing the liquid is at this time in the beam and the uncalibrated iris diaphragm, E, is adjusted for zero deflection of the galvanometer. The absorption cell is then removed from the beam.

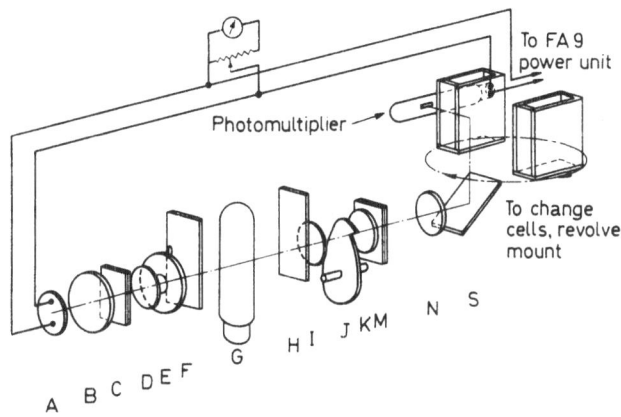

Colour filters can be inserted between liquid cells and photocell or at M

Figure 12.17. The modifications necessary to convert the absorptiometer into a fluorimeter. Reproduced by courtesy of Hilger & Watts Ltd.

A,	photocell	H,	heat absorbing filter (when in use)
B,	window	I,	window,
C,	filter	J,	variable shutter
D,	lens	K,	lens
E,	iris diaphragm	M,	filter (u.v.)
F,	heat absorbing filter (when in use)	N,	lens
G,	mercury lamp	S,	reflector

More light now falls on photocell, O, thus producing an out-of-balance current in the galvanometer. Equality of the photocell outputs is restored by adjusting the calibrated measuring disc. The optical density of the liquid is then read off from the disc markings.

The modifications to convert the absorptiometer into a fluorimeter are shown in *Figure 12.17*. Since the photocell A is easily fatigued, a special lamp housing (not shown on the diagram) is used. A pair of shutters in the

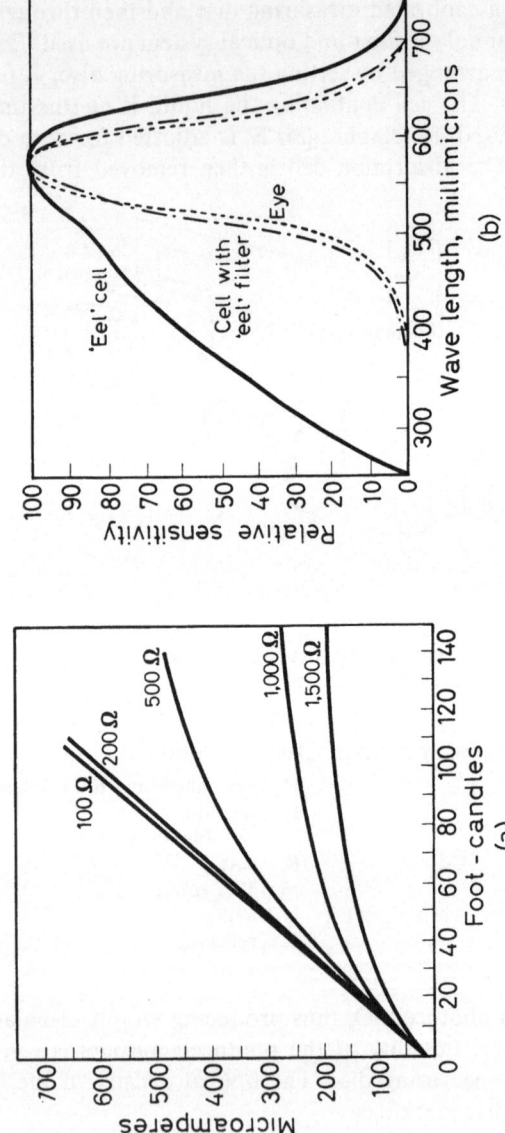

Figure 12.18. Characteristics of a selenium photocell. (*a*) shows the current output as a function of incident illumination for various load resistors. Only when the external load resistance is low is there good linearity. (*b*) shows the spectral response of a selenium photocell and the correction, with a suitable filter, to make it correspond to the response of the eye. Both diagrams are reproduced by courtesy of Evans Electroselenium Ltd., and refer to their cells

housing prevents the cell and the photomultiplier from being illuminated. When a reading is to be taken both shutters are opened simultaneously.

One advantage of the photovoltaic cell is that no power supply is required for energizing purposes since the cell generates its own output. The voltage from the cell is not a linear function of the light intensity. The current output, however, is almost linear provided the external circuit has a low resistance. *Figure 12.18* shows the relevant graphs. The current, therefore, may be used to operate a sensitive moving-coil microammeter directly. Many commercial light meters are merely photovoltaic cells connected to a meter that has a full-scale deflection of a few tens of microamperes. The meter is suitably calibrated for photographic purposes. The output is high when the cell is subject to blue-rich skylight. There is thus a spectral match with modern panchromatic films. For use in the laboratory the output from the cell can be used to operate a mirror galvanometer of the type described in Chapter 6 on measuring instruments (Pye Scalamp Galvanometer). Spectrophotometers, reflectometers, opacimeters and meters measuring the gloss of surfaces are instruments that use a galvanometer and appropriate optical arrangement with a photovoltaic cell. In the spectrophotometer head of Evans Electroselenium Ltd., the light from the lamp passes through a heat-absorbing glass and is focussed by a lens to form an approximately circular spot of light on the sample. Light reflected from the sample falls on the photocell and the current generated by the photocell is passed to the galvanometer unit. Data regarding the colour of a surface is obtained by obtaining an 'abridged' spectrum curve by measuring the reflectance in a limited number of regions of the spectrum which are isolated in turn by the use of narrow-band optical filters.

Phototransistors are devices that combine the photovoltaic effect with transistor amplification. Illumination of the base region creates electron-hole pairs there. The holes are swept into the collector region in a *pnp* phototransistor. The electrons are equivalent to the base signal current in a conventional transistor amplifier, and current amplification occurs as way described in Chapter 5. The basic circuits used with a phototransistor are shown in *Figure 12.19* along with the characteristics for the Mullard OCP71. The phototransistor has the advantages of small size, low power consumption and long life. Used with a supply voltage of 12 V, relay coils may be operated directly.

Silicon photovoltaic *pn* junctions have been described in the chapter on power supplies and therefore need no further description. Miniature *pn* junction cells with a tiny single lens optical system are often used as read-out cells in computers and data processing equipment. Here light passing through the punched tape energizes the cell, and the output is then processed in the associated electronic equipment.

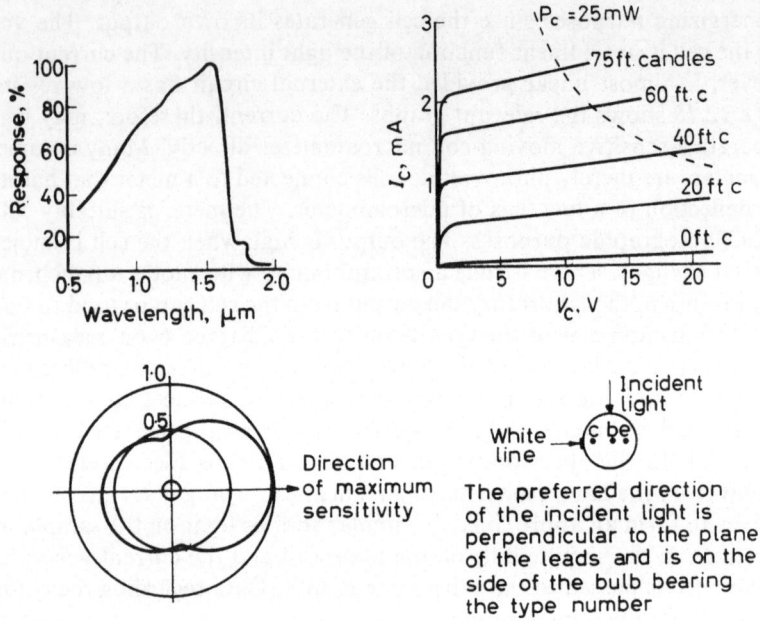

(a) Spectral response, I_c/V_c characteristics and polar diagram for OCP71 phototransistor

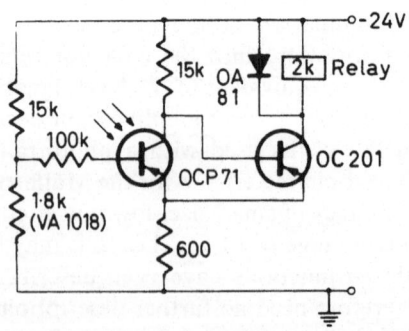

(b) A temperature-compensated photoelectric relay published by Mullard Ltd., for their OCP71 phototransistor. The VA 1018 is a thermistor

Figure 12.19. Details relating to a Mullard phototransistor

374

SUGGESTIONS FOR FURTHER READING

Useful publications on photometry

1. *Photometry* by J. W. T. Walsh, Constable, 3rd edn., 1958. A standard textbook with useful tables and definitions.
2. Kaye and Laby *Tables of Physical Constants*, 12th edn., Longmans, Green, 1959.
3. BS 233 : 1953 and BS 1991 : 1954.
4. *Units and Standards of Measurement Employed at the National Physical Laboratory: Part II Light (Photometry, Colorimetry and Radiometry).* H.M.S.O., 1952.

13

MISCELLANEOUS MEASURING
INSTRUMENTS

In this chapter we shall discuss the fundamental operating principles of some of the instruments commonly used by scientific workers. The range of measuring instruments employed in scientific investigations is very large, and it is not possible to discuss here every instrument that could be used. A selection has therefore been made of those instruments that are most likely to be used, and whose operation depends upon the electronic principles described in this book.

VALVE-VOLTMETERS

When measuring voltage in electronic circuits there are two factors that must receive special attention; firstly, a knowledge of the waveform of the voltage to be measured is required, and secondly, the impedance across which the voltage is developed must be known. Without these two items of information, any measurement is valueless.

Voltmeters that use pointer instruments as indicators are usually calibrated in terms of r.m.s. values of sine wave voltages; calibrations in terms of mean and peak values are also used. Any measurements made without a knowledge of the waveforms involved are worthless since the calibrations are valid only for sinusoidal waveforms. Where the waveform departs seriously from that of a sine wave, or is unknown, the safest course is to use a cathode-ray oscilloscope; the oscilloscope is then being used as a valve-voltmeter. With waveforms that are sinusoidal, or nearly so, it is cheaper and more convenient to use a pointer instrument.

Voltages in electronic circuits are frequently developed across high impedances, so if the circuit under examination is not to be unduly disturbed, the impedance of the measuring instrument must be very high indeed. The simple moving-coil voltmeter of the type discussed in Chapter 6 may have a resistance in the region of 20 k per volt if the meter movement is sensitive; the 100 V d.c. range then has a resistance of 2 M. Using a corresponding alternating voltage range the meter has a resistance that is less than 2 M. In fact many alternating voltage meters have resistances of only a few kilohms per volt. For many measurements such resistances prove to be satisfactory, but in electronic apparatus there are several occasions when the impedance of the voltmeter must be very much higher than the figures quoted above. Under these circumstances it is necessary to use a valve-voltmeter.

The pointer movement and the linear scale of a moving-coil meter make it attractive to retain this device as an indicating agency. It is necessary, however, to interpose some sort of impedance matching arrangement between the meter and the circuit under examination so that an extremely high impedance is presented to the circuit, thus causing negligible disturbance, whilst at the same time ensuring sufficient available power to drive the meter movement. The impedance matching arrangement consists of a suitably designed electronic amplifier; the combination of the electronic circuit and the moving-coil meter constitutes a valve-voltmeter.

Thermionic valves are frequently used in valve-voltmeters because of the ease with which a high input impedance and low output impedance can be arranged. This is, of course, the origin of the term 'valve-voltmeter'. High input impedance meters can be constructed using conventional transistors, but as yet it is not possible to devise transistor circuits with input impedances as high as those attainable with thermionic valves. Field-effect transistors are, however, able to compete with thermionic valves, and many newer voltmeters are using field-effect devices followed by conventional transistor circuitry. Several firms find it convenient to retain the term 'valve-voltmeter' for transistorized voltmeters. The use of the term is not incorrect since a transistor may be regarded as a crystal valve.

The simplest type of valve-voltmeter is shown in *Figure 13.1a*. It is merely a cathode-follower in which the output voltage is measured across the cathode load resistor. Although such a circuit meets the requirement of a high input impedance, it is unsatisfactory for measuring steady voltages because even when the input voltage is zero there is a steady output voltage; this voltage is due to the quiescent current through the load resistor. The circuit would operate with alternating voltages if (*a*) the steady voltage component were eliminated by having a series capacitor in the output line, and (*b*) the meter were replaced by an a.c. type (i.e. one using rectifiers.) The calibration, however, would depend upon such variable factors as the ageing of components and changes in the h.t. and heater voltages.

A better arrangement is to use a second valve to back off the steady component. In the circuit of *Figure 13.1b* two identical valves are used together with equal cathode resistors; the output voltage is developed between the cathodes. One of the grids is held at earth or zero potential whilst the signal voltage to be measured is presented to the other grid. In the absence of any input signal, the quiescent currents are equal. Slight mismatches between th valves and cathode resistors may give rise to a small output voltage when th input voltage is zero; the effect is eliminated however by adjusting the tap-

377

ping on the variable resistor in the anode loads. This variable resistor is in effect a 'zero set' control. If, after adjustment of the 'zero set' control, the h.t. or heater supply voltages vary, both valves are affected equally and the zero output condition is maintained. When a steady signal voltage is applied, or a signal having a steady mean value, the current in V1 changes. The voltage across the cathode resistor must also change and this results in a voltage

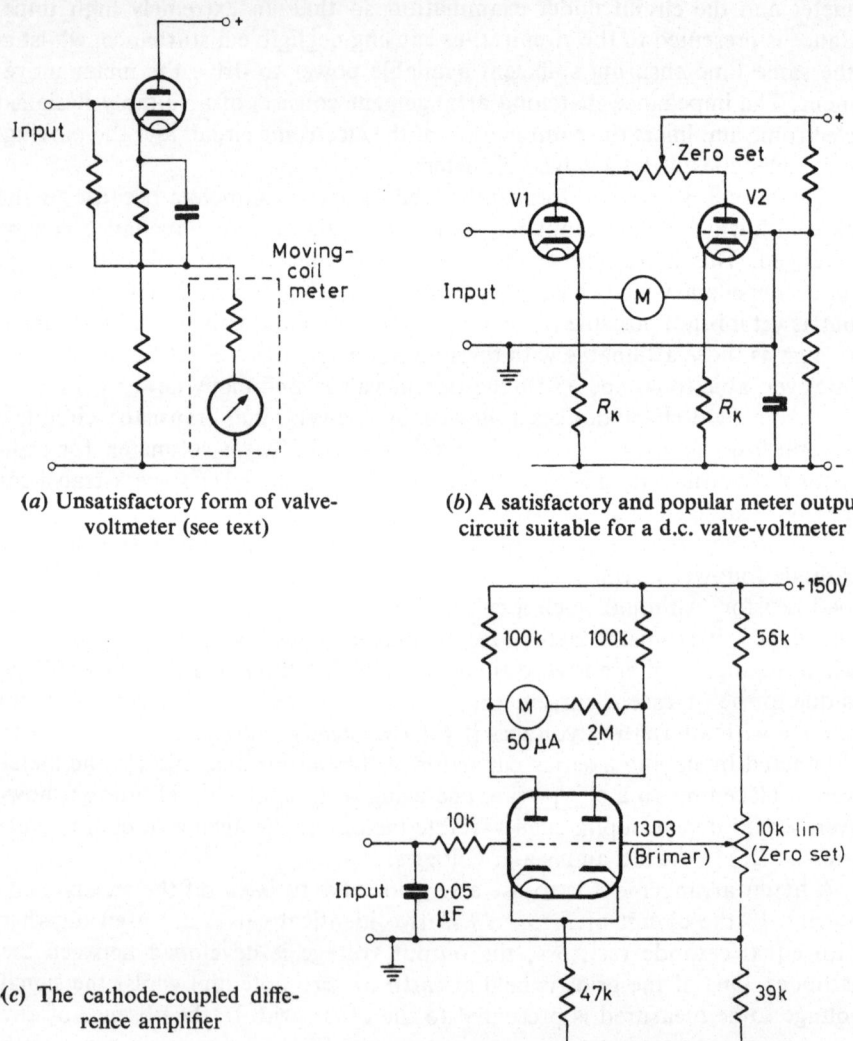

(a) Unsatisfactory form of valve-voltmeter (see text)

(b) A satisfactory and popular meter output circuit suitable for a d.c. valve-voltmeter

(c) The cathode-coupled difference amplifier

Figure 13.1. Various forms of valve-voltmeter. (*b*) and (*c*) are suitable only for steady voltages

378

difference between the two cathodes. This voltage difference is measured by the moving-coil meter, and is a linear function of the input voltage when the operating conditions have been correctly chosen.

An alternative to the cathode-follower amplifier of *Figure 13.1b* is the cathode-coupled difference amplifier shown in *Figure 13.1c*. This latter amplifier was discussed in Chapter 9; *Figure 9.51* and the related text showed that the voltage difference between the two anodes is directly proportional to the input voltage. With this circuit the advantages of using a balanced pair are retained in that the output voltage is not affected by changes in the supply voltages or ageing of the valve.

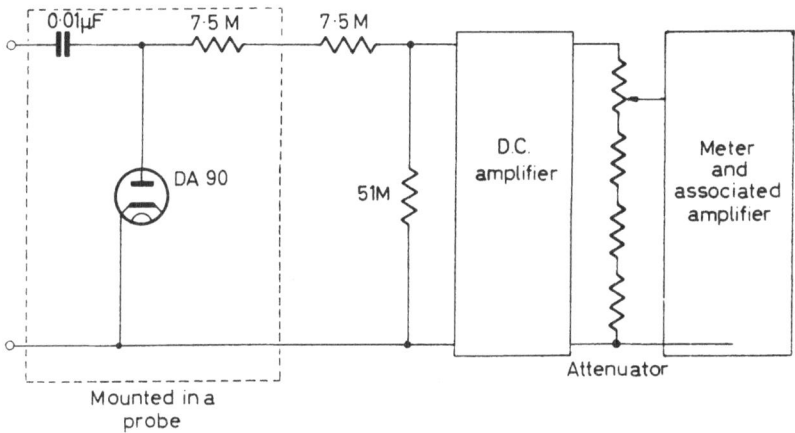

Figure 13.2. Basic circuit for a valve-voltmeter for use with alternating voltages

When measuring alternating voltages with a valve-voltmeter, some means of rectification must be provided. Rectifiers associated with the meter itself are not always satisfactory because of their limited frequency response. Only if the response is in the audio range will this method suffice. Various circuits use the rectifying properties of an amplifying valve (as, for example, in a reflex valve-voltmeter where the valve is biased almost to cut-off) but most manufacturers prefer to have a separate rectifying diode for the purpose. A small rectifying diode can easily be housed in a probe and brought right up to the circuit under examination. The frequency response of the small rectifiers used in this connection is excellent, extending up to hundreds of MHz. The basic valve-voltmeter circuit used for alternating voltages is shown in *Figure 13.2*. A diode rectifier and smoothing circuit give a steady output voltage that is a function of the input alternating voltage. This steady output voltage is then fed to a d.c. amplifier and meter stage similar to that of *Figure 13.1c*.

379

For sinewave input voltages the output from the rectifier is proportional to the peak value. The scale of the meter, however, is engraved for r.m.s. values since these are most commonly required. The presence of the rectifying section seriously reduces the input impedance of the valve-voltmeter. Whereas for steady voltages a valve-voltmeter may have an input impedance of 100 M, the corresponding input impedance for alternating voltages is often only a few megohms in parallel with 5–10 pF. However, for many purposes, this proves to be adequate. Continual development work has raised the input impedance on a.c. ranges of recent instruments to the region of 10 M, and further improvements are expected in the future.

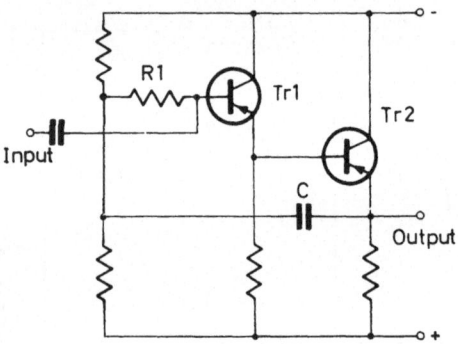

Figure 13.3. High input impedance stage for transistor voltmeters

Where measurements are to be taken of voltages of the order of millivolts then intermediate amplification must be employed. This is quite easily arranged. A single output stage, as in *Figure 13.1c*, is preceded by one or more balanced-pair stages. The voltmeter then consists of a rectifying input stage, a multistage directly-coupled balanced amplified and the balanced output stage.

Where the highest input impedance is not required voltmeters incorporating transistors are attractive. Provided the input stage is properly designed the input impedance can be at least 1 M with conventional transistors and considerably above this when field-effect devices are used. We must, however, tolerate a reduction in the frequency range of operation. This is due mainly to the input capacitance associated with the transistors. We have already seen that a Darlington pair (p. 251 and *Figure 9.28*) can provide a high input impedance. *Figure 13.3* shows the principle of an input stage suitable for a voltmeter. The function of C is to provide 'bootstrap' feedback. When the end of R1 connected to the base is driven negative by the signal, the emitters of Tr1 and Tr2 are also driven negative since they are connected in the common-emitter mode. The end of R1 remote from the base is, therefore, also driven negative so that the change of current through R1 is reduced. R1 therefore appears to be greater than it really is. We have met the same phenomenon when explaining the high input impedance of the cathode-follower. The name 'bootstrap' comes from the fact that the potentials at both ends of R1 rise simultaneously. These potentials are thus 'raised by their own bootstraps'. An alternative way of obtaining a high input impedance is to use a field-effect transistor in a circuit similar to that given in *Figure 9.34*.

For those who wish to build their own a.c. millivoltmeters, a design is given in *Wireless World* (June 1964, p. 264) which uses a Darlington pair input stage followed by an amplifier and metering circuit. A calibration multivibrator is added as an additional facility. The input impedance is 1 M plus 60 pF at the 3 mV range rising to 10 M plus 18 pF for input voltages greater than 3 V. There are 11 ranges going from 3 mV f.s.d. to 300 V f.s.d. and the cost of building is said to be £12 (in 1964). Full constructional details are given in the article.

Many other designs have been published over the years. Details of some of those that may be of interest to the reader can be obtained by consulting the references given at the end of the chapter.

Several laboratory instruments incorporate valve-voltmeters as the detecting and measuring device. Amongst these are the wave analyser, the Q-meter and the harmonic distortion meter.

The Wave Analyser

It is often of importance to know the composition of a complex periodic waveform in terms of the amplitude of the fundamental and the relative amplitudes of the harmonics. An instrument for making the necessary measurements is called a wave analyser. Commercially available analysers depend upon one of two principles.

In one type of analyser a valve-voltmeter is used, but instead of arranging a flat frequency response for the amplifiers, tuned circuits are employed. By varying the values of the components in the tuned circuits we can 'tune in' to the various harmonics of a complex periodic signal in much the same way as we select specific frequencies when we tune in to a radio transmitting station using a tuned r.f. receiver. The principle is shown in *Figure 13.4a*. The tuned amplifier consists of a highly selective frequency discriminating feedback amplifier that uses RC elements in the feedback line. The principles of this type of tuned amplifier have already been discussed in Chapter 9. The output from the amplifier is then measured with a valve-voltmeter circuit. Often the input signal presented to the tuned amplifier is adjusted to give a specific reading for the amplitude of the fundamental. Tuning to the harmonics will then give indications that are marked to show directly the percentage of the specific harmonic present. Alternatively a decibel scale may be used.

In the second system the wave analyser uses the superheterodyne principle. The i.f. (intermediate frequency) amplifier has a crystal filter input circuit; such circuits are capable of having very narrow bandwidths. Since the i.f. tuning depends upon crystals, we are not able to vary the frequency of the signal to the i.f. stage. The input signal is therefore mixed in a suitable mixer

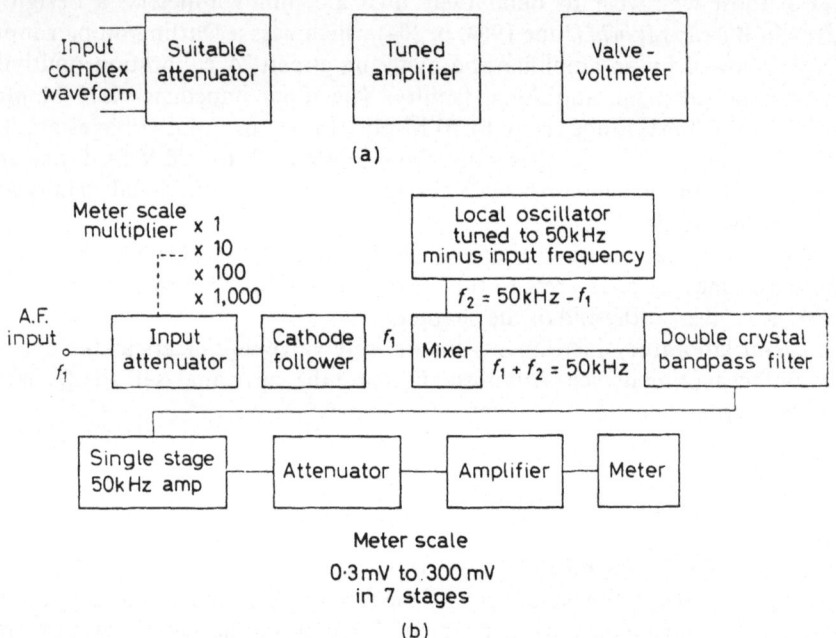

Figure 13.4. Two forms of wave analysis. (*a*) the straightforward tuned type and (*b*) the type operating on the superheterodyne principle. The block diagram shows the main stages of a Marconi TF 455E wave analyser

circuit with the output from a local oscillator, the frequency of which is variable. *Figure 13.4b* shows the principles in block diagram form of the Marconi Wave Analyser type TF 455E. This type allows an accurate evaluation of either the relative or absolute levels of the individual components of a complex waveform. This analyser has a pass band of only 4 Hz and can be tuned to any frequency in the range 20 Hz to 16 kHz. The input, after suitable attenuation, is fed to the mixer together with the local oscillator output which is adjusted to 50 kHz minus the frequency of the audio signal. The output from the mixer is then fed into the crystal filter. The mixer is a non-linear device so that when two frequencies f_1 and f_2 are fed into it the output contains com-

382

ponents with frequencies of (f_1+f_2) and (f_2-f_1). It is the frequency (f_1+f_2) that is selected by the crystal filter.

We can see how the sum and difference frequencies arise if we assume that the mixer has a square law response, i.e. the output $v_0 = Av_{in}^2$. Let $v_{in} = av_1 \sin \omega_1 t + bv_2 \sin \omega_2 t$ where $\omega_1 = 2\pi$ times the frequency of the input signal and $\omega_2 = 2\pi$ times the frequency of the local oscillator. Then

$$v_0 = A(av_1 \sin \omega_1 t + bv_2 \sin \omega_2 t)^2$$
$$= A[a^2v_1^2 \sin^2 \omega_1 t + b^2v_2^2 \sin^2 \omega_2 t + 2abv_1v_2 \sin \omega_1 t \sin \omega_2 t]$$

Now $\qquad 2 \sin \omega_1 t \sin \omega_2 t = \cos (\omega_2-\omega_1) t - \cos (\omega_2+\omega_1)t$

and thus we see that sum and difference frequencies are present in the output.

The double crystal 50 kHz bandpass filter discriminates severely against all but the wanted component (of frequency f_1+f_2). The output from the filter is then amplified by an amplifier tuned to 50 kHz, and the output from the amplifier is indicated on a directly-calibrated diode voltmeter. The local oscillator is of the tuned-anode type inductively coupled to the mixer.

The Distortion Factor Meter

When we wish to know the total harmonic component of a complex wave, as distinct from a knowledge of the individual components, we use a distortion factor meter. The readings obtained are then a measure of the total harmonic distortion. The distortion factor, D, of a voltage waveform is the ratio of the total r.m.s. voltage of all the harmonics to the total r.m.s. voltage.

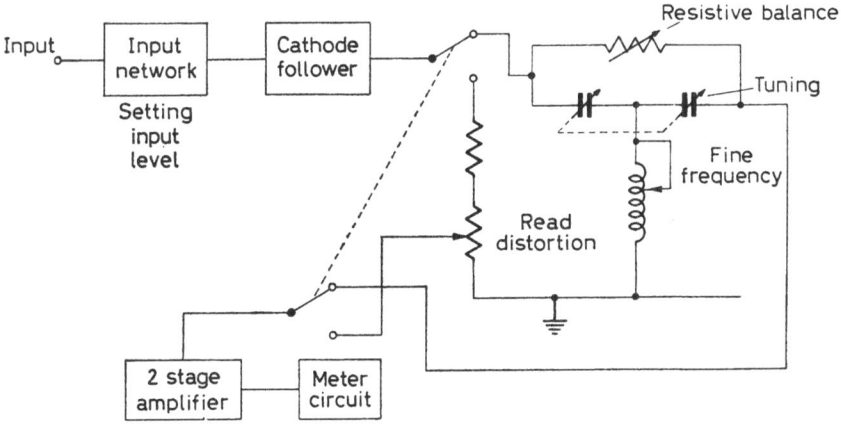

Figure 13.5. Block circuit diagram for a distortion factor meter

383

Expressed as a percentage

$$D = \frac{\sqrt{(E_2^2 + E_3^2 \ldots E_n^2)}}{\sqrt{(E_1^2 + E_2^2 + E_3^2 \ldots E_n^2)}} \times 100\%$$

Where E_1 is the amplitude of the fundamental and E_2, E_3 etc., the amplitude of the second, third, etc., harmonics.

Figure 13.5 shows the block diagram of the Marconi Distortion Factor Meter type TF 142F. The measurement is effected by suppressing the fundamental with a highly selective filter (the bridged-T RCL filter) and then comparing the amplitude of the residue with that of the total input. The comparison is semi-automatic in that controls are operated to bring the voltmeter pointer to a specified mark on the dial on two occasions. The percentage harmonic content is then read off directly from a calibrated dial.

The Q-meter

The ratio of reactance to resistance of a coil or capacitor is known as the Q-factor. Thus for a coil $Q = \omega L/R$, and for a capacitor $Q = 1/(\omega CR)$. R represents the resistive component of the coil or, in the case of the capacitor, the losses associated with the dielectric. In a high-quality coil (Q stands for quality), or in a good capacitor, R is low. The term Q-factor is also used in connection with a resonant circuit. In this case $Q = 2\pi$ times the energy stored in the circuit divided by the energy dissipated in the circuit during one cycle. In a resonant circuit the losses are almost entirely associated with the coil, the losses in a well-constructed capacitor being negligibly small. Very nearly then the Q of a resonant circuit is given by $Q = \omega L/R$.

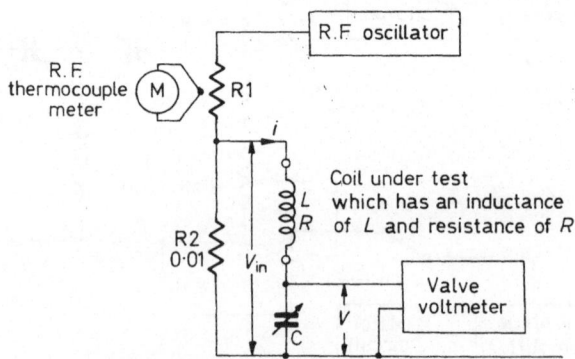

Figure 13.6. Principle of a Q-meter

The Q-meter, which is really a valve-voltmeter suitably calibrated, is used for measuring the Q-factor of a coil, or circuit. The principles of operation may be understood by considering *Figure 13.6*. The coil whose Q is to be determined is connected to the input terminals and forms a resonant circuit with the calibrated capacitor, C. At resonance $\omega L = 1/(\omega C)$ and the supply current, i, is determined entirely by R. If V_{in} is the supply voltage then $i = = V_{in}/R$. The voltage across C is $iX_C = V_{in}/\omega CR = V$(say). Then

$$\frac{V}{V_{in}} = \frac{1}{\omega CR} = \frac{\omega L}{R} = Q$$

The supply voltage V_{in} is taken from a potential divider R1, R2 supplied with current from an r.f. oscillator. For high-Q circuits the losses are very small and the impedance at resonance, R, is low. The resistance R2 must be much less than R and is commonly about 0·01 Ω. The resonant circuit does not then unduly load the potential divider R1, R2. The voltage across R2, viz. V_{in}, is monitored by means of an r.f. meter of the thermocouple type. For a constant current through R1, V_{in} is constant and the voltage V, as measured by the valve-voltmeter, is a measure of the Q. The valve-voltmeter can then be calibrated directly in Q values for various settings of the r.f. meter. In operation, C, a precision air-dielectric capacitor, is varied until the resonance condition is obtained (as detected by a maximum reading in the valve-voltmeter). The output from the r.f. oscillator is adjusted so that the r.f. meter gives a predetermined reading, and the Q is read off directly from the valve-voltmeter scale.

ELECTROMETER VALVES AND CIRCUITS

Although the valve voltmeters we have discussed have input resistances as high as 100 M ohms and are quite satisfactory for many measurements across high-impedance circuits, such voltmeters prove to be inadequate for certain applications. Whenever we wish to measure currents in the range 10^{-8} A to 10^{-15} A or wish to measure voltages from transducers that cannot supply currents larger than say 10^{-8} A, then it is necessary to employ electrometer techniques. Typical examples are the measurement of pH, the detection of sub-atomic particles, the measurement of current from a flame in a flame-ionization apparatus, and the determination of small electrostatic potentials. We shall be concerned here only with those electrometers that use electronic devices.

The unsuitability of the conventional valve when used as a very high impedance device is due to the random grid current that is present; such current is of the order of 10^{-8} A. Normally this is negligibly small, but when such currents flow in resistances of the order of 10^8 ohms, and higher, then the voltages developed across the resistors are several volts. The result is that conventional valve-voltmeters are useless when currents smaller than 10^{-8} A are involved since the effect of the random grid current makes any measurements invalid. Under these circumstances, we may use the conventional electronic circuit techniques, but must replace the ordinary input valve (or valves in a balanced arrangement) with devices that have been specially constructed for the purpose. Such devices are known as electrometer valves.

In conventional valves the presence of comparatively large random grid currents is due to

(a) the emission of positive ions from the cathode surface;

(b) the ionization of residual gases within the tube by electrons that travel from the grid to the anode under the influence of high anode voltages;

(c) the release of electrons from the grid due to the incidence of soft x-rays. Such rays are produced when high energy electrons strike the anode;

(d) leakage currents; and

(e) electrons reaching the grid by virtue of their finite emission velocity from the cathode surface.

Electrometer valves are specially constructed and operated under conditions that avoid the difficulties associated with excessive grid current. The cathodes of electrometer valves are always directly-heated filaments and receive special attention in their design and construction. The filament often forms part of a network that determines the d.c. operating points and, consequently, it is constructed from material having a low temperature coefficient. It is usual to mount the filament so as to have minimum contact area with the supports within the tube. Variable thermal losses are then minimized if the tube is subjected to mechanical disturbances. The filament is run at a reduced temperature so that no visible glow can be observed during operation. Photoemission from other electrodes and the walls of the tube is thus avoided; also the electron emission velocity is low. Pre-ageing before sale is undertaken to ensure that the emissive properties have been adequately stabilized.

It is impossible to avoid some residual gas within the tube and, therefore, low anode voltages (say up to 15 V) are used. The danger of ionizing the gas is thus avoided because the electrons never acquire sufficient energy to produce ionization. The production of soft x-rays is also avoided.

Leakages are reduced to a minimum within the tube by careful construction. The user can do much to reduce leakages on the exterior walls of the tube by ensuring that the surface is free of grease. Alcohol or ether are satisfactory cleaning solvents. Where the ambient humidity is likely to be high, the electrometer valve should be enclosed within a suitable housing and the air in contact with the walls kept dry with a suitable desiccant (e.g. P_2O_5 or silica gel). The housing also serves to exclude visible light. The leads to the grid should be kept short, and entry plugs should be adequately insulated with quartz or amphenol bushes.

Electrometer tubes may be triodes, tetrodes or pentodes. The multi-electrode tubes are often favoured because of their superior characteristics at low operating voltages. All electrometer tubes are extremely sensitive to mechanical vibrations and shocks, but the tetrodes are usually less sensitive to shock than the triodes.

Operation without range switching is usually preferred since switches introduce variable and undesirable leakage paths. Often, however, the range of currents we desire to measure is very large and logarithmic operation is desirable. The approximate number of decades possible with a triode of the Mullard ME1401 type is six whereas it is eight for their tetrode (ME1402). The introduction of the third electrode (ME1403) reduces the number of decades to seven.

Electrometer Circuits

Figure 13.7 shows a simple electrometer circuit that uses the slide-back principle. With the input short-circuited, RV2 is set so that the reading in the voltmeter shows zero. The reading on the galvanometer G is set to zero by adjustment of RV1. This counters the small anode current that is present when the input voltage is zero. If now a voltage is applied to the input, the reading on the galvanometer rises. By adjusting RV2 we can bring the galvanometer reading back to zero; the voltage indicated on the voltmeter is now the applied voltage. If a small current is to be measured then the procedure is exactly the same. Since the grid current is negligibly small, the signal current may be deduced by dividing the voltmeter reading by the resistance value of R_g.

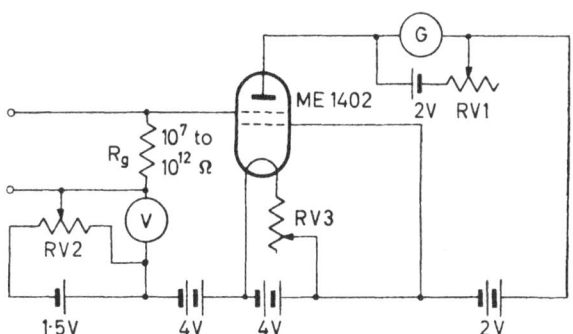

Figure 13.7. Slide-back voltmeter using the Mullard ME 1402 electrometer valve

Although adjustments must be made for every reading, the effects of valve and supply voltage variations are eliminated.

A simple electrometer circuit published by Mullard[2] is shown in *Figure 13.8*. The measurement circuit consists of a bridge, two arms of which are *npn* transistors. The output from the electrometer stage is fed to the base of one of the transistors upsetting the bridge and giving an out-of-balance reading. Full-scale deflections are obtained with input voltages of ± 25 mV, ± 250 mV and $\pm 2 \cdot 5$ V. The instrument is suitable as a simple pH meter and has also been used successfully as a flame-ionization detector for gas chromatography.[3]

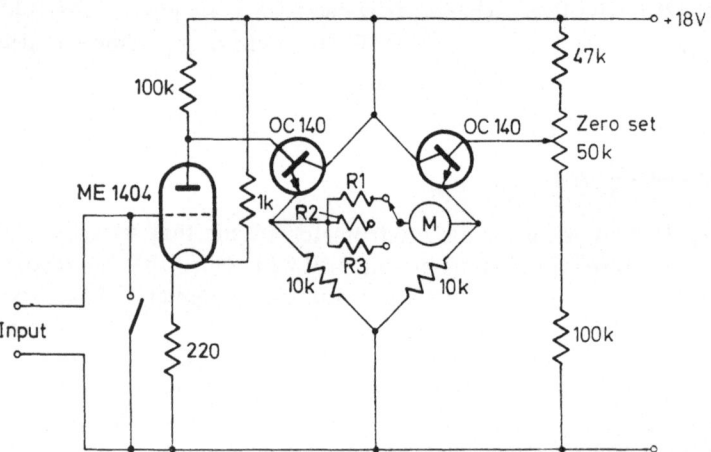

Figure 13.8. Electrometer circuit suitable for pH measurement, flame-ionization detection and low magnitude electrostatic potentials[2]. M = 50 μA meter. R1, R2, R3 chosen in connection with meter to give ± 25 mV, ± 250 mV and $\pm 2 \cdot 5$ V for full-scale deflection

A much more sophisticated instrument for gas chromatography is described by Gabriel and Morris[4]. The paper gives details of a flame ionization meter that uses the Mullard pentode electrometer type ME1403. The essentials of the circuit are shown in *Figure 13.9*. The instrument has six linear ranges with full-scale deflections of 10^{-10} A to 10^{-5} A, a logarithmic range of 10^{-10} A to 10^{-5} A, and three integrating ranges. The principles of operation are explained in the paper, and a photograph of the apparatus is given. The amplification of the ionization current is first carried out by the pentode electrometer valve (ME1403) whose more important function is that of producing the necessary high input impedance. The electrometer valve is then followed by a compound pair. We have already seen that such an arrangement has an extremely high input impedance, and, therefore acts as a satisfactory

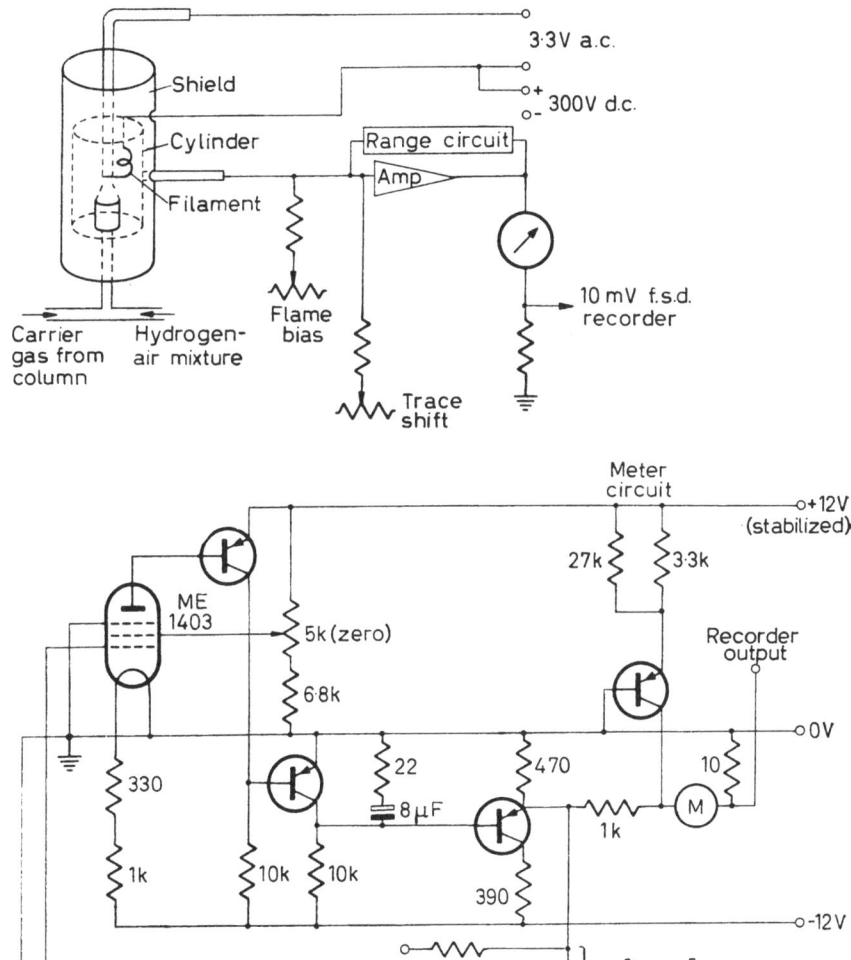

Figure 13.9. Main amplifier details of a flame ionization meter for gas chromatography designed by Gabriel and Morris[4]. Details of integral cancel circuit and bias voltages for flame bias and trace-shifting are omitted

load for the electrometer valve. The emitter-follower compound pair is follow-ed by a conventional amplifier stage. Accompanying the amplifier are the range setting components, not all of which are shown in *Figure 13.9*. Among the range-setting components are a pair of diodes, whose function is to pro-vide the logarithmic ranges. The characteristic of a *pn* junction is approxi-mately logarithmic. (It is shown in books on semiconductor devices that the current is given by $I = I_S\{\exp\ (eV/kT) - 1\}$ where V is the applied voltage, k is Boltzmann's constant, T the absolute temperature, e the charge on the electron and I_S the saturation current.) Two GEC type SX640 diodes are

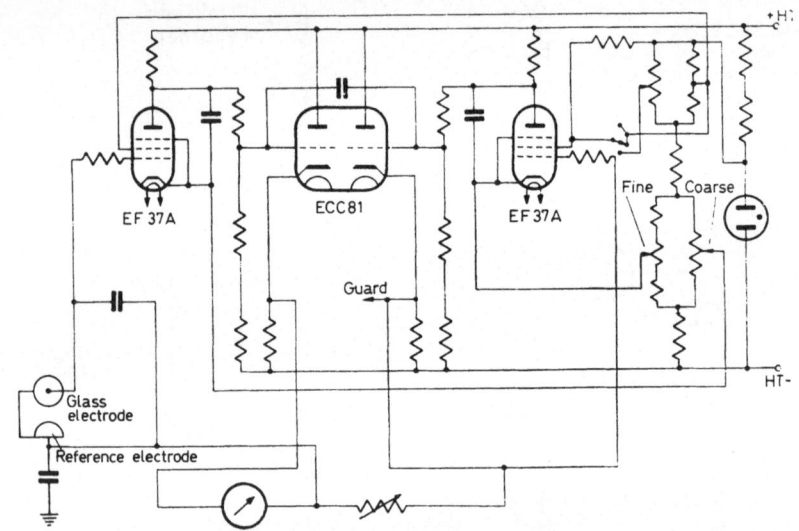

Figure 13.10. pH meter based on a valve-voltmeter design of Scroggie[7]. The EF37A valves are run under 'starvation' conditions and then behave as semiconductor valves

used connected in series and they yield approximately 0·2 V for each tenfold increase in input current in the range 10^{-10} A to 10^{-5} A. The function of the 1,000 M resistor is provide a stable zero since the diodes have an extremely high resistance at low input currents. The function of the capacitors is to pro-vide integration; we have already seen in Chapter 9 how a feedback line con-sisting of a capacitor can convert an amplifier into an integrator. Trace-shifting and flame bias controls, a meter circuit, and recorder output facilities are also provided.

Electrometer conditions can be approached using valves that are not pri-marily constructed for electrometer purposes[5, 6]. By operating under 'starva-tion' conditions acceptable results are obtained provided currents as low as say 10^{-11} A are not required to be measured. Starvation conditions means underrunning the heaters (using about 5·4 V instead of 6·3 V) and limiting the anode voltages to low values. Scroggie[7] has described how to use a Mullard EF86 in this way and has published a design for a valve voltmeter of extremely high input impedance. The design is an up-to-date version of an earlier one based on EF37A valves (*Wireless World*, Jan. 1952), which has an input resistance of 10^{10} ohms. A pH meter based on this circuit has been successfully constructed as a project by one of the author's students, and the basic circuit is given in *Figure 13.10*. Although EF37A valves are obsolete, they are cheap and readily available on the 'surplus' market. These valves are preferred to EF86 by the author because the grid is brought out to a top cap. Insulation difficulties associated with the grid circuit are, therefore, minimized.

MEASURING BRIDGES

Many of the instruments we use for measuring resistance, capacitance and inductance depend upon a circuit arrangement called a bridge. These bridges are all basically derived from the Wheatstone bridge used for measuring resistance. Although resistance can be measured with a moving-coil meter used as an ohmmeter, the accuracy of the determination is not very high and depends heavily upon the quality of the meter used. If we use a bridge, however, the meter or other null detecting device need not be highly accurate since it is only required to indicate a zero voltage condition. The basic resistance bridge has already been dealt with in Chapter 2 (p. 27 and *Figure 2.12*). The four resistors are arranged diagrammatically in the form of a square and the supply voltage bridges one of the diagonals whilst the detector bridges the other diagonal. When three of the resistors are accurately known standards the fourth can be calculated with precision.

A similar arrangement is possible with impedances, in which case we have what is called an a.c. bridge. *Figure 13.11* shows the basic arrangement for four impedances Z_1, Z_2, Z_3 and Z_4. The bridge must be supplied with an

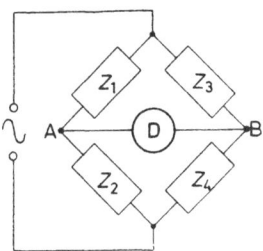

Figure 13.11. Basic a.c. version of Wheatstone bridge

alternating voltage from an oscillator or the mains; the detector may be a pair of headphones or some type of electronic a.c. meter. When we have a null condition $Z_1/Z_2 = Z_3/Z_4$. The production of a true null is often difficult with a.c. bridges because of the uncertain stray capacitances associated with the apparatus. We must often, therefore, be content with a minimum response in the detector. There is an additional complication when balancing a.c. bridges. To illustrate this, let us suppose that Z_1 and Z_2 are each resistors equal to R_1 and R_2, respectively, and Z_3 and Z_4 are inductors equal to $R_3 + j\omega L_3$ and $R_4 + j\omega L_4$, respectively. Substituting into the balance equation gives us

$$\frac{R_1}{R_2} = \frac{R_3 + j\omega L_3}{R_4 + j\omega L_4}$$

Therefore

$$R_1 R_4 + j\omega L_4 R_1 = R_2 R_3 + j\omega L_3 R_2$$

It follows at once from the fact that the ordinary and quadrature components in the balance equation can be separated that two balance equations are always obtained. The two balance equations enable two unknowns to be determined. Let us suppose that the coil $R_3 + j\omega L_3$ is the unknown. We can find the resistance of the coil from

$$R_1 R_4 = R_2 R_3$$

when R_1, R_4 and R_2 are known. R_1 and R_2 are usually two standard resistors or portions of an accurately known ratio arm and R_4 is the known resistive portion of a standard inductor. The inductance L_3 can be found by comparing the quadrature components so that $L_3 = L_4 R_1/R_2$. For convenience of manipulation, it is usual to arrange that only two components in a bridge should be variable and it is obviously desirable that each of the balance equations should contain one of the variables. The balance equations are then independent. With the example just quoted L_4 would be adjusted until a minimum condition was detected. R_4 would then be adjusted until a new and smaller minimum was obtained. L_4 would then be re-adjusted, followed by re-adjusting R_4, and so on until a final absolute minimum was obtained. This is the usual technique used for balancing a.c. bridges. When the balance conditions are independent three or four adjustments usually suffice. If the bridge is such that the balance equations are not independent, finding the balance is much more laborious.

Figure 13.12 shows three of the basic bridges that constitute the heart of many commercial instruments. The suffix x is used in conjunction with the unknown component that must be determined; the suffix s denotes the standard known component.

Bridges that use the Wheatstone bridge principle suffer from some disadvan-

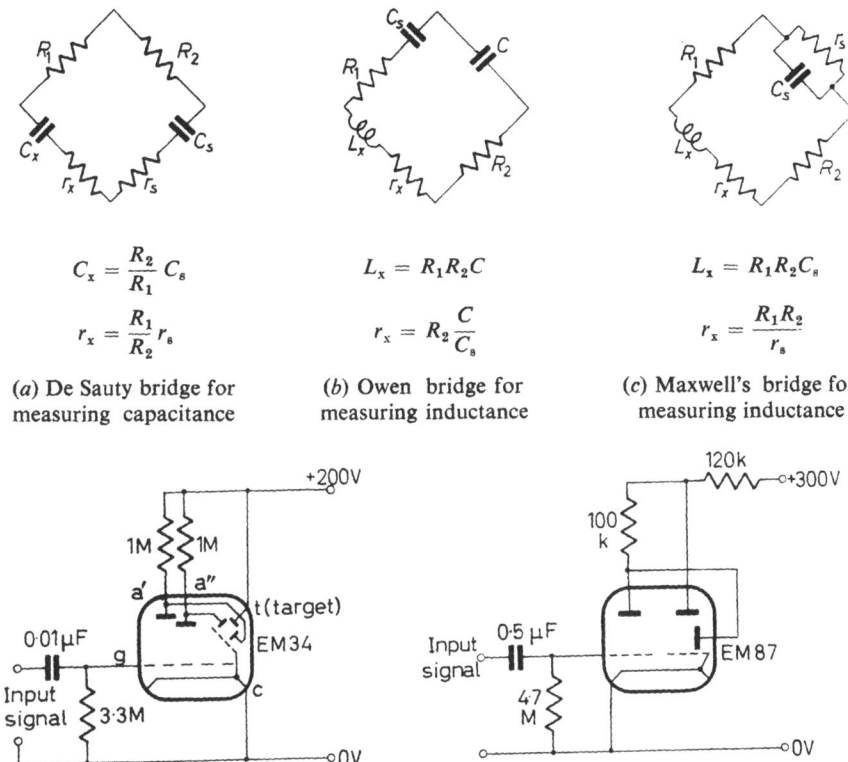

$$C_x = \frac{R_2}{R_1} C_s$$

$$r_x = \frac{R_1}{R_2} r_s$$

(a) De Sauty bridge for measuring capacitance

$$L_x = R_1 R_2 C$$

$$r_x = R_2 \frac{C}{C_s}$$

(b) Owen bridge for measuring inductance

$$L_x = R_1 R_2 C_s$$

$$r_x = \frac{R_1 R_2}{r_s}$$

(c) Maxwell's bridge for measuring inductance

Figure 13.12. Three of the standard bridges used by commercial firms for their measuring bridges. Magic-eye tuning indicators may be used as bridge detectors. Electrons from a heated cathode strike a fluorescent screen (target) which glows. A control electrode casts a shadow on the screen since it is placed in the electron beam path. The size of the shadow is determined by the voltage on the control electrode. This voltage is obtained from a triode mounted within the tube that acts as a high input impedance amplifier of the input signal. The EM34 uses two triodes and two control electrodes

tages, relying as they do upon the ratio of impedance standards. A restriction is introduced since the standards can be produced only over limited ranges of values. Very high and very low pure components are difficult to manufacture. Such standards are therefore expensive, especially since large numbers of them are required. Difficulties are also experienced over leads to the unknown; these leads must be short to avoid stray elements. Measurements with long cables are impossible to perform with accuracy, and it is usually highly inconvenient to make the calculations when the unknown has to be measured *in situ*. The transformer ratio-arm bridge overcomes these difficulties. The basic circuit is given in *Figure 13.13* and is essentially that used by

393

Wayne Kerr Laboratories Ltd. in their bridges. The balance equations are given in terms of admittances rather than impedances. It is not appropriate to go into the theory of this bridge here, but papers by B. Rogal[8] together with further information on the bridge will be sent to any interested reader whose request is forwarded to Wayne Kerr Ltd.

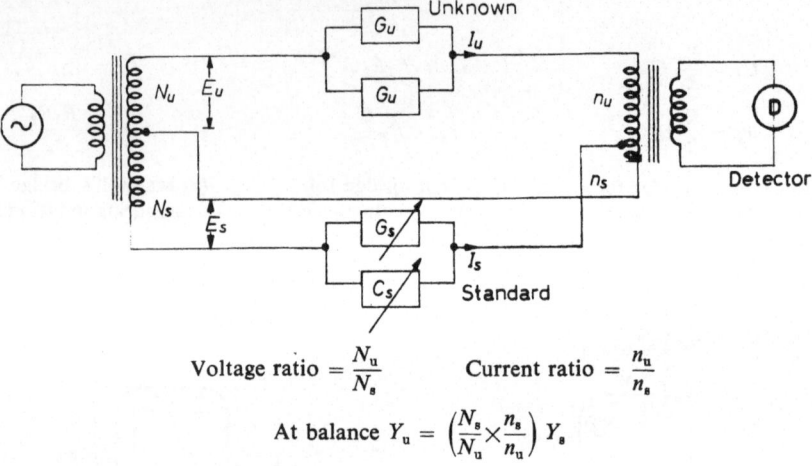

$$\text{Voltage ratio} = \frac{N_u}{N_s} \qquad \text{Current ratio} = \frac{n_u}{n_s}$$

$$\text{At balance } Y_u = \left(\frac{N_s}{N_u} \times \frac{n_s}{n_u}\right) Y_s$$

where Y_u and Y_s are the admittances (i.e. 1/impedance) of the unknown and standard respectively

Figure 13.13. The transformer ratio-arm bridge

SERVOMECHANISMS

A servomechanism (from the Latin *servus* meaning slave) is a machine designed to carry out orders. Unlike the open-loop control systems that make no direct check upon the output position or state of the system, servomechanisms are closed-loop error-correcting systems whose actions are continuously controlled by the difference between a desired and actual output state. Feedback is an essential feature of servomechanisms, or servos (which is the commonly used abbreviation); in servo systems some sort of power-amplifying device that drives the output is actuated by a feedback signal that is proportional to the difference or error between the actual and desired state of the output.

As an example of an open-loop control system let us consider *Figure 13.14a* in which, by setting a dial to some position, fuel is pumped from a tank into a furnace. The dial setting determines the rate of flow of oil and hence the temperature of the furnace. If we make no check on the temperature of the furnace our system is not likely to be an accurate one since so much depends upon the calibration of the input dial. If, however, a thermometer is used to determine the output condition (temperature in this case) then an operator may use his eyes to note the temperature. From a knowledge of the desired temperature and the observed temperature, he can note any difference or error involved and reset the dial in such a way as to reduce the error to zero. In detecting the error and feeding back the information to the input in such

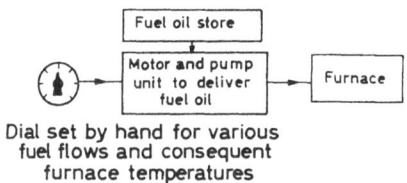

(*a*) Open loop control system

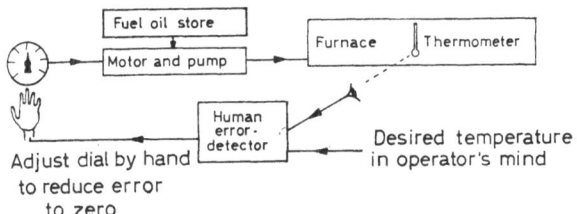

(*b*) Conversion of open-loop system to a closed-loop system, i.e. a servomechanism

Figure 13.14. Open and closed-loop control systems

a way as to reduce the error he is acting as a feedback path, and thus converts the open-loop system into a closed-loop one. Because a human operator is expensive to employ, is not equipped to detect very small errors or to react quickly, and cannot work for long periods without fatigue, he is often conveniently replaced by some other error-detecting device. Such a feedback system would then be called a servomechanism *(Figure 13.14b)*. Most readers will be aware that there is a growing need in industry for automatic controlling and regulating devices of which servos are an important part. Such devices were first invented to assist sailors to control the rudders of their ships. Nowadays servos are used for the automatic piloting of aircraft, for precision machining of metal parts from moulds or master templates, for measuring instruments such as potentiometric recorders and XY plotters, and for controlling rockets, guided missiles and industrial processes of many kinds.

395

Figure 13.15 shows a simple position servo. If we wished to rotate to a desired position a table supporting say a large telescope then a sufficiently powerful servomotor could be made to do the necessary work using the system shown. Two potential dividers are required, one attached to the input shaft and the other to the output shaft. The rotation of the input shaft through a given angle θ_i would cause a voltage to be applied to the input of the d.c. amplifier. The output of the amplifier would then drive the motor and gears so as to turn the table and associated potential divider. Once the table had rotated through an angle, θ_0, the input to the amplifier would be reduced. Eventually when $\theta_0 = \theta_i$ the input voltage to the amplifier would be zero and the motor would then ideally cease to produce any further rotation of the table. Unfortunately, is such a simple system, inertia of the moving parts would cause the table and output shaft to overshoot the desired position.

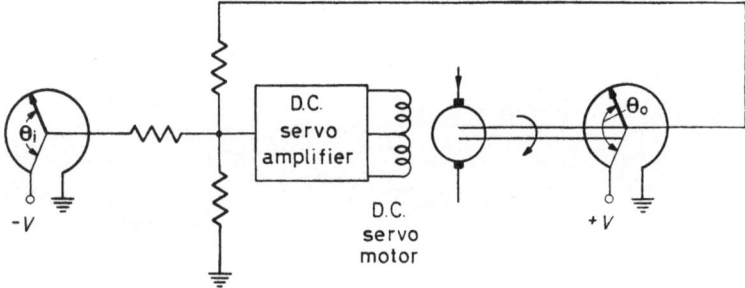

Figure 13.15. Basic position servosystem in which the angular position of the output shaft is determined by the position set on the input shaft

A signal would now be presented to the input of the amplifier such as to cause the motor to turn in the opposite direction in an attempt once again to make $\theta_0 = \theta_i$. Overshooting would take place again and the system would oscillate back and forth about the desired position; such oscillations in servo systems are termed 'hunting'. In mathematical terms the motor torque $I(d^2\theta/dt^2)$ is proportional to the angular difference in the input and output shaft angles, i.e.

$$I \frac{d^2\theta_0}{dt^2} = -K(\theta_0 - \theta_i)$$

Those familiar with the differential equation describing simple harmonic motion will recognize the form of the above equation and realize that the solution is

$$\theta_0 = \theta_i + A \cos(\omega t + \alpha)$$

where $\omega = \sqrt{(K/I)}$.

A is a constant of proportionality and I the moment of inertia of the rotating system. We see from the solution for θ_0 that the output shaft oscillates about the desired position with an amplitude, A. To eliminate the hunting and produce a useful system, we must introduce a damping term into our equation. The oscillations would then die away at a rate proportional to the damping term. One obvious method of producing damping in the system is to apply mechanical friction in the form of braking. Except for very low-powered systems such a solution is unsatisfactory because of the heat produced, the

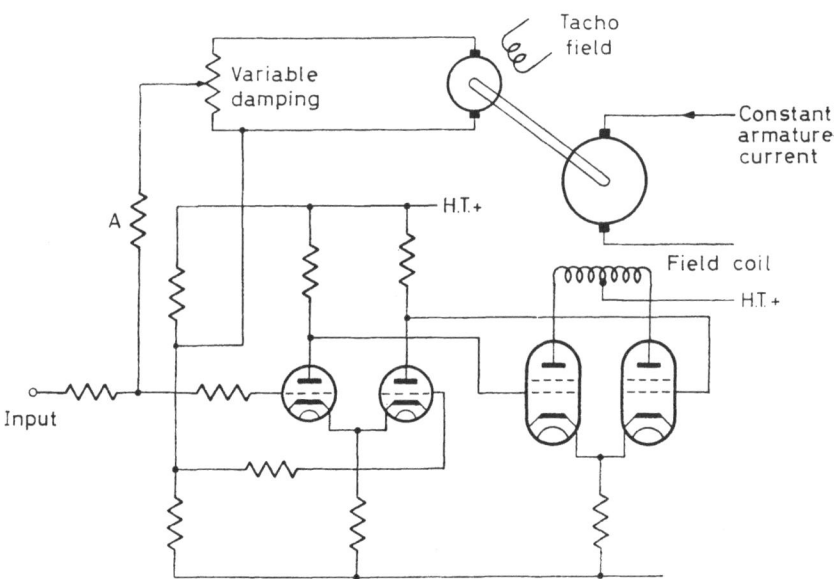

Figure 13.16. Basic velocity servosystem in which the rate of rotation of the servomotor is a function of the steady input voltage. Tacho feedback is employed for damping purposes. The tachogenerator must be connected to ensure that the voltage at point A is always in opposition to the input voltage

power lost and the increase in response time of the system. A better arrangement is to couple a small generator to the output or main servo motor. Such a generator is often called a tachogenerator or tachometer. The function of the tachogenerator is to supply a damping voltage proportional to the driving velocity. The damping voltage is then fed back to a convenient point in the amplifier so as to produce the desired effect. We then have

$$I(d^2\theta_0/dt^2) - r(d\theta_0/dt) = -K(\theta_0 - \theta_i)$$

where $r(d\theta_0/dt)$ is the damping term. The solution is

$$\theta_0 = \theta_i + A\,\exp(-rt/2I)\cos(\omega t + \alpha)$$

from which we see that the oscillatory term dies away. The decay of the oscil-lations is quite rapid in a well designed system. *Figure 13.16* shows one way of feeding back a damping voltage in a d.c. system. The voltage at the point A is always in opposition to the input error voltage. As the desired position is approached at full speed the error voltage is reduced below the opposing tachometer voltage. The torque on the main servomotor is, therefore, reversed and a braking action is thus exerted.

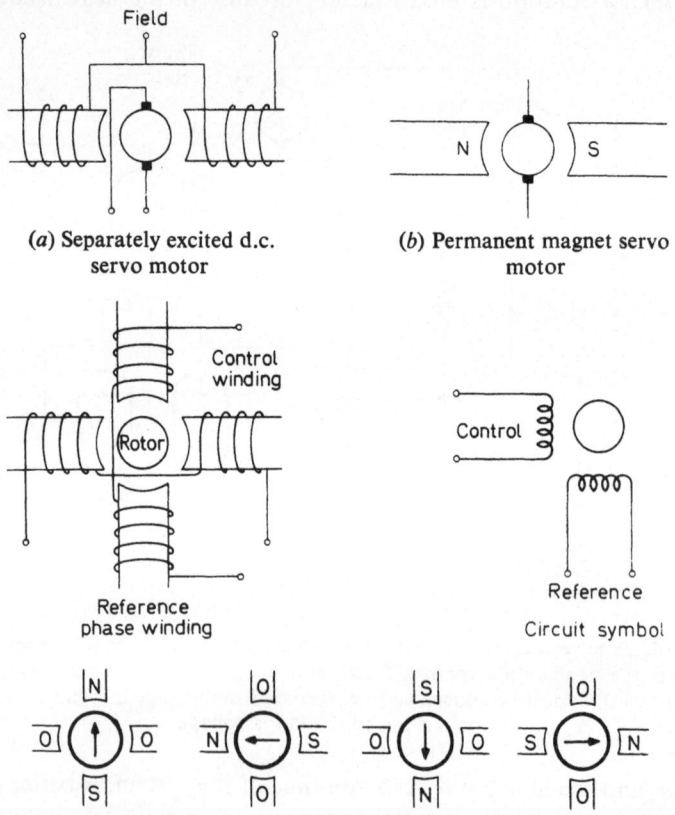

(*a*) Separately excited d.c. servo motor

(*b*) Permanent magnet servo motor

Reference phase winding

Circuit symbol

(*c*) A.C. induction 2-phase servomotor. The control winding is fed with a voltage that is 90° out of phase with the reference voltage. The smaller diagrams show the stages of rota-tion of the field. Starting with the control current at a maximum the reference current will be zero and the field is as shown. A quarter of a period later the reference current is a maxi-mum and the control current has fallen to zero. A further quarter of a period later the control current will be a maximum in the reverse direction whilst the reference current has fallen to zero. The reference current then reaches its maximum reverse value whereupon the whole cycle is repeated

Figure 13.17. Various forms of servomotor

398

Servomotors can be either separately excited d.c. motors or an a.c. induction type.

In the separately excited motor *(Figure 13.17a)* control can be effected by using the field or armature winding. A usual arrangement is to energize the armature winding from a constant current source and obtain control by supplying the field winding with the output of the servoamplifier (see *Figure 13.16*). Motor reversal is effected by reversing the direction of the field current. Such a system is useful in high-powered applications. For lower powers, it is possible to supply the field windings from a constant current source and effect control by variable currents in the armature winding. Feeding the armature with thyratron currents has already been mentioned in Chapter 8. The servoamplifier must then provide vertical control of the thyratron. The armature-controlled motor has the advantage of fast response time because of the lower armature inductance. The high power needed from the controller, however, is often a disadvantage. This type of motor is characterized by high starting and reversing torques.

For small powers in instrument work the main field may be provided by a permanent magnet. The armature is then supplied with current to obtain the necessary torque.

A two-phase a.c. induction motor *(Figure 13.17c)* is the usual choice for instrument servo systems. Two separate stator windings are provided and these are physically perpendicular to each other. The control winding is fed from the servo amplifier by a voltage that is 90° out of phase with the voltage supplied to the reference winding. The reference winding is connected directly to the source of power. When both windings are energized there is produced within the motor a rotating magnetic field. This field induces voltages in the rotor and the resulting rotor currents produce magnetic fields that interact with the rotating stator field. Such interaction produces rotation of the rotor in a direction determined by the phase of the control stator winding, a 90° lead over the reference voltage producing motion in one direction, and a 90° lag producing motion in the other direction. The speed of rotation is determined by the magnitude of the control winding voltage when the reference winding voltage is fixed. Obviously zero voltage on the control winding produces no revolving magnetic field and the rotor remains stationary. The advantage of this type of motor is that a.c. tuned amplifiers can be used. The servomotors operate at a fixed frequency, usually 50 Hz or 400 Hz, and by using amplifiers tuned to the operating frequency, it is easy to discriminate against noise and other unwanted voltages.

Tachogenerators for d.c. systems are straightforward dynamos that are coupled mechanically to the servomotor. The direct output voltage is strictly proportional to the shaft velocity, and reversals in direction of shaft rotation produce reversals in the polarity of the output voltage. The a.c. tachometer or generator has two windings one of which is connected to the a.c. power supply usually via a transformer since 115 volts is the common operating voltage. The rotor consists of a squirrel-cage arrangement with low resistance windings. These windings are shorted at the ends so that currents are produced in the rotor when the supply winding is energized. These currents produce fields that induce an alternating voltage in a second winding (which is perpendicular to the first), as the rotor moves. The rotor arm is connected to the servomotor so the speed of rotation is determined by that of the servomotor. The magnitude of the alternating voltage induced in the second coil is directly proportional to the angular velocity of the rotor.

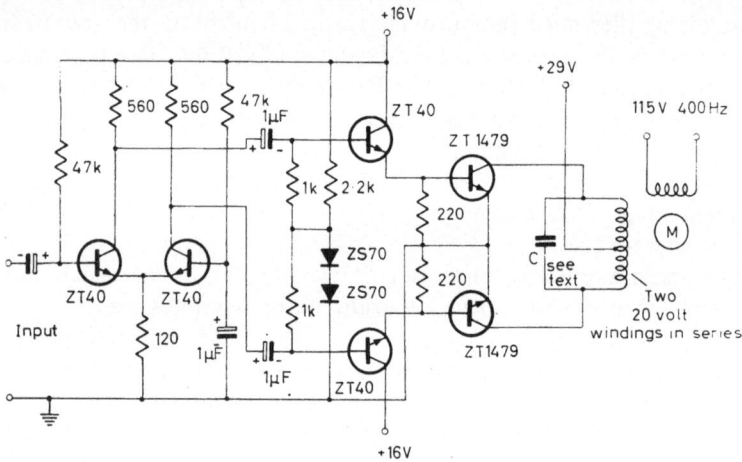

Figure 13.18. Complete circuit for a 6 watt transformerless servoamplifier using Ferranti silicon transistors[9]

Figure 13.18 shows the circuit for a 6-watt 400 Hz servoamplifier suitable for driving a small servomotor of the a.c. two-phase induction type.[9] A Class B push-pull output stage is used to energize the two 20 V motor control windings of a size 15 servomotor. The control windings are series connected. The output transistors are driven by two driver transistors that are fed from a conventional emitter-coupled phase-splitting stage. The value of the capacitor across the control windings must be determined by trial. The function of the capacitor is to tune the control windings to 400 Hz to give maximum stalled torque. The value required is typically 3·8 μF. The system is suitable for low-powered instrument work. No provision is made for electrical damping since for these powers sufficient damping is usually available as mechanical friction in the moving parts. The servosystem may be used as an a.c. pen recorder or XY recorder.

400

Potentiometric Pen-recorder

The simple system outlined above is not suitable for the measurement of small direct voltages, which is a frequent requirement in laboratories. For the recording of direct voltages, it is usual to use an arrangement that converts the steady voltage into an alternating one of suitable frequency. The alternating voltage can then be amplified in the usual way. The conversion of d.c. to a.c. is achieved by using a mechanical chopper operating at the frequency required by the servomotor. The principle of d.c. chopper amplifiers has already been outlined in Chapter 9. No demodulator is required at

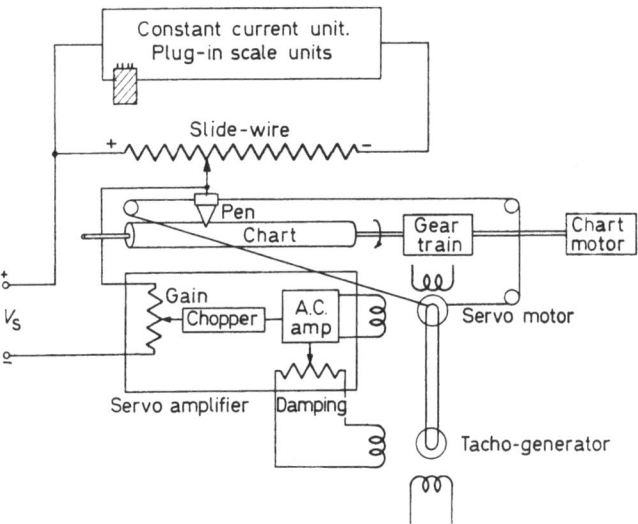

Figure 13.19. Schematic diagram showing the principle of a potentiometric pen recorder

the output since the a.c. is used directly to drive the servoamplifier. *Figure 13.19* shows the principle of the potentiometric pen recorder. The input to the chopper amplifier consists of the difference between the voltage to be measured and a portion of the steady voltage existing across an accurately manufactured, stable, linear resistance wire. When the input voltage changes there is an out-of-balance voltage input to the amplifier. This voltage is chopped at the operating frequency and amplified. The output from the amplifier then causes the servomotor to operate a pen arm. The pen arm moves to a new position and in doing so an associated sliding contact on the resistance wire moves until the voltage on the sliding contact is equal in magnitude, but opposite in sign, to the input voltage. When the voltages are equal in this way, there is no input to the amplifier and the servomotor stops. From a previous calibration carried out by the manufacturer, the movement of the pen along the chart is proportional to a given input voltage. The pen scale is thus a voltage scale. If the input voltage is proportional to some quantity, say temperature, then the chart may be calibrated directly in terms of the quantity being measured.

Suitable provision is made within the instrument to alter the scale. Usually a set of plug-in units is provided that alter the standard voltage across the slidewire. The wire is supplied with a current that remains constant in spite of variations of supply voltage. A constant current through the slidewire produces a known voltage across the wire, and it is a proportion of this voltage that is balanced against the input voltage to produce zero input to the amplifier.

Two preset controls are usually provided, one marked 'damping' and one 'sensitivity' or 'amplifier gain'. Altering the gain setting of the amplifier can have no effect on the scale recorded on the chart. Changing the gain serves to alter the time it takes to move the pen to the final position. With high gains the servomotor is made to operate with greater power thus shifting the moving parts more swiftly. Usually the use of excessive gain produces hunting, and a compromise must be achieved between maximum sensitivity and freedom from hunting. One procedure is to supply an input signal to give about half of the full-scale deflection and then to increase the gain or sensitivity, with the minimum damping, until oscillations are produced that are about 10 per cent of the full-scale deflection. The damping is then increased until, with the application of a step voltage, critical damping is obtained. (A step voltage is one which changes suddenly from zero up to a specified value and subsequently remains constant at that value.) If now, with the input voltage applied, the sliding contact or cord drive is moved manually to give first a deflection of about $\frac{1}{4}$in. to the right and then released the pen should return close to its original position. The difference between the original position and the position achieved after the drive cord is released is known as the 'dead-space' which should be less than 0·1 per cent of the full-scale deflection. If it is greater than this figure the sensitivity (gain) should be increased and the damping adjusted accordingly. Variable damping is effected by a pre-set control that determines the fraction of the tachogenerator voltage that is fed into the amplifier.

The XY Recorder

The potentiometric recorder described above serves to record a quantity that varies with time. The chart is made to pass the pen at a fixed speed determined by the sets of gears used in connection with chart motor. When, however, we have a quantity, Y, that varies with another quantity X, and X is not a function of time, we use an XY recorder. *Figure 13.20* shows the principles involved. Two servo systems, similar to those used in potentiometric recorders, cause a pen or other recording agency to be moved to any position on an area of paper. The movements of the pen produced by each servo system are mutually perpendicular. A voltage proportional to the X quantity is applied to the input to one servo system and this produces movements in what is defined as the X-direction. Voltages proportional to Y are fed into the second servo system which produces movements in the Y-direc-

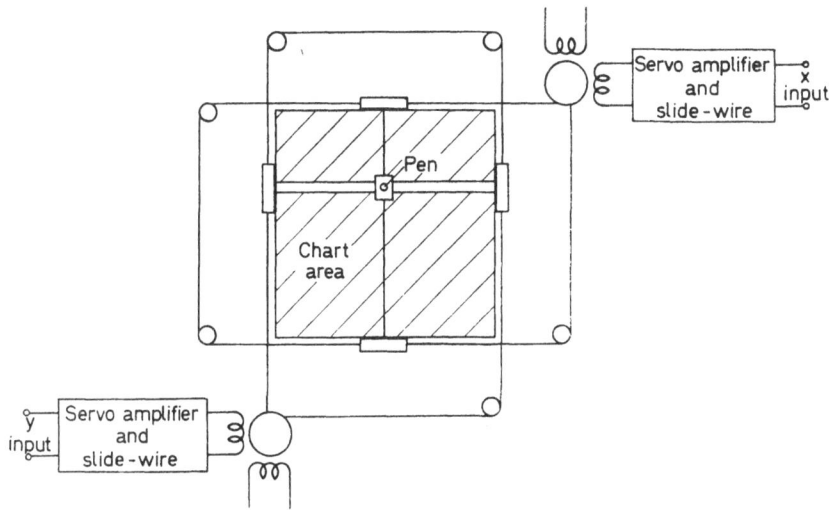

Figure 13.20. Principle of an X-Y recorder

tion. Because of the widespread use of the Cartesian co-ordinate system in graphical work the X- and Y-directions are mutually perpendicular. Adjustments of the servo systems follow the same lines as those used in a potentiometric recorder.

PHASE-SENSITIVE DETECTORS

Usually detectors or rectifiers yield outputs that are independent of the phase of the input voltage. In a power supply, for example, a full-wave rectifier yields a d.c. output and the polarity of the output voltage does not change if the phase of the a.c. input voltage changes. It is often necessary, however, for a change in phase of the input alternating voltage to cause a change in the polarity of the steady output voltage. This requirement entails the use of what is called a phase-sensitive detector. Servomechanisms and data transmission apparatus are typical systems that require phase-sensitive detectors. Let us consider, for example, the position servo shown in *Figure 13.15*. If we prefer to energize the slide wires with alternating instead of direct voltages then the system would not work even if we used an a.c. servo amplifier and an ordinary rectifier. This is because as the output shaft approaches the desired position the alternating voltage from the slidewire decreases to zero; any inevitable slight overshoot would mean a subsequent increase in the output voltage from the wire. Although a change of phase is experienced as we pass through the null point, a system that did not respond to

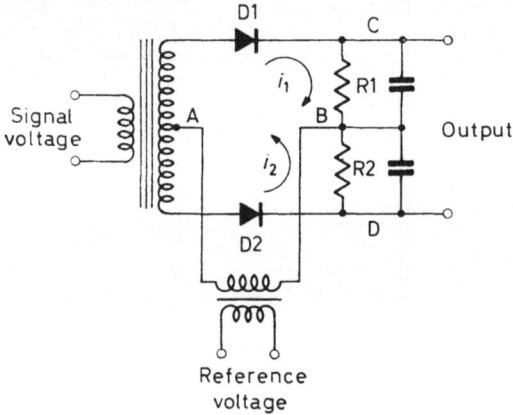

Figure 13.21. Basic phase-sensitive detector

phase changes would continue to move the output shaft away from the desired position. Clearly then we need some sort of device that can detect the change of phase and thus can 'sense' whether the output shaft is approaching or has overshot the desired position. A phase-sensitive detector is such a device.

After dealing with the principle of operation of this type of detector we shall be in a position to understand that another use of such a detector is the improvement of the signal-to-noise ratio in modulated systems. We therefore find phase-sensitive detectors being used to an increasing extent in scientific laboratories in such instruments as infra-red spectrometers and electron-spin resonance spectrometers. In fact, whenever a modulated system is being designed it is worth while considering the incorporation of a phase-sensitive detector to improve the signal-to-noise ratio.

Figure 13.21 shows the basic phase-sensitive detector. For descriptive purposes we shall assume that the signal and reference voltages are sinusoidal

404

although as a rule the reference voltage at least has a square waveform. Both the signal and reference voltages have the same frequency. In the absence of any signal voltage, diodes D1 and D2 conduct equally when the reference voltage is such that the point A is positive with respect to B. Assuming a balanced system with two identical diodes and balanced halves of the centre-tap transformer, the current through R1 will be equal to that through R2. The relative directions of the two currents is such that the points C and D are at the same potential with respect to the junction of R1 and R2. The result is that the output voltage is zero. Suppose now a signal voltage is applied which produces a voltage across the secondary of the signal transformer. During conduction both diodes conduct simultaneously, but unequally. Depending upon the phase of the signal voltage relative to the phase of the reference voltage, the voltage across one half of the secondary of the signal transformer is added to or subtracted from the reference voltage. (It is assumed that the signal voltage across half of the secondary is less than the reference voltage.) Let us suppose that the phasing is such that the voltage in the upper half of the secondary adds to the reference voltage. At the same time the voltage in the lower half subtracts from the reference voltage. D1 then conducts more current than D2, i.e. i_1 is greater that i_2. As a result the point C becomes positive with respect to point D by an amount which is proportional to the magnitude of the signal voltage. With the same magnitude of signal voltage, but a reversed phase, D2 conducts more current than D1, and C then becomes negative with respect to D. A phase reversal has thus produced a change of polarity of the output voltage.

The application of the detector to servo systems that use a.c. signals will be obvious enough. The detector is able to sense or which side of zero the system is lying by producing a voltage whose polarity is phase dependent. We may therefore use alternating signals and a phase detector to drive d.c. servo systems. (Alternatively, of course, for low powered systems, it is often convenient to use an a.c. servo amplifier with a two-phase servomotor. This enables us to dispense with the phase-sensitive detector. However if a d.c. servo system already exists and we wish to make use of it, the detector is useful when it is convenient to use alternating signals in the transducers.)

Let us see now how the detector discriminates against noise and other unwanted voltages. Depending upon the phase of the signal voltage there is a steady output voltage of a given polarity whose magnitude depends upon the magnitude of the input signal voltage. Superimposed upon the signal voltage it is assumed that there are noise and other unwanted voltages; the components of the noise have frequencies that differ in general from the reference voltage. Any noise component must then constantly be altering its phase relationship with respect to the reference voltage. The output voltage due to the noise component then has a polarity that is altering at the same rate as the phase alters with respect to the reference voltage. It is thus an easy matter to use a lowpass filter (similar to the filter used in a power pack) to filter out the alternating component due to the noise, but to allow the d.c. component due to the wanted signal voltage to remain. Had we used an ordinary linear detector or rectifier each noise component would have been rectified and a mean d.c. level would result that could not be satisfactorily filtered from the wanted direct voltage. The signal-to-noise ratio of a phase-sensitive detector is, therefore, superior to that of the conventional rectifier.

The detector is often used with chopper systems (e.g. chopped infra-red radiation) and thus we use square-wave reference and signal voltages. The transformers of *Figure 13.21* prove to be inconvenient for square waves, and therefore many practical circuits have been devised that do not require transformers. One very useful phase-sensitive detector has been designed by Faulkner and Stannett[10]. It was developed in the form of a general purpose laboratory instrument for use in the range 10 Hz to 100 kHz. Care has been taken to ensure that the operation of the instrument is not critically dependent on component values or valve characteristics. Their circuit, shown in *Figure 13.22*, is the subject of a British Patent Application No. 10,128/63 and is being used in a commercially available instrument. A photograph of the version built by the author is shown in Plate IV to illustrate the layout of electronic components in a conventional chassis arrangement.

Although the circuit is more complex than that of *Figure 13.21*, readers now have enough knowledge from previous chapters to recognize the major sections. In the main part of the detector circuit, the reference voltage, after amplification, is applied to a phase-splitter. The output from the phase-splitter is fed to a cathode-coupled pair of triodes (ECC82) via intermediate single-stage amplifiers. Each member of the cathode-coupled pair is alternately switched 'on' and 'off', i.e. into the conducting and non-conducting state, by the reference voltage. The common cathode load consists of a resistor and one of the triodes of an ECC81 double triode. This triode is fed from an amplifier driven by the signal. For a given phase of the signal causing the ECC81 to be in its low resistance state, one of the triodes of the ECC82 is in its 'on' state; this triode conducts harder than the other of the cathode-coupled pair, thus giving an output signal. It is the relative phase of the signal and reference voltages that determines which triode of the cathode-coupled pair conducts the greater current and this in turn determines the polarity of the output voltage. Further details of the operation of the circuit may be obtained from the paper quoted.[10]

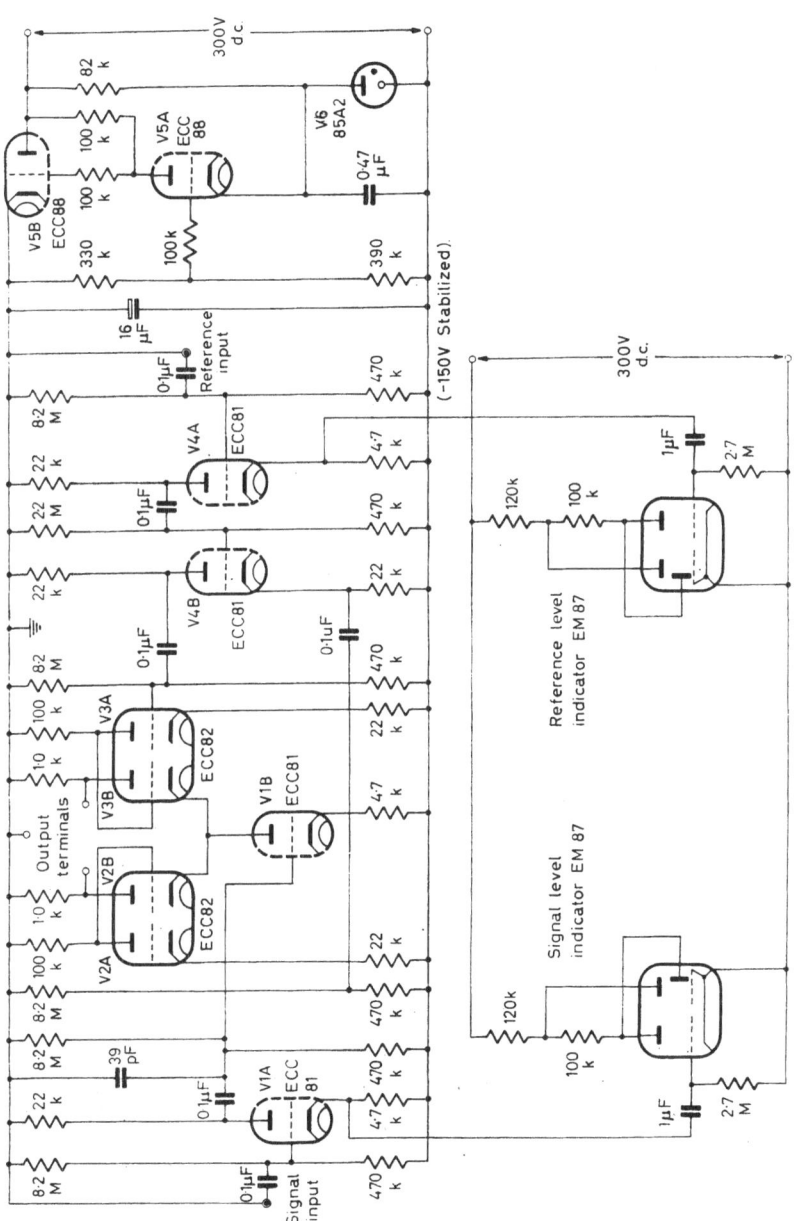

Figure 13.22. The general purpose phase-sensitive detector of Faulkner and Stannett[10]

COUNTING CIRCUITS AND DIGITAL INSTRUMENTS

Electronic circuits for counting events find many applications in nuclear physics experiments, industrial counting circuits, automatic control systems, rate meters, frequency meters, time measurers, etc.

We are accustomed by training to the counting system that counts in tens, i.e. the decimal system, but in electronic work reliability is greatly improved by using only two states (the 'on' and the 'off' state). We find, therefore, that the more usual electronic counters, especially in computers and automatic controllers, count in twos, i.e. they use the binary system. The discussion here, in the main, deals with decimal systems, and an account of the binary system appears at the end of Chapter 14 which deals with logic circuits. It was felt inappropriate to introduce the decimal counting tubes and rate meters in the chapter on logic circuits, whilst unnecessary duplication can be avoided by postponing a discussion on binary counters until after the logic gating circuits have been described. To obtain an overall picture of the various counting circuits used, it will be necessary to read this section in conjunction with the later parts of Chapter 14.

The starting point in any electronic counter is the transducer that converts the information to be counted into electrical pulses. The transducer may be a photoelectric cell where the energizing beam is interrupted by objects on a moving platform or belt; it may be a photomultiplier tube that responds to scintillations produced in suitable crystals by nuclear particles; it may be a geiger tube that produces pulses when a particle passes through the active volume; or it may be some sort of switch, electronic or mechanical, that is being operated by some process. In many cases the shape of the pulse is not suitable for the reliable operation of the counter. The pulse should be rectangular in shape and of suitable duration and amplitude. The usual procedure with sine waves is to amplify them and use biased diodes or zener diodes to clip the top and bottom of the waveform. In this way fast rise times can be achieved and the pulse height can be controlled by a suitable choice of bias voltages. For very short 'spiky' pulses such as are obtained in investigations of radioactive materials a different procedure is used. The main objection to the 'spiky' pulse is that the duration is too short to allow the counter to respond satisfactorily. The usual way of lengthening the pulse time and creating a pulse of satisfactory shape is to use a monostable multivibrator.

We have already met the astable or free-running multivibrator in the chapter on oscillators. By replacing one of the coupling capacitors with a suitable resistor we convert the free-running multivibrator into a monostable multivibrator, sometimes called the univibrator, one-shot multivibrator or 'flip-flop'. An example of the circuit is given in *Figure 12.23a* in which there is one stable and one unstable state. In the absence of any trigger pulse, the values of R1, R2 and R3 ensure that Tr2 is held off whilst Tr1 is in the conducting condition. On the receipt of a short positive pulse at the base of Tr1, that transistor is turned off. The collector voltage goes negative and this has the effect of turning Tr2 'on' by normal multivibrator action. The collector voltage of Tr2 therefore rises to 0V (near enough) and this voltage is transferred to Tr1 via C2 thus holding Tr1 in the 'off' condition. C2 discharges via

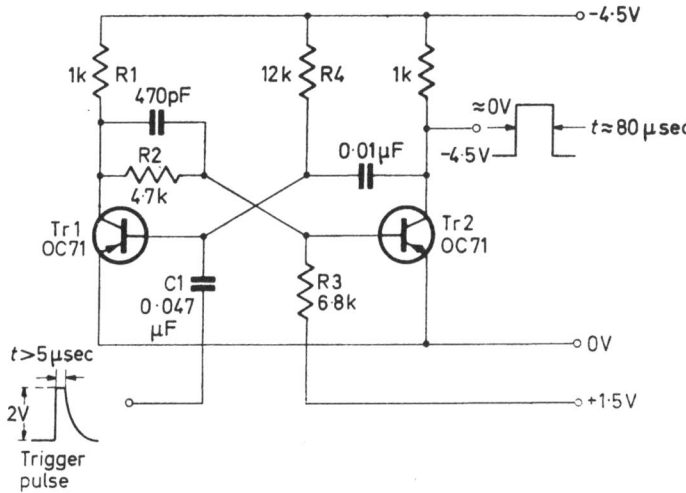

Figure 13.23a. A transistor univibrator used as a pulse shaper

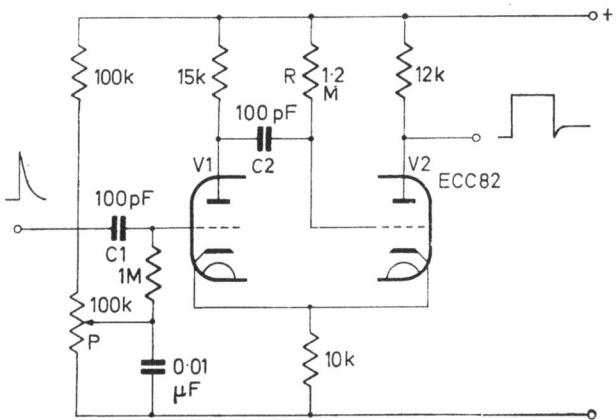

Figure 13.23b. The cathode-coupled univibrator used as a pulse shaper

409

R4 and Tr2, and eventually the base voltage of Tr1 becomes sufficiently negative for Tr1 to conduct again. As soon as conduction occurs in Tr1, Tr2 is driven to cut-off. In the absence of any further pulse the circuit remains in this stable state. The collector voltage of Tr2, therefore, returns to the negative supply voltage. The duration of the output pulse is controlled mainly by the time constant C2.R4. For the component values given the output pulse has a duration of about 80 μsec. The input pulse must have a duration that is long enough for the switching action to occur. For the given transistors the trigger pulse width must exceed 5 μsec. An improvement in the waveform can be effected by bypassing R2 with a small capacitor. The transition time between the unstable and stable states is thereby reduced.

The cathode-coupled univibrator of *Figure 13.23b* achieves the same purpose as the circuit of *Figure 13.23a* but the amplitudes of the waveforms are larger. In addition, variable control of the output pulse width is provided. In the stable state V2 is on whilst V1 is off. The large current drawn by V2 produces a large voltage across the common cathode resistor. The cathodes are therefore sufficiently positive to cut-off V1. The grid potential of V1 is set by the potential divider to a value sufficiently below the cathode potential to ensure that V1 is cut-off. When a large positive pulse is delivered via C1, to the grid of V1, this valve begins to conduct. The consequent fall in the anode voltage of V1 is transferred via C2 to the grid of V2 and V2 is thereby cut-off. The normal action associated with the cathode-coupled pair (described in the chapter on amplifiers) results in a transfer of current from V2 to V1. The circuit is thus driven into its unstable state. C2 now discharges until the voltage on the grid of V2 rises above the cut-off value. Once conduction recommences in V2 the grid-to-cathode voltage of V1 falls below the cut-off value. V1 is, therefore, cut-off and the current is transferred back to V2. The transfer is very rapid due to normal multivibrator action. The univibrator now remains in the stable state until the arrival of another input pulse. The time during which the univibrator is in the unstable state is determined by the time constant CR and the setting of the potential divider, P. Adjustments of P bring about variations of the duration of the output pulse since the adjustments determine the level of the bias voltage on V1. This voltage determines the current through V1 during the period in which that valve is conducting. This in turn sets the cathode potential which determines the time to be taken before the grid of V2 rises above the cut-off value.

The Ratchet Counter

Most readers will be familiar with the mechanical counter that is used on bicycles to count the number of revolutions of the wheel. Knowing the circumference of the wheel, it is possible to construct a gearing system that enables the counter to display distance directly. A similar principle is involved in motor-car distance indicators. Each time a mechanical impulse is delivered to the counter a wheel in the counter is made to rotate through a small angle. Each time the wheel makes a complete revolution an impulse is delivered to a second wheel, and so on for as many wheels as are necessary. A similar mechanical counter may be used in electronic systems provided the impulses are delivered at a rate not exceeding a few tens per second. *Figure 13.24* shows the principle of a very simple system in which a transistor is turned 'on' each time a pulse is received. The collector current energizes a sensitive relay which in turn operates the mechanical counter, and the number of pulses received are thereby counted.

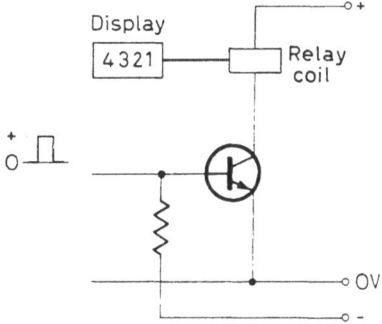

Figure 13.24. A simple electromechanical counter energized by a transistor. The transistor is held in the 'off' condition until the arrival of a positive pulse

The Gas-filled Decade Counting Tube (Dekatron)

The use of a mechanical counter imposes a severe limitation on the rate at which pulses may be counted. Where this rate exceeds about 50 impulses per second, it is necessary to use electronic counters. The electronic equivalent of the mechanical counter is the gas-filled stepping tube frequently referred to by the name dekatron (which is a trade name of Ericsson). The electrode assembly consists of a central disc anode surrounded by thirty equally spaced rod-like electrodes known as cathodes. Ten of the thirty electrodes are referred to as the main cathodes (k_0, k_1, k_2, etc.) and it is the glow discharge associated with any given cathode that indicates the number of the count. In a counting tube nine of the main cathodes are connected together electrically and brought out to a single pin on the base. The tenth cathode, k_0, is brought out to a separate connection. By including a resistor in the lead to the tenth cathode, an output voltage is developed across the resistor when the glow discharge rests on k_0. This voltage, when passed on to a second dekatron, results in a

411

division by ten of the incoming pulses. Each dekatron used, therefore, displays one decade of the count. Suitable bias voltages applied to the tenth cathode in each tube resets the count to zero.

The selector tube is identical to the counting tube in construction and mode of operation except that each main cathode is brought out to a separate pin in the base. Such a tube can be used for counting by connecting $k_1 - k_9$ together and returning them to the h.t. negative line. k_0 is still free and can produce output voltages. The advantage of the selector tube is that an output voltage can be produced with any cathode by the inclusion of a resistor in the lead to the selected cathode. This is useful in batching, for example, where an external circuit has to be operated after a count has been reached that is not a multiple of ten. The selector tube can also be used as an accurate timer when the input pulses are delivered at a steady known rate. Let us suppose that a circuit must be energized after a specified time (say 10·3 seconds). By feeding a chain of dekatrons with pulses from the mains (50 Hz), $50 \times 10\cdot3$, i.e. 515 pulses are received in the specified time. By selecting the fifth electrode in the third dekatron, the first in the second dekatron and the fifth in the first (or units) dekatron, three output signals can be made to energize relays via suitable circuitry; it is then arranged that an output voltage is produced only when all three relays are energized. (This is an example of an AND circuit described in the next chapter.) By suitable switching arrangements associated with the cathodes any time interval may be selected.

Interleaved with the ten main cathodes are two sets of subsidiary electrodes known as guide cathodes; each set contains ten cathodes. Those ten immediately following the main electrodes in a clockwise direction are called guide A cathodes by Mullard Ltd. and the ten electrodes following the guide A electrodes are called guide B cathodes. The guide A cathodes are all connected together electrically within the tube and brought out to a single pin in the base. The guide B cathodes are similarly connected and brought out to another pin. The arrangement is shown schematically in *Figure 13.25*. The function of the guides, as the name implies, is to guide the glow discharge from one main cathode to the next in a specified direction. In the absence of guides the glow may move forward or backward between the main cathodes in a random fashion, and then no reliance could be placed on the indicated count.

The electrode structure is enclosed in a glass envelope that contains inert gases at low pressure. During operation a glow discharge rests on one of the main cathodes and can be seen through the top of the tube. For convenience the tubes are usually mounted horizontally and the viewing end is circumscribed with an escutcheon engraved with the digits 1 to 9 and 0. For counting speeds up to 4,000 per second the filling gas is neon or neon/argon mixture. For high speed dekatrons operating up to 20 kHz, the filling gas is argon, often accompanied by hydrogen which acts as a quenching agent. In the Ericsson high speed dekatron thirty (instead of twenty) guide electrodes are used, but as the principle of operation is similar to the twenty guide electrode system we shall confine our explanation of the mode of operation to the latter type of tube.

Before describing the counting action of the tube, let us recall three facts about a glow discharge which were mentioned in connection with stabilizing tubes in Chapter 7. Firstly, when two electrodes are immersed in a gas at low pressure then the application of a sufficiently high potential difference between

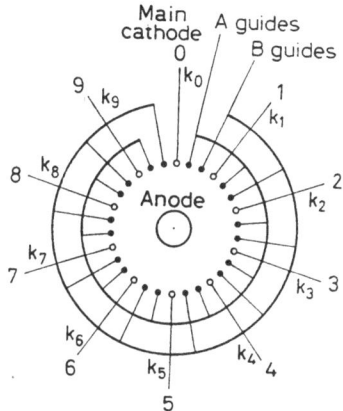

Figure 13.25. Schematic arrangement of the electrodes in a stepping tube. For ordinary counting purposes cathodes 1–9 are connected together; when the connection is within the tube the device is called a counter tube

the electrodes causes the gas to ionize and a glow discharge is established. The voltage necessary to initiate the discharge is called the striking voltage. Secondly, the voltage necessary to maintain the discharge is less than the striking voltage. Thirdly, during normal glow discharge the potential difference between the electrodes is almost constant. Any attempt to increase the voltage across the electrodes results in an increase of the discharge current. Since we normally use some sort of load resistor, any rise in the supply voltage results in a similar rise in the voltage drop across the resistor due to the increased discharge current; the voltage across the electrodes stays practically constant.

Let us consider now a basic circuit used for counting *(Figure 13.26)*. In the absence of any voltages on the guide electrodes the application of the main supply voltage results in a glow discharge being established on cathode k_0. This occurs because initially the voltage between the anode and k_0 is greater than that between the anode and any other cathode. k_0 is therefore the preferred cathode, and the glow is established there. Once the glow discharge is established a current passes through R_L and between the anode and k_0. The anode voltage, therefore, falls to a value that is determined by the maintaining or burning voltage. As this voltage is substantially lower than the striking voltage, glow discharges cannot be established on any other cathode. If for any reason the glow accidentally rests on a cathode other than k_0 when the apparatus is switched on, the application of a sufficiently high bias voltage to k_0 via a 'reset' switch ensures the transfer of the glow to k_0.

The glow discharge is transferred from k_0 to k_1 when a pulse arrives to be counted. The transfer is effected in the following way. When a pulse of sufficient magnitude and duration (120 V and 75 μsec for the Z502S tube) is

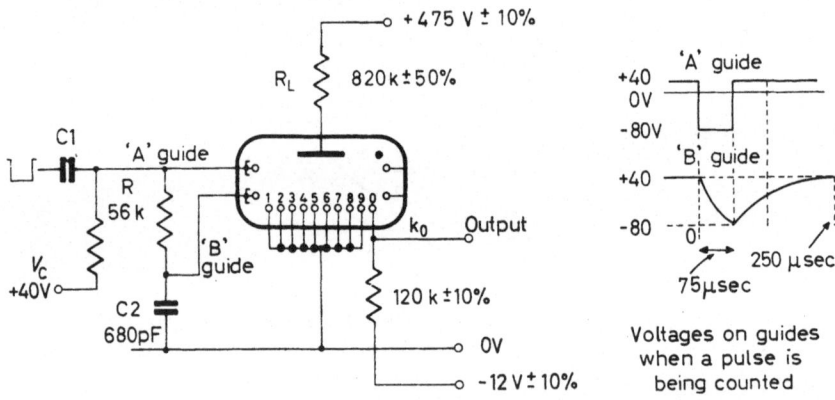

Figure 13.26. Basic circuit used for counting. The indicated voltages and component values are for a Mullard Z502S tube. The heavy dot indicates a gas-filled tube

applied to the A guides as shown in *Figure 13.26* the potential difference between the anode and the A guides exceeds the striking voltage. The A guide adjacent to k_0, however, is in a favoured position because of the presence of ions in the glow discharge on k_0. This particular guide is said to be primed and because its potential is now lower than k_0 the glow transfers from k_0 to the adjacent A guide. Once this occurs the anode voltage falls because the maintaining voltage between the anode and guide must be the constant value set by the gas ionization potential and electrode geometry. The voltage between the anode and k_0 is now below the maintaining voltage and the glow on k_0 is therefore extinguished. The next guide B in a clockwise direction is now primed because of the glow on guide A. The breakdown potential of this guide is therefore less than any other unprimed B guide. The arrival at guide

A of the pulse to be counted does not effect a change of the potential of guide B immediately because of the presence of C2. The potential of guide B falls exponentially at a rate determined by the time constant C2.R. After 75 μsec the pulse to be counted disappears and guide A returns to $+40$ V. Guide B, however, remains at -80 V because C2 does not lose its charge instantaneously. The primed guide B adjacent to k_1 therefore takes over the glow discharge. The potential on this guide B rises as C2 discharges. The anode voltage must rise simultaneously to satisfy the constant maintaining voltage condition. After approximately a further 75 μsec the voltage on guide B has risen above 0V. The voltage between the anode and k_1 is thus greater than that between the anode and guide B. Since k_1 is primed, the glow is transferred to k_1. The potential on guide B rises to $+40$ V and thus the glow on that guide is extinguished. We have transferred the glow from k_0 to k_1. Because the glow now rests on k_1 the adjacent B guide from which the glow came and the second A guide are both primed. However on the arrival of the next pulse to be counted the glow must continue to move in a clockwise direction because guide B does not receive the pulse until 75 μsec have elapsed from the time the pulse arrived at the A guides. We see, therefore, that having two intermediate guides between any pair of main cathodes ensures a clockwise rotation of the glow. This is guaranteed by applying a pulse to the A guides before it is applied to the B guides. Reversing the order in which the pulses are applied to the guides reverses the direction of rotation of the glow. This is useful when a counter must be reversible; such a counter can subtract or add the number of successive pulses received to the count already registered. The addition or subtraction process is selected by a switch that determines the order in which the pulses are applied to the guides.

Since it takes about 250 μsec for a reliable transfer of the glow discharge from one main cathode to the next, the maximum counting rate is 4,000 pulses per second.

The first counting tube in a chain is driven from an external source. It is necessary, therefore, to use a pulse shaping circuit to provide a negative pulse of the requisite amplitude and duration. The valve pulse shaper of the type already discussed is satisfactory for the purpose.

Sine waves can be used to drive dekatron tubes, which is useful when the tubes are to be used as timers. The sine waves may be obtained from the mains or from an oscillator operating at a known frequency. The circuit recommended by Mullard for their stepping tubes is given in *Figure 13.27*. The value of C should be 0·1 μF for 50 Hz operation, 0·005 μF for 1 kHz and 680 pf for 4 kHz. Further information on the operating principles and circuits for counting tubes may be obtained from the references to be found at the end of the chapter.[11, 12]

The output pulse from k_0 is not satisfactory for driving a further dekatron directly. The shape, magnitude and duration are unlikely to be correct and the pulse has the wrong phase (i.e. it is a positive-going instead of being negative-going). It is necessary, therefore, to use coupling circuits between the tubes. Many circuits have been devised some of which are discussed in the references just cited.

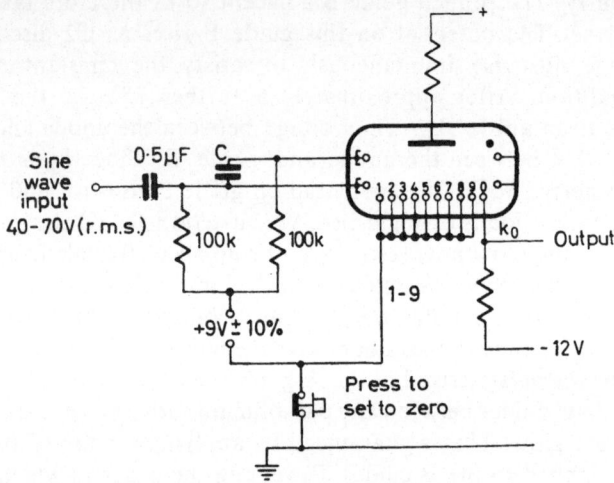

Figure 13.27. The input circuit for sine wave drive recommended by Mullard Ltd. in Technical Communications Vol. 4, no. 37

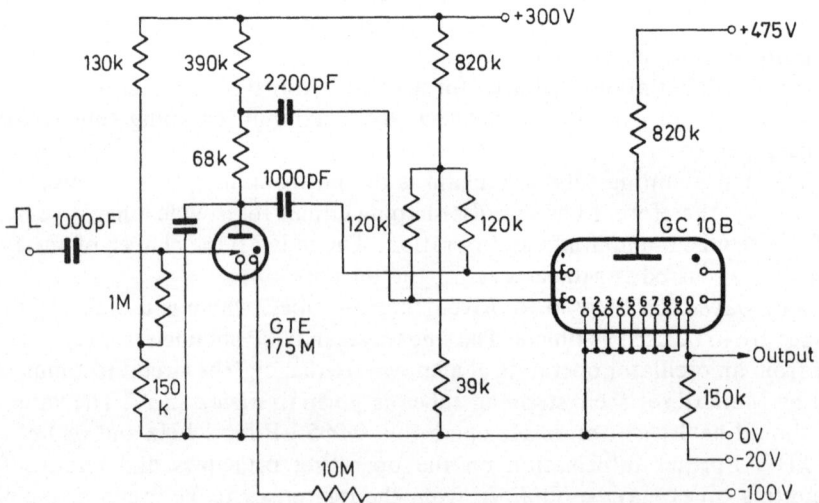

Figure 13.28. A coupling circuit recommended by Ericsson for their dekatron tubes, where the counting speed does not exceed 500 counts per second

416

A popular coupling circuit is given in *Figure 13.28*. It makes use of a trigger tube to convert the positive-going output pulse from one stage to the specially shaped negative-going pulses required to drive the guides of the subsequent stage. The trigger tube is a gas-discharge tube in which there is an anode, a cold cathode and an auxiliary electrode used for triggering purposes. The trigger electrode is closer to the cathode than to the anode and so the break-down or striking voltage between the trigger electrode and the cathode is substantially less than that between the anode and cathode. In operation, an anode-cathode voltage is applied that exceeds the maintaining or burning voltage for those electrodes, but which is less than the striking voltage. The trigger voltage is maintained just below the striking voltage for the trigger-cathode gap. When a pulse (of about 15–20 V and 100 μsec duration) is applied to the trigger electrode, breakdown occurs and ionization of the gas between the trigger and the cathode is established. This causes breakdown in the main anode-cathode gap, and so the tube strikes and a self-sustained discharge occurs that is unaffected by the decay of the trigger pulse. We have met a similar phenomenon when considering the action of the grid in a thyratron tube. To increase the reliability of firing, tetrode trigger tubes have been devised in which a subsidiary cathode is introduced. With this type of tube the main current builds up rapidly to its full value which is an advantage in triggering circuits. The function of the subsidiary cathode is to ensure a sufficient supply of available electrons within the tube when the main discharge is initiated.

Once the tube has fired, it can be extinguished only by reducing the anode voltage below the maintaining potential. This is achieved in *Figure 13.28* by placing a capacitor across the tube. Before striking, the capacitor charges through the anode load resistors. Once the main discharge is initiated the capacitor discharges via the tube and the anode voltage falls below the maintaining voltage. In the meantime, the fall of anode voltage provides the pulses at the A and B guides. For normal clockwise rotation the pulse delivered to B is delayed with respect to that delivered to A by using two RC couplings of different time constants.

The manufacture of transistors that withstand high voltages has led to the use of transistor coupling circuits. One such circuit devised by Frazer[13] is shown in *Figure 13.29*. The pulse received from k_0 is amplified by the transistor and, since a single-stage coupling is involved, there is a phase reversal. The transistor replaces the gas trigger tube and thus eliminates the need for critical trigger bias supplies.

Reversible counters are useful when it is desired to subtract the number of incoming pulses from the number already registered in the counter. As has already been explained, to achieve reversible counting, we must arrange that the two pulses to the guides are delivered in reverse order so that the A pulse is delayed with respect to the B pulse. A reversible counter has been described by Oxley.[14] The principle of a bidirectional driving circuit is shown in *Figure 13.30*. Both triodes are normally cut-off by the application of a sufficiently

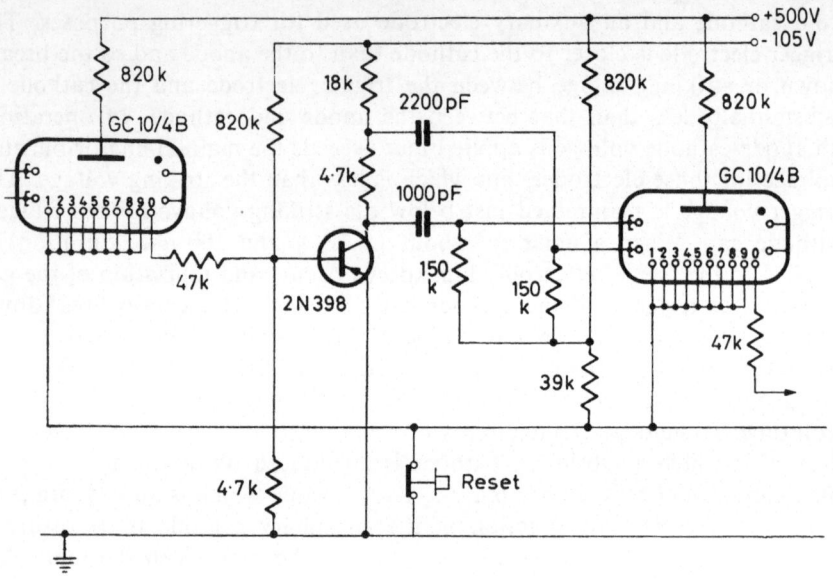

Figure 13.29. The transistor coupling circuit of Frazer[13]

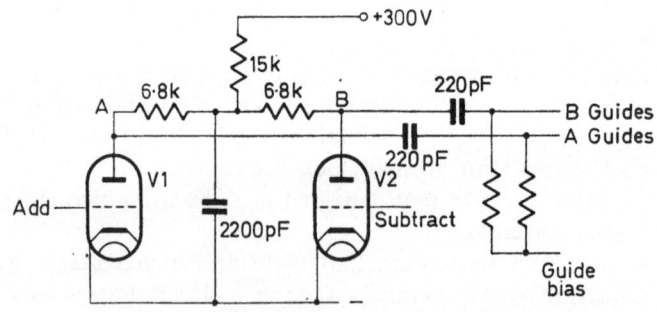

Figure 13.30. Basic circuit for bidirectional counting

large negative voltage to the grids. The points A and B are therefore initially at the same potential. When pulses are to be added they are delivered to the grid of V1. The pulses for this circuit must be positive-going. The arrival of the pulse causes conduction in V1 and an immediate drop in the anode potential. This drop is conveyed as a negative pulse via the coupling capacitor to the

418

A guides of the dekatron. There is a fall in the potential at the anode of V2 due to the current in the common 15 k load resistor. This fall is, however, delayed by the presence of the 2200 pF capacitor. When the pulse to be counted disappears both anode voltages return to 300 V. Throughout the duration of the pulse, the 220 pF capacitor feeding the A guides charges owing to the glow discharge current on the guide. The result is that although the anode potential of V2 is higher than that of V1, by the time the pulse ends,

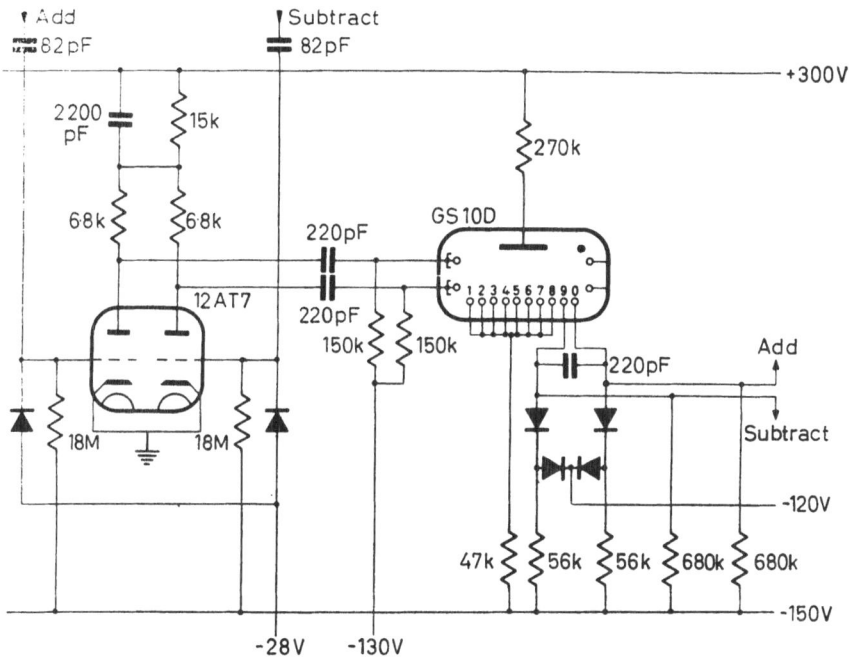

Figure 13.31. One decade of a 10 kHz reversible counter designed by Oxley[14]

and the two anode potentials become equal, the potential on the B guides is lower than that on the A guides. The glow, therefore, transfers from the A guide to the B guide. Subsequently when both guides return to a positive bias voltage the glow is transferred to the next main cathode. Because of the symmetry of the circuit we can see that an application of an incoming pulse to the grid of V2 causes a pulse to appear on the B guide before it arrives at the A guide. The glow therefore travels in an anticlockwise direction, which is equivalent to subtraction. One decade of a 10 kHz reversible counter due to Oxley is shown in *Figure 13.31*.

CIRCUITS FOR THE COUNTING OF RADIOACTIVE PARTICLES

Dekatron decade counters of the type described above are in common use in laboratories associated with radioactive materials. The detectors of radioactive particles deliver pulses of short duration and therefore the usual arrangement is to follow the detector with a pulse-shaping circuit and then on to the dekatron counter. A frequently used detector is the photoelectron-multiplier tube that detects the small scintillations produced in crystals such as sodium iodide by the radioactive particle. These tubes have already been described in the chapter on photocells.

A very common and well-known tube for the detection of radioactivity is the Geiger–Müller tube often referred to as a geiger tube. Such a tube, used for counting α and β particles, consists of a cylindrical tube with a very thin glass or mica window at one end. A coaxial wire (the anode) extends for most of the length of the tube. The tube is filled with gas at a pressure of a few centimetres of mercury. A typical filling is argon at 10 cm Hg pressure together with a quenching agent. The nature and reason for the quenching agent is discussed below.

The central wire anode is maintained at potentials from 300 V to 2,000 V above the outer tube or cathode. The particles enter the end window and a slight ionization of the gas is produced. The electric field between the cathode and anode accelerates the ions and gas multiplication takes place. For low anode voltages the ionization is restricted to the immediate vicinity of the primary ionization and does not spread along the tube. The resulting pulse height is directly proportional to the number of primary ions. Under these circumstances we are said to be using the tube in the proportional region, and as a proportional counter. On increasing the anode voltage, the ionization that results from the entry of a particle spreads right along the tube. The pulse height is then independent of the energy of the particle. We are now in the Geiger–Müller region. There is a range of anode voltage over which this region extends which is known as the G.M. plateau. If the anode voltage exceeds the value for operation on the plateau, continuous discharge takes place and the life of the tube is adversely affected.

Once a normal discharge has been initiated, the entry of a further particle cannot be detected until the tube has deionized. The time taken for the tube to return to its original state is known as the dead time. To control the dead time, we must arrange to quench the tube. Modern tubes include ethyl alcohol in the argon gas to quench the ionization. Ethyl alcohol dissociates and in doing so absorbs energy (which is mostly in the form of ultra-violet radiation) that would otherwise prolong the gas multiplication process. The life of a tube filled with ethyl alcohol is limited and therefore other quenching agents are used. These are normally halogens, notably bromine, which recombine after dissociation. The released energy of recombination is in the form of heat which does not prolong the gas multiplication process. An alternative method is to use electronic quenching, an example of which is given in *Figure 13.32*. Once the pulse is received at the grid of V1 a cumulative

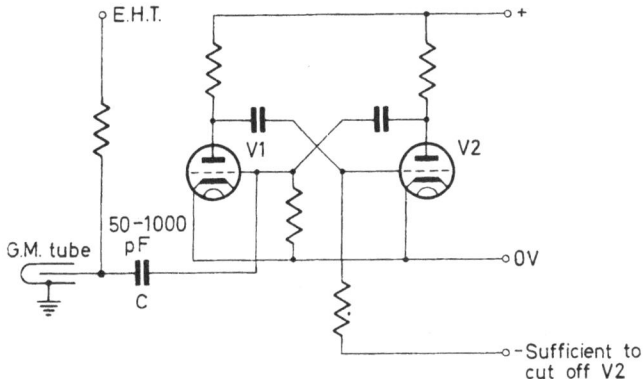

Figure 13.32. Electronic quenching of a G.M. tube

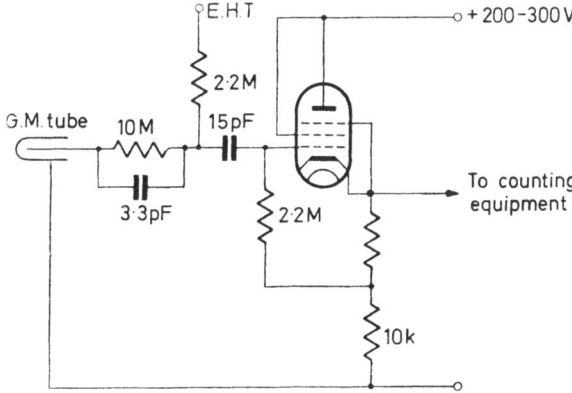

Figure 13.33. Cathode-follower circuit for use with halogen quenched tubes

action takes place in which the potential of the grid of V1 is rapidly reduced. This reduction is transferred via C to the anode of the tube thus reducing the anode voltage below that necessary to sustain gas multiplication. *Figure 13.33* shows a typical circuit used by 20th Century Electronics Ltd. for one of their halogen quenched tubes.

Solid state detectors are being used increasingly in the study of radio-active materials. When radiation enters a detector, electron-hole pairs are formed as described in Chapter 5. When these pairs are formed in the depletion layer of a *pn* junction a leakage current is produced the magnitude of which depends upon the number of pairs formed, i.e. on the energy of the incident radiation. The depletion layer width is a function of the applied reverse bias and so we have an easy means of varying the active volume of the detector. When an alpha particle enters the detector biased with only a

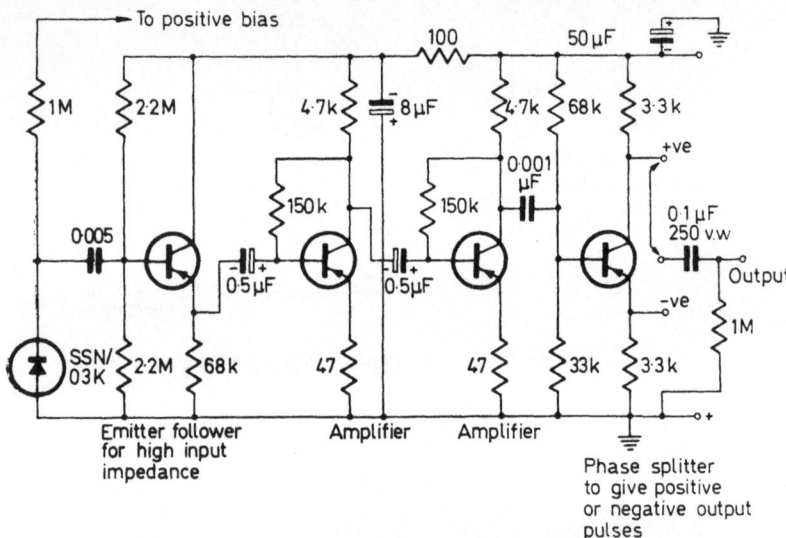

Figure 13.34. Amplifier for use with the 20th Century Electronics Ltd. solid-state detector for the detection of α-particles. All transistors are OC44 types

few volts, all its energy is converted into electron-hole pairs because alpha particles lose their energy within a distance of a few microns. Since the bias voltage is small the depletion layer is narrow and only alpha particles are detected. The electron-hole pairs formed are collected and passed to an amplifier circuit. The pulse formed is proportional to the energy of the incident radiation. These pulses may then be passed to a pulse shaper and then to a dekatron counter. Commercial counters, of course, incorporate the pulse shaper in their input circuits. A suitable amplifier published by 20th Century Electronics for use with their type SSN/O3K detector is given in *Figure 13.31*. The function of the individual sections of the amplifier is indicated on the diagram.

The Schmitt Trigger Circuit

This circuit is very useful when we wish to discriminate between noise or low amplitude pulses and pulses of larger amplitude. Although basically a voltage discriminator, and widely used as such, the circuit also acts as a pulse shaper and will operate only on input signals above a pre-determined voltage level. From the last section we recall that the pulse produced by the detector, and hence the amplifier, is proportional to the energy of the absorbed particle. If we lead the output pulses straight into a counter, all pulses are counted, and information regarding the energy of the particle is lost. However, if we interpose a Schmitt trigger the latter acts as a discriminator in that it rejects all pulses having a peak amplitude less than some chosen level, but accepts all those pulses with a peak value exceeding that level. Thus we can assess the energy of the incident particles.

Schmitt trigger circuits vary in detail depending upon the required rise time of the output pulse, the heights of the available input pulses and the operating voltages involved. A typical example of this trigger circuit is given in *Figure 13.35*. It bears a resemblance to the univibrator circuit of *Figure 13.23b*.

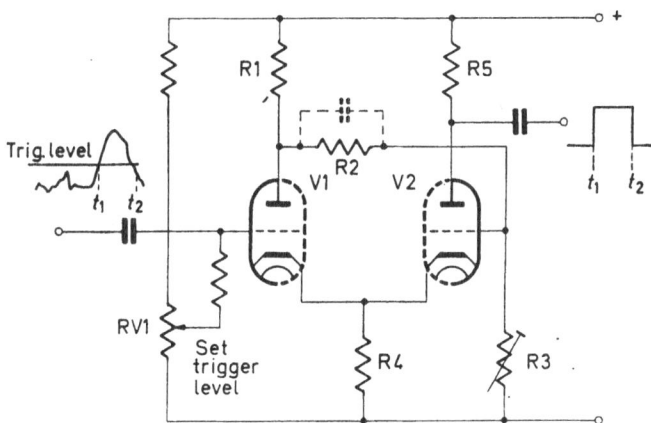

Figure 13.35. A typical Schmitt trigger circuit. Note that the turn-off time t_2 does not occur until the signal level has fallen below the trigger level (see text)

The univibrator circuit has only one stable state, however, and is triggered by the leading edge of the incoming pulse. The Schmitt trigger circuit on the other hand has two stable states and is, therefore, an example of a bistable multivibrator. The example given in *Figure 13.35* is a cathode-coupled bistable multivibrator. Unlike the astable or free-running multivibrator, that gives a continuous output of oscillations with a square waveform, the bistable multivibrator changes from one stable state to the other on the receipt of suitable signals. In the case of the Schmitt trigger, the leading edge of the input pulse changes the condition of the circuit from one stable state to the other and the trailing edge returns the circuit to its original state. In the case of the uni-vibrator, the circuit returns to its original state irrespective of the time delay or shape of the trailing edge of the input pulse.

The trigger of *Figure 13.35* works in the following way. In the absence of any input, V1 is cut-off and therefore the anode voltage of V1 is almost at the same potential as the h.t. positive line. The potential divider R1, R2, R3 maintains the grid of V2 at a value that ensures a fairly heavy current through V2. R3 is usually adjusted to make the grid of V2 about − 1 V with respect to the cathode. RV1 is then adjusted so that the voltage on the grid of V1 is below the cut-off value. The cathode of V1 (and V2) is usually 50 V or 60 V positive with respect to the h.t. negative line owing to the current through the common cathode resistor R4. Let us suppose that the cut-off voltage for V1 is − 10 V. If RV1 is set so that the grid is − 15 V with respect to the cathode then all incoming pulses with a peak height of less than 5 V will fail to make V1 conduct. These pulses are therefore rejected by the trig-ger; RV1 therefore sets the discriminator or trigger level. Positive pulses exceeding 5 V make V1 conduct, whereupon the anode voltage of V1 falls. This causes a reduction in the voltage on the grid of V2. Because V2 is coupled to V1 via the common cathode resistor a cumulative action occurs in which V1 is rapidly turned on whilst V2 is turned off. The action is often accelerated by the addition of a small capacitor (20–40 pF) connected in parallel across the coupling resistor R2. The rise time of the output voltage is therefore shorter than that of the input voltage. The trigger remains in this stable state with V1 conducting, for as long as the incoming pulse has a magnitude that exceeds the trigger level. As the input voltage diminishes, a point is reached where a further cumulative action returns the circuit to its original state. The input voltage at which this occurs is less than the pre-vious triggering level by a few volts. The phenomenon is called 'backlash' or 'hysteresis' and is due to the cathode-follower effect in V1 when that valve is conducting. During the rise of grid voltage on V1 due to the arrival of the input pulse, the cathode voltage is not affected because V1 is cut-off. When V1 is conducting, however, any fall in grid voltage tends to be followed by the cathode because of the cathode-follower effect. Before V1 can be cut-off, and a cumulative action started, the grid voltage of V1 must fall a few volts below the initial triggering level. Reliable operation as a discriminator can, therefore, be guaranteed only when the smallest pulses to be discriminated against have a peak voltage exceeding about 5 V. This then is the approximate threshold voltage with a normal Schmitt discriminator.

424

NUMERAL TUBES

Dekatrons are not the only types of tube available that give a visual indication of number. In instruments such as digital voltmeters, frequency meters and timers, it is convenient to have the number shown as a standard numeral. One such tube for performing this function is the cold cathode numeral tube known as a digitron. Basically, it resembles the dekatron but the cathodes, instead of being rod-like electrodes, are stacked elements in the form of metallic numerals. There are no guide electrodes. The application of a negative potential to the selected numeral with respect to a common anode results in a glow discharge, and the particular numeral is then displayed. The visible glow is larger than the size of the cathode numeral, and therefore the presence of the other metallic numerals does not interfere with the clarity of the display.

Optical projection displays are considered by some to give a superior visual display from the point of view of clarity. Whilst this is no doubt true such displays are more complicated to construct and therefore more costly to buy. In one system a set of lamps is used, each one representing the numeral or character to be displayed. The required lamp is selected by a suitable circuit, and the light from the lamp is then passed through a mask to form the given number's shape. Subsequently, a simple optical system causes the light to fall on a screen. The particular lamp may be selected by a subsidiary dekatron or more usually by a decoding circuit similar to that described at the end of the next chapter. (Sketches representing this form of tube and the digitron are shown in *Figure 13.36*.) An alternative to using a lamp for every

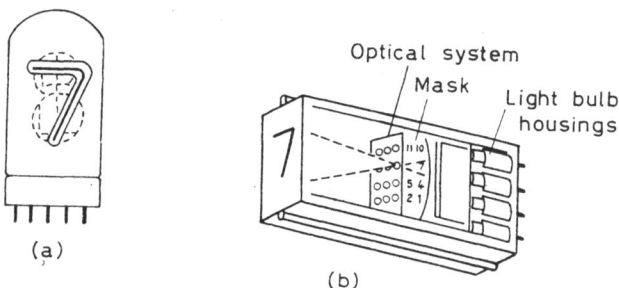

Figure 13.36. Two common forms of digital display. (*a*) The digitron cold cathode discharge tube and (*b*) the optical projection system

numeral is to use a single lamp together with a motor driven belt. The numerals or other characters are cut out along the belt which passes between the lamp and optical systems thus displaying the particular character that happens to be in front of the lamp at any given time. An electromagnetic switching system energizes the motor, and moves the belt when a mismatch exists between the code representing the character and the position of the belt. When the correct character is displayed the motor stops.

Another decade counting tube that displays its information in the form of numerals is the E1T tube. The tube is, in effect, a miniature cathode-ray tube in which a thermionic cathode emits electrons that are formed into a narrow beam by conventional methods. The beam passes between two deflector plates and is made to take up one of ten stable positions by being passed

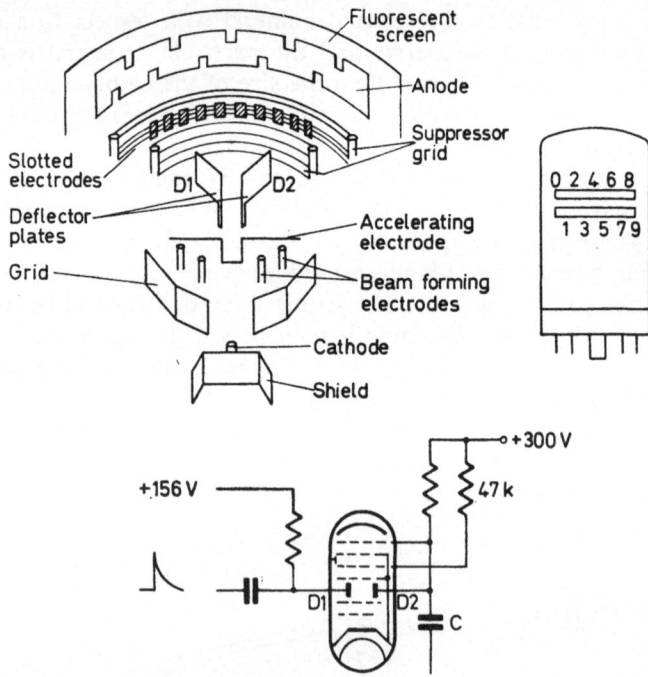

Figure 13.37. The Philips E1T beam switching counter tube and basic operating circuit

through a special arrangement of slotted electrodes *(Figure 13.37)*. On emerging from the anodes the beam strikes a fluorescent screen. The glow of the screen is suitably masked so that only the shape of a particular numeral is visible. Alternatively, the position on a ribbon-shaped screen is indicated by an adjacent numeral.

The operation of tube can be understood by considering *Figure 13.37* showing the electrode structure and the basic associated circuit. In the absence of any incoming pulse the beam passes between the slotted electrode and most of the beam current is collected by the slotted anode. A proportion passes

426

through to illuminate the screen. If the beam attempts to move out of the stable position, it is intercepted by the slotted electrode. Let us suppose that an attempt is made by the beam to move to the left, e.g. because of a gradual increase in the potential of D1. The beam is cut-off from the anode, the anode current falls and so the anode potential rises. The anode and D2 are connected together, so the potential of D2 rises thus bringing the beam back to its original position. This position is not quite symmetrical in relation to the slot. The beam is actually biased towards the left-hand edge of the slot. Any tendency for the beam to move to the right is, therefore, accompanied by an increase in anode current. This results in a fall in the potential of the anode and D2, which results in a movement of the beam back to its former position.

The behaviour of the beam is different, however, when pulses having a fast rise time and suitable shape are applied to D1. Owing to the presence of C the potential on the anode and D2 cannot change instantaneously. Although C is small, and comparatively slow changes of anode voltage are possible, any sudden increase in the potential of D1 is not immediately accompanied by an increase in the potential of D2 and the anode. The beam is, therefore, deflected into the next stable position. Suitable pulse-shaping circuits control the amplitude and duration of the pulse presented to D1 and, together with a careful choice of C, movement past the adjacent stable position is prevented. A reset anode is provided at the extreme left-hand position of travel. This is used for zeroing the indication and providing a carry-pulse for the next tube. Operational frequencies up to 30 or 40 kHz are possible with the Philips E1T tube, which constitutes the main advantage of this tube over the conventional dekatron. Other tubes working on similar principles to the E1T tube can count pulses with repetition rates up to 1 MHz. For counting speeds in excess of this, it is necessary to use fast binary counters as explained in the next chapter.

THE DIODE PUMP

We often need to know not the total number of pulses received by a transducer, but the rate at which those pulses are arriving. Counters can give us this information when the total number of pulses are delivered in a known period of time. Such an elaborate piece of equipment is unnecessary, however, when only a simple indication of the rate is required. One popular ratemeter is based on the diode pump, the circuit for which is given in *Figure 13.38*.

The function of the amplifier and pulse shaper is to present the diode circuit with pulses of constant amplitude and duration. On the arrival of the first positive pulse C1 charges via R and D1, which conducts because it is forward-biased. D2 is reverse-biased so no current passes into the meter or C2. The current that flows, i_1, depends upon the pulse height and width as well as

C1 and R. After the first pulse has disappeared the input voltage becomes zero and the input is effectively shorted. The charge on C1 now flows as i_2 via D2 and the parallel combination of the meter and C2. The polarity of the voltage on C1 is such that when it is discharging D2 is forward-biased and D1 is reverse biased. By choosing the value of C_1 to be much less than that of C2, practically the whole of the initial charge on C1 is lost to the meter/C2 combination. As C1 is practically completely discharged, it is in its former state to receive the next pulse. We therefore have the position where the charge delivered to C1 by the pulse is 'pumped' into the 'reservoir' C2. The meter

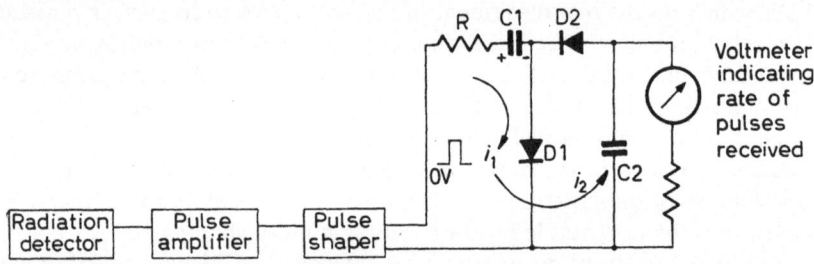

Figure 13.38. The diode pump ratemeter

receives current from C1 during the pumping period and C2 during the period that a pulse is being delivered to the input terminals. C2 therefore averages out the meter current. For given values of circuit constants and pulse shape the current through the meter, and hence the indication, is a function of the rate at which pulses arrive.

The circuit outlined above is called a ratemeter when it is used in connection with radiation or other transducers that yield pulses that are delivered at a varying rate. If the output from an oscillator, or other device operating at a fixed frequency, is fed into the pulse shaper the ratemeter performs the same function, but is now called a frequency meter.

DIGITAL INSTRUMENTS

When measuring voltage, time and frequency it is becoming increasingly popular to dispense with analogue instruments such as moving-coil meters and potentiometers and to use digital instruments as alternatives. Such instruments deliver their information in the form of a numerical display using digitrons or optical projection indicators. These instruments are capable of a high degree of accuracy and, because of the nature of their operation, it is possible to obtain digital information in a form suitable for feeding into a digital computer or data processing apparatus. It is not possible here to do more than sketch the outlines of the principles of operation of a digital voltmeter, frequency meter and timer. For those seeking further information a useful introductory book is *Digital Instruments* by K. J. Dean (Chapman and Hall, 1965).

Digital frequency and time measuring instruments rely on gating circuits for their operation. *Figure 13.39* shows a gating circuit which, although not

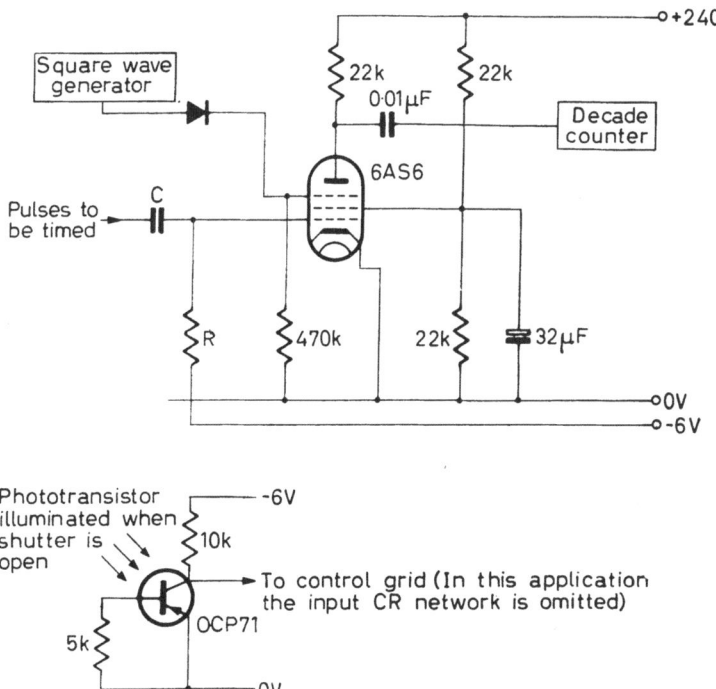

Figure 13.39. Gating circuit using the 6AS6 pentode. The circuit can be used as a timer or digital frequency meter (see *Figure 13.40*). As shown the author has used the circuit to time camera shutter speeds. A simple phototransistor circuit feeds pulses into the first grid which are then timed as described in the text

used in modern digital instruments, serves to illustrate the principles. The modern instruments use semiconductor devices to achieve the same ends. Semiconductor gating circuits are often referred to as AND gates, which use a type of logic circuit that is discussed in Chapter 14.

The circuit of *Figure 13.39* uses a specially constructed pentode in which the suppressor grid and first grid both act as control grids. It is necessary for the voltage on each grid to reach an appropriate level before the valve con-

429

ducts. Let us consider now the action of the circuit as a timer. When some event has to be timed the appropriate signal is applied to the first control grid. For example, if we wished to time the shutter speeds of a camera, a photocell can be arranged behind the shutter. A beam of light is made to pass through the shutter and to illuminate the photocell when the shutter is open. During the period that the photocell is illuminated the voltage developed across the photocell load is applied to the 6AS6 so as to drive the first control grid sufficiently positive with respect to the bias voltage to 'open the gate'. Any positive pulses that arrive simultaneously on the second control grid cause a substantial rise in the anode current. Each pulse delivered to the second control grid, therefore, gives an output pulse at the anode. The output pulses are then fed to a decade counter. Knowing the frequency of the pulses fed to the second control grid, and the count, we can calculate the period during which the gate was open. If, for example, the oscillations from the oscillator had a period of 10 μsec then a count of N on the decade counter represents a camera shutter speed of 10 N μsec. With no input voltage to the first control grid, that electrode is biased back sufficiently to 'close the gate'. No output pulses are then delivered to the counter even though pulses are being applied to the second control grid.

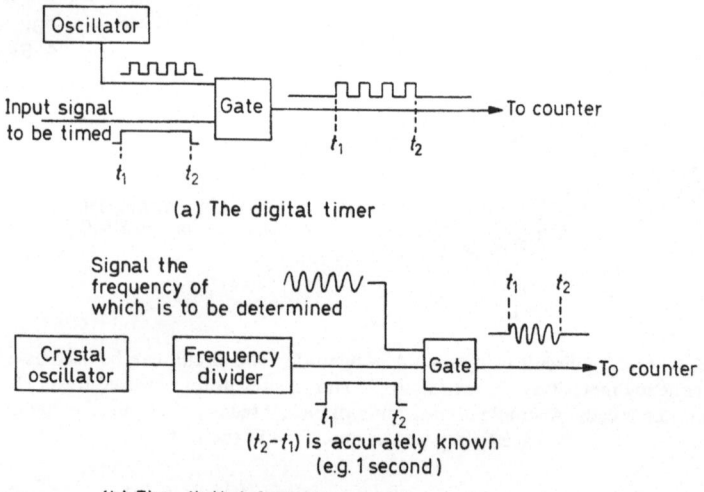

(a) The digital timer

(b) The digital frequency meter

Figure 13.40. The use of an electronic gate to measure time and frequency. The display of the information is in digital form

To use the circuit as a frequency meter *(Figure 13.40b)*, highly stable oscillations from a crystal oscillator, operating at an accurately known frequency, are first fed to a chain of multivibrators that act as a frequency divider (see

Chapter 11, *Figure 11.7*). Each multivibrator is designed to oscillate at 1/10th the frequency of the previous multivibrator. By using each multivibrator as a trigger for the next one in the chain, and by triggering the first multivibrator with the output from the crystal oscillator all multivibrators are kept under rigid control with respect to their frequency of operation. The output from the last multivibrator is applied to the gate. If the duration of the output pulse from the frequency divider is 1 second then the indicated count gives the frequency of the signal applied to the second control electrode directly. Special circuitry is used to reset the counter during the intervals between the arrival of the 1 second pulses.

Digital voltmeters use one of several principles and we shall briefly describe two of them here. A further type employing bistable controlled transistor switches, logic gates and decoding matrices is described in the next chapter.

A ramp function digital voltmeter makes use of a high gain amplifier to compare the unknown voltage with an internally generated reference voltage. This reference voltage has a linear ramp waveform (i.e. sawtooth waveform) and its production has already been discussed in the chapter on relaxation oscillators and also in connection with cathode-ray oscilloscopes. For a digital voltmeter special precautions must be taken to ensure that the rise of voltage with time is precisely linear. At the commencement of the rise of the ramp voltage a logic gate is opened and pulses from a stable oscillator, operating at an accurately known frequency, are counted in the way described in the previous section. Once the ramp voltage reaches the unknown voltage the logic gate is closed. (The way in which a transistor logic gate works is explained in Chapter 14.) Counting then ceases. We have, in effect, measured the time it takes for the ramp voltage to reach the unknown voltage. Since the ramp voltage rises linearly with time this is equivalent to measuring the particular ramp voltage that is needed to close the gate, i.e. the unknown voltage has been measured. It is, of course, necessary for the manufacturer to design his equipment so that with a particular frequency of operation of the oscillator, and with a chosen ramp slope the indicated count represents voltage directly.

A second method uses a servo system and potentiometer slide wire similar to those used in a potentiometric recorder. The chart and pen are superfluous. The unknown voltage is compared with that developed along a high-grade potentiometer wire. The number of revolutions of the servo motor necessary to effect a balance is counted and the count represents the input voltage. Suitable gearing ensures that the digital display shows the voltage directly.

REFERENCES

1. References for valve-voltmeters.
   (a) Bull, C. S. 'The diode as an a.c. voltmeter'. *J. Sci. Instrum.* Oct. 1947.
   (b) *Philips Tech. Rev.*, Jan. 1950. p. 206 gives details of the Philips GM 6006 voltmeter.
   (c) Amos, S. W. 'Simple valve voltmeter'. *Wireless World*, 1950, **56**, No. 12, 430 (Dec.).
   (d) Waddington, D. E. O'N. 'Silicon transistor multivoltmeter'. *Wireless World*, 1966, **72**, No. 3, 111 (March).
   (e) A transistor d.c. voltmeter with an inaccuracy of less than 1%, voltage ranges from 200 μV to 32 V in eight ranges and an effective input resistance of 1 M per volt is described in a Mullard publication *Educational Electronic Experiments—No. 2. 1 M/Volt d.c. Voltmeter*.
   (f) A particularly simple reflex type valve-voltmeter is described by Scroggie, M.G. in *Wireless World* for October, 1949, p. 401 and also in his book *Radio Laboratory Handbook* (Iliffe).
2. 'A low voltage electrometer' *Educational Electronic Experiments No. 8* published by Mullard.
3. Hughes, D. E. P. 'Flame-ionization detector for gas chromatography'. *J. Chem. Educ.* Aug. 1965, p. 450.
4. Gabriel, W. P. and Morris, R. A. 'A flame ionization meter for gas chromatography'. *J. Sci. Instrum.* 1966, **44**, No. 2, Feb. p. 104.
5. Crawford, K. D. E. 'H.F. pentodes in electrometer circuits'. *Electronic Engng*, July 1948, p. 227.
6. Hay, G. A. 'Receiving valves suitable for electrometer use'. *Electronic Engng*, July 1951, p. 258.
7. Scroggie, M. G. *Radio Laboratory Handbook* (Iliffe) see chapter on indicators.
8. Rogal, B. 'Recent advances in three-terminal bridge techniques'. *Proc. Inst. Elect.* **4**, No. 2.
9. 'A 6-watt transformerless servo-amplifier using silicon transistors'. *Ferranti Application Note No. 13*.
10. Faulkner, E. A. and Stannett, R. H. O. 'A general-purpose phase-sensitive detector'. *Electronic Engng*, 1964, **36**, No. 433, Mar. p. 159.
11. Jeynes, G. F. 'Decade stepping tubes and their operation'. *Mullard Tech. Comm.* 1959, **4**, No. 37, Feb.
12. Barker, C. S. 'Operating principles of the Z504S stepping tube'. *Mullard Tech. Comm.* 1965, **8**, No. 75, March.
13. Frazer, H. J. 'Transistorized dekatron driving circuits'. *Electronic Engng*, 1962, **34**, No. 407, Jan.
14. Oxley, A. J. 'Some simple dekatron coupling circuits'. *Electronic Engng*, 1964, **36**, No. 434, April.

14

INTRODUCTION TO LOGIC CIRCUITS

The progress of modern industry depends to a large extent on our ability to introduce an increasing amount of automatic control into our industrial processes. Profitable fields for the introduction of automation techniques are the assembly of machine parts, the precision drilling and cutting of raw materials, the automatic sequencing of sorting, weighing, checking and storaging of raw materials and the automatic control of chemical and other processes. To an increasing extent the routine, and often mundane, work once performed by human beings is now being undertaken by machines. The digital computer is a good example of a sequencing machine that performs commercial and scientific calculations. We are already at the stage where machines are superior to man in many fields because of their ability to work to precision limits for long periods without fatigue. The speed of computer calculations is such that no human being could compete successfully with an electronic digital computer.

With all industrial control apparatus we must consider the problem of reliability. If, between two specified limits, a machine must distinguish accurately between two closely spaced states then the selection is not usually reliable. For example, let us suppose that a process depends upon measuring the anode current of a control valve. If the limits of the current were 0 and 10 mA and within these limits we had accurately to distinguish between 5 mA and 6 mA, we would find that the control process was not a very reliable one. Several factors, such as ageing of components and variations of the power supplies, could introduce errors into our system. For this reason control systems are designed to have only two possible states, namely 'on' and 'off'. The system has then to detect only when a signal is in the 'on' or the 'off' state; it is easy to do this with a high degree of reliability.

The best known on/off device for controlling industrial processes is the electromechanical relay. For over one hundred years relays have been the standard equipment used for routeing telephone signals. During the Second World War the design of switching networks for the selection and routeing of signals became very complicated. In an attempt to optimize a given system (i.e. use the minimum number of relays to perform a specified task) it was realized that the techniques involved were almost identical to those devised by the Irish mathematician, George Boole, for the symbolic representation of logic. His original efforts were directed to finding a symbolic algebra that could be used to represent arguments in philosophy and logic. His paper, published in 1847, dealt with the Mathematical Analysis of Deductive Reasoning and formulated relationships between the true and false propositions of his contemporary logicians. His ideas were not seen to have any practical application until Shannon in 1938 published a paper entitled 'Symbolic Analysis of Relay and Switching Circuits'. In the following years it was realized that digital computers were a practical possibility.

The starting point in the design of any control system is the specification setting out the operations that are to be performed. The specifications for each operation or set of operations is met by designing a suitable unit. We shall see later why these units should be electronic 'building bricks'. Logical design consists of deciding how these units can best be fitted together to meet the specification. The design is efficient if it uses the minimum possible number of 'bricks'. In the design of complicated systems a mathematical representation is convenient since it provides a shorthand by which the operational processes can be described. Such a shorthand can be very important because the design can follow mathematical lines and lead to simplifications that cut the cost of the installation. In many cases, these simplifications are by no means obvious from a consideration of the circuit configuration alone.

Boolean algebra is the mathematical method used in the design and optimization of control systems. We shall not be using the techniques of Boolean algebra in this book, but content ourselves with an explanation of the symbolism used. This will enable readers to understand the terminology and logical representations used by manufacturers and designers. Even where the various theorems are not used to optimize a system, descriptions of various control systems in books, papers and manufacturers' literature use Boolean symbolism in their descriptions. If the reader is going to use, design or discuss control systems, he must obviously learn the basic language first. He can then proceed to some of the specialist books that are entirely devoted to the subject.

BOOLEAN ALGEBRA

In this form of algebra the signs used are borrowed from ordinary algebra, but we must rid our minds entirely of all the ordinary mathematical meanings. This form of algebra is concerned with the relationship between classes or sets and each symbol used represents a class. We may make statements about any class, and these statements must be true or false. No half-truths or shades of meaning are allowed. This corresponds to our on/off arrangement. A switch is either 'on' or 'off'. We cannot have any intermediate state such as 'half-on', or 'nearly-off'. Each statement is given a truth value depending upon whether the statement is true or false. If it is true the statement is given a value of 1; if it is false we assign a value 0 to the statement. Let us take some examples to clarify what has been said. If we consider the statement 'women marry men' we are dealing here with the class that consists of human beings who are women and who marry. We are not concerned with the class of all women since clearly some women do not marry. If we let A represent the statement 'women marry men' then, since this is true, we write

$$A = 1$$

If B represents the statement 'women marry women' then $B = 0$ because the statement is false.

We may consider more than one statement simultaneously and decide whether the combination is true or false. Two statements may be joined to form one proposition by the use of the connectives AND, OR and NOT. For example let

$$A = \text{Rain is wet}$$

$$B = \text{Normal women have two legs}$$

therefore A AND B is true, that is, it is true to say that rain is wet AND normal women have two legs. We write this

$$A \cdot B = 1$$

where the dot signifies the AND combination. Often where no ambiguity arises the dot may be omitted. The AND combination of two logical propositions is known as the 'logical product' (hence the dot) and is true only if both propositions are true.

The connective OR may be used as follows:

$$A = \text{Rain is wet}$$

$$C = \text{Snow is white}$$

therefore 'rain is wet OR snow is white' is true. We write this

$$A + B = 1$$

The OR combination of two propositions is called the 'logical sum' and is true if either or both propositions are true. It is, therefore, an inclusive OR. (Exclusive ORs where one or the other, but not both, propositions are true need not concern us here.) The use of the plus sign for the OR connective rather than the AND can be confusing to the beginner. However, by using the signs as described above, familiar associative and distributive laws are obeyed.

If we wish to negate a statement we write a bar over the symbol, for example

$$C = \text{Snow is white}$$

therefore

$$\overline{C} = \text{Snow is not white}$$

We thus have

$$C = 1, \quad \overline{C} = 0 \quad \text{and}$$

$$A + \overline{C} = 1$$

i.e. 'rain is wet OR snow is not white' is true. The bar over the symbol represents the NOT connective i.e. NOT C is written $\overline{C}$. A bar over the combined statement negates the whole statement. For example, if $E \cdot \overline{F} = 1$ then $\overline{E \cdot \overline{F}} = 0$, i.e. if E and NOT F is true then NOT (E and NOT F) must be false, because $\overline{1} = 0$. We may illustrate this by having

$$E = \text{all Englishmen are Europeans}$$
$$F = \text{all Eskimos are black}$$

then $E \cdot F = 0$, i.e. it is false to say that all Englishmen are Europeans AND all eskimos are black. However

$$E \cdot \overline{F} = 1$$

because it is true that 'all Englishmen are Europeans AND all Eskimos are not black'. The negation of this statement must be false i.e. $\overline{E \cdot \overline{F}} = 0$, because it is not true to say that 'it is not true that all Englishmen are Europeans AND all Eskimos are not black'.

Most readers will understand how easy it is to become confused when considering logical propositions. It is precisely for this reason that we find it so convenient to attempt a mathematical formulation of the problem. It frequently is the case that for the majority of us mathematics can be a powerful aid when we are mentally lazy or merely unable to solve a given problem. (No doubt readers can remember from their schooldays expressing verbal problems in the form of algebraic equations and then proceeding to use simple techniques to arrive at the solutions of the equations and hence the problems). It would take a whole book to describe and prove the validity of the many theorems and techniques of Boolean algebra that are used to design control systems, and so we cannot go into them here. However, it is necessary for those interested in these systems to understand how the terminology and symbolism originates and to be quite happy about such statements as $A + B = 0$ or $\overline{C} = 1$.

436

RELAY REPRESENTATION OF THE CONNECTIVES

We commence our introduction to control system logic by considering the effect of opening and closing switches in simple circuits. In *Figure 14.1a* when the relay coil is energized, contacts are made to close thus starting some system, say a motor that pumps liquid into a tank. Before the relay coil is energized two conditions must obtain, namely switch A AND switch B must be closed. Switch A may be controlled by a safety cage and switch B by a temperature sensitive device. When switches A and B are open they are said to be in the '0' state, and when they are closed they are in the '1' state. Both A and B must be in the '1' state for the relay to be energized, i.e. in the '1' state. Thus $A.B = 1$ represents the position when the control system is set in motion. If $A = 0$ or $B = 0$ then the system will not operate. When $A = 0$ it is false to say that the safety cage is closed. Similarly $B = 0$ means that the temperature is not correct. A table, known as a truth table, is often used to enable us to see at a glance the output condition for any combinations of states of A and B. This is shown in *Figure 14.1a*. There are only four different combinations possible with the two variables A and B. Only when A and B are 1 can the output $(A.B)$ be 1.

In *Figure 14.1b* we have the OR combination. If either or both switches are in the 1 (i.e. 'on') state the control system is energized. As in the former case we are assuming that with the relay not energized the contacts are normally open i.e. in the 0 state; only when the relay coil is energized will the relay contacts close and come into the 1 state. The appropriate truth table is shown alongside.

Figure 14.1c shows the NOT combination. The relay contacts are normally closed, i.e. in the 1 state, when A is in the 0 state. On closing switch $A(A = 1)$ then the output contacts are opened, i.e. NOT closed.

Although not a logical connective, the MEMORY or HOLD circuit of *Figure 14.1d* is a useful and basic one in control systems. In this operation we require that an impulse on the SET switch should energize the relay and that the relay coil should remain energized after the pulse has disappeared. This is achieved by having a second set of contacts on the relay wired across the SET switch. On closing the SET switch the relay is energized and both relay contacts are closed. Now when the SET switch opens there is still a path for the energizing current via the auxiliary relay contacts. In order to reset the system to its former state, it is necessary to have a RESET switch, which is normally closed, in the position shown. On pressing the RESET switch the relay coil is no longer energized and the contacts open. There is now no longer a path via the SET switch or auxiliary contacts, and so the relay remains de-energized.

As a simple example to show how the circuits of *Figure 14.1* may be used

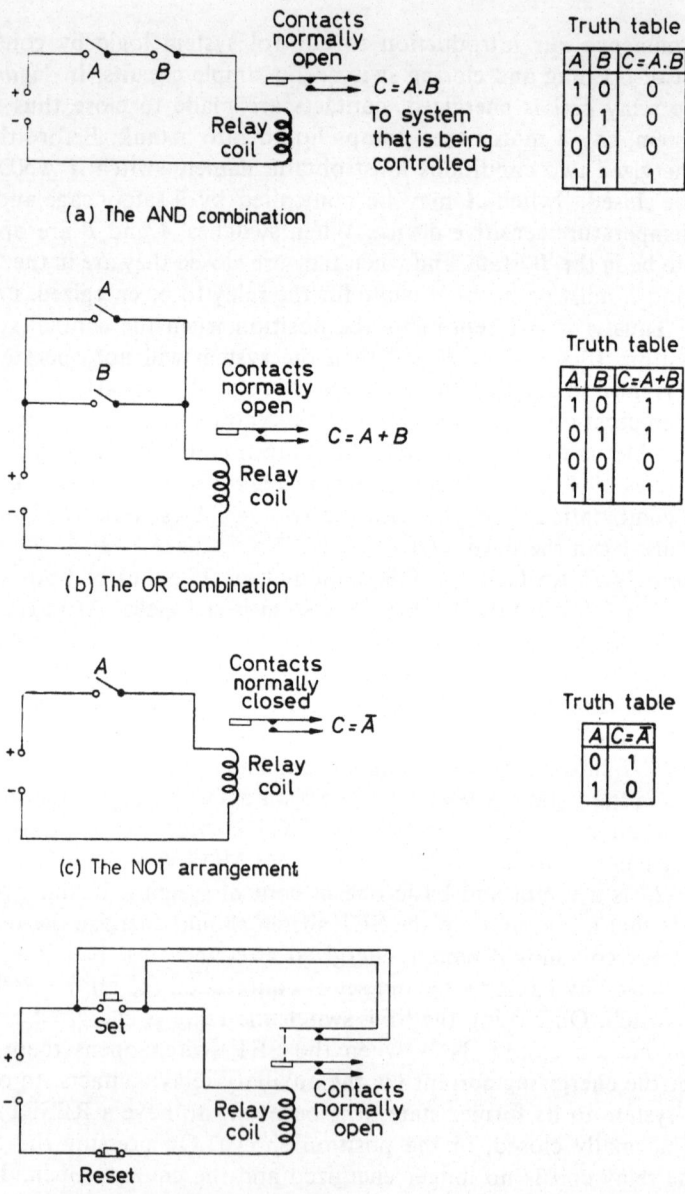

Truth table

A	B	C=A.B
1	0	0
0	1	0
0	0	0
1	1	1

(a) The AND combination

Truth table

A	B	C=A+B
1	0	1
0	1	1
0	0	0
1	1	1

(b) The OR combination

Truth table

A	C=Ā
0	1
1	0

(c) The NOT arrangement

(d) The MEMORY or HOLD circuit

Figure 14.1. The basic logic circuits—electromechanical relays

let us suppose that a three-man committee wishes to employ an electric circuit to record a simple majority vote. The circuit must be so designed that each member can push a button for his 'yes' vote (not push for a 'no') and a signal light must go on if a majority of the committee members vote 'yes'. A solution to the problem is shown in *Figure 14.2*. We require three AND circuits and an OR circuit. The outputs from the AND circuits are respectively $A.B$, $A.C$ and $B.C$. Thus if $A.B = 1$ or $A.C = 1$ or $B.C = 1$ then the output from the OR circuit will be 1 and the lamp will be illuminated. In *Figure 14.2a* we have the actual relay circuit. In this simple case the

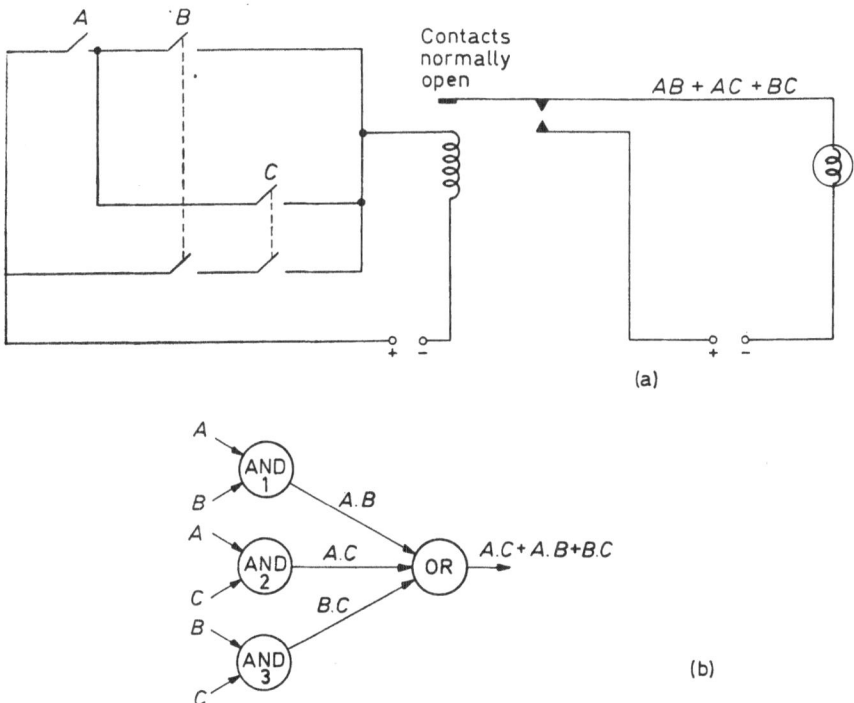

Figure 14.2. The solution to the voting problem. (*a*) shows the relay circuit while (*b*) shows the logical symbolic representation of the same circuit

operation of the system is easily understood from a consideration of the circuit. In more complicated systems, however, there is unnecessary confusion when the whole circuit is drawn. The combination of the various logical bits or bricks is more easily understood if the individual AND and OR switching units are represented by a simple symbol, as shown, in *Figure 14.2b*. Here it is assumed that a 1 input is available on the A lines when a button is pressed. When the button is not being pressed there is no voltage on the *A* lines and this corresponds to the 0 state. If *A* and *B* buttons are pressed simultaneously some arbitrary voltage (e.g. 10 V) is impressed on the *A* and *B* lines. The output from AND 1 goes from 0 V to 10 V where 10 V corresponds to the 1 state. This operates the OR unit, the output of which goes to 10 V and the lamp lights. The AND and OR units in a control system such as this are often referred to as 'gates'. The name can be understood by considering the AND 1 gate. The 1 signal 'gets through' the unit only when the gate is open. The gate is open only when *A* and *B* are in the 1 state.

TRANSISTOR SWITCHES

Switches that depend upon electromechanical relays are suitable only for simple and slow systems. Such switches require considerable power and, being mechanical, it is inevitable that their moving parts suffer from wear. The inertia of the moving parts limits the maximum permissible switching speed to values below a few hundreds of operations per second. Because of their physical size, slow speed, limited life and somewhat unreliable performance, electromechanical relays are being replaced in control systems by transistor switches. Modern high speed computers would not be possible if they had to rely on relays for their operation. By comparison transistors are small, light and cheap. They have no moving parts to wear out and their life is indefinitely long. Very fast switching speeds are possible with transistors that have been made specially for switching purposes. Computing transistors can switch at rates up to hundreds of megacycles per second.

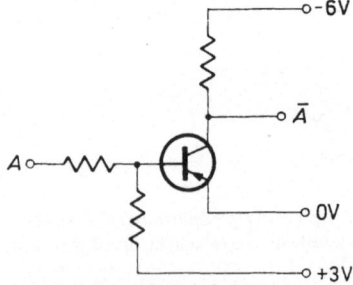

Figure 14.3. The transistor as a switch. The voltages are only an example of those that may be used. Different manufacturers choose voltages to suit their units

A simple transistor switch is shown in *Figure 14.3*. In the absence of any input at *A* or when *A* is connected to the 0 V line, the base is held positive with respect to the emitter. The base emitter junction is reversed biased and no current passes through the transistor. There can be no drop of voltage across the collector load and, consequently, the output voltage is -6 V. If we define 0 V as the 0 state and -6 V as the 1 state then when the input is 0 the output is 1. Conversely when the input is connected to -6 V sufficient electrons flow into the base to allow the transistor to go into the saturation condition. A large collector current is established and the collector-to-emitter voltage falls to only a few hundred millivolts. For practical purposes in logic circuits the output is at 0 V i.e. in the 0 state. This state is unaffected by changes in collector current provided saturation current conditions are maintained; this is ensured by the circuit designer when he selects the value of the input resistor for given supply voltages. The transistor, therefore, changes from the cut-off to the saturation state, i.e. from 'off' to 'on' with very great reliability and speed. The limiting switching speed depends upon how quickly the large number of charge carriers can be swept out of the base region. For specially constructed silicon transistors this process takes place very rapidly (in about 1 nsec, i.e. 10^{-9} sec).

The circuit of *Figure 14.3* is a NOT circuit because the output is always different from the input. We must remember that there are only two permissible states and so there are no intermediate states. If the input is 0 then the output is 1 and vice versa. Let us now consider *Figure 14.4* in which we have

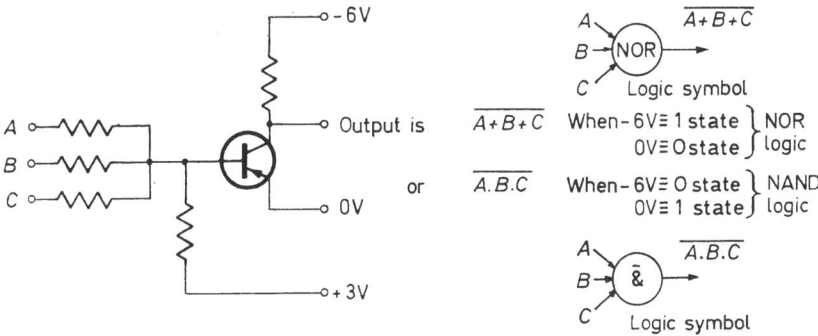

Figure 14.4. The basic logic unit that can be used for NOR or NAND logic systems. The transistor and associated resistors are represented by a circle and appropriate symbol in logic circuits

multiple inputs (three in this case). If all three inputs are at 0 V, i.e. in the 0 state, then if A or B or C goes to -6 V, i.e. the 1, state then the output changes. On our definition of 0 and 1 the circuit of *Figure 14.4* is a NOT OR gate, which we refer to as a NOR gate. It is a NOR gate because if A or B or C is 1 then the output is 0, i.e. NOT 1. A whole system of logic can be built upon this type of gate and we are then said to be using NOR logic. The actual gates can be bought in micromodule form, an example of which are the

441

NORBITS of Mullard Ltd. Let us now look at *Figure 14.4* from a different point of view. Suppose we define −6 V as 0 and 0 V as 1. Then if the three inputs are connected to the −6 V line they are all in the 0 state. With this connection the transistor will be conducting and, therefore, the output will be in the 1 state, i.e. at 0 V. If *A* and *B* are connected to the 0 V line they are now in the 1 state, but the transistor still conducts because base current is drawn via the *C* input. We see that *A* and *B* and *C* must be 1 for the output to change to 0. With this definition of line voltages we may call the circuit of *Figure 14.4* a NOT AND gate or more usually a NAND gate. Elliott-Automation Microelectronics Ltd. use this system in their control and computer systems. They give the trade name 'Minilog' to the small modules or bricks that can be built into control systems. We see that whether we use NOR or NAND logic is purely a matter of definition. Precisely the same electronic circuit may be used as a NOR or a NAND gate. The choice between the two logics depends upon the control system envisaged. In certain cases, a system may be built with the minimum number of bricks using NOR logic whereas other systems can be more efficiently constructed using NAND logic.

For a given system the choice depends largely upon whether AND gates or OR gates are in the majority. *Figure 14.5* compares the number of 'bricks'

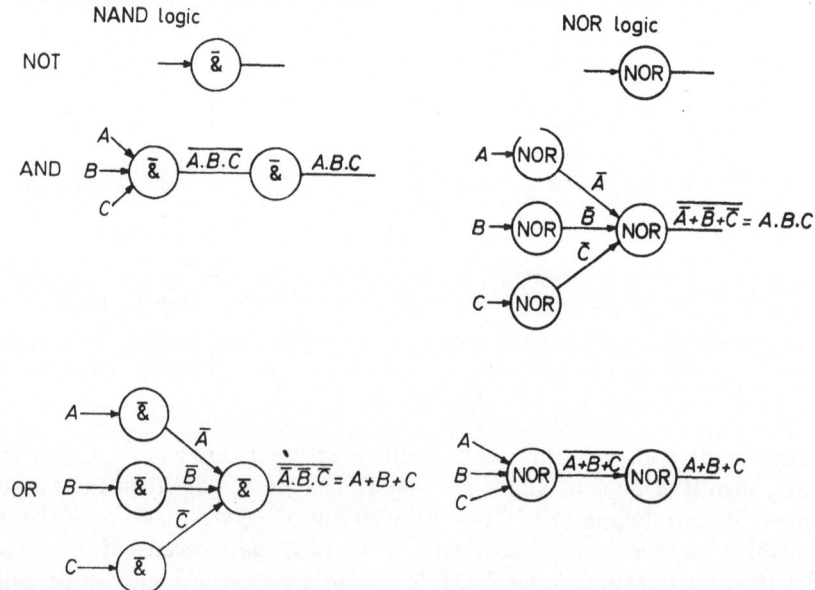

Figure 14.5. A comparison of the NOR and NAND systems showing the number of basic 'bricks' needed for each of the logic functions

required for the basic logical processes in the two systems. The circuit of *Figure 14.4* showing the transistor and associated resistors is not drawn in full each time, but is represented by a circle. Within the circle is the symbol Ī for a NOR unit or $\overline{\&}$ for a 'not and', i.e. NAND unit. The reader should confirm for himself that the combinations do, in fact, perform the respective logical functions. For example, if the 'OR' combination in the NAND logic system is considered we note the voltages at the input and output of each gate. Assuming A and B and C are all in the 0 state their respective voltages must be −6 V. The output voltage from each initial gate is, therefore, 0 V (corresponding to the 1 state). The final output from the combination must be −6 V, i.e. 0. If A or B or C goes to the 1 state, i.e. 0 V, the output voltage of the appropriate input gate goes to −6 V. If any one of the outputs from the initial gates goes to −6 V the transistor in the final gate conducts and the output falls to 0 V, i.e. to the 1 state. Thus the output must be $A+B+C$.

Two further combinations are found to be useful in logic systems. They are the MEMORY or HOLD combination and the DELAY combination. *Figure 14.6* shows the MEMORY combinations using the NAND system.

Memory or hold circuit

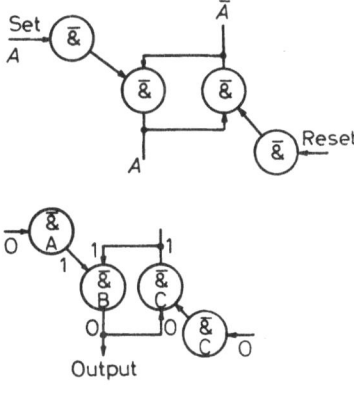

(a) Initial position throughout the circuit

(b) 'Set' pulse received. Even after the 'set' pulse disappears and the output from A goes back to 1 the output from B will remain 1. Further 'set' pulses produce no change in the output conditions

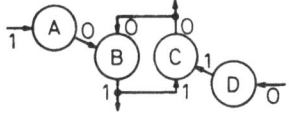

(c) When the 'reset' pulse is received the output of C changes to 1 which resets the output from B to 0. After the reset pulse disappears the output from C will remain in the 1 state and the position has reverted to that shown in (a)

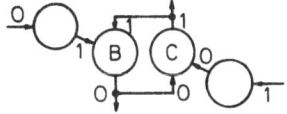

Figure 14.6. The memory or hold circuit showing the changes in state that occur when 'set' and 'reset' pulses are applied to their respective input units

The 'memory' or 'hold' function has already been mentioned. The function of the memory is to provide an output when a given signal input is received and to continue to provide that output even when the input signal is removed. There must also be provision for resetting the memory to its original condition in readiness for the next input signal pulse. Because a descriptive account in the text can be confusing, the various changes throughout the circuit when set and reset pulses are received are shown with the main circuit diagram in *Figure 14.6*.

The DELAY unit, as the name suggests, is a unit that gives an output voltage of the same kind as the input voltage, but which is delayed in time with respect to it. We require delay units in control systems to guarantee that a certain sequential process is reliable. If two input signals arrive at the input of a unit simultaneously then the unit is not able to decide which signal to process first. When the processing order is important then one input must be delayed for a short time so that the correct sequence is maintained. The arrangement and circuit diagram is shown in *Figure 14.7*. In the absence of an input signal pulse Tr1 and Tr2 are both conducting thus Tr3 is cut off and the collector voltage of Tr3 is -10 V. Using NAND logic the -10 V level is defined as the 0 state. When the leading edge of an input signal is presented to the input of the delay unit Tr1 is cut off sharply. The collector voltage of Tr1 becomes -10 V. Tr2 is, however, already conducting so there is no change in the output voltage of Tr2 or Tr3. After a time C1 charges via R1 and the base of Tr1 returns to its original value. Tr1 then conducts. The time taken to go through this operation is proportional to the capacitance of C1. As soon as Tr1 conducts the collector voltage falls to zero and this voltage is conveyed to the base of Tr2. Tr2 is now cut off which makes the output of Tr3 go to 0V, i.e. the 1 state. C2 charges via R2 and once the voltage on the base of Tr2 is restored to its former value Tr2 conducts and the output from Tr3 goes to -10 V, i.e. the 0 state. We have thus produced a short output pulse whose leading edge is delayed by a time determined by C1. The output pulse width is smaller than the input pulse width and is determined by the values of C2 and R2. When we shorten a pulse in this way we are said to obtain a logical differentiation of the input signal. The circuit shown is, therefore, a delay and differentiation circuit. By using this arrangement even if the input voltage remains in the 1 state, the output will still be a short delayed pulse. This is frequently an advantage in control systems.

The terms 'fan-in' and 'fan-out' used in connection with logic units refers to the number of inputs and outputs respectively that can be reliably handled by the units. There is a limit to the maximum number of inputs that can be fed into the base of a transistor. Most commercial units can accommodate up to six inputs although the manufacturer should always be consulted about his particular products. The number of outputs is under a more severe

444

limitation. It is the fan-out that determines the number of logic units that can be driven by any particular unit. Where a resistive load is to be driven (as, for example, where we lead the output of one gate to the input resistor of another logic unit) most units will drive at least two others. Where a fan-out greater than two is required, it is usual to employ a special driving unit that

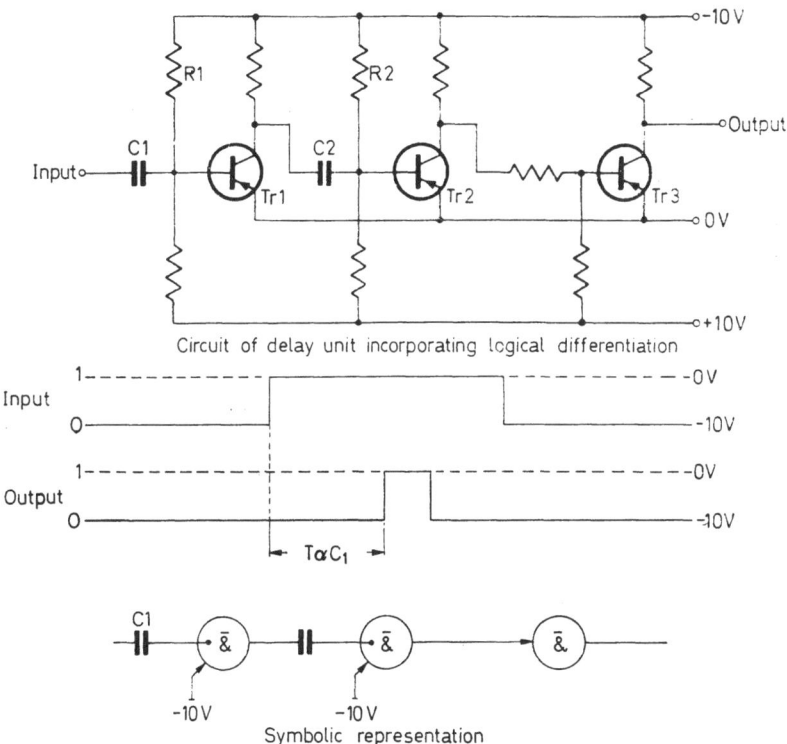

Circuit of delay unit incorporating logical differentiation

Symbolic representation

Figure 14.7. The delay circuit incorporating logical differentiation. In the symbolic circuit the capacitors are shown connected directly to the base by having the lead to the unit terminating with a dot within the symbol. The arrow to the symbol signifies a connection to an input resistor in the usual way

incorporates a transistor with larger power handling capabilities and which is operated in the common emitter mode. Where the load is an indicator lamp quite high currents are required. Special units are also available for such inductive loads as electromagnetic relays.

DESIGN CONSIDERATIONS

In all logic problems, we must start with a specification outlining the requirements of the system. If the system is not too large we may then draw out a block diagram and subsequently use this diagram to produce the most efficient system. (For complicated problems involving many hundreds of units it is convenient to use Boolean algebraic techniques to optimize the system, but we are not considering this method here.)

In order to illustrate the procedure let us consider a simple problem and obtain a solution. We may suppose that a large bath of electrolyte must have its contents pumped into a reaction chamber by an electrically driven pump. The pumping action must not commence until

- (A) the pH has reached a specified value, *and*
- (B) the temperature of the bath is correct, *and*
- (C) the volume of electrolyte has reached a specified value, *and*
- (D) the start switch is on.

The pumping action must stop

- (E) immediately any one of several emergency buttons is pressed, *or*
- ($\bar{B}$) if the temperature drops to too low a value, *or*
- ($\bar{A}$) the pH becomes too low (or high, depending upon the circumstances), *or*
- (F) 5 seconds after the level of liquid in the bath has fallen below a specified level.

We may obtain the control signals in the following way. The pH can be measured with a standard glass electrode and a switch operated when the value is correct. Alternatively, if suitable chemical indicators are available the change of colour on reaching the correct pH value may be detected by a photocell and appropriately coloured filter. The indicator initially has a certain colour to which the photocell-filter combination is insensitive and the switch operated by the photocell will be 'off' As soon as the required pH is reached the indicator changes colour, the photocell responds and the switch is moved into the 'on' position. The signal corresponding to the temperature may be obtained from a thermostat or a servo-operated switch controlled by a thermocouple. The levels of the liquid may conveniently be detected by light beams operating photocells or more simply by a couple of platinum electrodes held in a block of insulating material. When the electrodes are immersed in the electrolyte a small alternating current passes which energizes an a.c. relay. For the top level the relay contacts are normally open until the platinum electrodes become immersed in the electrolyte. After immersion the relay is energized and the contacts are closed. For the lower level the contacts are normally open when the relay is energized and closed once the level of the electrolyte falls below the platinum electrodes. By detecting the levels of the electrolyte we control the volume of liquid pumped into the reaction chamber.

Figure 14.8 shows the block diagram solution to the problem. Conditions *A*, *B*, *C* and *D* must be correct (i.e. in the 1 state) before the output from the AND gate sets the memory and starts the pumping motor. The memory can

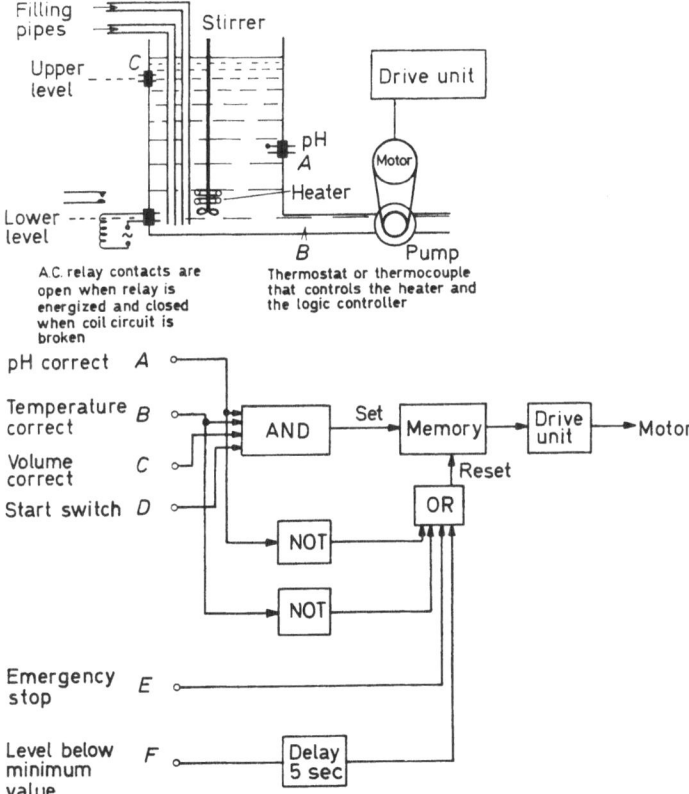

Figure 14.8. Block diagram showing the solution to the control problem described in the text

be reset via the OR gate, thus stopping the motor, if conditions *A* or *B* change to the 0 state. The memory can also be reset if any one of several emergency stop buttons is pressed or 5 seconds after the level of the liquid has fallen below a specified level. If we use NAND logic then we must consult *Figures 14.5, 14.6* and *14.7* for the basic arrangements. The block diagram can now be converted into a circuit diagram *(Figure 14.9)* using the basic NAND

447

bricks, each brick consisting of a transistor switch similar to that shown in *Figure 14.4*.

We must now optimize our controller so as to reduce the number of NAND units to a minimum and still produce the required control signals. For this

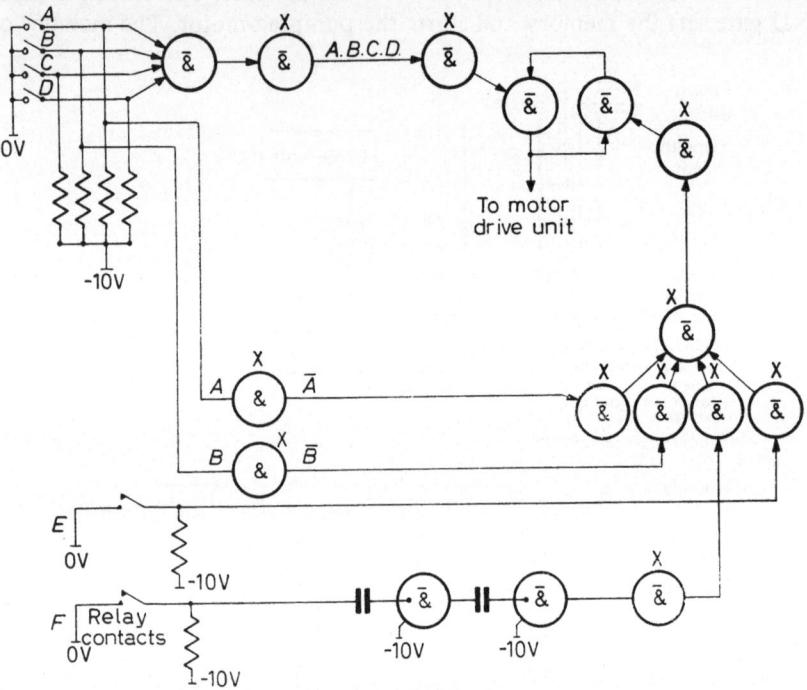

Figure 14.9. The logic unit interpretation of the block diagram. Those units with a cross above them may be eliminated because they are redundant (see *Figure 14.10*)

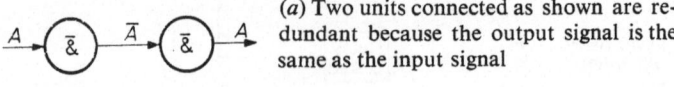

(*a*) Two units connected as shown are redundant because the output signal is the same as the input signal

(*b*) Three units in the left-hand configuration may be replaced by a single unit as can be seen by comparing the outputs

Figure 14.10. Redundant units

simple system we need to recognize only two forms of simplification. When two NAND units are linked by a single line as in *Figure 14.10a* they are redundant and may be eliminated without upsetting the logic. This is because a double inversion takes place, the output from the two units being the same as the input. It is equivalent to saying that a statement that is not 'not true' is true. *Figure 14.10b* shows a further example of redundancy. By eliminating redundant units of *Figure 14.9* we arrive at the final solution of *Figure 14.11*.

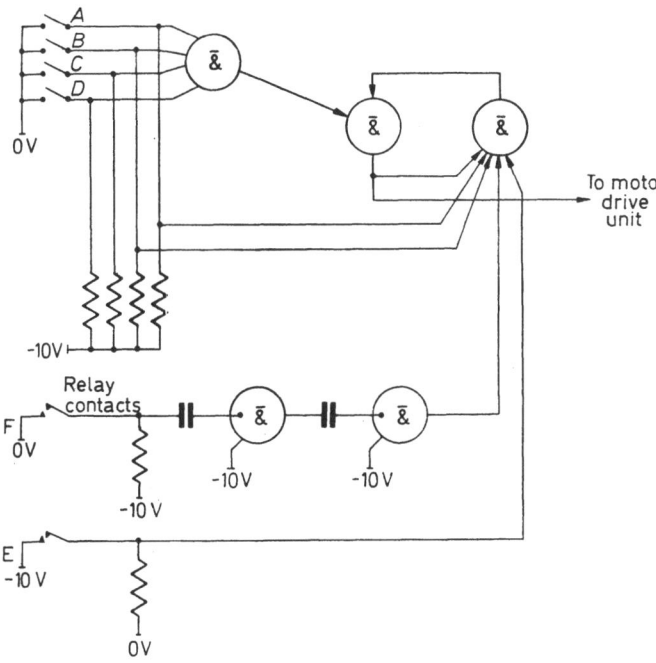

Figure 14.11. The final solution of the control problem. Note the change in voltage associated with the emergency stop switch (E). By reversing the polarities it was possible to eliminate the unit in the OR gate associated with the emergency stop switch. We now have a controller that uses the minimum number of logic units to perform the given specified functions

BINARY COUNTERS

An industrial process often depends upon counting events; counting is also the central function involved in the operation of a digital computer. The decade counters described in the last chapter are too slow for many purposes and much too slow for computer work. When the counting rate must approach or exceed 10^6 per second, it is necessary to dispense with counting tubes and use electronic circuitry. As already explained, electronic circuitry operates most reliably in the counting field when it is used as an on/off device. Instead of using the more familiar decade scale, therefore, it is necessary to count in the binary scale. This scale permits only two digits, 0 and 1. When 1 has to be added to 1 we cannot call this 2. We must carry one to the next column. We do the same sort of thing in the decimal scale using columns for units, tens, hundreds, etc. Once the count has reached 9 in any column the addition of a further 1 brings the digit in that column back to 0, and 1 is carried to the next column. The binary and decimal scales are compared below.

Binary	Decimal Equivalent
0000	0000
0001	0001
0010	0002
0011	0003
0100	0004
0101	0005
.	.
.	.
.	.
1001	0009
1010	0010

Whereas the columns in the decimal scale are units, tens, hundreds and thousands (i.e. 10^0, 10^1, 10^2 and 10^3) the columns in the binary scale are 2^0, 2^1, 2^2, 2^3, etc.

Binary counters often use bistable multivibrators as counting elements. (Computer men often refer to bistable multivibrators as 'flip-flops'. This use of the term is different from that adopted in Chapters 10 and 13. Many people agree that 'flip-flop' should mean a monostable multivibrator.) *Figure 14.12(a)* gives the circuit of one stage of a binary counter. This circuit is the transistor equivalent of the original Eccles-Jordan bistable multivibrator based on thermionic valves. The form of the circuit bears a resemblance to the astable or free-running multivibrator. The coupling capacitors, however, are replaced by resistors, and so the flip-flop is directly coupled. (The small capacitors in parallel with the coupling resistors serve to speed up the transition period from one state to the other. Usually they are only a few hundred pF.)

Let us suppose that the circuit is originally in the state where Tr1 is conducting and Tr2 is not. The collector of Tr1 is, therefore, almost at zero potential whereas that of Tr2 is almost at the potential of the negative supply line. This is a stable state because the base of Tr1 is connected to a point that is sufficiently negative to ensure that base current is drawn thus keeping Tr1 conducting. So long as Tr1 conducts the base of Tr2 is connected to a

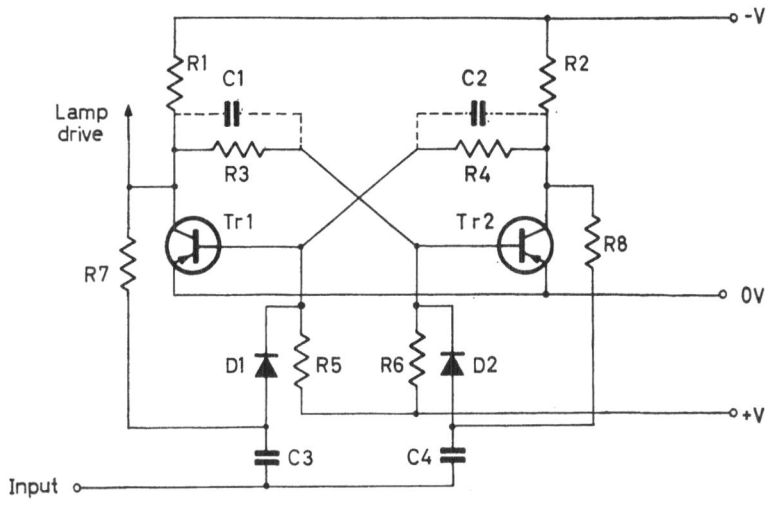

Figure 14.12a. One stage of a binary counter

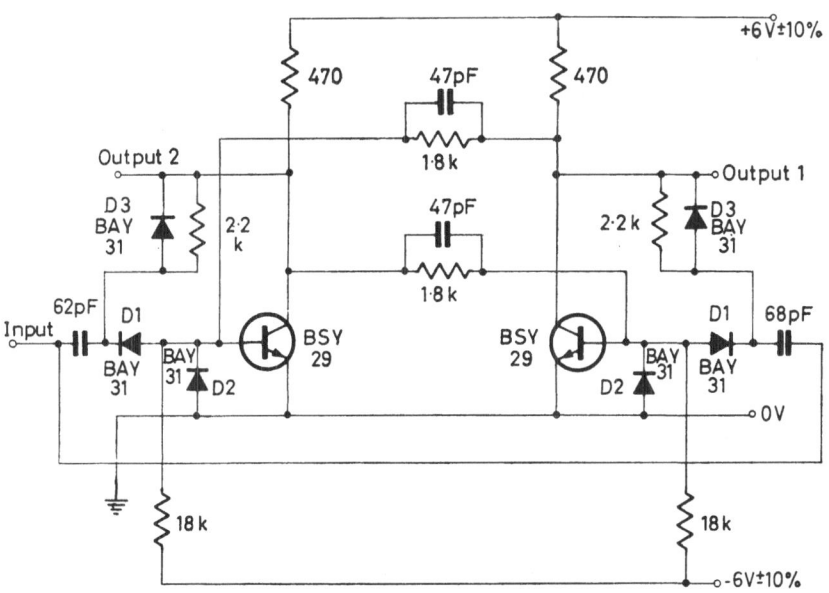

Figure 14.12b. The circuit diagram of a binary divider suitable for counting speeds up to 10 MHz. At input frequencies of 5 MHz or less diodes D2 may be omitted and at frequencies below 2MHz diodes D3 may also be omitted. The circuit is essentially that published by Standard Telephones & Cables in their publication MK/176X describing 10 MHz counting circuits

point which is almost at zero potential. Tr2, therefore, remains cut off. If now a short positive pulse appears at the input, one of short duration appears at the bases of the two transistors via C3 and C4 and the two isolating diodes. The positive pulse at the base of Tr2 does not affect that transistor because it is already cut-off, but the positive pulse at the base of Tr1 reduces the collector current. The collector voltage of Tr1, therefore, goes negative and base current is then drawn by Tr2. The collector voltage of Tr2 falls, and a regenerative action ensues in which Tr1 is cut-off and Tr2 is now conducting. When a further pulse is injected into the input terminal a similar action takes place and the circuit resumes its original state. It therefore requires two input pulses to make the circuit go through a complete cycle. The output is taken via a capacitor, from one of the collectors. For every two input pulses there is one output pulse. The circuit, therefore, divides by two. By cascading n of these circuits (i.e. by connecting the output of one unit to the input of the next for a chain of n units) we divide by 2^n.

A binary counter consists of n units in cascade. An indicator lamp associated with each unit is often used to indicate the state of the unit. One of the collectors is connected to a suitable lamp-driving circuit which in turn energizes the lamp. A reset line is connected to all the corresponding bases in each binary counting unit. By giving the reset line a suitable potential all units are brought to the same state. All the lamp drives are connected so that all of the lamps are extinguished. The arrival of the first pulse switches the state of the first unit and the corresponding lamp is illuminated. The output pulse is negative going and, therefore, does not affect the second unit because the isolating diodes are driven into reverse bias conditions. The diodes thus

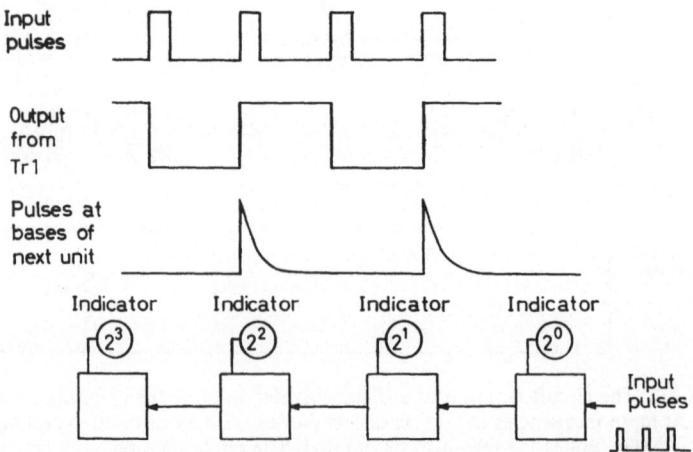

Figure 14.13. A binary counter. Each rectangle represents a single stage of the form of *Figure 14.12a* or *b*

452

block the signal from the bases of the second unit. Only when the output from the first unit is positive going will the state of the second unit be affected. This happens on the arrival of the second pulse. The lamp on the first unit is then extinguished whilst that on the second unit is illuminated. As successive pulses arrive the units up the chain are affected and the binary number is displayed on the corresponding set of lamps. *Figure 14.13* shows the binary counter in schematic form. Each rectangle represents the circuit of *Figure 14.12a or b.*

By redrawing the essential parts of *Figure 14.12a* as shown in *Figure 14.14* we see how the single stage binary counter can be represented by logic gates. R1, R2, etc. correspond in the two diagrams. *Figure 14.15* shows a symbolic logical representation of a binary counter. It is not an exact counterpart of

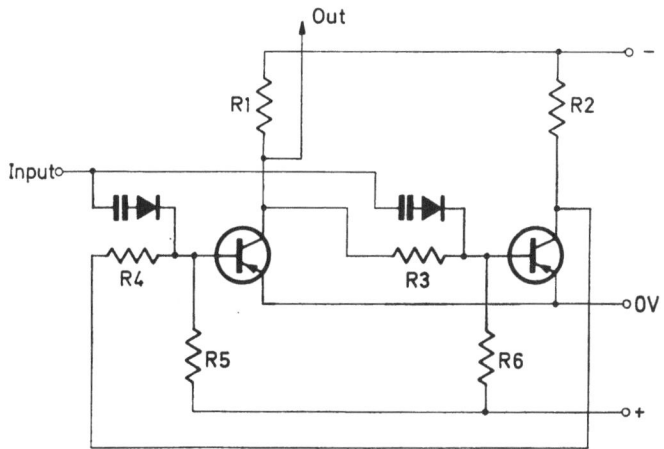

Figure 14.14. A redrawing of the essential parts of *Figure 14.12* to show the two logic units involved

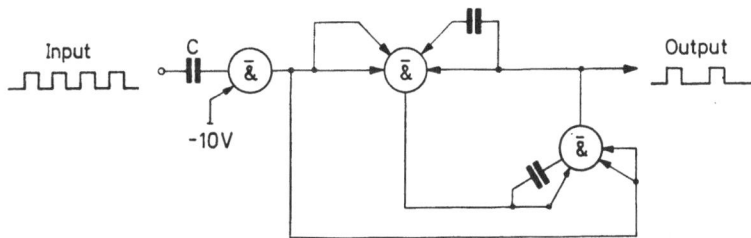

Figure 14.15. A binary counter circuit represented in logic form. This is not the exact equivalent of *Figure 14.14* but operates on the same principles

453

Figure 14.14 but the general principles involved are the same. The diode input circuit has been replaced by a logic unit that gives an output pulse the width of which is determined by C and not by the width of the input pulse.

DECODING MATRICES

It is unlikely that the lamp display representing a binary number can be interpreted readily since we have been educated in the decade system of counting. Although the electronic portions of a counter, computer or digital voltmeter are most reliable when counting in the binary system, it is convenient to translate the binary number into its decimal equivalent. One way of doing this electronically is to use a decoding matrix. These matrices are based on

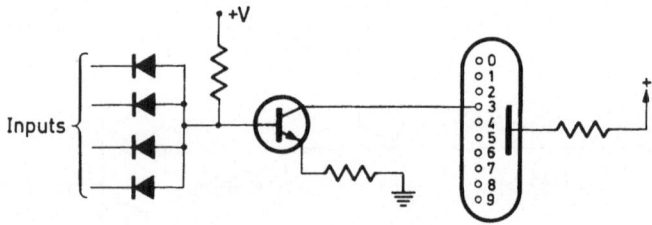

Figure 14.16. Part of the decoding system necessary to display a binary number in the form of a decimal number

diode transistor logic units (DTL units). Such units perform in the same way as the resistor-transistor logic (RTL) units already described, but for faster operation the input resistors are replaced by diodes. Also the energy loss associated with the input resistors is eliminated.

Let us consider the circuit of *Figure 14.16*. Here we have a cold cathode numeral tube displaying numbers on the decimal system. Associated with each cathode is an *npn* transistor, only one of which is shown in the diagram. When any transistor is turned on, there is a low resistance path to earth for the particular cathode and a glow discharge is associated with that cathode. With the transistor turned off a high resistance path results and the cathode is not illuminated. In our circuit, the numeral 3 is displayed only when the transistor inputs are all positive. (The change in polarity compared with previous circuits should be noted and results from the use of an *npn* transistor. The *npn* transistor is necessary in this circuit because we are using a glow discharge indicator. The collector is, therefore, connected to a source of potential that is positive with respect to the emitter when the glow discharge is established.) When any one (or more) of the inputs is connected to a line on which the potential is zero the appropriate diode conducts and its resistance is practically zero. The base of the transistor is, therefore, at zero volts and the transistor is off. Only when all inputs are connected to a source of positive voltage is the transistor turned on. Suppose now we wished to convert the binary number representing 3 (i.e. 0011) into the decimal number 3 and display this number. Using four flip-flops their states would be as shown in *Figure 14.17*. The first two units working from right to left have to be 'on' and the next two 'off'. If the flip-flop has one output 'on' the voltage on the ON terminal must be positive for our system. The voltage at the OFF terminal is zero. If, however, the flip-flop is 'off' the voltage at the ON terminal will be zero, but that at the OFF terminal will be positive. That this is so can be understood when we consider the two collector voltages in any flip-flop

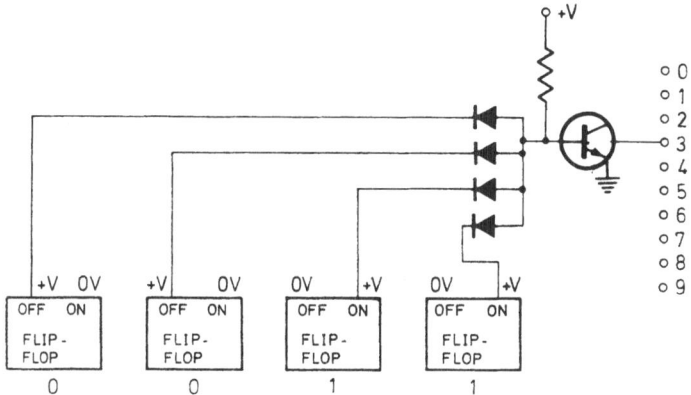

Figure 14.17. When the four flip-flops are representing the pure binary number 0011 the polarities of the output terminals are as shown. Connection to the transistor via the four diodes will cause cathode number 3 to be displayed

unit. To energize the number 3 cathode, therefore, we need to connect our diodes as shown. All diode cathodes are now positive and the transistor conducts. All other numbers are selected in a similar way. To avoid confusion on the circuit diagram, all the diodes are drawn on a network, called a matrix, as shown in *Figure 14.18*. P. R. Adby has shown[1] that a big reduction in the number of diodes can be effected by using a modified circuit. This circuit is not given here since those interested can easily consult the reference given.

The diode matrix given in *Figure 14.18* does not decode the simple binary code already given. The simple binary code is liable to error when transmitted from point to point within computers and data processing apparatus. If we refer to the simple binary scale on p. 450 it will be seen that as we

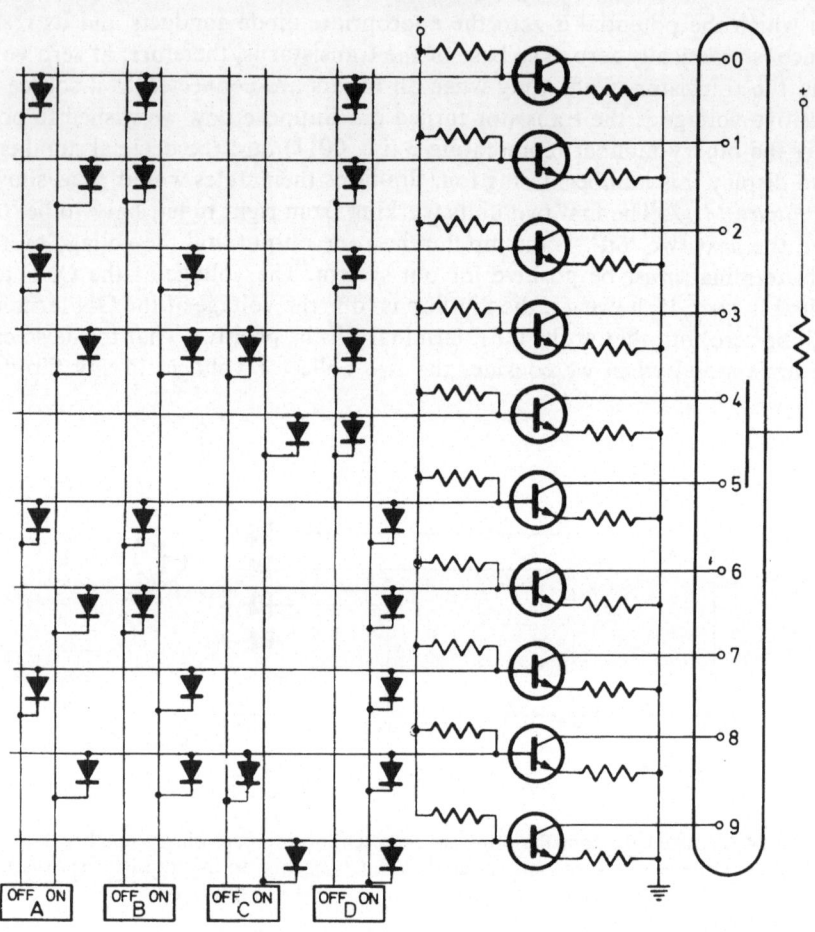

Decoding table for a
5.1.2.1. counter

	A	B	C	D
0	0	0	0	0
1	1	0	0	0
2	0	1	0	0
3	1	1	0	0
4	1	1	1	0
5	0	0	0	1
6	1	0	0	1
7	0	1	0	1
8	1	1	0	1
9	1	1	1	1

Figure 14.18. Decoding matrix for a 5.1.2.1 counter, but see Adby[1]
for important economies

progress from 0001 to 0010 (i.e. decimal 1 to decimal 2) then two bits of the binary number must change (a *bit* in computer jargon means *binary digit*). The transition from decimal 3 to decimal 4 involves a change of three bits in the simple binary equivalent numbers. The transfer of data by the simple binary code is not therefore so reliable as when other codes are used. Such codes are still binary in nature employing only the digits 0 and 1, but increased reliability is possible. (The book by Dean[2] gives further information on this point.) The circuit of *Figure 14.18* decodes the 5121 BCD code. (BCD = Binary Coded Decimal.)

Digital voltmeters frequently use flip-flops and a decoding matrix to yield a suitable display in conventional numbers. In the Solartron voltmeter, for example, the input signal is compared with a stable reference voltage by a bank of precision resistors. Each resistor is sequentially connected to the reference supply through bistable controlled transistor switches so that the proportion of reference voltage applied to the summing points is increased by discrete and known increments until the input signal is balanced out. The binary representation of the signal voltage is now held in the bistable circuits. The outputs from the bistables are applied to a decoding matrix which drive cold cathode numerical indicator tubes to display the voltage in the familiar decimal digits.

REFERENCES

1. Adby, P. R. 'Binary read-out circuits using transistors and gas-discharge numeral indicator tubes'. *Electronic Engng*, 1966, **38**, No. 457 (March.).
2. Dean, K. J. *An Introduction to Counting Techniques and Transistor Circuit Logic*, Chapman and Hall, 1964. See also his book *Digital Instruments*, Chapman and Hall, 1965.

15

PRACTICAL CONSIDERATIONS

When dealing with electronic equipment there is no substitute for practical experience. Every opportunity should therefore be taken to extend one's capabilities in this respect. For most readers of this book there are two aspects to consider, namely, the building of equipment, and the servicing and maintenance of electronic apparatus.

MAINTENANCE AND SERVICING

Reliable though most equipment is, there never has been up to the present an electronic instrument that is guaranteed never to go wrong. Manufacturing faults, accidents, carelessness, misuse and old age are the causes of failure in electronic circuits. It is not always convenient to return the apparatus to the manufacturer for repair, especially if the instrument is in constant use. Long delays in returning the apparatus can be very frustrating, especially in a research laboratory. There is no reason why those who have studied this book should not feel confident enough to look into the electronic boxes and correct minor faults at least. Contrary to the opinion of many, one need not be an electronic engineer to replace a faulty valve or transformer. Diagnosing the fault is a little more difficult, but an intelligent and methodical approach will often uncover the reason for failure of the equipment.

Unlike those in mechanical apparatus, faults in electronic equipment cannot often be found by visual inspection. It is, therefore, necessary to have some servicing equipment. There are many cases where some 'magician' of an electronic engineer corrects a fault in a short time using little more than a screwdriver, piece of wire and a multimeter. Although giving the impression that 'there was absolutely nothing to the job' we must not forget that he probably brings years of experience to bear, some of them perhaps on the very equipment he is repairing. Also there is no way of telling how many possible causes of failure are quickly rejected because his experience and logical approach enables him to see rapidly that certain explanations of the trouble could not possibly be valid. Only by cultivating a methodical approach and gaining the necessary experience will the more humble worker become an expert. Even for the expert, however, a certain minimum of test gear is essential for diagnosing the majority of faults. In order of priority the test instruments required are a multimeter, a signal generator, a resistance-capacity bridge and a cathode-ray oscilloscope. Vague descriptions of apparatus can be irritating in a chapter such as this, so firm recommendations are made. Naturally the recommendations reflect the opinion of the author, and there is no implication that other equipment is not equally as good as that mentioned.

Excellent service over several years has been obtained from an Avo 8 universal multimeter. This has a sensitivity of 20 kΩ/volt on the d.c. ranges, which may be regarded almost as a minimum specification for servicing transistor equipment. Chapter 6 outlines the main points to be considered when buying and using meters of this kind. The signal generator is a little more difficult to specify since much depends upon the nature of the apparatus to be serviced. However, for general low-frequency work an Advance H1 oscillator gives good service. Both square and sine waves are available at the output terminals with a frequency range of 15 Hz to 50 kHz. The Solartron CO.1004.2 has a range of 10 Hz to 1 MHz and gives two sine wave outputs, one being 180° out of phase with the other. Both firms will give advice on the oscillator to buy for a specific purpose. In the high-frequency range the oscillators of Advance and Marconi have given very good service. The RC bridge used by the author is a Marconi TF2700. It is very convenient to use and has a range that is more than adequate for general service work. The cost of the Avo 8, Advance H1 oscillator and Marconi RC bridge is approximately £20, £50 and £80, respectively. The cathode-ray oscilloscope can be classified as almost essential. Details are given in Chapter 11 of some of the available instruments, but for a single recommendation for general purpose work the Solartron CD1400 with appropriate plug-in units is very satisfactory. If the money is available, it is worthwhile considering the purchase of an insulation tester, a valve-voltmeter with a.c. and d.c. ranges, a valve tester and a transistor tester. It must be stressed however, that, provided one is prepared to apply deductive thought to the problem, a great deal of servicing can be done with a single item of equipment, namely the multimeter.

In addition to the test equipment, other equipment and tools will be needed especially if apparatus is to be built. These are

(a) appropriate power supplies, stabilized if necessary,

(b) a low wattage (< 50 W) soldering iron with a supply of resin cored solder, 60% tin/40% lead (Ersin Multicore or Savbit solder),

(c) diagonal wire cutters, long-nosed pliers, large and small screwdrivers, wire strippers (Bib type of Multicore Solders Ltd.), and

(d) for printed circuit boards, a magnifying glass for the detection of small cracks, a soft wire brush to remove solder, and a needle-point probe for testing the circuit.

No matter how hard-pressed for time one is, an attempt should be made to test every new piece of equipment that is bought. In particular, a record should be kept of the direct voltages at various points in the circuit, together with sketches of oscilloscope traces obtained in equipment known to be in working order. Manufacturers' figures are often only nominal or average for a particular piece of equipment. Those responsible for looking after and using the equipment should insist on having the circuit diagram, specifications and block diagram for each item, together with the appropriate handbook. These should be stored in a safe place and be readily available.

The methodical procedure for fault-finding is to eliminate successively the portions of the circuit not affected by the fault. As stated in the introduction to this book, all equipment may be regarded as being built up of basic units. In servicing the equipment, therefore, we should eliminate each basic unit in turn until the faulty unit is located. Further investigation into the basic unit should reveal the faulty components. Sometimes it is obvious where to look. If, for example, a cathode-ray oscilloscope gives a well-focussed vertical line that varies in length with the peak-to-peak value of the input signal and can be shifted in the Y direction, it is obvious that the power supplies, the tube and the Y-amplifier are working. We must look to the X-amplifier and timebase to locate the fault.

Because the basic units may be assembled in many different ways to form the composite equipment, it is difficult to take a typical example. However, let us illustrate the principles of servicing by considering a fault in the equipment represented by the circuit of *Figure 15.1*. Let us suppose that when a known signal is applied to the input terminals the servomotor does not work although previously the apparatus was known to be in working order. With all due respect to the reader the author's experience leads him to suggest that the first thing to check is that the mains plug is inserted in the socket, the switch is on and that the mains fuses are sound. Then, with a multimeter the steady voltages should be checked. With the negative lead of the meter on the h.t. negative line and the appropriate setting on the meter switches, the points A, B, C and D should be checked in turn. If A shows no voltage or a low voltage then clearly the early part of the power supply is probably at fault. A short across the h.t. lines anywhere in the circuit will also produce this symptom. The h.t. + line at B should be broken and the voltage at A noted again. If this is now correct a short in the later circuit exists. If the voltage is still down the electrolytics may be faulty, although if this is the case any protecting fuse in position F, for example, should have blown and should be checked. If the fuse has blown it should be replaced with a similar one. If that blows too, we must suspect the diodes or an electrolytic capacitor. A good check for the rectifiers, and all junction diodes, is to unsolder one end and measure the diode resistance with the multimeter. With the meter leads connected one way the resistance will be low, but with the leads connected in the reverse order, the resistance should be very high. (The lead of a multimeter that is normally regarded as positive is in fact negative when the meter is arranged as an ohm-meter.) If the diode is faulty the resistance will be low in both directions. The electrolytic capacitor should be isolated and connected in series with a milliammeter across a variable voltage supply. The supply

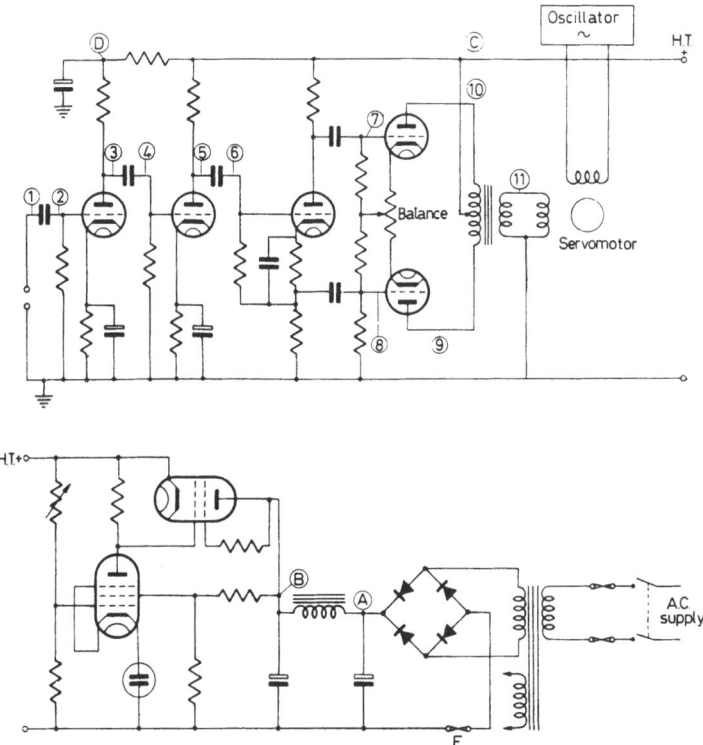

Figure 15.1. Checking points used in fault finding

must be connected with the correct polarity and the voltage increased from zero up to the working voltage of the capacitor. Excessive leakage will be evident at low applied voltages if the capacitor is faulty. If the meter is an Avo 8 it has internal protection against excessive currents. If the meter is not so protected, the double-diode protecting circuit described in Chapter 6 may be temporarily connected.

If the h.t. voltage proves to be satisfactory at B, but not at C, then the fault is located in the stabilizing section. At this point, if a valve tester is available, the valves of the stabilizer should be tested. It is a good idea to test the remaining valves at the same time. The most prevalent source of trouble in all thermionic valve equipment is due to faulty valves. If checking facilities are not available in the laboratory a local radio dealer will carry out the check for a very small, or no, charge. If testing facilities are not available, a suspect valve should be replaced by one of the same type. If the h.t. voltages and the valves prove to be satisfactory, it will then be necessary to check the voltage at each anode and cathode. It is very likely that the faulty section will now have been located, and thus we see that considerable progress can be made by using a multimeter alone.

If the fault has not yet been revealed, the waveform and signal voltage should be checked at the points 1 to 11 in turn. We presumably have a signal at 1 since we have assumed the transducer to be operating; alternatively, a suitable signal may be obtained from an oscillator. If we now lose the signal at any point, the location of the fault is obvious. Let us assume that all is well right up to point 11 then we must check the oscillator of the apparatus under-going examination. If this is satisfactory the servomotor must be at fault. The best way to check this is to replace it with a similar one. This is generally true of all apparatus connected to the electronic equipment although each case must receive appropriate consideration.

Once we have located the basic section which is at fault we must next ex-amine the individual components in that section. As has already been stated valves are the likely components to be faulty. The main causes of trouble are heater failures, heater-to-cathode shorts, low cathode emission, the gradual loss of vacuum, interelectrode shorts and fracture or disconnection of the leads from the pins to the electrodes. The smaller valves on B9A or B7G bases often suffer from bent pins. Care should be taken when straightening the pins not to fracture the metal-to-glass seal.

Resistors may develop a very high resistance, go open-circuit, or fracture, if faults elsewhere lead to excessive power dissipation in the component. Burn-ing can usually be detected by a preliminary visual check of the apparatus before testing has begun. The resistor itself may have been faulty originally, but it is always wise to suspect other reasons for the excessive current that caused the damage; for example the bias voltage on the valve may have dropped to zero or gone positive due to excessive leakage of the coupling capacitor to the previous stage.

Capacitors may develop short circuits with age due to failure of the dielec-tric. By assuming short circuits, or near short circuits, the effect on the rest of the circuit can be deduced. Overheating is often evident as a discolouration of the casing or a melting of the wax coating. It may be that the physical posi-tion of the capacitor in the apparatus is not satisfactory and the ambient temperature there is high. Well-designed equipment should not, of course, develop this symptom.

462

Coils in transformers or inductors may go open-circuit due to a melting of the wire from which the coil was made. A simple continuity test will reveal this fault. Short circuiting between adjacent turns is more difficult to locate. Often there is an excessive rise of temperature in the component, in which case smoke and smell soon reveal that something is wrong.

Looking for faults in transistor equipment is usually more difficult. Power failure is not immediately obvious as it is in the case of valves in which the heaters can be seen not to be glowing. Also most, if not all, transistor equipment is assembled on printed circuit boards. The steady-voltage checks can easily be performed as can the tracing of the signal through the circuit. It is in the replacing of components that the main difficulty arises. Location of the faulty component is not always easy because of the difficulty of isolating it from the circuit. Once a component is proved to be faulty, it should be cut away and the leads removed. Any surplus solder can then be brushed away after first getting it into the molten state with the soldering iron. (Excessive heat for a prolonged period may cause the copper to delaminate, i.e. come away from the board.) With the holes clear, the new component can then be soldered in place.

Unlike thermionic valves, transistors are soldered directly into the circuit. Excessive heat from the soldering iron may cause irreparable damage to the device. The emitter and/or collector connections may become loose, and the transistor material itself may diffuse into the base region thus destroying transistor action. If complete damage is not sustained a damaged base region usually gives rise to increased noise. This is evident in audio equipment as an audible 'hiss'; in other equipment the c.r.o. trace reveals the presence of noise. With no signal in the equipment the horizontal trace should be a clean straight line, or almost so. If the trace has the appearance of a cross-section of a grassy lawn, noise is present. Even correctly operating equipment makes some noise however, and this must be taken into consideration. The soldering iron may give rise to further trouble in that current surges from the mains (via a capacitive effect) can easily damage transistors that are connected to earth. To avoid all of these difficulties the transistor should be soldered quickly into the circuit using a low melting-point solder. A thermal shunt, which may be a pair of cool pliers, should be clamped across the lead between the transistor and the soldering iron. Soldering should never be undertaken when the circuit is energized and, if the apparatus is connected to earth, the earth lead should be temporarily disconnected. If the transistor has a black lacquered coating this should not be scratched because the entry of light, especially if it is modulated as is the case with a fluorescent lamp, can upset the transistor action. It is an easy matter to repaint the damaged coating. A check should be made to ensure that the leads are correctly wired to the appropriate points in the circuit. Mistakes can be fatal for the transistor since connecting it round the wrong way, with emitter and collector leads reversed, for example, is equivalent to connecting the battery incorrectly. The emitter junction may not be able to dissipate the power normally dissipated by the collector junction in which case irreparable damage may be sustained. Voltage measurements should be made with voltmeters having sensitivities of at least 10 kΩ/volt, otherwise the sum of the meter current and transistor current may be too great for the device. The transistor should always be sited sensibly to avoid high ambient temperatures. If necessary, heat sinks should be used to prevent excessive temperature rises.

With insulated-gate field-effect transistors, it is very easy to induce charges at sufficient voltage to damage the insulating layer irreparably. Just rubbing the free leads on a nylon overall is sufficient to ruin the device. F.E.T.s should never be soldered with a mains electric soldering iron. Destruction is almost certain if the gate lead is inserted into its socket with the main equipment 'floating' with respect to earth. Since these devices may cost anything up to £3 each such mistakes are expensive.

In spite of what has been said no one should gain the impression that transistors are too delicate to handle. They do, however, require care during installation and such points as have been made above may not be obvious to the newcomer or those brought up on valve circuits. Once installed, transistors exhibit the advantages described in Chapter 5. Apart from the differences outlined above, 'trouble shooting' in transistor equipment follows similar principles to those already described for the valve circuit.

Soldering

Poor soldering techniques can lead to considerable trouble when repairing or building electronic equipment. Dry joints are particularly difficult to locate since they are often not evident on a visual inspection. They are caused by layers of grease, dirt or undiffused flux between the component lead and the main solder blob. Noisy or indifferent performance usually results, or in severe cases the apparatus may not function at all because of the high resistance which is introduced into the circuit. The original cause of the trouble is usually the use of an iron whose temperature is too low, or the application of the iron for too short a period. For electronic work the iron should be about 60 deg C above the liquefying point of the solder. Too low a temperature gives a plastic type fluid whilst too high a temperature is evident by 'spitting' of the flux. High temperatures reduce the life of the bit. The bit, which should be small, must be kept clean and properly tinned. Only non-corrosive fluxes should be used, i.e. those specifically prepared for electronic work. Where the flux is incorporated as a core in the solder this is automatically ensured. When

464

using cored solders the iron should be applied to the joint and then the solder may be applied. If the solder is carried to the joint on the iron then the resin flux is burnt off and a poor joint results. Frequently, since one has not got three hands, it is not possible to avoid carrying the solder on the iron. Under these circumstances, a separate flux (e.g. Coraline red flux) must be previously applied to the joint, any surplus being removed after soldering with a rag moistened with a solvent. Carbon tetrachloride is a suitable solvent, but care should be taken not to inhale the fumes. All wires, tags, etc., that are to be soldered together must be clean and free from grease. For very fine enamelled wire scraping with a razor blade or knife is not satisfactory because of the possibility of damaging the wire. Such wires may be cleaned by heating in a methylated spirit flame, plunging them into liquid methylated spirits and then drying.

BUILDING EQUIPMENT

The process of constructing an actual piece of apparatus from a theoretical circuit diagram is not always obvious to the beginner because the physical layout of the components rarely follow the layout of the circuit diagram. Confusion can often arise when trying to decide the best layout. The prospective constructor is well advised to inspect the layout of apparatus built by firms with a reputation for high quality workmanship. Neatness and tidiness are often indications of well-thought-out arrangements.

When a new circuit is to be built and the performance is not known, it is often convenient to make a temporary assembly on a universal type chassis. Such chassis are easy to make from aluminium sheets using a simple bending machine or large vice. Physical appearance is not the prime concern at this stage. Once the circuit is known to function correctly, it must be constructed in a more permanent form. Professional engineers make suitable drawings from which the equipment can be assembled, but most readers will not want to go to this trouble. Perfectly satisfactory results may be obtained by taking the actual components and arranging them on a sheet of paper. The components can then be rearranged to produce the most satisfactory layout. In general, leads to grids, or bases, should be kept as short as possible as should all signal-carrying wires. It may be desirable to use screened wires (e.g. television coaxial cable) to avoid unwanted electrostatic voltages being induced in the signal lines. The outer sheathing is then earthed or connected to the h.t. negative line. If screening is employed one should not overlook the effect of the additional capacitance to earth that is introduced.

One of the most prevalent sources of induced noise and 'hum' is due to what are called 'earth loops'. Although it may seem strange, not all points on the chassis, earth or h.t. negative line are at the same potential. The result is that small voltage differences (of microvolts) can give rise to large currents round any low resistance loop that may be present. Such loops are formed by connecting the h.t. negative line and associated components to several points on the chassis. The interconnection of several different pieces of apparatus can also cause earth loops if the earthing points are numerous. Within a single item of equipment earth loops are avoided by using a single thick wire bus-bar to which the h.t. negative supply and the earthy leads of components are connected. This bus-bar should be connected to the chassis at one point only, preferably at the signal input socket. It is easy to forget that electrolytic capacitors are connected to the chassis electrically by means of the retaining clip holding the outer can. The author always prefers to wrap insulating tape around the component thus isolating the outer can from the chassis. A separate connection is then made between the outer can and the bus-bar.

The elimination of 'hum' in sensitive equipment can often be difficult. Screening is effective only against electrostatic fields unless it is of mumetal or similar high permeability alloy. Such alloy screens are expensive and out of the question in the majority of cases. The troublesome magnetic field is most frequently due to the mains transformer and/or smoothing choke. Rotation of the transformer, where possible, often reveals a position where 'hum' is a minimum. Wherever possible, it is a good idea to build the transformer and power pack on a separate chassis from the other equipment. In high amplification stages, it is often an advantage to connect all the earthy leads of any given stage to one point on the bus-bar. Heater leads should be tightly twisted together so as to reduce to a minimum the external magnetic field due to the varying heater current. The leads should be sensibly sited and as close to the metal chassis as possible. One should particularly avoid having them near to the input socket. The input and output sockets should not be adjacent. It is usual to reduce 'hum' due to the heater connections by earthing, or connecting to the h.t. negative line the centre tap on the heater secondary of the transformer. Often, however, improved results are obtained by earthing one lead of the secondary winding rather than the centre-tap. The better lead to connect to earth is found by experiment. Earthing at the transformer secondary is not essential. One of the heater pins of the input valve may prove to be a more satisfactory earthing point.

It is a great help in locating faults if some sort of colour code is adopted for the wiring. Although not universally accepted, the following code is used by several manufacturers. Leads to the following electrodes and associated components should be of the following colours:

Anode—blue
Cathode—yellow
Grid—green

Screen grid—orange
Heaters—brown
H.T. positive—red
H.T. negative—connected to chassis or earth—black
H.T. leads more negative than chassis or earth—purple
Transistor base—green
Emitter—yellow
Collector—white

In Great Britain a.c. leads to the mains are (a) red—live; (b) black—neutral; and (c) green—earth. One should always check equipment from other countries before connecting to the mains supply as it may originate in one which does not follow the same code.

Instability, especially in very high-gain equipment, is often a problem for the constructor. Apart from exercising care in the layout, he must also ensure adequate decoupling of stages, i.e. each stage must be fed via a resistor and have its own large smoothing capacitor across its h.t. supply lines. One is apt to forget the earth line resistance in equipment built on Veroboard or printed circuit board. Take, for example, the circuit of *Figure 12.14*. Although the upper supply line is adequately decoupled, this is effective only if the common chassis line is everywhere at the same potential. A high-resistance line here can be very troublesome. It may be necessary to break the line between each stage and bring a separate wire back from each stage to a common chassis point, thus ignoring the copper strip as a common 'earthy' line.

Stable results, with low noise, cannot usually be obtained in high-gain equipment unless the whole assembly is enclosed within a metal box that is connected to the 'earthy' line.

APPENDIX 1

The Simple Differentiating Circuit

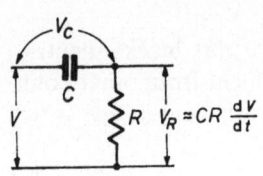

$$V_C = \frac{1}{C} \int i\, dt$$

$$\therefore\ i = C\,\frac{dV_C}{dt}$$

$$V_R = iR = CR\,\frac{dV_C}{dt}$$

By making C and R very small $X_C \gg R$ and almost the whole of the signal voltage appears across C. ($X_C = 1/2\pi fC$ and is the reactance of the capacitor.)

$$\therefore\ V_C \approx V$$

and
$$V_R \approx CR\,\frac{dV}{dt}$$

i.e. the voltage across R is proportional to the differential coefficient of the input or signal voltage.

The Simple Integrating Circuit

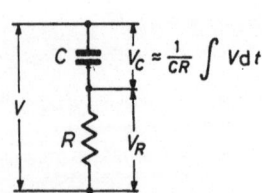

In this case the output voltage is taken from across the capacitor. By making C and R large, $R \gg X_C$ and the current, i, is given very nearly by

$$i = V/R$$

$$V_C = \frac{1}{C} \int i\, dt = \frac{1}{CR} \int V\, dt$$

i.e. the output voltage is proportional to the time integral of the input or signal voltage.

APPENDIX 2

DECIBEL SCALES

$$N(\text{db}) = 10 \log_{10} \frac{P_1}{P_2}$$

where P_1 and P_2 are the ratios of two powers. For powers dissipated in the same resistances (or equal resistances)

$$A(\text{db}) = 20 \log_{10} \frac{V_1}{V_2}$$

where V_1 and V_2 are the voltages across the resistors.

db loss when comparing V_2 with V_1	$\dfrac{V_1}{V_2}$	db gain when comparing V_2 with V_1	$\dfrac{V_1}{V_2}$
0	1	0	1
1	1·122	1	0·891
2	1·259	2	0·794
3	1·414	3	0·707
6	1·995	6	0·501
10	3·162	10	0·316
20	10·0	20	0·1
30	31·62	30	0·0316
40	100·0	40	0·01
60	1000·0	60	0·001

APPENDIX 3

CALCULATION OF C AND L VALUES FOR SIMPLE POWER SUPPLIES

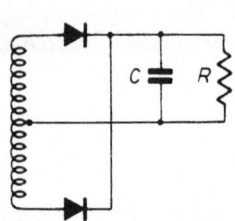 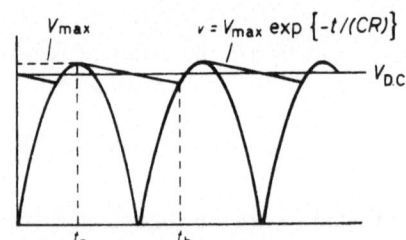

Using the mean value theorem

$$V_{DC} = \frac{\displaystyle\int_{t_a}^{t_b} V_{max} \exp -\{t/(CR)\}}{t_b - t_a}$$

Since the charging period is short $t_b - t_a \approx \tau$ (the half period) $\tau = 1/2f$ where f is the supply frequency.
Performing the integration

$$V_{DC} = \frac{V_{max}}{\tau} \cdot [\exp \{-t/CR)\}]_{t_a}^{t_b} \cdot (-CR)$$

$$= V_{max} \, 2fCR \left[\exp \{-t_a/(CR)\} - \exp \{-t_b/(CR)\}\right]$$

Taking t_a as our starting point, then $t_a = 0$.

$$\therefore \ V_{DC} = V_{max} \, 2fCR \left[1 - \exp \{-t_b/(CR)\}\right]$$

$$t_b - t_a \approx t_b \approx \tau = \frac{1}{2f}$$

The Exponential Theorem states that

$$e^{-x} = 1 - x + \frac{x^2}{2!} - \frac{x^3}{3!} \cdots$$

470

Using this and substituting t_b/CR (i.e. $1/2fCR$) for x

$$V_{DC} = V_{max}\, 2fCR \left(1-1+\frac{1}{2fCR}-\frac{1}{2!\,(2fCR)^2}+\cdots\right)$$

$$= V_{max}\left(1-\frac{1}{4fCR}\right)$$

In practice with the values of C R and f encountered later terms in the series may be neglected.

Example. Suppose we wished to supply 50 mA at 300 V to a load. This implies a load of 6 k. Let us have a steady output voltage of 95% V_{max} so that the amplitude of the ripple voltage is only about 5% V_{max}.

$$\frac{V_{DC}}{V_{max}} = 0{\cdot}95 = 1 - \frac{1}{4fCR}$$

$$\therefore \quad \frac{1}{4fCR} = 0{\cdot}05 \quad \text{i.e.} \quad C = \frac{1}{4fR(0{\cdot}05)}$$

Since $\qquad R = 6\text{ k}, \quad C = 1/\{4\times 50(6\times 10^3)(5\times 10^{-2})\}$

$$\approx 16\ \mu\text{F}$$

To cut the ripple down by a factor of 100 we may regard L and C as a potential divider so far as the 100 Hz ripple is concerned. We have, therefore, $X_L \approx 100 X_C$ i.e. $\omega L = 100/\omega C$

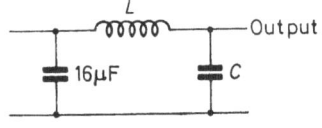

$$\therefore\ LC = \frac{100}{4\pi^2 f^2} \approx \frac{1}{4\times 10^3} = 250\times 10^{-6}$$

A reasonable value for C would be 32 μF.

$\therefore\ L \approx 8$ H. In practice a 10 H choke would be used.

The only serious precaution to be taken in designing power packs is to avoid exceeding the manufacturer's limit on the value of the reservoir capacitor. Otherwise, economic considerations apart, the larger the value of the choke and smoothing capacitor, the better will be the power supply.

APPENDIX 4

RANGE OF PREFERRED VALUES OF RESISTORS

20%	Tolerance 10% (Silver)	5% (Gold)
10	10	10
		11
	12	12
		13
15	15	15
		16
	18	18
		20
22	22	22
		24
	27	27
		30
33	33	33
		36
	39	39
		43
47	47	47
		51
	56	56
		62
68	68	68
		75
	82	82
		91
100	100	100

Larger values are obtained by multiplying the values given in the table by an appropriate multiple of 10.

APPENDIX 5

MATRIX ALGEBRA

There are two main reasons for including an appendix on matrix algebra in a book written for those not aiming to become specialist electronic engineers. First, the large number of parameters associated with transistors is bewildering to the beginner because he may not realize that there is a simple, basic pattern from which many of the parameters are derived. Secondly, several results have been quoted without proof in the book, and some readers may wish to know how the formulae were derived.

Many arrangements of linear components may be regarded as four-terminal networks, such networks being represented by a 'black box' as shown in *Figure 1*.

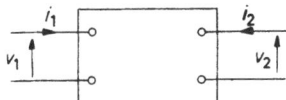

Figure 1. 'Black box' representation of a circuit network

In the chapter on a.c. theory we dealt with circuit arrangements that can be reduced to two-terminal networks. With these circuit arrangements we found it very convenient to describe such a network by the simple expression $R + jX$. Unfortunately, such a simple expression is inadequate for four-terminal networks because of the increased number of variables involved. It is obviously of great value to find an expression, involving the minimum number of parameters, that adequately describes a four-terminal network. Matrices are such expressions. A matrix is a set of coefficients arranged in an orderly array of rows and columns. The number of coefficients need not be limited in a

---

* In electronics it is often useful to disregard the individual properties of the circuit elements comprising a given network, such a network being judged solely by external observations. When the workings within the network are imperfectly understood, or are of no immediate interest, the network may be called a 'black box'. The transistor is a good example of a 'black box'. Although the way in which a transistor works may be difficult to understand, it is still possible to use the device as a circuit element when its external behaviour is known.

mathematical sense, but for the majority of four-terminal networks encountered in electronics, we need consider only the following forms:

$$\begin{pmatrix} v_1 \\ i_1 \end{pmatrix} \quad \begin{pmatrix} a_{11} & a_{12} \\ a_{21} & a_{22} \end{pmatrix}$$

The first expression is known as a column matrix whilst the second is a two-by-two square matrix (i.e. having two rows and two columns). The coefficients v_1, a_{11}, a_{12}, etc. are known as the elements of the matrix. When suitable expressions for the elements are found, it is possible to describe four-terminal networks that contain linear circuit components. A suitable combination of matrices can then be found to enable us to express a set of algebraic simultaneous equations in a very concise form. Matrix algebra is the manipulation of these matrices in an orderly manner so as to obtain solutions of the equations and other useful results.

Let us see now how we may express in matrix form the simultaneous equations that arise in the description of the four-terminal network of *Figure 1*. The four variables v_1, i_1, v_2, i_2 give rise to six different ways of describing the external behaviour of the network; we need only consider four of these ways. Obviously v_2 and i_2 depend upon v_1, i_1 and the contents of the 'black box'. One way of expressing this dependence is

$$v_1 = a_{11}v_2 - a_{12}i_2 \tag{1}$$

$$i_1 = a_{21}v_2 - a_{22}i_2 \tag{2}$$

(In general, the coefficients of the first line of any set of equations expressed in this form are all a_1's, the first being a_{11}, the second a_{12}, etc. In the second line they are all a_2's with a_{21} being the first and a_{22} the second etc.) In matrix form this pair of simultaneous equations is expressed thus

$$\begin{pmatrix} v_1 \\ i_1 \end{pmatrix} = \begin{pmatrix} a_{11} & a_{12} \\ a_{21} & a_{22} \end{pmatrix} \begin{pmatrix} v_2 \\ -i_2 \end{pmatrix} = [A] \begin{pmatrix} v_2 \\ -i_2 \end{pmatrix} \tag{3}$$

Knowing the rules for multiplying the right-hand side the original equations can always be recovered. In many cases, however, this is not necessary; and with practice, it is possible to think in the language of matrices rather than in the more familiar language of networks and algebraic equations.

Rules of Matrices

(1) Two matrices are equal if, and only if, they are both column or both square and the elements in corresponding positions are equal.

(2) Two matrices of the same kind can be added to give a third matrix, as follows

$$\begin{pmatrix} a_{11} & a_{12} \\ a_{21} & a_{22} \end{pmatrix} + \begin{pmatrix} b_{11} & b_{12} \\ b_{21} & b_{22} \end{pmatrix} = \begin{pmatrix} a_{11}+b_{11} & a_{12}+b_{12} \\ a_{21}+b_{21} & a_{22}+b_{22} \end{pmatrix}$$

We merely add the elements in corresponding positions.

(3) The multiplication of a matrix by a number merely multiplies each element by the number, i.e.

$$m \begin{pmatrix} a_{11} & a_{12} \\ a_{12} & a_{22} \end{pmatrix} = \begin{pmatrix} ma_{11} & ma_{12} \\ ma_{21} & ma_{22} \end{pmatrix}$$

(4) Two matrices are multiplied as follows

$$\begin{pmatrix} a_{11} & a_{12} \\ a_{21} & a_{22} \end{pmatrix} \begin{pmatrix} b_{11} & b_{12} \\ b_{21} & b_{22} \end{pmatrix} = \begin{pmatrix} a_{11}b_{11}+a_{12}b_{21} & a_{11}b_{12}+a_{12}b_{22} \\ a_{21}b_{11}+a_{22}b_{21} & a_{21}b_{12}+a_{22}b_{22} \end{pmatrix}$$

i.e. $\qquad (A)(B) = (C) \quad$ where $\quad (A) = \begin{pmatrix} a_{11} & a_{12} \\ a_{21} & a_{22} \end{pmatrix}$, etc.

The reason for choosing this seemingly odd method of multiplication need not concern the reader except to note that a consistent mathematical system is defined that produces useful results. Matrix multiplication is not in general commutative, i.e. [A] [B] $\neq$ [B] [A]. It is, however, associative, i.e. [A]([B]$\times$ [C]) = ([A]$\times$[B])[C]. (Matrix addition, on the other hand, is both associative and commutative.)

The significance of the multiplication rule becomes apparent when we consider the cascading of two four-terminal networks (*Figure 2*).

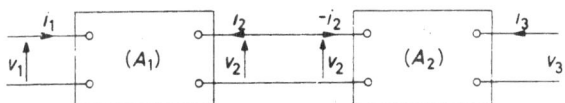

Figure 2. The cascading of two 'black boxes'

$$\begin{pmatrix} v_1 \\ i_1 \end{pmatrix} = (A_1) \begin{pmatrix} v_2 \\ -i_2 \end{pmatrix} \quad \text{where} \quad A_1 = \begin{pmatrix} a_{11} & a_{12} \\ a_{21} & a_{22} \end{pmatrix}$$

i.e., the matrix describing the first network.
However,

$$\begin{pmatrix} v_2 \\ -i_2 \end{pmatrix} = [A_2] \begin{pmatrix} v_3 \\ -i_3 \end{pmatrix}$$

Hence

$$\begin{pmatrix} v_1 \\ i_1 \end{pmatrix} = [A_1] [A_2] \begin{pmatrix} v_3 \\ -i_3 \end{pmatrix}$$

Therefore the combination of the two networks can be regarded as a single quadripole network that can be described by a matrix $[C]$ such that

$$(C) = (A_1)(A_2).$$

Let us exploit this rule by first finding the matrix elements for some common circuit arrangement and then deducing the matrix for the combination of these arrangements.

From *Figure 3*

$$v_1 = a_{11}v_2 - a_{12}i_2 \qquad (1)$$

$$i_1 = a_{21}v_2 - a_{22}i_2 \qquad (2)$$

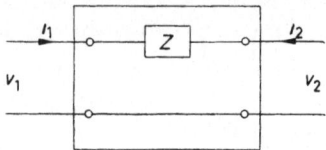

Figure 3. A four-terminal network in which an impedance Z is connected between one input and one output terminal as shown

When the output is open circuited, $i_1 = -i_2 = 0$ and $v_1 = v_2$.

$$\therefore \quad v_1 = a_{11}v_2 \quad \text{i.e.} \quad a_{11} = 1,$$

and $\qquad\qquad 0 = a_{21}v_2 \quad \therefore \text{ since } \quad v_2 \neq 0, \quad a_{21} = 0.$

When the output is short circuited,

$$v_2 = 0, \quad i_1 = -i_2 = \frac{v_1}{Z}.$$

$$\therefore \quad a_{12} = Z \quad \text{and} \quad a_{22} = 1.$$

Hence

$$\begin{pmatrix} v_1 \\ i_1 \end{pmatrix} = \begin{pmatrix} 1 & Z \\ 0 & 1 \end{pmatrix} \begin{pmatrix} v_2 \\ -i_2 \end{pmatrix} \qquad (3)$$

By similar reasoning we may build the shown in table *Figure 4*.

Note that from the equation $v_1 = a_{11}v_2 - a_{12}i_2$, when the output is open circuit $v_1 = a_{11}v_2$ i.e. the transfer function $v_2/v_1 = 1/a_{11}$. This is a most useful and important result.

In combining networks the example of the Wien bridge circuit may be taken (*Figure 5*).

Let $\qquad\qquad R_1 + \dfrac{1}{j\omega C_1} = Z \quad \text{and} \quad \dfrac{1}{R_2} + j\omega C_2 = Y$

Network   Matrix

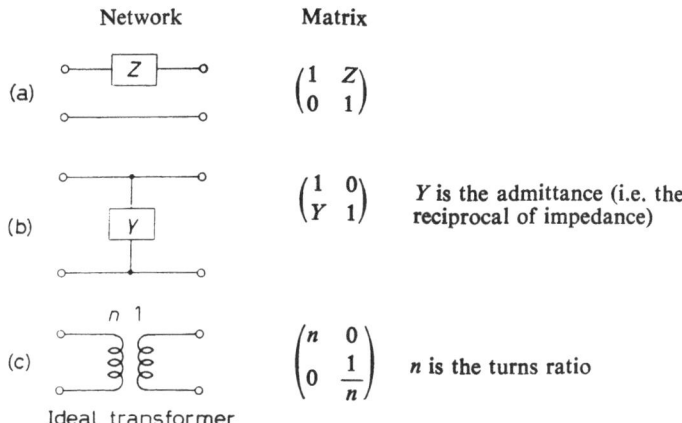

(a) $\begin{pmatrix} 1 & Z \\ 0 & 1 \end{pmatrix}$

(b) $\begin{pmatrix} 1 & 0 \\ Y & 1 \end{pmatrix}$ Y is the admittance (i.e. the reciprocal of impedance)

(c) $\begin{pmatrix} n & 0 \\ 0 & \dfrac{1}{n} \end{pmatrix}$ n is the turns ratio

Ideal transformer

Figure 4. Some matrices of common circuit arrangements

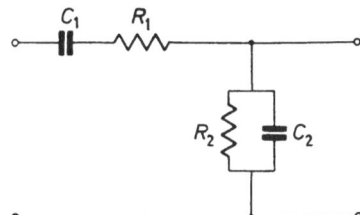

Figure 5. The Wien bridge drawn as a four-terminal network which may be regarded as two four-terminal networks [(a) and (b) of *Figure 4*] in cascade

The matrix for the network, [A], can be obtained by considering the cascading of the Z and Y portions, i.e.

$$[A] = \begin{pmatrix} 1 & Z \\ 0 & 1 \end{pmatrix} \begin{pmatrix} 1 & 0 \\ Y & 1 \end{pmatrix} = \begin{pmatrix} 1+ZY & Z \\ Y & 1 \end{pmatrix}$$

We see immediately upon inspection that

$$\frac{V_2}{V_1} = \frac{1}{1+ZY}$$

In fact if we were wishing to obtain only the transfer function it would not be necessary to proceed with the whole of the multiplication—obtaining the a_{11} element of [A] would suffice

Proceeding, $$\frac{V_2}{V_1} = \frac{1}{1 + \left(R_1 + \dfrac{1}{j\omega C_1}\right)\left(\dfrac{1}{R_2} + j\omega C_2\right)}$$

477

If $R_1 = R_2 = R$, $C_1 = C_2 = C$ the transfer function becomes

$$\frac{v_2}{v_1} = \frac{1}{1+1+j\omega CR + \dfrac{1}{j\omega CR}+1}$$

$$= \frac{1}{3+j\left(\omega CR - \dfrac{1}{\omega CR}\right)}$$

It is clear that when $\omega^2 = 1/C^2R^2$ i.e. $f = 1/2\pi CR$, v_1 and v_2 are in phase and the attenuation is $1/3$. In the chapter on oscillators the section dealing with the Wien bridge oscillator makes use of these results.

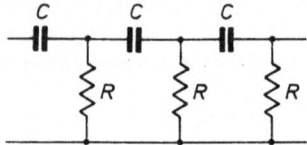

Figure 6. Three-section RC phase-shifting network used in phase-shifting oscillators

In the section on the phase-shift oscillator, it will be recalled that we required to know the attenuation of the network at a frequency which gave a phase-shift of 180° between the output and input voltages. In trying the ordinary Kirchhoff approach to the three-section network of *Figure 6*, a set of simultaneous equations is obtained that proves to be an algebraic handful. It is in these circumstances that the power of the matrix method is revealed. Matrix methods may not necessarily involve any great reduction in labour (though they often do); the advantages lie in the organization and clear procedure that is laid down.

In *Figure 6* the first CR section has a matrix given by

$$\begin{pmatrix} 1 & Z \\ 0 & 1 \end{pmatrix}\begin{pmatrix} 1 & 0 \\ Y & 1 \end{pmatrix} \quad \text{where} \quad Z = \frac{1}{j\omega C} \quad \text{and} \quad Y = \frac{1}{R}$$

$$= \begin{pmatrix} 1+ZY & Z \\ Y & 1 \end{pmatrix} = [M]$$

The transfer function of the whole network is the reciprocal of the a_{11} element of $[M]^3$. The reader may care to find the a_{11} element for himself. It is

$$1+6ZY+5Z^2Y^2+Z^3Y^3$$

which yields

$$1+\frac{6}{j\omega CR}-\frac{5}{\omega^2C^2R^3}-\frac{1}{j\omega^3C^3R^3}$$

478

For a phase-shift of $180°$ the j term $= 0$. This will occur when $6\omega^2C^2R^2 = 1$, i.e. at a frequency given by $f = 1/\{2\pi\sqrt{(6)}CR\}$. At this frequency the a_{11} element of $[M]^3 = -29$. Thus the gain of the amplifier must be at least 29.

Equations 1 and 2 lead to the A-matrix of equation 3. The A-matrices, as we have seen, are extremely useful when analysing circuits such as *Figures 5* and *6*, in which the elements are in a cascade or chain arrangement.

Circuits are not always, however, made up of networks connected in cascade as above. We may have networks connected in parallel (e.g. the parallel-T filter) or in series at their inputs and outputs, or we may have the inputs in series and the outputs in parallel. In such cases the A-matrix may not be the best description available. Other matrices are possible and, especially for certain transistor work, these alternatives have distinct advantages over the A-matrix. For those readers who are just starting on transistor work, it is hoped that this appendix shows how some of the parameters associated with transistors arise; perhaps they will then take some comfort in the fact that some order does indeed exist among the bewildering number of parameters that can be used.

The A-matrix and other matrices are all obtained by using one central idea or theme. From the quantities v_1, i_1, v_2, i_2 in *Figure 1* we may regard any pair as known and the other pair as unknown. With v_2 and i_2 known, v_1 and i_1 can be expressed as in equations 1 and 2. From these equations the A-matrix is defined. Similarly the following matrices (which are most useful in circuit analysis) may be obtained:

$$i_1 = y_{11}v_1 + y_{12}v_2 \tag{4}$$

$$i_2 = y_{21}v_1 + y_{22}v_2 \tag{5}$$

which in matrix form is

$$\begin{bmatrix} i_1 \\ i_2 \end{bmatrix} = \begin{bmatrix} y_{11} & y_{12} \\ y_{21} & y_{22} \end{bmatrix} \begin{pmatrix} v_1 \\ v_2 \end{pmatrix} = [Y] \begin{pmatrix} v_1 \\ v_2 \end{pmatrix} \tag{6}$$

The elements of the Y-matrix are known as the y-parameters since each has the dimensions of an admittance. Y-matrices are particularly useful when analysing circuits consisting of two networks in parallel at their input and output terminals as in *Figure 7*. The overall matrix is given by $[Y] = [Y_1] + [Y_2]$ i.e.

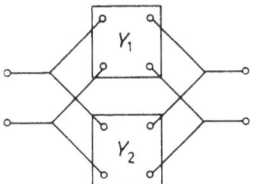

Figure 7. The overall Y-matrix of two four-terminal networks connected as shown is
$$(Y) = (Y_1) + (Y_2)$$

the elements in corresponding positions in Y_1 and Y_2 are merely added. An example of an analysis using this technique is given later.

If we take as our starting point

$$v_1 = Z_{11}i_1 + Z_{12}i_2$$
$$v_2 = Z_{21}i_1 + Z_{22}i_2$$

which in matrix form is

$$\begin{array}{c} v_1 \\ v_2 \end{array} = \begin{pmatrix} Z_{11} & Z_{12} \\ Z_{21} & Z_{22} \end{pmatrix} \begin{pmatrix} i_1 \\ i_2 \end{pmatrix} = (Z) \begin{pmatrix} i_1 \\ i_2 \end{pmatrix}$$

we obtain the Z-matrix for the network. The Z-matrices are most useful in the analysis of networks arranged in series as in *Figure 8*. The individual elements of a Z-matrix all have the dimensions of an impedance.

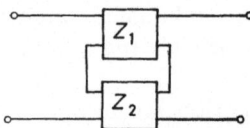

Figure 8. $(Z) = (Z_1) + (Z_2)$ when two four-terminal networks are connected as shown

The last matrix we need consider is the h-matrix. It is this matrix that is most useful when considering transistors. The individual elements of this matrix are the hybrid or h-parameters often quoted by manufacturers of transistors. The matrix is defined from the following equations

$$v_1 = h_{11}i_1 + h_{12}v_2$$
$$i_2 = h_{21}i_1 + h_{22}v_2$$

which in matrix form is

$$\begin{array}{c} v_1 \\ i_2 \end{array} = \begin{pmatrix} h_{11} & h_{12} \\ h_{21} & h_{22} \end{pmatrix} \begin{pmatrix} i_1 \\ v_2 \end{pmatrix} = [H] \begin{pmatrix} i_1 \\ v_2 \end{pmatrix}$$

In circuit analysis the h-matrices are most useful in the configuration of *Figure 9*.

The overall h-matrix is given by $[H] = [H_1] + [H_2]$.

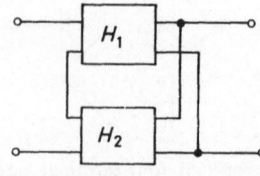

Figure 9. $(H) = (H_1) + (H_2)$

Consider now the transistor manufacturer's dilemma when wishing to publish design data on his products. The matrices described so far each have their advantages when applied to suitable circuit configurations, but from the manufacturer's point he wishes to describe his transistors with parameters that can easily be measured accurately. Unfortunately the Z- and y-parameters do not permit of easy determination because of the way in which they are defined. An examination of the h-parameters shows that they can all be determined easily and accurately. This is why the h-parameters are now widely used in manufacturers' data sheets, in textbooks and articles.

It is essential to point out that it will be necessary to be able to convert the h-parameters into the corresponding Z-, y- and A-parameters depending upon the particular circuit arrangement it is wished to examine. Since no one can rapidly convert one set into another from the first principles every time a conversion is required, it is usual to refer to a table. There is nothing complicated about the use of the tables. Such a table is given on p. 482. Only one symbol used has not been defined so far and that is the one used for the determinant of a matrix. Taking the A-matrix as an example we have

$$A = \begin{pmatrix} a_{11} & a_{12} \\ a_{21} & a_{22} \end{pmatrix} \quad \text{det. } A \text{ (written } |A|) = a_{11}a_{22} - a_{12}a_{21}$$

Applications

When using matrices in circuit analysis we must first determine the most useful matrix form to use. As an example let us take the case of the parallel-twin-T network shown in *Figure 10*.

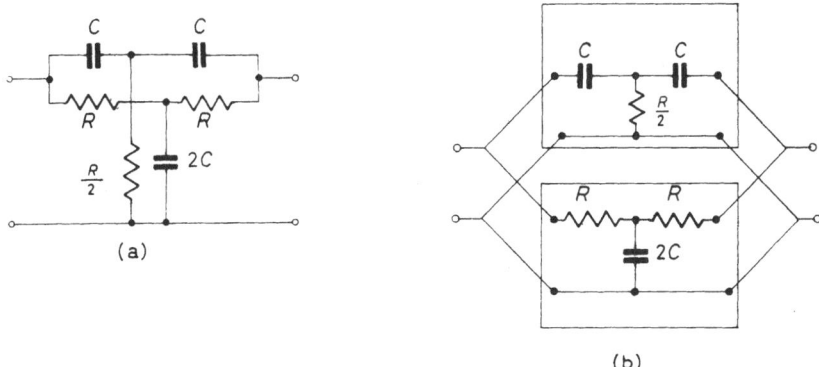

(a)

(b)

Figure 10. The parallel-twin-T network of (a) may be regarded as being made of two T-networks connected in parallel at their input and output terminals as in (b)

Conversion table

to

Conversion from

	$z_{11} = \dfrac{a_{11}}{a_{21}}$ $\quad z_{12} = \dfrac{\lvert A\rvert}{a_{21}}$ $z_{21} = \dfrac{1}{a_{21}}$ $\quad z_{22} = \dfrac{a_{22}}{a_{21}}$	$y_{11} = \dfrac{a_{22}}{a_{12}}$ $\quad y_{12} = \dfrac{-\lvert A\rvert}{a_{12}}$ $y_{21} = \dfrac{-1}{a_{12}}$ $\quad y_{22} = \dfrac{a_{11}}{a_{12}}$	$h_{11} = \dfrac{a_{12}}{a_{22}}$ $\quad h_{12} = \dfrac{\lvert A\rvert}{a_{22}}$ $h_{21} = \dfrac{-1}{a_{22}}$ $\quad h_{22} = \dfrac{a_{21}}{a_{22}}$	$\lvert A\rvert = a_{11}a_{22} - a_{12}a_{21}$
A-matrix parameters				
z-parameters	$a_{11} = \dfrac{z_{11}}{z_{21}}$ $\quad a_{12} = \dfrac{\lvert z\rvert}{z_{21}}$ $a_{21} = \dfrac{1}{z_{21}}$ $\quad a_{22} = \dfrac{z_{22}}{z_{21}}$	$y_{11} = \dfrac{z_{22}}{\lvert z\rvert}$ $\quad y_{12} = \dfrac{-z_{12}}{\lvert z\rvert}$ $y_{21} = \dfrac{-z_{21}}{\lvert z\rvert}$ $\quad y_{22} = \dfrac{z_{11}}{\lvert z\rvert}$	$h_{11} = \dfrac{\lvert z\rvert}{z_{22}}$ $\quad h_{12} = \dfrac{z_{12}}{z_{22}}$ $h_{21} = \dfrac{-z_{21}}{z_{22}}$ $\quad h_{22} = \dfrac{1}{z_{22}}$	$\lvert z\rvert = z_{11}z_{22} - z_{12}z_{21}$
y-parameters	$z_{11} = \dfrac{y_{22}}{\lvert y\rvert}$ $\quad z_{12} = \dfrac{-y_{12}}{\lvert y\rvert}$ $z_{21} = \dfrac{-y_{21}}{\lvert y\rvert}$ $\quad z_{22} = \dfrac{y_{11}}{\lvert y\rvert}$	$a_{11} = \dfrac{-y_{22}}{y_{21}}$ $\quad a_{12} = \dfrac{-1}{y_{21}}$ $a_{21} = \dfrac{-\lvert y\rvert}{y_{21}}$ $\quad a_{22} = \dfrac{-y_{11}}{y_{21}}$	$h_{11} = \dfrac{1}{y_{11}}$ $\quad h_{12} = \dfrac{-y_{12}}{y_{11}}$ $h_{21} = \dfrac{y_{21}}{y_{11}}$ $\quad h_{22} = \dfrac{\lvert y\rvert}{y_{11}}$	$\lvert y\rvert = y_{11}y_{22} - y_{12}y_{21}$
h-parameters	$z_{11} = \dfrac{\lvert h\rvert}{h_{22}}$ $\quad z_{12} = \dfrac{h_{12}}{h_{22}}$ $z_{21} = \dfrac{-h_{21}}{h_{22}}$ $\quad z_{22} = \dfrac{1}{h_{22}}$	$y_{11} = \dfrac{1}{h_{11}}$ $\quad y_{12} = \dfrac{-h_{12}}{h_{11}}$ $y_{21} = \dfrac{h_{21}}{h_{11}}$ $\quad y_{A2} = \dfrac{\lvert h\rvert}{h_{11}}$	$a_{11} = \dfrac{-\lvert h\rvert}{h_{21}}$ $\quad a_{12} = \dfrac{-h_{11}}{h_{21}}$ $a_{21} = \dfrac{-h_{22}}{h_{21}}$ $\quad a_{21} = \dfrac{-1}{h_{21}}$	$\lvert h\rvert = h_{11}h_{22} - h_{12}h_{21}$

482

To obtain the overall matrix of this configuration, we first note that two T-networks are connected in parallel at their input and output terminals. The Y-parameters are therefore used, since simple addition of the Y-elements for each T-network gives us the overall Y-matrix. If only the A-matrix parameters are known we would adopt the following procedure

The A-matrix of the lower T-network of *Figure 10b* can be found from the individual matrices of the components by the methods already discussed. The result is given by

$$(A_1) = \begin{pmatrix} 1+2j\omega CR & 2R+2j\omega CR^2 \\ 2j\omega C & 1+2j\omega CR \end{pmatrix} = \begin{pmatrix} \alpha & \beta \\ \delta & \alpha \end{pmatrix}$$

(It should be noted in passing that $|A| = 1$, a property of the A-matrices of passive linear symmetrical networks.) Using the conversion table

$$(Y_1) = \begin{pmatrix} \alpha/\beta & -1/\beta \\ -\dfrac{1}{\beta} & \dfrac{\alpha}{\beta} \end{pmatrix}$$

Similarly the A-matrix of the upper network of *Figure 10b* is given by

$$(A_2) = \begin{pmatrix} 1+ZG & 2Z+Z^2G \\ G & 1+ZG \end{pmatrix} \quad \text{where} \quad Z = \frac{1}{j\omega C}$$
$$G = 2/R$$

$$= \begin{pmatrix} \alpha' & \beta' \\ \delta' & \alpha' \end{pmatrix}$$

Thus

$$(Y_2) = \begin{pmatrix} \alpha'/\beta' & -1/\beta' \\ -1/\beta' & \alpha'/\beta' \end{pmatrix}$$

The overall y-matrix for the parallel T-network is therefore

$$(Y_3) = \begin{vmatrix} \dfrac{\alpha}{\beta}+\dfrac{\alpha'}{\beta'}, & -\left(\dfrac{1}{\beta}+\dfrac{1}{\beta'}\right) \\ -\left(\dfrac{1}{\beta}+\dfrac{1}{\beta}\right), & \left(\dfrac{\alpha}{\beta}+\dfrac{\alpha'}{\beta'}\right) \end{vmatrix}$$

From equation 5 the transfer function is $-y_{21}/y_{22}$ i.e. $(\beta+\beta')/(\alpha\beta'+\alpha'\beta)$. For infinite attenuation $\beta+\beta' = 0$. This occurs when $\omega = \omega_0$ say.

$$\therefore \quad 2R+2j\omega_0 CR^2 + \frac{2}{j\omega_0 C} - \frac{2}{\omega_0^2 C^2 R} = 0$$

$$\therefore \quad \omega_0 = \frac{1}{CR}$$

31*

Infinite attenuation, therefore, occurs at a frequency of $1/(2\pi CR)$ which is the equivalent result quoted in the chapter on amplifiers (page 257).

Transistor Parameters

In designing transistor amplifiers there are four quantities of particular interest: the current gain A_i, the voltage gain A_v, the input impedance Z_i and the output impedance Z_o. These quantities may be obtained by using a suitable equivalent circuit (e.g. as on page 247) but today most designers are using the hybrid matrix parameters (or h-parameters) as a starting point for their calculations. Manufacturers prefer to publish the performance details of their transistors in h-parameter form since these parameters (unlike the A-, Y- and Z-parameters) can all be evaluated in the laboratory with reasonable accuracy. If a designer finds that a given analysis would proceed with greater ease if he used the y-parameters, for example, it is quite easy for him to use the table to convert the h- into y-parameters.

There are four h-parameters for each of the three basic transistor amplifier configurations. A transistor in any given configuration may be regarded as a four-terminal 'black box'. The h-parameters are, therefore, defined from the equations

$$v_1 = h_{11}i_1 + h_{12}v_2$$
$$i_2 = h_{21}i_1 + h_{22}v_2$$

i.e.

$$\begin{pmatrix} v_1 \\ i_2 \end{pmatrix} = [H] \begin{pmatrix} i_1 \\ v_2 \end{pmatrix}$$

where v_1, v_2, i_1 and i_2 are the alternating or small signal quantities. The first parameter, h_{11}, is the input impedance with the output terminals shorted i.e. $v_2 = 0$. $h_{12} = v_1/v_2$ when the input is open-circuited to a.c. since under these conditions $i_1 = 0$. This parameter is, therefore, the reverse voltage transfer ratio. h_{21} is the forward current transfer ratio when the output is short-circuited to a.c. Like h_{12} it is a dimensionless ratio. In the common-base configuration h_{12} is the current gain often denoted by α, whilst in the common-emitter configuration h_{12} is equal to β. To distinguish between the two configurations many writers prime the h-parameter to indicate that the common-emitter mode is being used; thus $h'_{12} = \beta$. Alternatively, the American system may be used as described below. The last parameter, h_{22} is the output admittance with the input open-circuited to a.c. It is measured in reciprocal ohms, i.e. mhos. It can be seen now why these parameters are referred to as 'hybrid', since one is in ohms, another in mhos, whilst the other two are dimensionless ratios. Because these parameters are not all in one compatible set of units, difficulties arise when circuit analyses are undertaken. However, as has already been mentioned, conversion to a compatible set of units is possible by using the table. In this way the equivalent impedance, or Z-parameters, or the admittance, i.e. the Y-parameters, may be obtained. Indeed conversion

484

formulae have been published[1, 2] to convert to other parameter systems such as the modified Z-parameters and the T-parameters. As this book does not deal to any great extent with circuit analysis, the matter will not be pursued here. Interested readers may consult the given references.

In the Institute of Radio Engineers nomenclature, adopted by the majority of American writers, there are two subscripts for each hybrid matrix parameter. The first subscript may be i, f, r, or o meaning input, forward, reverse and output respectively. The second subscript refers to the mode of operation and may be b, e or c referring to the common-base, common-emitter, or common-collector configuration, respectively. Thus $h_{11} = h_{ib}$, $h_{12} = h_{rb}$, $h_{21} = h_{fb}$ and $h_{22} = h_{ob}$. $h'_{11} = h_{ie}$, $h'_{12} = h_{re}$, $h'_{21} = h_{fe}$ and $h'_{22} = h_{oe}$. This system is also recommended by the British Standards Institution in B.S. 3363 : 1961.

Finally let us take one example to show how matrices are used to derive formulae. Suppose we wished to know the voltage gain and input impedance of the simple amplifier of *Figure 11*.

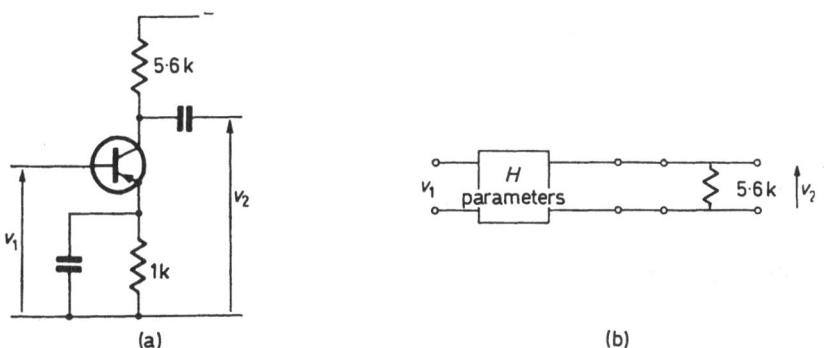

(a)   (b)

Figure 11. The simple transistor amplifier of (a) may be regarded as the transistor in cascade with the load resistor. (b) shows the arrangement for circuit analysis in which only alternating quantities are involved

Regarding *Figure 11a* as two four-terminal networks in cascade we find it convenient to use A-matrices. We first convert the h-parameters supplied by the manufacturer to A-parameters, using the table. The resulting A-matrix is then multiplied by that representing the load. The overall A-matrix for *Figure 11b* is then given by

$$[A] = \begin{bmatrix} \dfrac{-|h'|}{h'_{21}} & \dfrac{-h'_{11}}{h'_{21}} \\[2mm] \dfrac{-h'_{22}}{h'_{21}} & \dfrac{-1}{h'_{21}} \end{bmatrix} \begin{bmatrix} 1 & 0 \\[2mm] G & 1 \end{bmatrix} \quad \text{where} \quad G = \dfrac{1}{R_L}$$

By applying the rules for multiplying matrices and noting that the voltage gain is $1/a_{11}$, where a_{11} is the appropriate element of $[A]$, we obtain

$$\text{Voltage gain} = \frac{-h'_{21}}{|h'| + Gh'_{11}} \quad \text{(The minus sign shows a phase reversal)}$$

The input impedance is given by a_{11}/a_{21}

i.e.
$$-\left(\frac{|h'| + Gh'_{11}}{h'_{21}}\right) \div \left(\frac{-h'_{22}}{h'_{21}} - \frac{G}{h'_{21}}\right)$$

$$= \frac{|h'| + Gh'_{11}}{h'_{22} + G}$$

Suppose, therefore, that we take typical low frequency h-parameters and have $h'_{11} = 1 \, \text{k}$, $h'_{12} = 3 \times 10^{-4}$, $h'_{21} = 50$ and $h'_{22} = 50 \, \mu\mho$ then $|h'| = 35 \times \times 10^{-3}$. The voltage gain for a load of $5\cdot6 \, \text{k}$ is $-50/(35 \times 10^{-3} + 1/5\cdot6) = 235$. The input resistance is $[10^3 + (35 \times 10^{-3})5\cdot6 \times 10^3]/(1 + (50 \times 10^{-6})5\cdot6 \times 10^3) = = 930$ ohms (approximately).

REFERENCES

1. Hakim, S. S., *Junction Transistor Circuit Analysis*. Iliffe Books Ltd., 1962.
2. *Mullard Reference Manual of Transistor Circuits*, published by Mullard Ltd., 1960.
3. Tilsley, D. N., 'Transistor h parameters' *Wireless World*, 1964, **70**, No. 5. May, p. 229.
4. Olsen, G. H., 'Matrix algebra' *Wireless World*, 1965, **71**, Nos. 3 and 4, March and April.

INDEX